AF810046

PARTICLE CONTROL FOR SEMICONDUCTOR MANUFACTURING

PARTICLE CONTROL FOR SEMICONDUCTOR MANUFACTURING

R. P. DONOVAN
Center for Aerosol Technology
Research Triangle Institute
Research Triangle Park, North Carolina

MARCEL DEKKER, INC. **New York and Basel**

Library of Congress Cataloging-in-Publication Data

Donovan, R. P.
Partical control for semiconductor manufacturing / Robert P. Donovan.
p. cm.
Includes bibliographical references.
ISBN 0-8247-8242-9 (alk. paper)
1. Semiconductors--Design and construction. 2. Semiconductors--Defects. 3. Particles. I. Title.
TK7871.85.D647 1990
621.381'52--dc20 89-49025
CIP

This book is printed on acid-free paper

Copyright © 1990 by MARCEL DEKKER, INC. All Rights Reserved

Neither this book nor any part may be reproduced or transmitted in any form or by any means, electronic or mechanical, including photocopying, microfilming, and recording, or by any information storage and retrieval system, without permission in writing from the publisher.

MARCEL DEKKER, INC.
270 Madison Avenue, New York, New York 10016

Current printing (last digit):
10 9 8 7 6 5 4 3 2 1

PRINTED IN THE UNITED STATES OF AMERICA

PREFACE

As used in this book, the term particle refers to a stable or quasi-stable agglomeration of molecules, forming a discrete liquid or solid entity, typically 2 nm to 1 mm in size. Particles exist everywhere in nature, as both living and inorganic entities and are generated in most industrial processes. The ambient air and water of the planet contain huge concentrations of particles. Only with painstaking difficulty can a surface, be it a lens, mirror or silicon wafer, be made particle-free.

Large particles—those exceeding 50-100 μm in size—can be readily seen by the naked eye. On the other hand, small submicrometer particles require high-quality microscopes or electron microscopes to be detected. At the lower end of the size range specified in the particle definition, the distinction between a particle and a large molecule begins to blur. The concern of the semiconductor industry has been with smaller and smaller particles as device geometry itself has become smaller and smaller. Thus, the emphasis in this book is on submicrometer particles. Particles as small as 0.1 μm (100 nm) are now widely recognized as potential "killer" particles. Because of their growing importance, that size particle and those in the size decade below 0.1μm are of special concern.

Because of the ubiquitous nature of these particles and because their dimensions are of the same order of magnitude as the dimensions of the devices on today's silicon chips, particles constitute a serious threat to chip yield, introducing unplanned and undesired deviations in the designed structure. These deviations can come about primarily as alterations in either the artwork photolithographically transferred to a wafer surface or in the bulk electrical/structural properties of the chip because of impurities introduced by the particle or both. In any event, the relationship between particulate contamination and yield, while not necessarily simple, is direct, and no debate exists regarding the destructive role of the extraneous particles found on wafer surfaces during processing.

This book is an introduction to the measurement and control of the contaminating particles that plague the manufacturing of silicon chips. These particles introduce unwanted, unmodeled effects into device performance, often resulting in chip failure. Their presence has been estimated to account for anywhere from one-third to two-thirds of the failures detected at wafer probe. The trend toward chip designs of ever decreasing dimensions promises increased sensitivity to particle control practices.

While no letup in particle vulnerability is in sight—indeed quite the reverse—increased awareness of the particle threat at all levels of manufacturing has resulted in significant progress in combatting the particle menace. Particle detection methods—in gases, in liquids, and on surfaces—have improved dramatically; particle generation as a factor in the design of wafer processing equipment is now universally recognized; management now generally supports increased resources and personnel for particle control.

This book reviews this recent progress. It is an introduction to the particle problems of silicon device manufacturing and their solution. While it is written primarily for the particle control novice and includes considerable background fundamentals, it also describes many state-of-the-art practices so that the practitioner can also find information of value in responding to his daily challenges. The goals of the book are to summarize what has been learned over past years in understanding particle problems and what has been successful in minimizing the deleterious effects of particles on device yield.

This book is an updated revision of the course notes distributed at a short course entitled Particle Control in Semiconductor Manufacturing that was presented for the first time on April 19-21, 1988, in Research Triangle Park, North Carolina. The course was organized by researchers at the Research Triangle Institute who have participated in Semiconductor Research Corporation (SRC)-sponsored research in manufacturing science over the past 5 years. The course syllabus included much of this SRC-sponsored research and the course staff included many of these researchers. In addition, however, the staff also included a university professor, an instrument vendor, a gas vendor, a chemical vendor, and three semiconductor technologists who, through the eyes of a chip manufacturer, have developed an appreciation for the importance and the subtleties of particles and particle problems to the successful manufacturing of today's silicon chips.

The course began with definitions of particles and particle properties (Chapters 1 to 3), and progressed to general methods of measuring and controlling particles (Chapters 4 to 7) and on to discussions and descriptions of particle sources in a typical silicon fabrication facility (Chapters 8 to 12). The second day of the course was dedicated to lectures on particle measurement methods and control strategies customized for semiconductor operations (Chapters 13 to 20). Wafer-cleaning methods (Chapters 21 to 23)—the removal of particles that, in spite of the best preventive measures, have found their way onto a silicon surface—concluded the second day.

The final half-day provided practical guidelines for semiconductor operations (Chapter 24), discussed the impact and provisions of the recently revised Federal Standard 209D (Chapter 25), presented data relating particle defects to electrical properties (Chapter 26), and allowed some time for course critiques by all participants (not reproduced here).

It is hoped that the added circulation provided by this updated text will serve to increase the technology transfer impact of this activity. This book is also planned to be used as the core of course notes to be handed out at future short courses on this topic.

In addition to the authors, acknowledged with a byline at the beginning of each chapter, others contributed to the preparation of this volume. Ms. Jan Shirley, Word Processing Supervisor, led the word processing group who converted diverse author inputs into a uniform, readable format and produced the camera-ready copy used to reproduce the commercial product. RTI editors Richard Straw and Erin Newton painstakingly reviewed all text for clarity, style, and consistency.

R. P. Donovan

CONTENTS

CONTRIBUTORS

Jeffrey M. Davidson
The BOC Group, Inc.
Group Technical Center
100 Mountain Avenue
Murray Hill, NJ 07974

Keith Dillenbeck
Ashland Chemical Company
Electronics and Laboratory Products Division
P.O. Box 2219
Columbus, OH 43216

R. P. Donovan
Center for Aerosol Technology
Research Triangle Institute
P.O. Box 12194
Research Triangle Park, NC 27709-2194

Gilbert Drab
Ashland Chemical
8201 S. Central Expressway
Dallas, TX 75239

David S. Ensor
Center for Aerosol Technology
Research Triangle Institute
P.O. Box 12194
Research Triangle Park, NC 27709-2194

Wayne G. Fisher
Texas Instruments, Inc.
P.O. Box 84, MS884
Sherman, TX 75090

David Leith
Department of Environmental Sciences and Engineering
University of North Carolina
Chapel Hill, NC 27514

Venugopal B. Menon
Sematech
2706 Montopolis Drive
Austin, TX 78741

Les Nichols
Ashland Chemical
8201 S. Central Expressway
Dallas, TX 75239

Gilmore J. Sem
TSI, Incorporated
500 Cardigan Road
Box 64394
St. Paul, MN 55164

Jim Tichich
Ashland Chemical
8201 S. Central Expressway
Dallas, TX 75239

B. Tullis
Hewlett-Packard Co.
3500 Deer Creek Rd.
Building 25L
Palo Alto, CA 94304

Andrew S. Viner
3M Company
3M Center
Building 260
St. Paul, MN 55144-1000

Toshiaki Yamamoto
Center for Aerosol Technology
Research Triangle Institute
P.O. Box 12194
Research Triangle Park, NC 27709-2194

1. PARTICLE INTERACTION WITH INTEGRATED CIRCUITS

WAYNE G. FISHER Texas Instruments, Inc., Sherman, Texas

INTRODUCTION

This chapter is a qualitative and quantitative overview of the effects of particles on integrated circuits. First is a review of the various ways that particles can cause yield loss on integrated circuits. In particular, there are several ways that particles can cause damage that have only recently been recognized as important by particle control technologists. By some of these mechanisms, the circuit may be damaged by particles so small that they were previously thought to be of little concern.

Particle-integrated circuit interactions will then be reviewed quantitatively. Basic information on yield models will be presented, along with the effect of nonuniform distributions of defects on the yield model.

An excellent introduction to the problems caused by particles in integrated circuit manufacturing has been given by Cooper (1986), and a qualitative description of the various ways that particles can cause defects on integrated circuits is reviewed in a previous article by the author (Fisher, 1987). One example of a defect caused by a particle is where the particle itself is the defect and causes failure of the circuit by disruption of the pattern or by bridging separated features. Particle-caused defects include the shorting of two conductors by a particle and open circuits in the same features.

Another example of a defect caused by a particle is where the particle causes a defect that remains even after the particle is gone. Defects of this type commonly occur when the particle masks an area of the circuit that should not have been masked by photoresist or some other masking layer. An area may also be unmasked that should have been masked if the particle leaves a hole in a photoresist pattern. If an area is masked prior to an etching step, it may not be etched as intended. If the area is unintentionally exposed, material may be etched away where it was supposed to remain. Prior to an implantation step, a masked area may not be doped as intended. An exposed area may be implanted when it should not be.

In the two examples just cited, a particle is itself a defect or a particle causes a defect that remains even after the particle is gone. Both of these cases result from a physical or mechanical interaction of particles with the integrated circuit. In these cases, the size of the particle relative to the circuit feature size is a major factor in determining whether the particle causes a defect or not. A rule of thumb often applied is that the particle must be some fraction of the minimum circuit feature size to be large enough to cause a killing defect. Commonly this fraction is taken to be 1/10. Because the minimum feature sizes on the most advanced products are near 1 μm, particles near 0.1 μm are now of concern. Feature sizes will undoubtedly decrease, so ever smaller particles will be a problem.

DEFECTS CAUSED BY PARTICLES < 0.1 μm

In addition to the two familiar physical or mechanical defect-causing mechanisms described above, particles much smaller than the minimum circuit feature size may cause killing defects in other ways through both physical or mechanical and chemical interactions. These include particle interactions with thin layers, chemical interactions, and particle-induced defects where the resulting defect is much larger than the particle itself.

If a particle causes a defect in a thin layer, it is the particle size relative to the layer thickness that determines if the defect is killing, rather than the particle size relative to a circuit feature size. Much smaller particles can cause killing defects in circuits by disruption of thin layers than by the previously described mechanisms because layer thicknesses may be much less than minimum feature sizes in the plane of the wafer. For example, the gate or capacitor oxide thicknesses in MOS devices may be as thin as 100 Å (0.01 μm) in circuits with minimum feature sizes of 1 μm. A particle with a diameter of only 100 Å is large enough to bridge this oxide. The interaction of small particles with thin layers may be either physical or chemical.

Particle defects occur through chemical interactions when the particle chemically reacts with circuit features. One example is the formation of a conducting region in an insulator when the particle reacts with it, such as with silicon dioxide during growth of a gate oxide. A particle much smaller than the gate oxide thickness may cause a defect in the oxide by inhibiting or enhancing the formation of the oxide during thermal oxidation. Or, silicon may be unintentionally doped by particles containing elements that are electrically active in silicon. An example is the formation of conductive regions between the collector and emitter of bipolar transistors that short the two regions together.

When a particle causes a defect by chemical means, the chemical identity of the particle is more important than the particle size. This is in contrast to physical or mechanical defect-causing mechanisms where the particle size is more important than the chemical identity. The relative importance of the size and composition for the two mechanisms is summarized in the following table.

	Particle size	**Chemical composition**
Physical interaction	More important	Less important
Chemical interaction	Less important	More important

A final case of defect formation by a very small particle occurs when a defect is formed in such a manner that the defect is much larger than the particle. An example is when a small particle masks the etching of a layer that is much thicker than the diameter of the particle. If the etching is very anisotropic, the layer under the particle will not be etched, leaving a pillar of unetched material much larger than the particle. The interaction to cause this type of defect is physical. Another example is the formation of a large defect in a silicon epitaxial film nucleated by a small particle in the film during its growth.

PROBABILITY OF DEFECT FORMATION

The probability that a given particle causes a killing defect depends on:

- Interaction mechanism—physical or chemical,

- Size of the particle,
- Chemical identity of the particle,
- Where the particle falls on the circuit, the sensitive area, and
- Point in the process where the particle falls on the circuit, and how long it stays there.

For each interaction mechanism, some particles are so large that they will always cause a defect, and some are so small that they will never cause a defect. The general shape of the curve describing the probability of failure as a function of the defect size is shown below. (Also see Stapper, Armstrong, and Saji, 1983.)

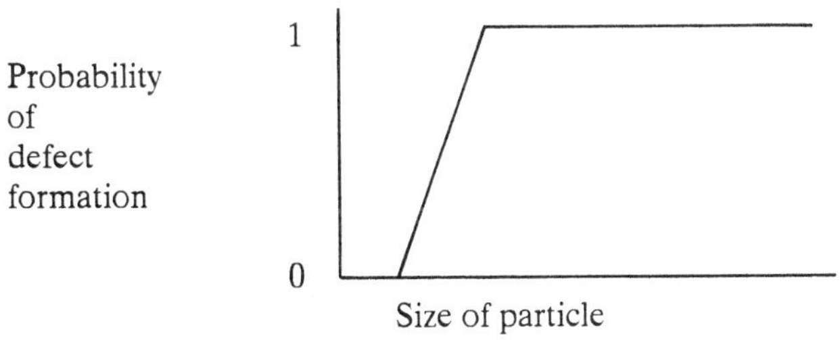

The exact shape of this curve depends on the circuit design and specifics of the manufacturing process. The exact sizes of particles that never or always cause defects depend on the interaction mechanism. For each circuit, process, and interaction mechanism, there is a different curve.

Each circuit has some areas that are insensitive to particles. A particle in these areas is very unlikely to cause a killing defect. These include the bonding pads, alignment marks, and test structures on the circuit. Other insensitive areas on the wafer include unused area at the wafer circumference and scribe line areas.

Finally, there are many points in an integrated circuit manufacturing process where particles may come and go without causing a killing defect. One example is a particle that is part of a photoresist masking layer that is removed with the resist when the resist pattern is no longer needed. Another is a particle that is deposited during one process step, then removed during a cleaning step that immediately follows.

QUANTITATIVE EFFECT OF PARTICLE DEFECTS ON YIELD

Qualitative descriptions of the effect of particle defects on integrated circuit yield allow us to judge the probability that a given particle causes yield loss or not. The quantitative dependence of the yield of integrated circuits on the defect density of the circuit may be described by yield models. The referenced article by Stapper, Armstrong, and Saji (1983) summarizes the derivation of several yield models in detail.

Yield models are studied primarily because of their economic importance. They may be used to predict the yield of new technologies or to forecast the results of process improvements. They illustrate that yield decreases rapidly as defect density rises, and that yield decreases rapidly as chip area increases.

We must first distinguish between point defects that may be caused by particles, and gross or area defects that have other causes. Area defects are those that cause failures of all the chips in areas that are large compared to the area of an individual chip, such as a whole wafer or half a wafer, or of several wafers in a lot of wafers. These failures usually result from gross misprocessing, such as mask misalignment or the use of a wrong mask, a missing or double ion implantation, an incorrect film thickness, over or under etching, or incorrect substrate resistivity or type. The yield loss from broken or lost wafers may also be included in this type of yield loss.

Yield loss due to gross or area defects is not a function of the point defect density, so they are not accounted for by models that assume point defects. In practice, the yield loss due to incorrectly processed wafers or lots of wafers is accounted for by a multiplicative factor to determine the overall yield of a processing line.

One model for the dependence of yield on the point defect density assumes a random distribution of the defects, where the probability of the occurrence of a defect on a chip is independent of the location of any of the other point defects (a Poisson distribution). This distribution of defects leads to the expression for the yield:

$$Y = e^{-DA} , \tag{1-1}$$

where

Y = yield, or fraction of circuits that do not have a defect,
D = defect density, and
A = area of the circuit.

Note that the product of the defect density and the circuit area, DA, is the number of defects per chip. This model predicts an exponential decrease of yield with defect density and with chip area, so that the yield falls dramatically as either of these two functions increases.

Fortunately, the distribution of defects on circuits is generally not random, and this nonrandom distribution leads to real device yields that are higher than those predicted using Poisson statistics. Variations in the defect density occur within individual wafers, from wafer to wafer within a lot of wafers, and from lot to lot. These variations occur because it is unlikely that any two parts of a wafer, any two wafers in a lot, or any two lots of wafers experience exactly the same conditions during processing. For example, in a cassette of wafers being loaded into a processing machine, an end wafer may be exposed for a longer period of time to clean room air, and the end wafer may shield adjacent wafers from particle deposition from the air.

The yield is higher when the defect distribution is not random than when it is random because there is increased probability, with the same total number of defects, that a given chip will have multiple defects. This means that some other chip is more likely to be free of defects and to contribute to the yield. The form of the yield model will depend on the shape of the distribution of the defects. Some distribution shapes may also be characterized by a variance. For the same distribution shape, a high variance will produce a higher yield than a low variance. When the defect density is not random, the yield depends not only on the defect density (now a mean defect density) and circuit area, but also on the shape of the defect distribution and the deviation (or variance) of the distribution.

The shapes of several distributions that may characterize defect densities and the yield relationships that result for each are summarized in Figure 1-1 (from Koehler and Beadle, 1985). In this figure are shown distributions of defect densities for the delta, triangular, rectangular, exponential, Gaussian, and Gamma functions. The delta, triangular, rectangular, and exponential distribution functions result in yield functions that depend on the defect density and chip area. The variance in these cases is fixed and defined by the function. For the Gaussian and Gamma functions, the variance is also a variable in the yield relationship.

The Gamma function is particularly interesting because it simplifies to the delta function when the variance is zero and to the exponential functions when the variance is one. It closely approximates the triangular function when the variance is about 0.2. Compared to the Gaussian function, the Gamma function does not allow negative values for the defect density distribution. The Gamma function predicts some zero defect densities when the variance in the distribution is large, a condition generally observed in real integrated circuits. Note that this means a production line should see wafers with 100 percent yield when the gross yield losses are under control and when the mean defect density is decreased toward zero.

Figure 1-2 (Koehler and Beadle, 1985) compares the yield as a function of the defect density-chip area product for the yield relationship obtained from the Gamma function for various values of the variance. Table 1-1 shows the yield for a circuit with an area of 0.4 cm^2, a size close to that of present 256k DRAM memory chips. The yield is calculated using the yield relationship obtained from the Gamma function for variances ranging from 0 to 2, and defect densities ranging from 0.5 to 10 cm^{-2}.

As noted above, for a zero variance the exponential function is obtained. As the variance increases, the yield at a given defect density increases. This increase is particularly dramatic for the larger defect densities. When the defect density is very small, corresponding to high yields, the difference between the different yield relationships is not great.

What is the real yield model? The answer is that the defect density distribution depends on the type of device being manufactured and the operation of the manufacturing line. So, the yield relationship can vary from device type to device type and from processing line to processing line. For each manufacturing line, the yield model should be chosen that best matches data obtained from devices made on that line. The fit is made empirically by adjusting the parameters in the yield model to the yields and areas of devices manufactured there. In the three parameter yield models, the variance provides an extra degree of freedom for the fit above that in models that depend on only the defect density and circuit area.

DISTRIBUTION OF D	$\frac{Y}{Y_0}=$	$\lambda =$
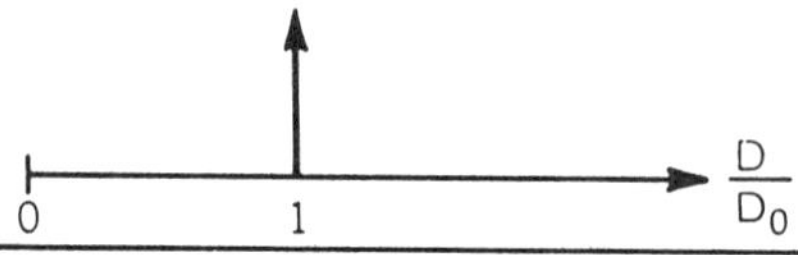	$e^{\;D_0 A}$	0
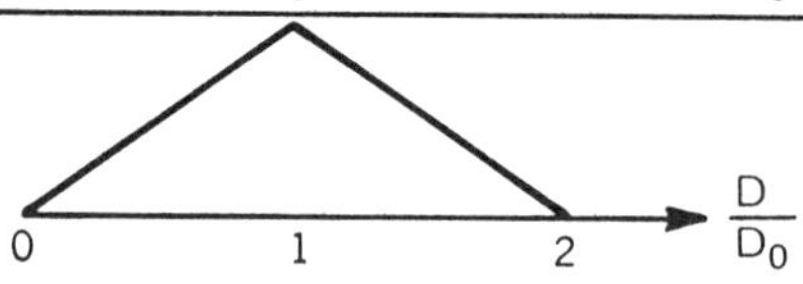	$\left(\frac{1-e^{-D_0 A}}{D_0 A}\right)^2$	0.22 (±0.02)
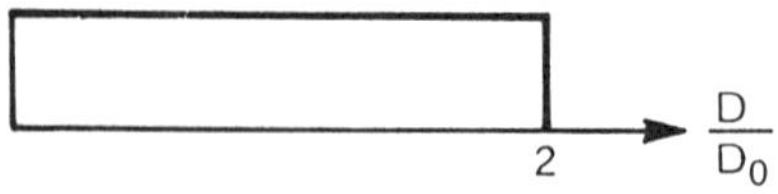	$\frac{1-e^{-2D_0 A}}{2D_0 A}$	0.5 (±0.1)
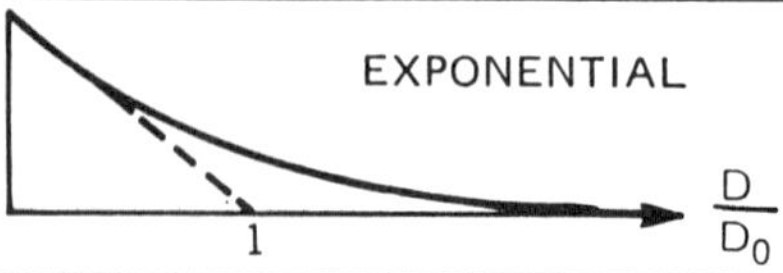	$\frac{1}{1+D_0 A}$	1
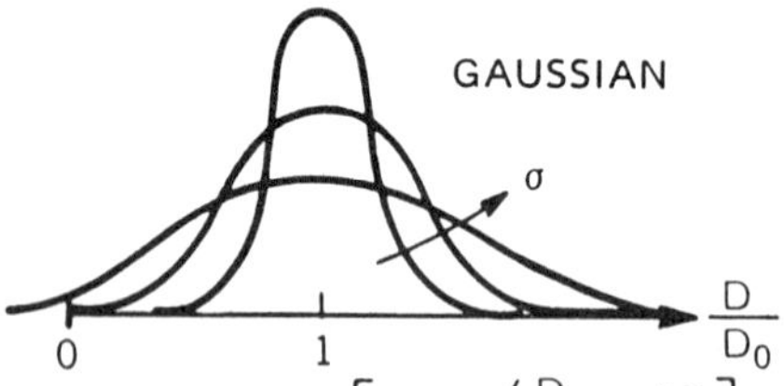 $\frac{1}{\sqrt{2\pi}\sigma}\;\exp\left[-\frac{1}{2}\left(\frac{\frac{D}{D_0}-1}{\sigma}\right)^2\right]$	$\frac{1}{2}\exp\left(-AD_0+\frac{\sigma^2 AD_0}{2}\right)\left\{1+\mathrm{erf}\left[\frac{1}{\sqrt{2}}\left(\frac{1}{\sigma}-\sigma D_0 A\right)\right]\right\}$	$\approx \sigma^2$ (for small σ)
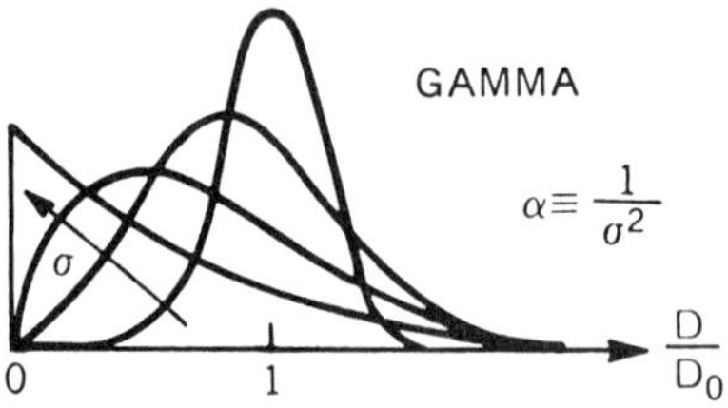 $\frac{\alpha}{\Gamma(\alpha)}\left(\alpha\frac{D}{D_0}\right)^{\alpha-1}\exp\left(-\alpha\frac{D}{D_0}\right)$	$\frac{1}{(1+\sigma^2 D_0 A)^{1/\sigma^2}} \equiv \boxed{\frac{1}{(1+\lambda D_0 A)^{1/\lambda}}}$	σ^2

σ = standard deviation normalized with respect to D_0

Source: Koehler and Beadle, 1985.

Figure 1-1. Yield models.

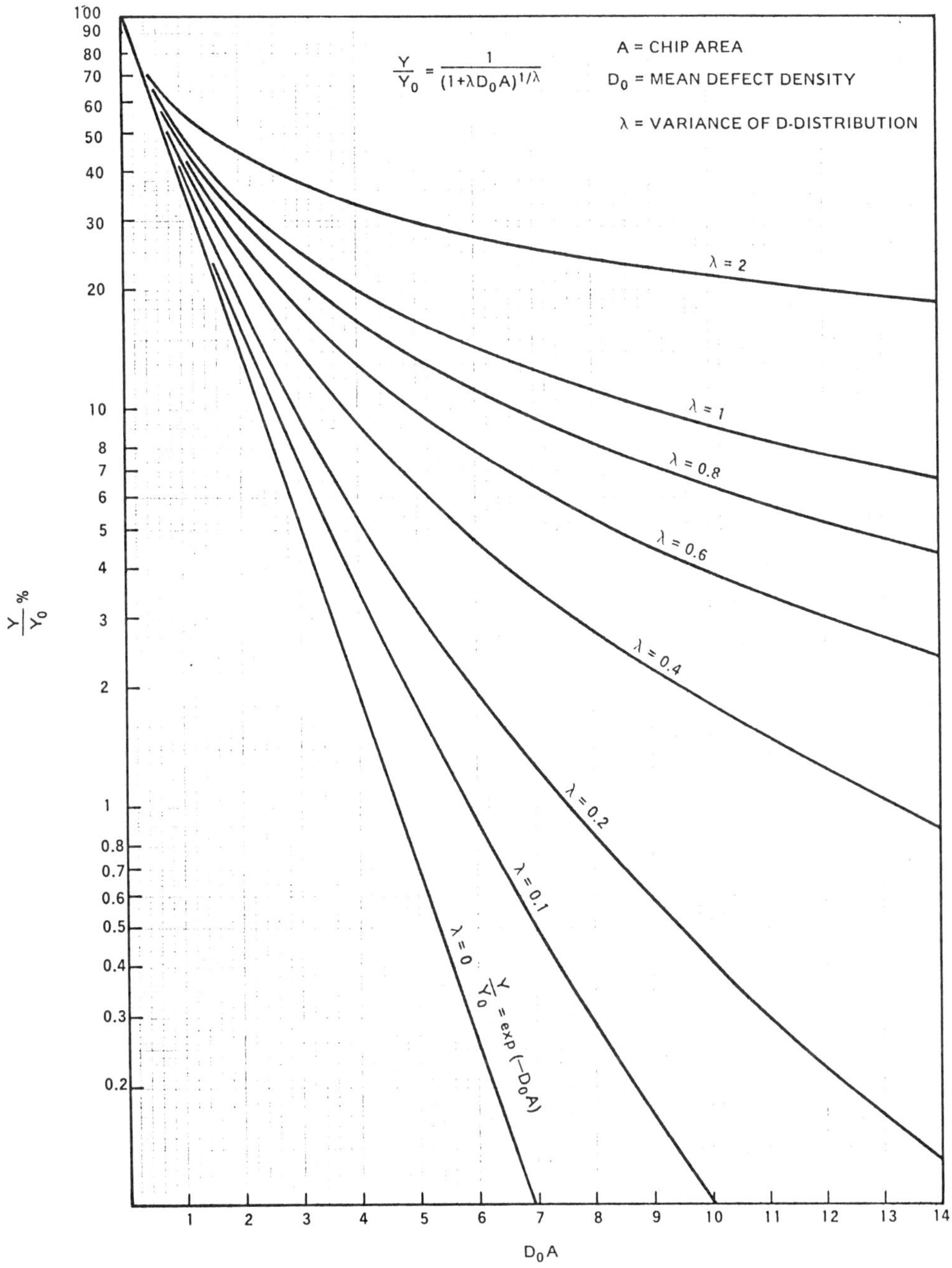

Source: Koehler and Beadle, 1985.

Figure 1-2. Three-parameter yield model.

Table 1-1. Yield for a Circuit with an Area of 0.4 cm^2 for Various Values of the Defect Density and Variance in the Defect Density Distribution

	Defect density (cm^{-2})						
Variance	0.5	1.0	2.0	4.0	6.0	8.0	10.0
0.0	0.819	0.670	0.449	0.202	0.091	0.041	0.018
0.1	0.820	0.676	0.463	0.223	0.116	0.062	0.035
0.2	0.822	0.681	0.476	0.250	0.141	0.084	0.053
0.4	0.825	0.690	0.500	0.290	0.186	0.127	0.092
0.6	0.828	0.699	0.520	0.326	0.226	0.168	0.130
0.8	0.831	0.707	0.539	0.357	0.262	0.204	0.166
1.0	0.833	0.714	0.556	0.385	0.294	0.238	0.200
2.0	0.845	0.745	0.620	0.488	0.415	0.368	0.333

SUMMARY

Particles can cause defects on integrated circuits in a variety of ways. A particle may itself be a defect, or it may cause a defect that remains even after the particle is gone. Particles may also cause defects by interaction with thin layers or by chemical interactions with circuit features. Defects caused by particles may be much larger than the particles that caused them.

The probability that a particle causes a defect depends on how it interacts with the circuit, the size of the particle and its chemical identity, and where and when the particle falls on the circuit. The yield of good circuits decreases dramatically when the defect density increases or when the circuit area increases. This decrease is generally not as fast as predicted by a random distribution of defects because the defects are generally not uniformly distributed. The shape and variance of the defect density distribution must be determined empirically.

REFERENCES

Cooper. 1986. "Particulate Contamination and Microelectronics Manufacturing: An Introduction." *Aerosol Science and Technology*, *5*(3):287-299.

Fisher, W. G. 1987. "Particle Monitoring in Clean Room Air with the TSI 3020 Condensation Nucleus Counter." *TSI Journal of Particle Instrumentation*, *2*(1):3-19.

Hinds, W. C. 1982. *Aerosol Technology*, Wiley, New York, NY.

Koehler, D., and W. E. Beadle. 1985. "IC Structures and Wafer Properties." In: *Quick Reference Manual for Silicon Integrated Circuite Technology*, W. E. Beadle, J.C.C. Tsai, and R. D. Plummer, eds. Wiley, New York, NY.

Stapper, C. H., F. M. Armstrong, and K. Saji. 1983. "Integrated Circuit Yield Statistics." *Proceedings of IEEE*, *71*(4):453-470.

2. INTRODUCTION TO PARTICLE SCIENCE

DAVID LEITH University of North Carolina, Chapel Hill, North Carolina

This chapter will introduce some ideas basic to understanding particles. Books have been written on this subject, so the treatment here is necessarily incomplete. The purpose of this chapter is to explain some phenomena important to the particle problems of the microelectronics industry and at a level commensurate with the balance of this book. For further information, or a deeper discussion of the ideas presented here, you may wish to consult:

Fuchs, N.A., *The Mechanics of Aerosols*, Pergamon, Oxford, 1964.

Hinds, William, *Aerosol Technology*, Wiley Interscience, New York 1982.

Reist, Parker C., *Introduction to Aerosol Science*, Macmillan, New York, 1984.

Hidy, George M., *Aerosols, An Industrial and Environmental Science*, Academic Press, New York, 1984.

INTRODUCTION

A particle is a small piece of material that is larger than a cluster of a few molecules, but small enough to remain suspended in a fluid for a time. It has the same physical properties as the bulk material of which it is a piece.

When particles are suspended in a gas, the result is an aerosol. Note that for an aerosol to be present, both particles and gas must exist together. Aerosols composed of particles in air are around us every day as dust, haze, visible smog, mist, rain, and fog. Understanding the properties of aerosols can help explain natural phenomena such as why sunsets are red and clouds are white, as well as why particles collect on the wafers used in the microelectronics industry, why they are difficult to remove once they deposit, how particles can be removed from liquids or gases, and how they can be detected.

PROPERTIES OF PARTICLES

The most important property of a particle is its size because many properties, such as particle velocities, adhesion forces, and ability to scatter light, depend strongly on size. Particle size is expressed in micrometers, abbreviated μm, where

$$\begin{aligned} 10^6\ \mu\text{m} &= 1\ \text{m}, \\ 10^4\ \mu\text{m} &= 1\ \text{cm}, \\ 25{,}400\ \mu\text{m} &= 1\ \text{in.} \end{aligned} \tag{2-1}$$

The smallest dot that can be seen with the unaided eye is about 50 μm in diameter. A human hair is roughly 100 μm in diameter. The period at the end of this sentence is several hundred micrometers in diameter. The particles of concern to the microelectronics industry are less than 10 μm in size; most are less than 1 μm. If all particles in an aerosol are the same size, the aerosol is said to be monodisperse. If particles with different sizes are present together, the aerosol is polydisperse. Almost all aerosols in nature are polydisperse.

Other particle properties can be important as well. Density affects particle dynamics. The shape of a particle affects how it scatters light. A particle's chemical identity helps determine the force with which it adheres to a surface or to another particle in an agglomerate. Overall, though, these other particle properties are not as important as size.

Ways to Express Particle Size

Because spheres are easy to characterize, particle size is often expressed as a *diameter*. That is, one speaks of a 0.5-μm particle as though it were actually a sphere with a diameter of 0.5 μm. Actually, particles are rarely spheres. Although very small solid particles from combustion sources may be spheres of a few hundredths of a micrometer in diameter, these usually form aggregates that are more chainlike than spherical. Solid particles that form in other ways can be jagged or crystalline.

As a practical matter, one is usually more interested in the behavior or characteristics of a particle than in its physical dimensions. For this reason, one often characterizes a particle by comparing it to a sphere that behaves the same way as the particle in question.

For example, particle size might be characterized through its equivalent light-scattering diameter, where

Light-scattering diameter = diameter of a sphere with known refractive index that scatters light like the particle in question. (2-2)

With this definition, one can speak of a particle with a light-scattering diameter of 1-μm without actually knowing the actual shape or size of the particle. Knowing that it scatters light like a 1-μm sphere with known refractive index may be sufficient to characterize it.

Similarly, particle size can also be characterized in terms of an *aerodynamic diameter*, the diameter of a sphere with density of 1 g/cm^3 that falls with the same velocity as the particle in question. Aerodynamic diameter would be an appropriate way to characterize particle size if one were primarily interested in particle motion, whereas light-scattering diameter would be an appropriate way to characterize size if one were primarily interested in a particle's ability to scatter light.

Although these definitions are convenient and often used, particles characterized through an equivalent spherical diameter such as light-scattering diameter or aerodynamic diameter need not actually be spherical. For example, a fiber with an equivalent light scattering diameter of 0.5 μm would be larger than 0.5 μm in some dimensions and smaller than 0.5 μm in other dimensions. This distinction can be important if one property of a particle, such as its ability to scatter light, is used as a surrogate for another particle property such as its ability to bridge the conducting pathways on a semiconductor chip.

The Frank Chart

Figure 2-1 lists properties of particles that change with particle size. Note that for spheres between 0.01 and 10 μm in diameter, the range of interest here, many particle properties change substantially.

PROPERTIES OF AEROSOLS

The most important property of an aerosol is its particle concentration. Concentration can be expressed in different ways. In the microelectronics industry, number concentration is the method most frequently used:

$$\text{Number concentration} = \text{number of particles/volume of gas} \ . \qquad (2\text{-}3)$$

Number concentration can be expressed in different units, such as particles per cubic foot or particles per cubic meter or per cubic centimeter. Clearly, the numerical value of number concentration will vary with the units used, as 1 particle/cm^3 = 28,317 particles/ft^3.

Particle concentration can also be expressed on a mass basis:

$$\text{Mass concentration} = \text{mass of particles/volume of gas} \ . \qquad (2\text{-}4)$$

Again, the value of mass concentration will vary with the units used. Frequently, mass concentration is expressed in micrograms per cubic meter. The mass concentration of particles in outdoor air is generally from tens to a few hundreds of micrograms per cubic meter.

Particle concentrations cannot be converted from a number to a mass basis or vice versa unless the sizes of the particles are known. For example, the cubic meter of clean air that contains a bowling ball has a number concentration of 1/m^3 (the ball) and a mass concentration of about 7 x 10^9 μg/m^3. If the same cubic meter of clean air contains a single particle of unit density (1 μg/cm^3) that is 1.0 μm in diameter, its number concentration is still 1/m^3 but its mass concentration is 7.9 x 10^{-7} μg/m^3, a difference of 16 orders of magnitude in mass concentration. Because an aerosol has a low number concentration does not necessarily mean that its mass concentration is low, as this rather silly example shows. An aerosol comprised of 0.1-μm particles has a mass concentration three orders of magnitude less than an aerosol with the same number concentration made up of 1-μm particles. Two aerosols with similar number concentrations may have considerably different mass concentrations if the sizes of the particles in the two aerosols differ.

If one is concerned with the density of particles on the surface of a wafer, the density can be analyzed in terms of number of particles present per surface area, analogous to a number concentration; or in terms of the mass of particles present per surface area, analogous to a mass concentration. Usually in microelectronics work, particle density is expressed in terms of number of particles per unit area, with the idea that the density of surface defects is more strongly related to the number of particles present than to the mass of particles present. Again, the number density of particles on a surface is related to the mass density of particles on that surface through the sizes of the particles present.

The properties of an aerosol depend on the properties of the gas as well as on particle properties. Some properties of air at 20 °C are in Table 2-1.

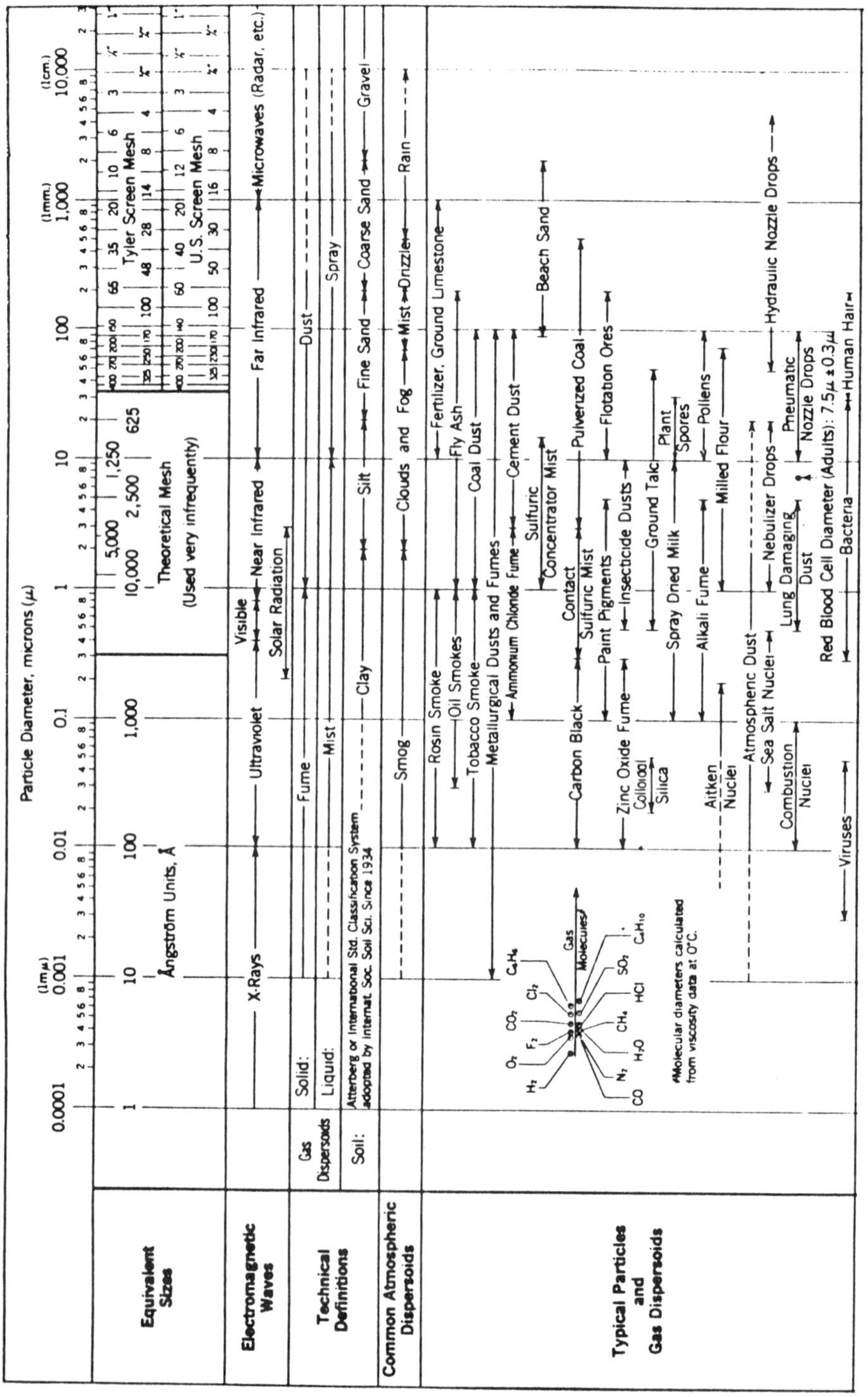

Particle Diameter, microns (μ)
0.0001
0.001 (1mμ)
0.01
0.1
1
10
100
1,000 (1mm.)
10,000 (1cm.)
Equivalent Sizes
Ångström Units, Å
Theoretical Mesh (Used very infrequently)
Tyler Screen Mesh
U.S. Screen Mesh
Electromagnetic Waves
X-Rays
Ultraviolet
Visible
Near Infrared
Far Infrared
Microwaves (Radar, etc.)
Solar Radiation
Technical Definitions
Gas Dispersoids
Solid:
Fume
Dust
Liquid:
Mist
Spray
Soil:
Atterberg or International Std. Classification System adopted by Internat. Soc. Soil Sci. Since 1934
Clay
Silt
Fine Sand
Coarse Sand
Gravel
Common Atmospheric Dispersoids
Smog
Clouds and Fog
Mist
Drizzle
Rain
Typical Particles and Gas Dispersoids
Rosin Smoke
Oil Smokes
Tobacco Smoke
Metallurgical Dusts and Fumes
Ammonium Chloride Fume
Fertilizer, Ground Limestone
Fly Ash
Coal Dust
Cement Dust
Sulfuric Concentrator Mist
Beach Sand
Carbon Black
Contact Sulfuric Mist
Pulverized Coal
Paint Pigments
Flotation Ores
Zinc Oxide Fume
Colloidal Silica
Insecticide Dusts
Ground Talc
Spray Dried Milk
Plant Spores
Pollens
Alkali Fume
Milled Flour
Aitken Nuclei
Atmospheric Dust
Sea Salt Nuclei
Nebulizer Drops
Hydraulic Nozzle Drops
Combustion Nuclei
Lung Damaging Dust
Pneumatic Nozzle Drops
Red Blood Cell Diameter (Adults): 7.5μ ± 0.3μ
Viruses
Bacteria
Human Hair
Gas Molecules*
*Molecular diameters calculated from viscosity data at 0°C.

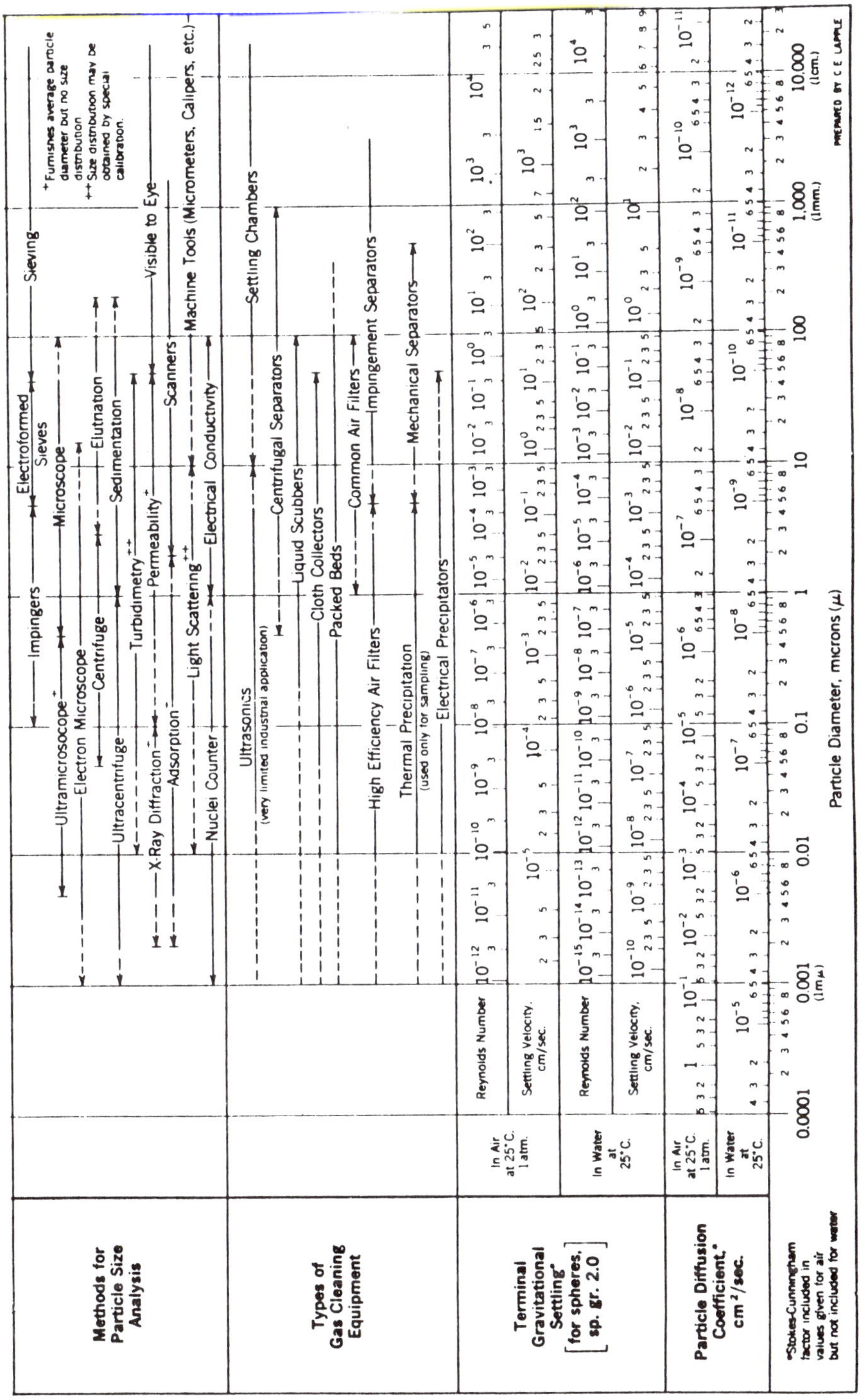

Reprinted from *Stanford Research Institute Journal*, Third Quarter, 1961. Single copies 8½ by 11 inches free or $15 per hundred. 20- by 26-inch wall chart $15 each. Both charts available from Dept. 360.

Figure 2-1. Characteristics of particles and particle dispersoids.

Table 2-1. Properties of Air at 20 °C

Property	Symbol	Value
Density	ρ_g	1.2×10^{-3} g/cm^3
Viscosity	η	1.8×10^{-4} g/(cm•s)
Mean free path	λ	0.066 μm
Molecular weight	M	29

PARTICLE STATICS AND DYNAMICS

Particle motion in an aerosol, and the adhesion of particles to the surfaces on which they deposit, are determined by the forces that act on these particles. Although these forces are the same as those that act on the larger objects with which we are familiar in everyday life, the relative magnitude of these forces can be very different. Forces such as adhesion that are relatively small for large-scale objects can become relatively large for submicrometer particles. For example, the adhesion force that attracts a 1-cm marble to the ceiling is orders of magnitude smaller than the gravity force that pulls it down so that the marble falls; however, for a 0.1-μm particle, adhesion forces are many orders of magnitude greater than gravitational force. As a result, small particles can be difficult to remove from a surface once they deposit there.

The forces of primary importance to determining the characteristics of particles in the microelectronics industry are drag force, gravitational force, inertial force, diffusion, electrostatic forces, and adhesion forces. Some additional concepts, such as interception and particle agglomeration, are also important. These forces and concepts, and their interactions, are discussed in the next sections.

Drag Force

As a particle moves through a fluid, either gas or liquid, drag will act on it in a direction opposite to the direction of its motion. This drag arises from two sources: the first, form drag, comes from pushing fluid molecules out of the way so the particle can get through; the second, friction drag, comes from friction between the surface of the particle and the fluid molecules it passes through. For small particles moving slowly, drag force can be expressed through Stokes' Law,

$$F_d = 3 \pi \eta \, d \, v / C_c \, , \qquad (2\text{-}5)$$

where η is fluid viscosity, d is particle diameter, v is the relative velocity between the particle and the fluid, and C_c is the "Cunningham slip correction factor." Strictly speaking, Stokes' Law is valid only in the creeping flow regime, where Reynolds number is substantially less than unity, as given in Equation (2-6),

$$Re = d \, v \, \rho_g / \eta \ll 1 \, , \qquad (2\text{-}6)$$

where ρ_g is gas density. As a practical matter, Equation (2-6) will almost always be satisfied for the particles of concern to the microelectronics industry. It will not be satisfied, and Equation (2-5) will give substantial errors, for large particles moving quickly; for example, for a bullet, a thrown baseball, or a moving automobile.

If Re > 1, drag force depends upon the projected area of the object, its kinetic energy, and an empirical term called the drag coefficient, C_D. For a sphere,

$$F_d = (\pi d^2/4)(\rho_g v^2/2)C_D \quad . \tag{2-7}$$

Drag coefficient varies with Reynolds number as shown in Figure 2-2. If Re < 0.1, C_D = 24/Re and Equation (2-7) simplifies to Stokes' Law, Equation (2-5) given above. For larger values of Re, the dependence of C_D on Re is complex, and no simple equation describes the relationship. Refer to the resources cited at the beginning of this chapter for further information on this subject.

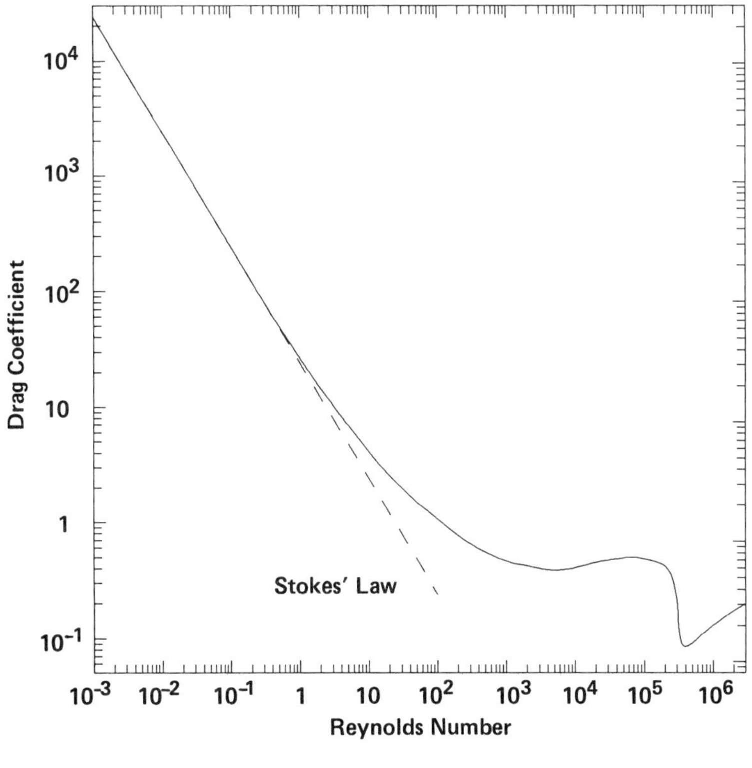

Figure 2-2. Drag coefficient, C_D, vs. Reynolds number, Re, for spheres.

The Cunningham slip correction factor, C_c, is sometimes necessary to correct Stokes' Law for its assumption that the fluid around the particle is continuous. Although $C_c \rightarrow 1.0$ for particles larger than a few micrometers in diameter, $C_c > 1$ for submicrometer particles.

$$C_c = 1 + [\lambda/d]\,[2.514 + 0.800\,\exp(-0.55\,d/\lambda)] \;, \tag{2-8}$$

where λ is the mean free path of the gas molecules, which depends on the temperature and pressure of the gas. For air at 1 atm and 20 °C, λ is 0.066 μm and Equation (2-8) reduces to

$$C_c = 1 + 0.16/d_{\mu m} \;, \tag{2-9}$$

where $d_{\mu m}$ is particle diameter in micrometers. Equation (2-9) shows that the slip correction factor for a 1-μm particle will be 1.16, so that drag force on this particle will be 1/1.16 or 0.86 as much as determined from Stokes's Law uncorrected. Figure 2-3 is a plot of slip correction factor against particle diameter for particles from 0.01 to 10 μm in diameter. The slip correction factor becomes very important for particles smaller than 0.5 μm or so in size.

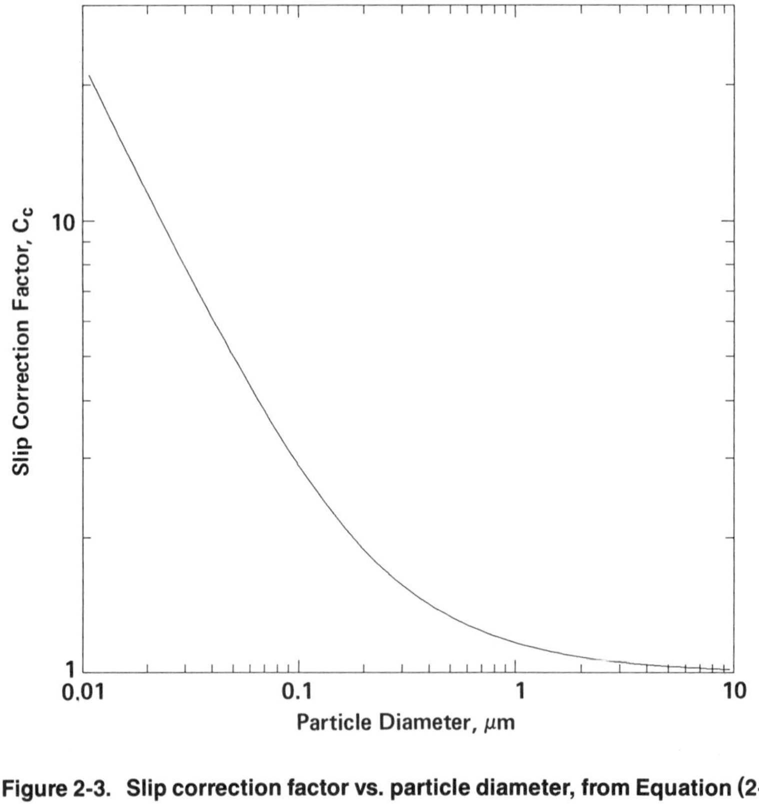

Figure 2-3. Slip correction factor vs. particle diameter, from Equation (2-8).

Gravity Force

Gravity, among all forces that can act on aerosol particles, is especially important because it acts on all particles all the time. It is the product of particle mass (volume times density) multiplied by the acceleration of gravity:

$$F_g = [\pi\, d^3/6]\, (\rho_p - \rho_f)\, g\ , \tag{2-10}$$

where $(\rho_p - \rho_f)$ is the difference between particle density, ρ_p, and the density of the fluid through which it moves, ρ_f.

The terminal velocity with which a particle falls by gravity, v_t, can be determined by setting the drag force from Equation (2-5) equal to the force of gravity from Equation (2-10).

$$v_t = [d^2\, (\rho_p - \rho_f)\, C_c/(18\, \eta)]\, g = \tau\, g. \tag{2-11}$$

The bracketed term in Equation (2-11) is the particle relaxation time, τ, and has the dimensions of time. Relaxation time is an important, fundamental property of aerosol particles and will appear again in our discussion of particle properties.

Figure 2-4 is a plot of terminal settling velocity against particle aerodynamic diameter as determined from Equation (2-11). Although gravity causes large particles to fall quickly, particles of submicrometer aerodynamic diameter fall very slowly and at a very low rate as compared to the velocity of the air currents in a room. Although gravity will always add to the number of particles that collect on a wafer surface as long as particles are present in the air above that surface, the deposition rate will be low if the particles are all small.

Inertia

In accordance with the laws of physics, a particle projected in any direction will tend to continue traveling in that direction. Outside forces such as gravity or drag will alter the direction and velocity of the particle. Imagine that a small particle is thrown with velocity v_o into a room filled with still air. If the particle is small enough, the gravitational force that acts on it will be negligible; nevertheless, drag will continue to act on the particle and slow it down. Eventually the particle will slow to the point that it stops. A force balance can be set up for this particle to describe its motion. Integration of the resultant differential equation allows calculation of how far the particle will travel before it stops. This distance is called the particle "stopping distance," S:

$$S = [d^2\, \rho_p\, C_c/(18\, \eta)]\, v_o = \tau\, v_o. \tag{2-12}$$

Again, the particle's relaxation time, τ, is important to determine the particle's behavior.

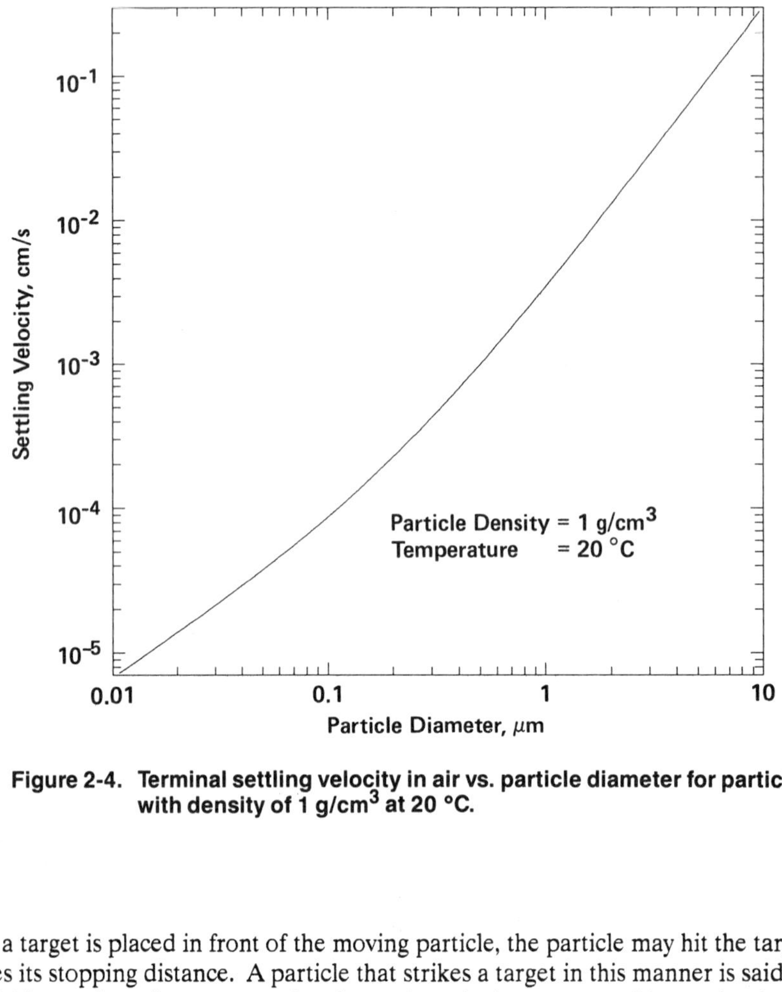

Figure 2-4. Terminal settling velocity in air vs. particle diameter for particles with density of 1 g/cm³ at 20 °C.

If a target is placed in front of the moving particle, the particle may hit the target before it reaches its stopping distance. A particle that strikes a target in this manner is said to collect by inertial impaction. A parameter that determines the importance of impaction is the ratio of stopping distance, S, to a characteristic dimension of the target, D, a quantity known as the Stokes number, Stk:

$$\mathrm{Stk} = S/D = [d^2 \rho_p C_c/(18\,\eta)]\, v_o/D = \tau\, v_o/D \ . \qquad (2\text{-}13)$$

Impaction is an important mechanism by which filters remove particles from gases and liquids. For these applications, characteristic dimension, D, is usually taken to be the diameter of the collecting element, often a fiber. Equation (2-13) shows that impaction is important for large particles (d) moving quickly (v_o) and where the target diameter (D) is small. It is less important than diffusion (described below) for collection of submicrometer particles and is usually unimportant as a mechanism for collection of particles on wafers.

Brownian Motion and Diffusion

Just as large ships pass through the ocean unaffected by the action of individual waves, so large aerosol particles are unaffected by collisions with individual molecules of the fluid that surrounds them. Nevertheless, the same waves that leave a ship unaffected can toss about a small dinghy; similarly, bombardment by fluid molecules greatly affects the motion of particles that are sufficiently small. Particle motion due to bombardment by fluid molecules is called Brownian. Brownian motion becomes important for particles less than 1 μm or so in diameter.

The average distance, x, that a particle travels due to its Brownian motion depends on its diffusivity, D, and on time, t,

$$x = (2 D t)^{1/2} \tag{2-14}$$

where

$$D = C_c k T/(3 \pi \eta d) \ . \tag{2-15}$$

Here, k is Boltzmann's constant, 1.38 x 10^{-16} dyne cm/°K; T is absolute temperature, °K; and the other quantities are as defined before. Note that particle diffusivity increases for smaller particles, so that smaller particles will, on average, travel farther over time by Brownian motion than will larger particles. Figure 2-5 is a plot of distance traveled after 1 min against particle diameter for Brownian motion and for gravitational settling, from Equations (2-14) and (2-11).

Gas molecules undergoing Brownian motion that collide with a surface will rebound. In contrast, small particles that collide with a surface generally stick. As a result, the particle concentration is lower near the surface than in the bulk gas. A gradient in concentration becomes established near the surface. This gradient produces a net flux of particles from the gas to the surface, according to Fick's second law, through the process of diffusion. The rate that particles collect on a surface by diffusion is

$$J = n_o [D/(\pi t)]^{1/2} \ , \tag{2-16}$$

where J is the diffusion flux in particles/(area x time) and n_o is the particle concentration in the gas far from the surface in particles per volume. This equation is based on the assumptions that the surface is suddenly immersed in aerosol that has particle concentration n_o, and that the particle concentration far from the surface remains n_o.

The number of particles per unit area that collect by diffusion on a surface in time t, N(t), can be found by integrating Equation (2-16) over the time that diffusional deposition occurs, assuming the concentration in the gas far from the surface remains constant at value n_o. The result is Equation (2-17). Here, time is measured from the instant when the surface is first immersed in the aerosol:

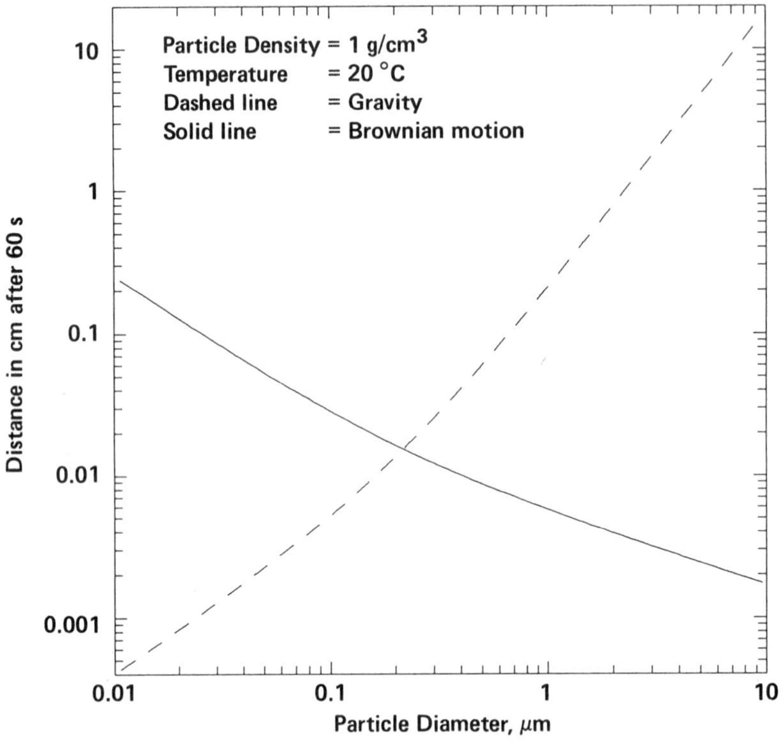

Figure 2-5. Distance traveled after 60 s due to gravitational settling and by Brownian motion.

$$N(t) = 2\, n_o\, (D\, t/\pi)^{1/2}. \tag{2-17}$$

Diffusion is an important process whereby small particles collect on surfaces. Like gravity, it always occurs. As shown in Equation (2-17), the only ways to reduce the degree of surface contamination by diffusion are to reduce the particle concentration in the gas, n_o, and to reduce the time the surface is exposed, t. Because particle diffusivity increases with temperature as shown in Equation (2-15), diffusional deposition will increase with temperature as well. Whereas surface contamination by gravitational settling is most important for large particles, contamination by diffusion is most important for small particles. Figure 2-6 is a plot of the number of particles that collect per square centimeter of horizontal surface due to gravitational settling and due to diffusion.

The above discussion is concerned with diffusion caused by Brownian motion. Although this mechanism causes particles to move only small distances, on the order of

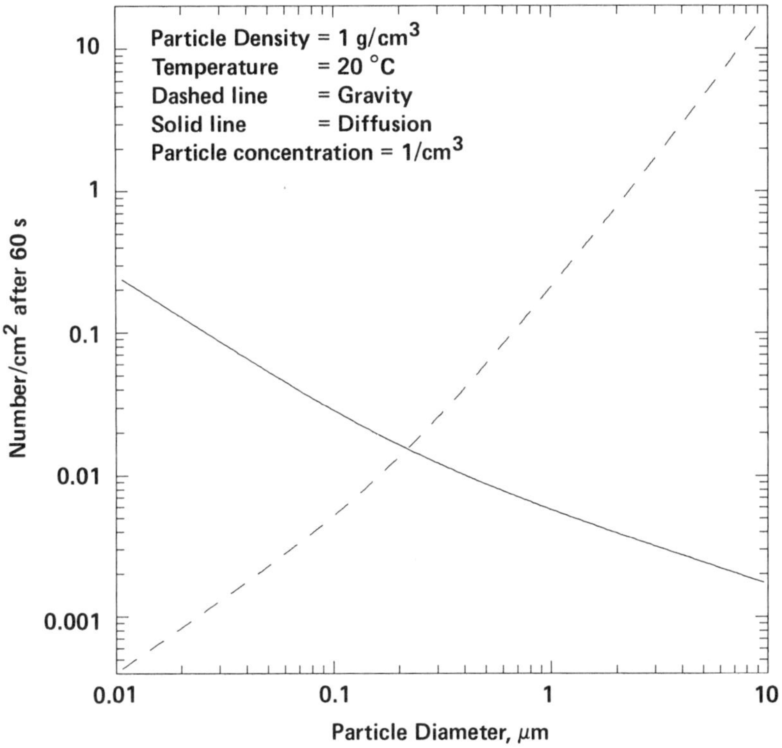

Figure 2-6. Number of particles collected per square centimeter of horizontal surface vs. particle diameter for collection by gravity and by diffusion.

millimeters or less, it is important because it can bring particles through the thin boundary layer that surrounds a surface to collect on that surface.

Particles move larger distances, on the order of meters, by convective diffusion, defined as particle motion caused by large-scale flow of the aerosol. For example, as air moves from one side of a room to the other, driven by thermal gradients or by the ventilation system, the particles in that air move along with it by convective diffusion. Convective diffusion is important to bring aerosol from the place where it is generated to a place where the particles in the aerosol can collect on a surface by Brownian diffusion, gravitational settling, or another collection process.

Electrostatic Forces

Particles can carry an electrical charge, acquired by contact with ions that are naturally present in the air with a concentration of about $10^3/cm^3$. These ions arise because cosmic rays continuously bombard the air with energy high enough to ionize air molecules that they strike.

Because both positive and negative ions are present in the air, particles can have either positive or negative charge. Larger particles are likely to acquire multiple charges.

For particles larger than about 0.02 μm, a normal distribution describes the probability, f_n, that a particle of diameter d has "n" charges, either positive or negative:

$$f_n = [2/(\pi\, d\, k\, T)^{1/2}]\, \exp[-n^2 e^2/(d\, k\, T)] \ , \tag{2-18}$$

where k is Boltzmann's constant, 1.38×10^{-16} dyne cm/°K; T is absolute temperature, °K; and e is the charge on a single electron, 4.8×10^{-10} statcoulombs. The distribution of particle charges given in Equation (2-18) is called the Boltzmann equilibrium charge distribution and applies to particles not charged intentionally. If particles are charged intentionally, they can carry many more electronic charges than the number specified by Equation (2-18). As a practical matter, very few particles smaller than about 0.01 μm carry any charge at all. The average number of electronic charges carried by a particle varies with particle size, and for particles larger than about 0.1 μm the average is approximated by

$$n = 2.37\, (d_{\mu m})^{1/2} \tag{2-19}$$

where "n" is the average number of charges calculated without regard to sign, and $d_{\mu m}$ is particle diameter in micrometers. Figure 2-7 shows the average number of electronic charges on a particle plotted against particle diameter. Both the actual average number of charges determined from the Boltzmann distribution and the number of charges calculated using Equation (2-19) are plotted.

Electrostatic force attracts opposite charges and repels like charges according to Coulomb's Law:

$$F = (n_1 e)(n_2 e)/r^2 \, . \tag{2-20}$$

Here, n_1 and n_2 are the number of electronic charges on objects one and two that interact to cause the electrostatic force, e is again the charge per electron, and r is the distance between the electronic charges. Note that cgs units must be used in Equation (2-20); if other systems of units are used in which electronic charge is not measured in statcoulombs, a constant must be used in the equation to preserve dimensional consistency.

Interception

The discussion above has considered particles to be points in space on which drag, gravity, or electrostatic forces act; however, actual particles have physical dimensions. As a result, a particle can deposit on a surface even if its center remains away from that surface if particle shape is such that some portion of the particle extends away from its center to touch the surface. Collection by this means is called interception.

The effect of interception is to increase particle collection above that predicted by consideration of forces that act on points in space. The importance of interception increases as particle size increases and as particle shape becomes more extreme.

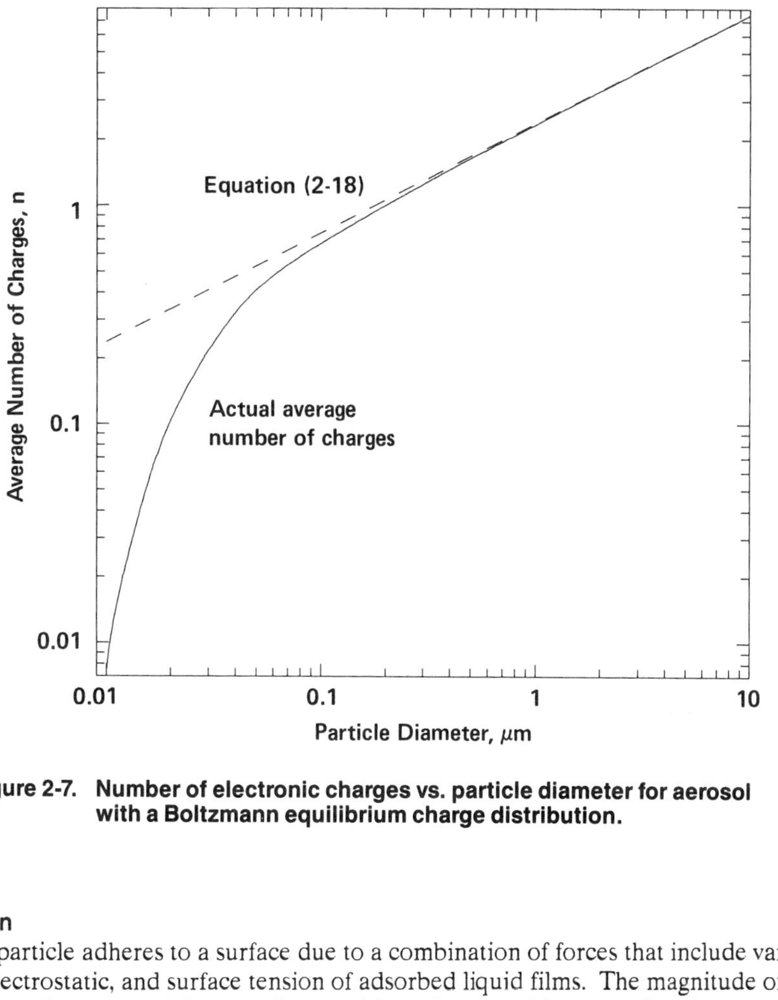

Figure 2-7. Number of electronic charges vs. particle diameter for aerosol with a Boltzmann equilibrium charge distribution.

Adhesion

A particle adheres to a surface due to a combination of forces that include van der Waals, electrostatic, and surface tension of adsorbed liquid films. The magnitude of adhesion forces depends on material properties, particle and surface history, and environmental conditions. Although theory is available to predict the adhesive force for a sphere of known properties under ideal conditions, this situation rarely arises. As a result, theory is not adequate to describe particle adhesion adequately for practical situations. For hard materials and clean surfaces, the following equation describes adhesive force based on measurements conducted at 25 °C:

$$F_{adh} = 150\, d\, [0.5 + 0.0045\, (\%RH)]\,, \qquad (2\text{-}21)$$

where adhesive force is in dynes, particle diameter d is in centimeters, and %RH is the relative humidity in percent.

The adhesive force can be much greater than other forces for small particles. Figure 2-8 is a plot of adhesive force and gravitational force against particle diameter. For example, this figure shows that adhesion calculated for relative humidity of 50 percent is over three orders of magnitude greater than the gravitational force for particles 100 μm in diameter, and is nearly ten orders of magnitude greater for a 0.1 μm particle. Clearly, once particles become attached to a surface, they can be very difficult to remove.

Agglomeration

The particles within an aerosol may collide as they undergo Brownian motion. If they do, they are likely to stick together, forming an agglomerate. If all particles are the same size, agglomeration is described by

$$1/n(t) = 1/n_o + (4\,k\,T\,C_c\,t)/(3\,\eta)\,. \tag{2-22}$$

Here, n(t) is concentration after time t, n_o is concentration at time zero, and the other terms are as discussed in conjunction with Equation (2-15). If particle concentration is low, as in the

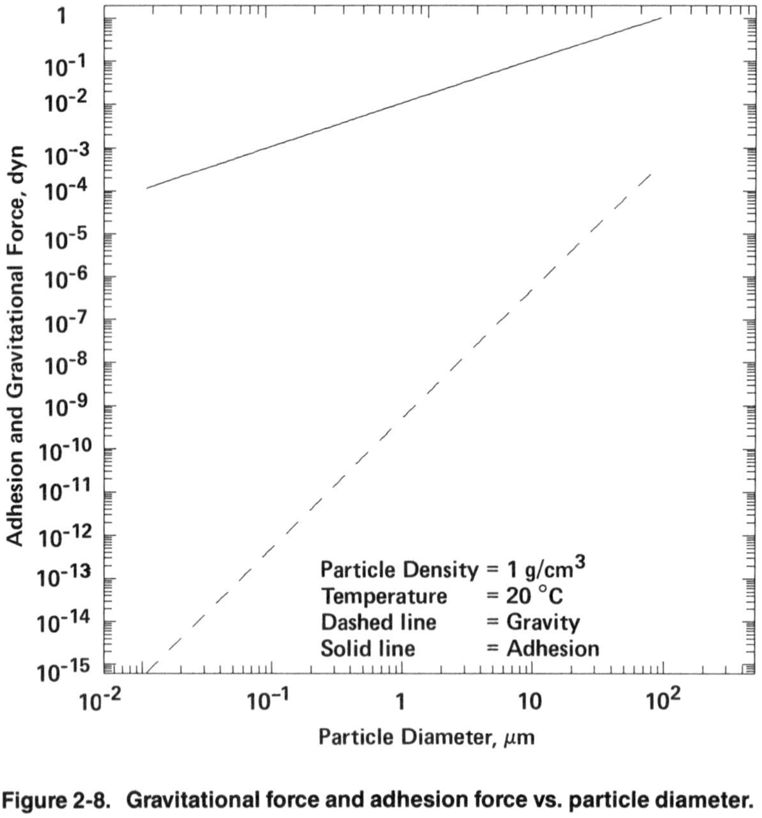

Figure 2-8. Gravitational force and adhesion force vs. particle diameter.

clean rooms of the microelectronics industry, Equation (2-22) suggests agglomeration will occur so slowly that particle concentration in the room will be stable unless particle sources or collection sinks are present.

SUMMARY

The behavior and properties of particles in aerosols or attached to surfaces can be described and understood using principles of Newtonian physics. For particles with sizes of interest to the microelectronics industry, generally 1 μm or smaller, inertia has less effect on particle deposition than diffusion and gravity. Once small particles collect on a surface, they can be very difficult to remove because adhesion forces are large relative to dislodgement forces.

HOMEWORK PROBLEM

The concentration of 0.5 μm particles in a clean room is 100 per cubic foot of air.

- How long will it take before the number concentration of particles on a clean wafer reaches 0.01/cm^2 if the clean wafer is placed on a horizontal surface in this room? (Answer: 48 minutes, collection primarily by gravitational settling.)
- How long will it take for the number concentration of particles on the surface to reach the same value if the wafer is stored vertically (edge up) in the same room? (Answer: 115 days, collection primarily by diffusion.)

3. PARTICLE SIZE DISTRIBUTIONS

DAVID S. ENSOR Research Triangle Institute, Research Triangle Park, North Carolina

INTRODUCTION

Significance of Particle Size Distribution

The physical behavior of a particle is determined largely by its size. The settling velocity of the particle, light scattered from the particle, and adhesion of a particle to a surface are examples of particle-size-dependent physical properties important in contamination control. It follows that selection of the measuring instruments, cleaning techniques, and understanding detrimental effects from particles are critically dependent on the particle size.

However, populations of particles normally found suspended in a volume of gas are rarely all of one size or monodisperse. Monodispersed particles are rarely found in nature, and the generation of monodispersed test aerosols in the laboratory is the result of decades of research. Populations of particles usually contain a range of sizes and are classified as polydisperse.

The concentration of particles in a gas is also an important property. A particle size distribution is the concentration of particles as a function of particle size. The concentration may be expressed as a number, surface, mass, or other property per volume of gas. The concentration measurement method is selected to describe the properties of the aerosol in a way directly related to the problem of interest.

It has been observed that populations of particles tend to conform to distinct distributions that can be approximated by equations. At the present time, no theoretical explanations exist for the selection of a particular model over another. Particle size distributions are a result of the physical phenomena forming the particle, such as nucleation and abrasion, as well as the events modifying particle size distributions, such as condensation, coagulation, sedimentation, surface deposition, and filtration.

This chapter is an overview of the analysis of size distribution data and the application of size distribution statistics. Also, model distributions are introduced with a discussion of areas of application.

Particle size distribution statistics is a well-developed field in environmental sciences, meteorology, and industrial hygiene. The understanding of particle size distributions in modern clean areas is currently a developing topic. For more detailed information, the reference books by Hinds (1982), Friedlander (1977), and Dennis (1976) should be consulted.

Application

Size distribution statistics permit one to describe data in compact ways. It is important to select the representation that reflects the limitations of the measurement method and the planned application of the results. One of the complications in the aerosol field has been the limitations in instrumentation. Often the available instruments do not measure directly the property of the particles of interest.

The usual way of expressing particle size data in clean rooms is the "class" or total number concentration (number per cubic feet of air) greater or equal to 0.5 μm. This approach may have been developed from the concept that a limiting small particle size exists that will not damage the process or is smaller than a critical tolerance in a precision machine. In semiconductors, the critical particle size is one-tenth the lateral design rule. However, it has been suggested by Larrabee (1987) that the dimensions in the vertical direction such as thin films on the order of nanometers in thickness may be damaged by particles much smaller than 0.1 μm. It is possible that, as the dimensions of semiconductors become even smaller, all particles present in the gas surrounding the chip may be potentially damaging. If this is the case, the entire size distribution will need to be characterized.

In addition, the reporting of class does not provide sufficient information needed to understand the transport and behavior of contaminants in processes. The detailed features of the particle size distribution and computation of other size distribution formats such as the differential distribution are needed to document and ultimately control the formation, transport, and damaging effects of particles.

PARTICLE SIZE DATA

Discrete Distributions

Particle size distribution data are obtained by measuring the concentration of particles as a function of size. Many ways are available to measure particle size distributions ranging from automatic counting to microscopic examination with manual counting and logging. A common feature of all approaches is that the particle concentration is sorted into channels or "bins" of finite particle diameter. A large number of individual particle measurements can be summarized in a compact histogram form. An example is shown in Table 3-1, which is a summary of particle counts measured with a laser spectrometer in an office area. The instrument sorts the signal of pulses of light scattered from the particles into 16 channels from 0.09 to 5 μm.

The data in Table 3-1 are plotted as a histogram in Figure 3-1 in linear coordinates. The size ranges of the channels are used to segregate the particle counts, and this case corresponds to the characteristics of the particular instrument. In this case, roughly a logarithmic increase in the width of the particle diameter channels has been used to classify the counts. The reason for selecting narrower channel widths for small particles is to resolve features of the curve in this rapidly changing region. Airborne particle size distributions are usually skewed to the small particle side of the spectrum.

One limitation of this practice lies in interpretation of histogram plots of raw data. The plot may be misleading because relatively more particles will be counted in this channel as the increment becomes wider. This may be significant when different instruments are compared. Also, much of the insight into particle size distribution formation and behavior is obtained by looking at the plotted data. Bias introduced by channel widths could be misleading.

Table 3-1. Tabulated Size Distribution Data

Particle diameter range (μm)	(Number/channel) (gas volume) (number/cm^3)
0.090 - 0.110	1.96×10^{3}
0.110 - 0.150	3.12×10^{3}
0.150 - 0.200	2.43×10^{3}
0.200 - 0.250	9.92×10^{2}
0.250 - 0.300	5.10×10^{2}
0.300 - 0.400	4.09×10^{2}
0.400 - 0.500	1.23×10^{2}
0.500 - 0.650	6.37×10^{0}
0.650 - 0.800	1.35×10^{0}
0.800 - 1.00	5.97×10^{-1}
1.00 - 1.25	1.87×10^{-1}
1.25 - 1.50	5.54×10^{-2}
1.50 - 2.00	2.50×10^{-2}
2.00 - 2.50	7.50×10^{-3}
2.50 - 3.00	8.30×10^{-4}
3.00 - 5.00	4.89×10^{-4}

Histogram data are usually normalized by the channel widths to eliminate instrumental artifacts. The data are normalized by dividing channel particle counts by the diameter increment of the size channels. This practice is given by:

$$n_i = \Delta n / \Delta dp_i \ , \tag{3-1}$$

where:

n_i = normalized concentration
Δn = raw concentration
Δdp_i = channel width.

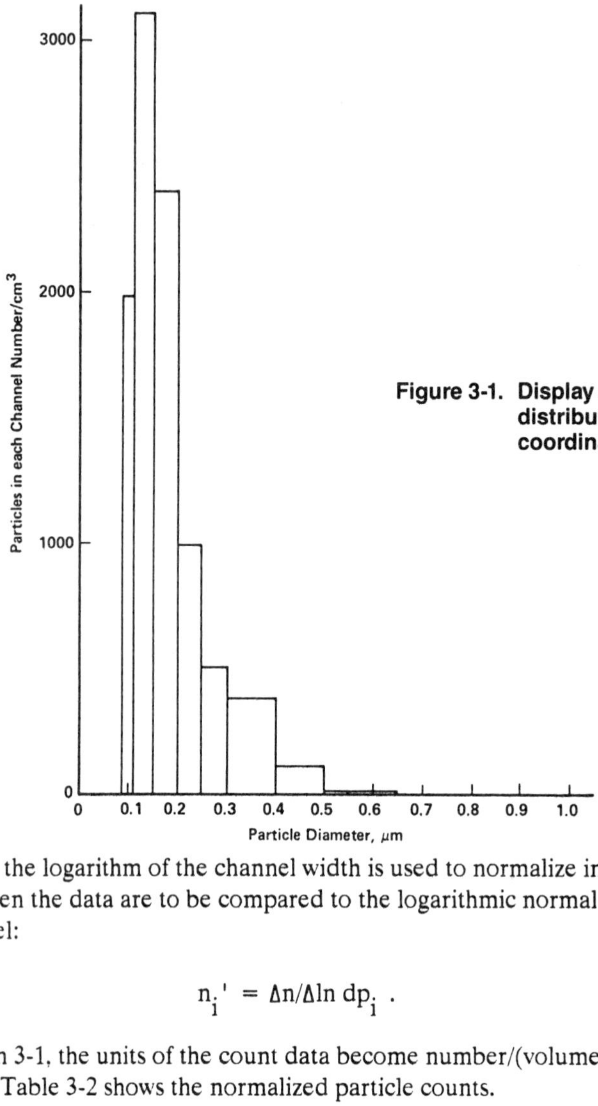

Figure 3-1. Display of particle distribution in linear coordinates.

Sometimes the logarithm of the channel width is used to normalize instead of the linear channel width when the data are to be compared to the logarithmic normal (log-normal) size distribution model:

$$n_i' = \Delta n / \Delta \ln dp_i \ . \tag{3-2}$$

In Equation 3-1, the units of the count data become number/(volume μm); in Equation 3-2, number/vol. Table 3-2 shows the normalized particle counts.

The normalized histogram is shown in Figure 3-2 in linear coordinates and in Figure 3-3 in logarithmic coordinates with a logarithmic normalization. Interpreting the data is dependent on these steps. For example, in Figure 3-3 the normalized distribution appears to peak at 0.12 μm. It is a good possibility that the data follow the log-normal distribution (to be discussed later). Therefore, a mirror image of the right-hand side of the curve may exist on the left-hand side below the size resolution of the laser spectrometer.

The total number of particles is given by:

$$N = \sum_i n_i \Delta dp_i \ , \tag{3-3}$$

Table 3-2. Normalized Size Distribution Data

Particle diameter range (μm)	$\Delta n_i/\Delta dp_i$ (number/cm^3 μm)	$\Delta n_i/\Delta \ln dp_i$ (number/cm^3)
0.090 - 0.110	9.83×10^4	9.81×10^3
0.110 - 0.150	7.80×10^4	1.00×10^3
0.150 - 0.200	4.82×10^4	8.39×10^2
0.200 - 0.250	1.58×10^4	4.45×10^2
0.250 - 0.300	1.02×10^4	2.80×10^2
0.300 - 0.400	4.09×10^3	1.42×10^2
0.400 - 0.500	1.23×10^3	5.52×10^1
0.500 - 0.650	4.25×10^1	2.43×10^1
0.650 - 0.800	9.00×10^0	6.50×10^0
0.800 - 1.00	2.98×10^0	2.68×10^0
1.00 - 1.25	7.48×10^1	8.38×10^{-1}
1.25 - 1.50	2.21×10^{-1}	3.24×10^{-1}
1.50 - 2.00	5.00×10^{-2}	8.68×10^{-2}
2.00 - 2.50	1.50×10^{-2}	3.36×10^{-2}
2.50 - 3.00	1.65×10^{-3}	4.56×10^{-3}
3.00 - 5.00	2.44×10^{-4}	9.77×10^{-4}

where:

n_i = normalized histogram
Δdp_i = size increment
N = total number concentration of particle.

Figure 3-3 with a log increment is:

$$N = \sum_i n_i' \, \Delta \ln dp_i \, . \tag{3-4}$$

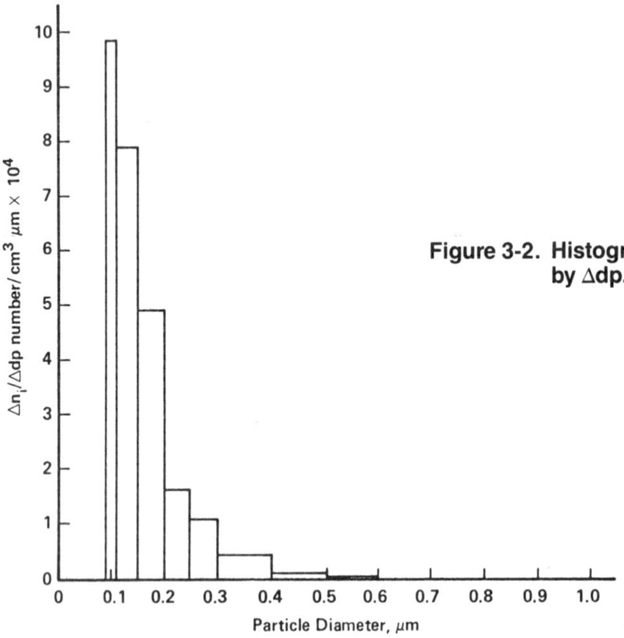

Figure 3-2. Histogram normalized by Δdp.

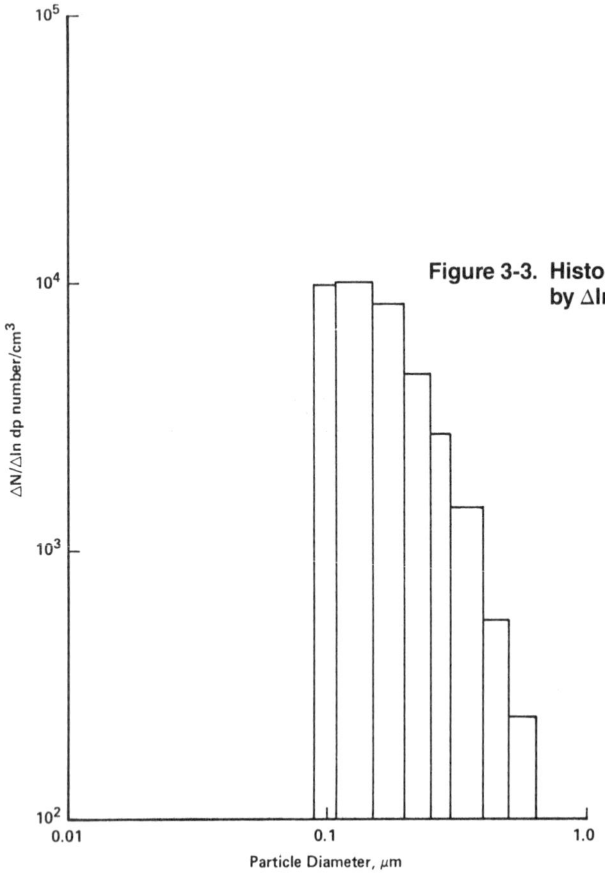

Figure 3-3. Histogram normalized by Δln dp.

An additional normalization is often made by dividing each increment by the total concentration, thus producing a fractional histogram. A fractional histogram allows direct comparison of the shapes of distributions taken under various conditions:

$$f_i = n_i/N \quad , \tag{3-5}$$

$$f_i' = n_i'/N \quad , \tag{3-6}$$

$$\Sigma f_i = \frac{\Sigma n_i \, \Delta dp_i}{N} = 1.0 \quad , \text{ and} \tag{3-7}$$

$$\Sigma f_i' = \frac{\Sigma n_i' \, \Delta \ln dp_i}{N} = 1.0 \quad , \tag{3-8}$$

where

N = total number concentration of particles.

The shape of the distributions is shown in Figures 3-2 and 3-3 depending on the choice of log or linear size increment.

Continuous Functions

One further generalization of particle size distribution is to assume a continuous function. This approach parallels statistical developments in other fields that consider velocity and mass distributions. Such a practice appears to be valid when sufficient numbers of counts have been made. However, when concentrations are very low, the counts may be random due to Poisson sampling statistics and may not describe a smooth curve. The channel may be characterized by the arithmetic mean computed by:

$$d_{am\,i} = \frac{(d_i + d_{i+1})}{2} \quad , \tag{3-9}$$

or the geometric mean given by:

$$d_{gm\,i} = (d_i \bullet d_{i+1})^{1/2} \quad . \tag{3-10}$$

Our example is shown in Figure 3-4 in which a continuous function is formed by joining the geometric mean of each increment.

This development leads to a number of definitions. The fraction of the total number of particles df having diameters between dp and dp + ddp is:

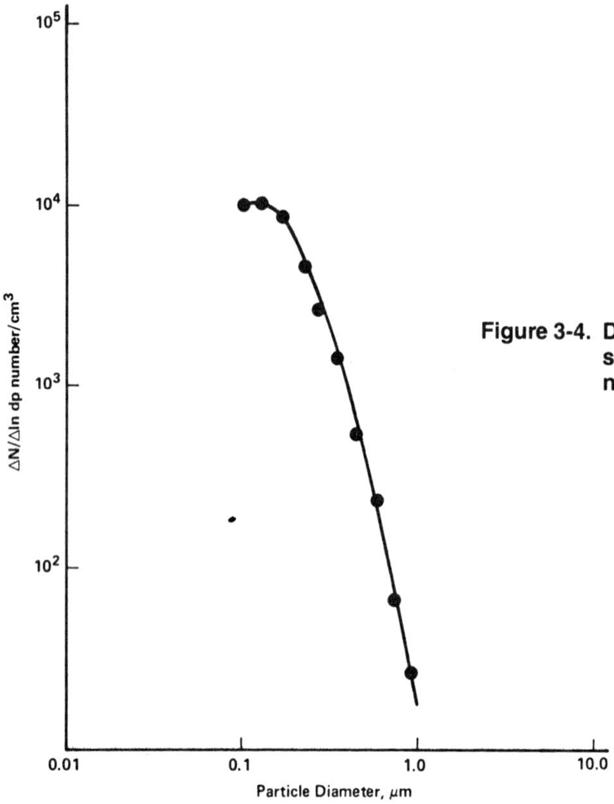

Figure 3-4. Differential particle size distribution normalized by Δln dp.

$$df = f(dp)\ ddp \quad , \tag{3-11}$$

where:

f(dp) = frequency function
ddp = a differential of particle size.

An analogous function can be developed for the log increment given by:

$$df' = f'(dp)\ d\ \ln dp \quad . \tag{3-12}$$

The area under the curve is:

$$\int_0^\infty f\ (dp)\ ddp = 1.0 \tag{3-13}$$

and

$$\int_0^\infty f'\ (dp)\ d\ \ln dp = 1.0 \quad . \tag{3-14}$$

The fraction of particles between diameters a and b can be computed for a small slice of the distribution:

$$fab = \int_a^b f(dp)\, ddp \quad . \qquad (3\text{-}15)$$

The fraction of particles for a single diameter is given by:

$$faa = 0 = \int_a^a f(dp)\, ddp \quad . \qquad (3\text{-}16)$$

The fraction of particles at a single diameter is zero because the incremental width approaches zero.

Cumulative Distributions

Smaller than

The cumulative distribution of particles smaller than a specific diameter, a, is related to the differential of the fractional distribution. $F_s(dp)$ is given by:

$$F_s(a) = \int_0^a f(dp)\, ddp \quad . \qquad (3\text{-}17)$$

The "smaller than" distribution is often used in air pollution studies because the small particles cause the most serious problems such as health or visual effects. Also, cascade impactors backed up with filters often used to obtain size-segregated particle samples are more accurate for the small particles than the large particles.

Larger than

A similar function can be defined for the distribution of particles larger than a specific diameter. The F_l is defined by:

$$F_l(b) = \int_b^\infty f(dp)\, ddp \quad . \qquad (3\text{-}18)$$

The "larger than" distribution is used in fields such as contamination control and protection of turbine blades in hot gas cleanup because of the need to protect critical parts from large particles. It is often used in sieve analysis because the size of large particles retained on the sieve are known with better certainty than the size of small particles falling through the sieve.

Relation to differential distributions

The fractional curves are formed from the slope of the cumulative curves as given by:

$$f(dp) = dF(dp)/ddp \quad . \qquad (3\text{-}19)$$

A practical application of these distributions is the analysis of noisy data. Often, the data can be smoothed as a cumulative curve and then the curve is numerically differentiated to obtain the differential curve. An example of cumulative distributions is tabulated in Table 3-3 and is shown in Figure 3-5.

Table 3-3. Tabulated Cumulative Size Distributions

Particle diameter range (micrometers)	"Smaller than" fractional	"Larger than" fractional
0.090 - 0.110	0.262	0.738
0.110 - 0.150	0.530	0.47
0.150 - 0.200	0.754	0.246
0.200 - 0.250	0.872	0.128
0.250 - 0.300	0.946	0.0535
0.300 - 0.400	0.984	0.0156
0.400 - 0.500	0.9991	0.0009
0.500 - 0.650	0.9997	0.0003
0.650 - 0.800	0.9999	0.0001
0.800 - 1.00	1.0	0.0
1.00 - 1.25	1.0	0.0
1.25 - 1.50	1.0	0.0
1.50 - 2.00	1.0	0.0
2.00 - 2.50	1.0	0.0
2.50 - 3.00	1.0	0.0
3.00 - 5.00	1.0	0.0

PARTICLE STATISTICS

Measure of Central Tendency

The display of the distribution is useful, but it may also be important to compute statistical properties of the distribution. A number of different statistics are often computed for size distributions depending on the application.

The arithmetic mean of a distribution is given by:

$$d_{nm} = \frac{\Sigma\, n_i dp_i}{\Sigma\, n_i} = \int_o^\infty dpf(dp)\, ddp \quad , \tag{3-20}$$

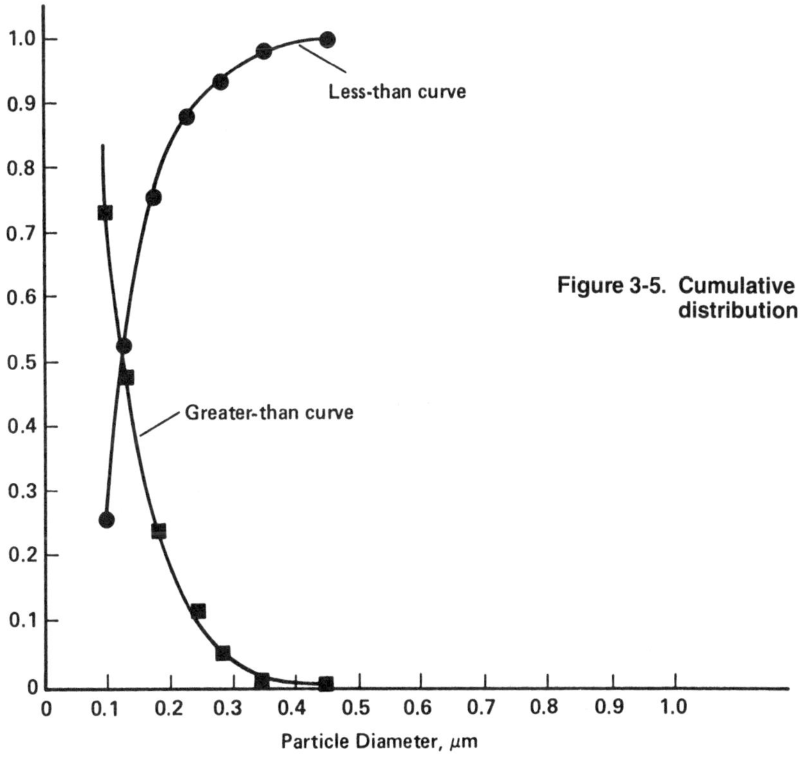

Figure 3-5. Cumulative size distribution.

where:

n_i = number of particles in size i
dp_i = diameter of channel i, either mean or geometric mean.

Other statistics often used to describe central tendency are median (cumulative point of 50 percent with corresponding diameter) and mode (most frequent size).

Particle size distributions are often skewed toward smaller sized particles as shown in the example. The geometric mean better represents the properties of these distributions.

The geometric mean for grouped data is given by:

$$d_{gnm} = (dp1^{n_i} \bullet dp2^{n_2} \bullet dp3^{n_3} \bullet dpn)^{1/N} \quad . \tag{3-21}$$

Taking the logarithm of the above equation:

$$\ln d_{gnm} = \frac{\Sigma\, n_i \, \ln dp_i}{N} \quad . \tag{3-22}$$

Measures of Diversity

The width of the distribution is important to determine the spread of the distribution. In many cases, this can greatly alter the effective properties of the aerosol.

The standard deviation is given by:

$$S_d = \frac{(\Sigma\ (dp_i - d_m)^2)^{1/2}}{N-1} \quad . \qquad (3\text{-}23)$$

In a similar manner, a geometric standard deviation can be computed:

$$\ln S_{dg} = \frac{[\Sigma\ (\ln dp_i - \ln d_m)^2]^{1/2}}{N-1} \quad . \qquad (3\text{-}24)$$

DISTRIBUTIONS BASED ON DIFFERENT PHYSICAL PROPERTIES

Background

The use of a number distribution may not be sufficient to address a specific problem. Because electric charge depends on the surface area distribution of the aerosol, insights into aerosol charge behavior may be obtained by computing surface distribution. However, the desired property may not be directly measured. For example, number may be determined by an optical particle counter, and elemental analysis on size-segregated samples may be collected with a cascade impactor. The two sets of data may need to be compared to assess a particular effect.

Transformation Between Various Types of Distributions

Distributions can be computed by multiplying each increment of the distribution with an appropriate weighing factor. For example, the mass distribution can by computed by the following formula:

$$m_i = n_i\ Mp_i \quad , \qquad (3\text{-}25)$$

where:

m_i = mass in each increment
n_i = number in each increment
Mp_i = mass of average particle in each increment.

In a similar way, surface area can also be computed. A note of caution--errors in the original distribution can be amplified during these calculations.

Moments of the Distribution

One of the important aspects of particle size distribution theory is that the moments have physical significance.

For example, the number mean as shown before is:

$$d_{nm} = \Sigma\, dp_i/N \quad . \tag{3-26}$$

The surface mean is given by:

$$d_{sm} = \Sigma\, s_i dp_i/S \tag{3-27}$$

where:

s_i = surface per increment
S = total particle surface.

And the mass mean is given by:

$$d_{mm} = \Sigma\, m_i dp_i/M \tag{3-28}$$

where:

m_i = mass in each increment
M = total mass.

In Equations (3-27) and (3-28), d_{sm} is the surface mean diameter, d_{mm} is the mass mean diameter, s_i is the incremental surface, and m_i is the incremental mass.

A generalized rule can be developed:

$$(d_{qm})_{\bar{p}} = \left[\frac{(\Sigma\, n_i\, d_i^{\,q}\, d^p}{\Sigma\, n_i d_i^{\,q}}\right]^{1/p} \quad . \tag{3-29}$$

When q = 0, the weighing is by number; when p = 1 and q = 0, the arithmetic mean is computed; when p = 1 and q = 3, the mass mean diameter is computed.

The distribution moments are (Friedlander, 1977):

0th = Concentration
1th = Diameter
2nd = Surface

3rd = Volume or mass
4th = Total surface area settling from a stationary fluid
5th = Mass flux of material settling
6th = Total light scattering when the particles are smaller than the wavelength of light.

DISTRIBUTION MODELS

Use of Models

Mathematical models provide a guide in interpretation of size distribution data. In addition, theoretical simulations of the effects of particles can be computed with an appropriate analytical expression.

Normal Distribution

The normal distribution results from the action of many random additive and subtractive effects on an initially homogeneous population. In particle statistics, normal distributions are only found in cases of a vary narrow size range. Examples are nearly monodisperse calibration aerosols. The normal distribution is given by:

$$f(dp)\,ddp = A \exp\left[\frac{-(dp-d_{nm})^2}{2\,S^2}\right] ddp \quad , \qquad (3\text{-}30)$$

where:

$$A = \frac{1}{S\sqrt{2\pi}}$$

d_{nm} = mean diameter by number
S = standard deviation
f(dp) = fractional size distribution
dp = particle diameter.

Logarithmic Normal (Log-Normal) Distribution

The log-normal distribution results when an initially homogeneous population is subjected to random effects that change the sizes by a multiplication process. As demonstrated in Figure 3-1, particle size distributions found in nature are skewed toward the smaller sizes, and often the log-normal distribution will fit the data very well. Examples are distributions formed by condensation and grinding. In addition, logarithmic normal distributions have properties that allow simple transformations between number, surface, and mass distributions.

The log-normal distribution is given by:

$$f(d) \quad d\ln dp = A \quad \exp\left[\frac{-(\ln dp - \ln d_{gnm})^2}{2\,(\ln S_g)^2}\right] d\ln dp \quad , \qquad (3\text{-}31)$$

where:

$$A = \frac{1}{\sqrt{2\pi}\ \ln S_g}$$

d_{gnm} = geometric mean diameter
S_g = geometric standard deviation
$f(d)'$ = fractional size distribution, log increment.

The equation forms the normal distribution in logarithmic coordinates as shown in Figure 3-6, plotted with the example data shown previously. Additional data taken with a differential mobility size analyzer (DMSA) covering the particle size range from 0.014 μm to 1.0 μm confirmed the conjecture that the particle size distribution was log-normal in form (Owen et al., 1988).

The transformations between different kinds of log-normal distributions are given by the Hatch-Choate equations for which the relationships for number, surface, and volume are presented. For more information, see Hinds (1982). As an example, the relationship between number mean diameter and mass mean diameter is given by:

$$d_{gnm} = d_{gnm} \exp\left(3 \ln^2 Sg\right) \ . \qquad (3\text{-}32)$$

The geometric standard deviations of the various distributions remain the same, while only the geometric mean diameter changes for the distributions.

The geometric mean diameter and geometric standard deviation are computed by plotting a cumulative distribution on logarithmic probability paper, as shown in Figure 3-7. The 50 percent point is the geometric mean diameter. The slope of the line is the geometric standard deviation as given by the following:

$$Sg = dp(84\%)/dp\ (50\%) \quad . \qquad (3\text{-}33)$$

Curved lines indicate a deviation from the log-normal equation or the possibility of two or more distributions with different mean diameters and standard geometric deviations.

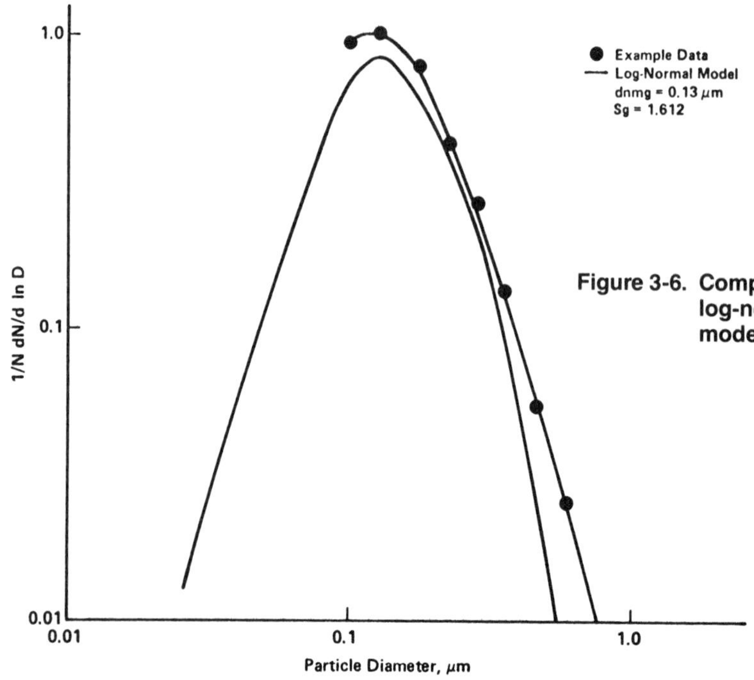

Figure 3-6. Comparison of the log-normal size distribution model to example data.

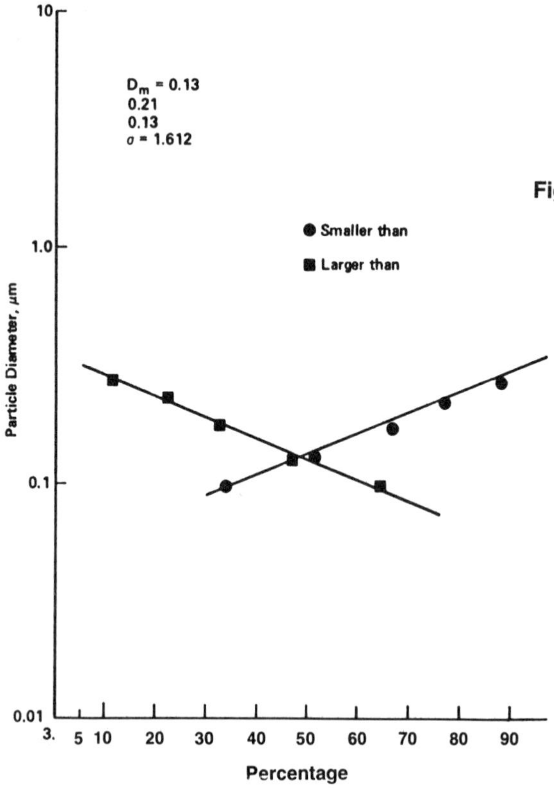

Figure 3-7. Cumulative size distribution on a log-probability.

Power Law Distribution

The power law distribution has the form:

$$f(d) = C\, dp^{-\beta} \,, \qquad (3\text{-}34)$$

where C and β are constants obtained by fitting the trend of particle concentration size data on logarithmic coordinates. This relationship is only valid between the small and large particle diameters in the data base.

Atmospheric differential size distribution data have been described by power law distributions (Junge, 1963).

The clean room class curves described in Federal Standard 209D are power law relationships and are shown in Figure 3-8. These curves are cumulative "larger than or equal to stated size," which are equal to the "larger than" relationships described earlier in the limit of very small channel widths. The slopes of the curves were obtained by averaging size distribution data from a wide range of indoor and industrial locations during the initial development of the standard. The particle size distribution actually found in clean rooms may be quite different, as reported by Donovan et al. (1984) and shown in Figure 3-9. Compliance to a class is determined by summing the particle counts greater than or equal to 0.5 μm with an optical particle counter.

SUMMARY

Particle size distribution or the concentration as a function of particle size is critical in the understanding of deposition rate, filter performance, and cleaning. Particle size distributions can be reported in many different ways depending on the use. Changing the way we interpret the size definition of killer particles may change the way particle concentrations are used in the future. Particle data can be reported as histograms, differential curves and cumulative distribution. Differential curves are used to compute potential transport properties of the particles.

Distribution models allow us to simplify the way we look at data and include normal, log-normal, and power law distributions. The class specification curves of Federal Standard 209D are cumulative distributions greater than or equal to stated particle size assuming a power law model. Actual particle size distributions in a modern clean room may be quite different than 209D depending on the activity.

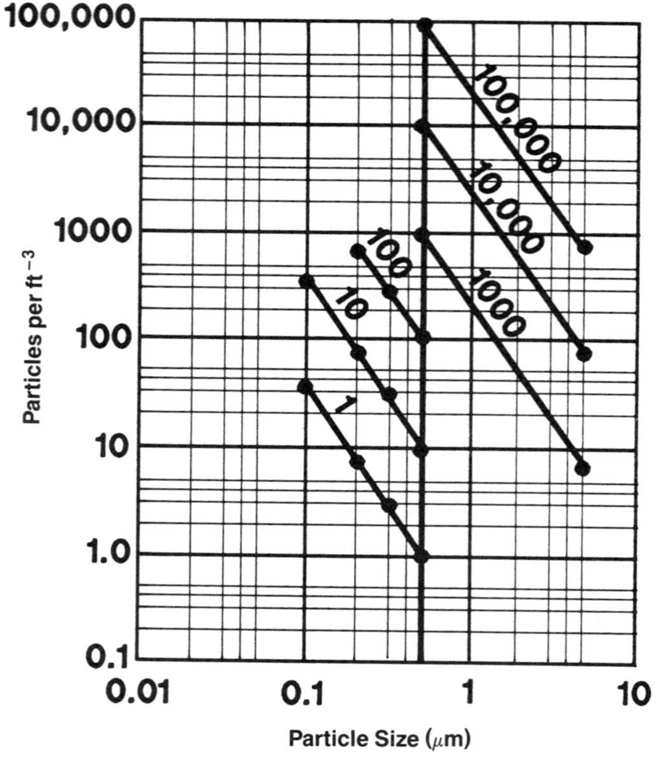

Figure 3-8. Class limits in particles per cubic foot of size equal to or greater than particle sizes shown.

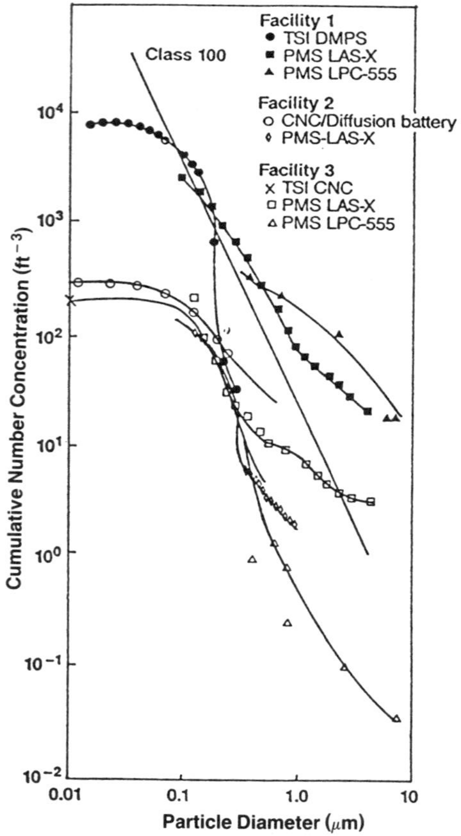

Note: To convert #/ft^3 to #/m^3 multiply the curves by 35.3.

Figure 3-9. Cumulative size distribution curves.

REFERENCES

Dennis, R. 1976. *Handbook on Aerosols*. TID-26608. National Technical Information Service, U.S. Department of Commerce: Springfield, VA.

Donovan, R. P., B. R. Locke, D. S. Ensor, and C. M. Osburn. 1984. The Case for Incorporating Condensation Nuclei Counters into a Standard for Air Quality. *Microcontamination*, (Dec.):39-44.

Federal Standard No. 209C. 1987. Federal Standard Clean Room and Workstation Requirements, Controlled Environment. General Services Administration: Washington, DC.

Friedlander, S. K. 1977. *Smoke, Dust and Haze*. Wiley: New York, NY.

Hinds, W. C. 1982. *Aerosol Technology*, Wiley: New York, NY.

Junge, C. E. 1963. *Air Chemistry and Radioactivity*. Academic Press: New York, NY.

Larrabee, G. B. 1987. Contamination Control in Semiconductor Device Manufacturing. *Microcontamination Control: The 21st One-Day Videoconference Seminars via Satellite*. IEEE, Piscataway, NJ.

Owen, M. K., D. S. Ensor, L. S. Hovis, W. G. Tucker, and L. E. Sparks. 1988. Particle Size Distributions of an Office Aerosol. Submitted to *Aerosol Science and Technology*.

4. LIGHT SCATTERING THEORY

DAVID LEITH University of North Carolina, Chapel Hill, North Carolina

Principles of light scattering are used in many instruments that measure the presence and properties of particles in air, in liquids, and on surfaces. For particles with sizes of interest in the microelectronics industry, theory that describes light scattering is complex. The objective of this chapter is to introduce the important concepts of light-scattering theory in a way that represents the principles that govern this topic, without extensive use of the complex equations that govern these concepts. As a result, the discussion here will be qualitative, although illustrated with figures developed using the full theory. For more complete discussion of this topic, you may wish to consult:

Hinds, William, *Aerosol Technology*, Wiley Interscience, New York 1982.

Kerker, M., *The Scattering of Light and Other Electromagnetic Radiation*, Academic Press, New York, 1969.

Reist, Parker C., *Introduction to Aerosol Science*, Macmillan, New York, 1984.

Van de Hulst, H.C., *Light Scattering by Small Particles*, Wiley, New York, 1957.

INTRODUCTION

A particle illuminated from a given direction will scatter light in all directions. The intensity of the scattered light is proportional to the intensity of the incident light, and depends strongly on the size of the particle. It also depends on the refractive index of the particle, on the wavelength and polarization of the incident light, and on the angle at which the particle is viewed.

Figure 4-1 shows an illuminated particle. The rays of incident light and scattered light, which intersect at the particle, define the "scattering plane." The angle between the incident and the scattered light rays is the "scattering angle." If the scattering angle is close to 0°,

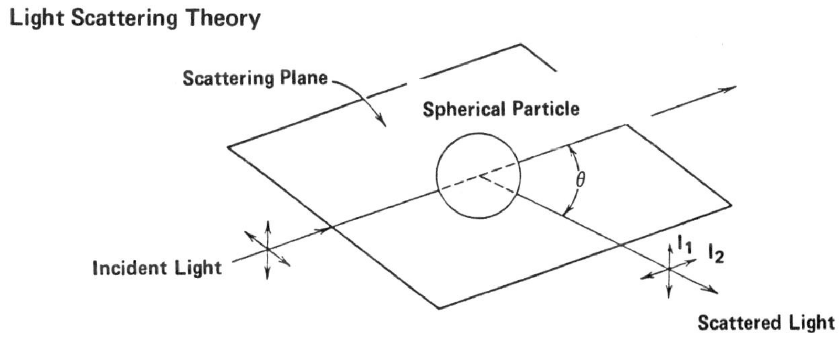

Figure 4-1. Light scattered by an isolated sphere.

scattering is said to be in the "forward" direction; that is, nearly aligned with the direction of the incident light. Conversely, if the scattering angle is close to 180°, scattering is said to be in the "backward" direction.

Normal, unpolarized light can be broken into two polarized components. The I_1 component is perpendicular to the scattering plane, whereas the I_2 component lies in the scattering plane. If the incident light is unpolarized, the scattered light will have both an I_1 and an I_2 component. If the incident light is polarized, the scattered light will be polarized in the same way; that is, if only the I_1 component is present in the incident light, then the scattered light will possess only an I_1 component.

The amount of light a particle scatters depends on its size. Particle size in this field is often expressed in terms of a dimensionless number called the size parameter, α, the ratio of particle circumference to light wavelength:

$$\alpha = \pi d/\lambda \ . \tag{4-1}$$

Here, d is particle diameter and λ is the wavelength of the illuminating light. Visible light has a wavelength of from about 0.4 μm (violet) to 0.7 μm (red). The wavelength of an HeNe laser used in many instruments to measure the optical properties of particles is 0.6328 μm, in the "red" region.

The refractive index, m, is the ratio of the speed light travels in a vacuum, c, to its speed through the material of concern, v_p. Light travels fastest in a vacuum, so m is always greater than one:

$$m = c/v_p \ . \tag{4-2}$$

For particles not in a vacuum, a relative index of refraction, m_r, can be used, defined as the ratio of the velocity of light in the fluid or medium, v_m, to the velocity in the particle, v_p:

$$m_r = v_m/v_p \ . \tag{4-3}$$

Because the velocity of light in air is practically the same as its velocity in a vacuum, $v_m \rightarrow c$ for aerosols so that $m_r \rightarrow m$. For particles in liquids, the value of v_m is significantly less than c so that m_r must be used.

The index of refraction is a complex number whose imaginary part is zero for nonabsorbing particles. Most particles that are electrical insulators, such as glass or sodium chloride, are nonabsorbing. In contrast, absorbing particles, whose imaginary part is nonzero, include materials that are electrically conductive, such as metals or carbon. The index of refraction for a material is not necessarily constant, as it may change with the wavelength of the incident light. Table 4-1 lists refractive indices for several materials.

Table 4-1. Indices of Refraction for Common Materials at Wavelength 0.589 μm

Material	Index of Refraction
Vacuum	1.0
Air	1.00025
Water	1.3330
Ice	1.309
Glass	1.50 - 1.72
Sodium chloride	1.544
Quartz	1.544
Polystyrene latex	1.5905
Diamond	2.417
Carbon	1.59 - 0.66i
Iron	1.51 - 1.63i
Copper	0.62 - 2.63i

Source: Hinds, 1982.

LIGHT SCATTERING BY ISOLATED SPHERES

Exact solutions to the equations that describe light scattered by isolated spheres have been available for many years. These relationships are important because they determine the performance of instruments that count and size particles in a gas or in a liquid. Lord Rayleigh was first to describe mathematically the light scattered by particles much smaller than the wavelength of light, $\alpha << 1$. For particles whose sizes are around the wavelength of light, $\alpha \cong 1$, the Mie solutions must be used to describe light scattering. For very large particles, larger than those of interest in the microelectronics industry, principles of geometric optics apply.

Light Scattering by Small Particles, $\alpha << 1$, Rayleigh Scattering

Rayleigh theory can be used to describe light scattered by small particles, $\alpha << 1$; $d < 0.05$ μm. The intensity of light scattered under this condition at angle θ is

$$I_1(\theta) = [I_o \pi^4 d^6/(8 R^2 \lambda^4)] [(m^2-1)/(m^2+2)]^2, \tag{4-4}$$

$$I_2(\theta) = [I_o \pi^4 d^6/(8 R^2 \lambda^4)] [(m^2-1)/(m^2+2)]^2 [\cos^2 \theta] \tag{4-5}$$

Here, I_o is the intensity of the incident light, R is the distance from the particle to the observer, and the other terms are as before. Note that $I_2 = I_1 \cos^2 \theta$ so that

$$I(\theta) = [I_o \pi^4 d^6/(8 R^2 \lambda^4)] [(m^2-1)/(m^2+2)]^2 [1 + \cos^2 \theta] \tag{4-6}$$

for the total intensity of light scattered. Equation (4-6) shows that the intensity of the scattered light depends very strongly on the wavelength of the incident light, λ^4, and on the diameter of the particles, d^6, for light scattered in the Rayleigh region.

Figure 4-2 is a plot of I_1, I_2, and $I = I_1 + I_2$ against scattering angle, θ, from Equations (4-4) to (4-6). It shows that the amount of light scattered perpendicular to the scattering plane, I_1, is independent of the scattering angle, as can be seen by inspection of Equation (4-4). In contrast, the amount of light scattered in the scattering plane, I_2, varies from a value identical to I_1 in the forward and backward directions to a value of zero for light scattered at 90°.

Figure 4-3 is a plot of $I = I_1 + I_2$ at 0° against particle size for particles from 0.01 to 0.05 μm in diameter. It shows the extreme dependence of light scattering on particle size within the Rayleigh region. As particle diameter increases by a factor of five, I increases by a factor of over 10^4.

Light Scattering by Particles in the Mie Region

For particles not significantly smaller than α, the equations that describe light scattering become more complicated.

$$I_1(\theta) = I_o\lambda^2 i_1/(8\pi^2 R^2) \ , \tag{4-7}$$

$$I_2(\theta) = I_o\lambda^2 i_2/(8\pi^2 R^2) \ , \tag{4-8}$$

$$I(\theta) = I_o\lambda^2 (i_1 + i_2)/(8\pi^2 R^2) \ , \tag{4-9}$$

where i_1 and i_2 are Mie intensity parameters that depend on α, particle refractive index, and scattering angle. The equations used to determine i_1 and i_2 are complex enough to require a computer for solution. Some solutions are given in the figures below.

Figures 4-4 through 4-9 are plots of relative intensity for I_1, I_2, and I against scattering angle for spheres 0.1, 0.3, 0.7, 1.0, 3.0 and 7.0 μm in diameter, when illuminated by an HeNe laser. All these figures are for a particle with refractive index of 1.544, a reasonable value for glass and the correct value for sodium chloride and quartz. Relative intensities are plotted in these figures rather than absolute intensities, which will depend on I_o and R.

The 0.10-μm particles that are the subject of Figure 4-4 are small enough to be near the Rayleigh region. The light-scattering pattern for these particles is similar to that shown in Figure 4-2. I_1 is nearly constant for all angles; I_2 falls off sharply as scattering angle approaches 90°.

Figure 4-5 shows the scattering pattern for particles 0.3 μm in diameter. Again, I_1 shows less variation with scattering angle than I_2, but is noticeably higher for scattering in the forward direction than for scattering in the backward direction. I_2 again shows a dip, but the angle at

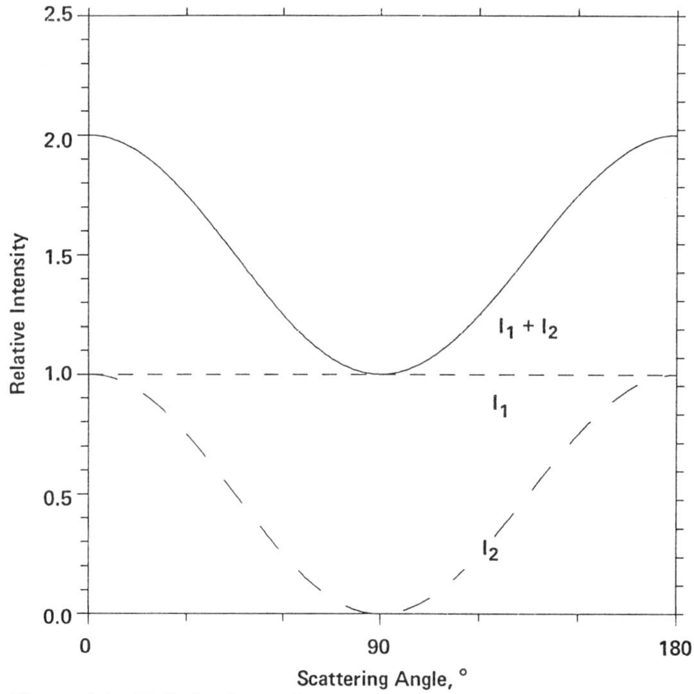

Figure 4-2. Relative intensity vs. scattering angle for a sphere smaller than about 0.05 μm in diameter (Rayleigh region).

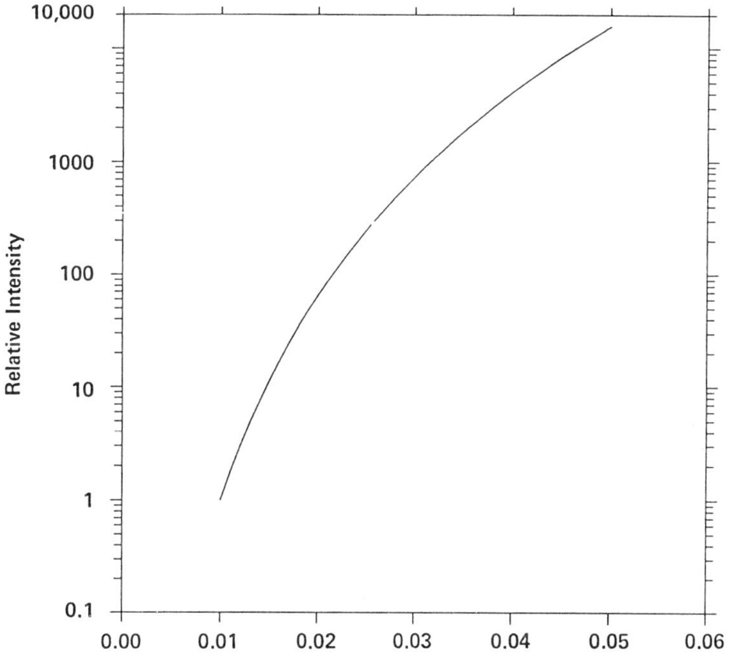

Figure 4-3. Relative intensity vs. sphere diameter for light scattering in the Rayleigh region.

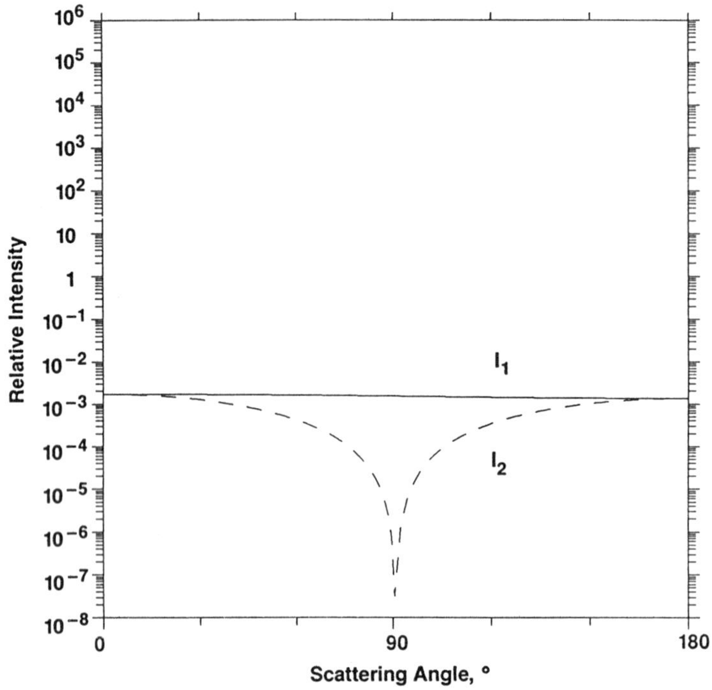

Figure 4-4. Relative intensity vs. scattering angle for a sphere 0.10 μm in diameter.

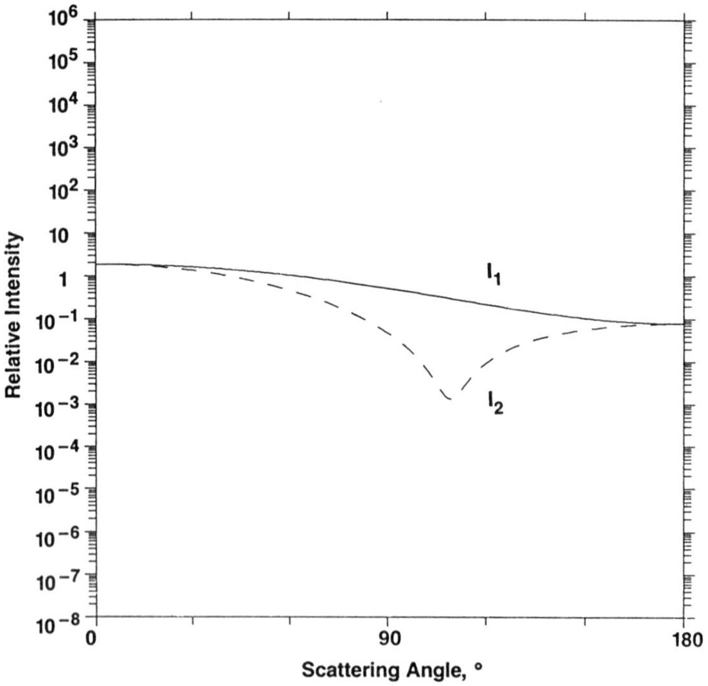

Figure 4-5. Relative intensity vs. scattering angle for a sphere 0.30 μm in diameter.

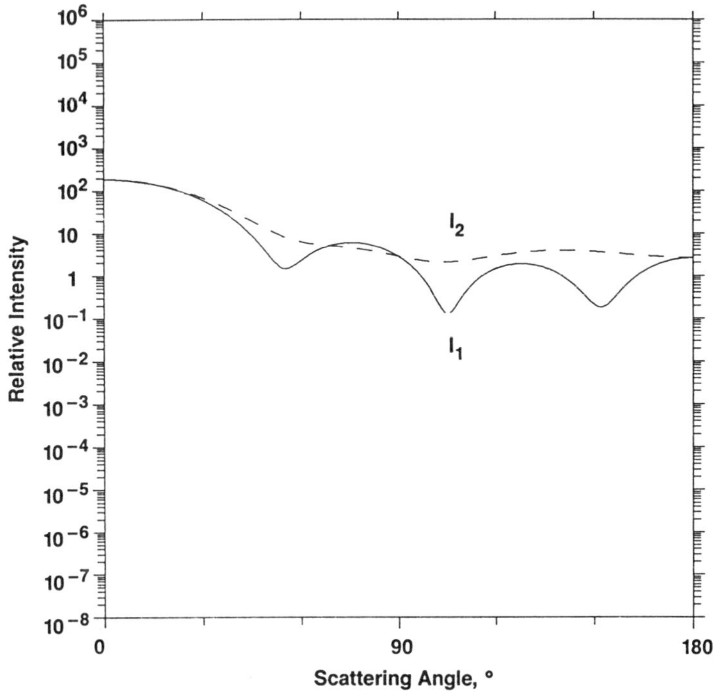

Figure 4-6. Relative intensity vs. scattering angle for a sphere 0.70 μm in diameter.

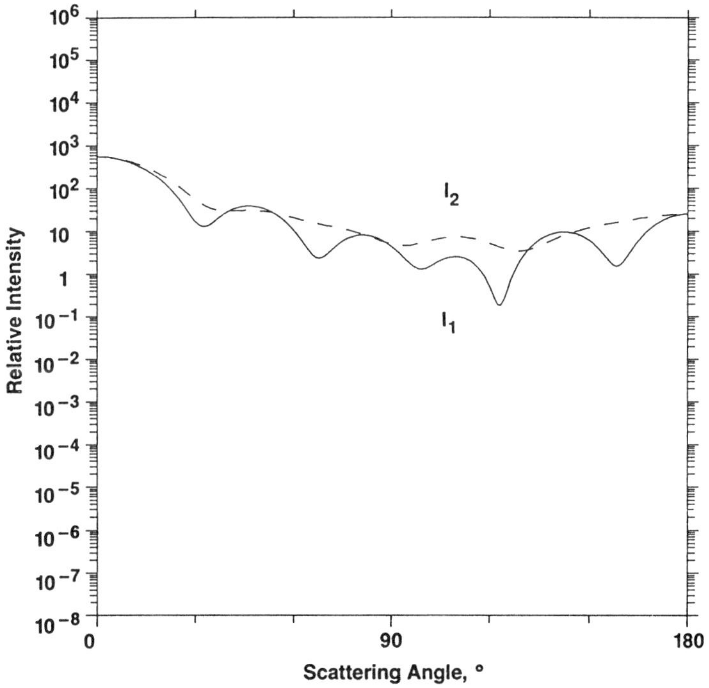

Figure 4-7. Relative intensity vs. scattering angle for a sphere 1.0 μm in diameter.

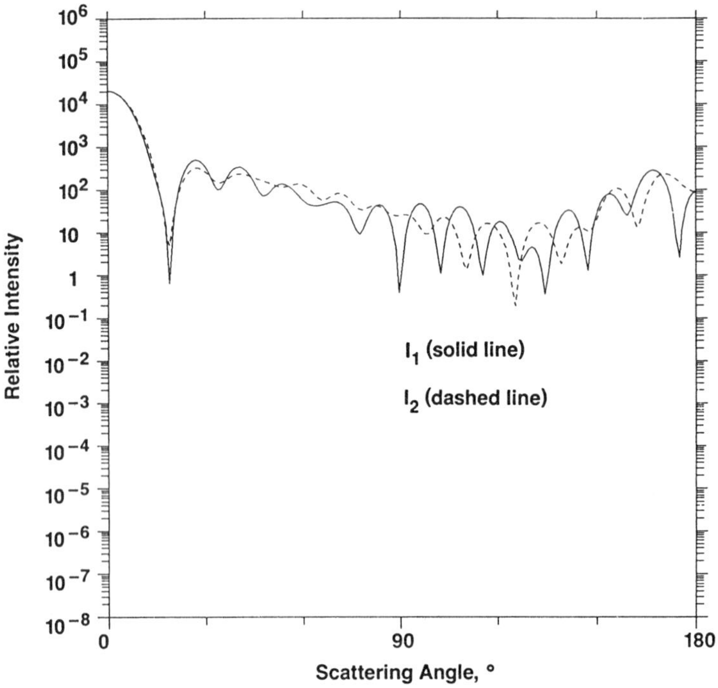

Figure 4-8. Relative intensity vs. scattering angle for a sphere 3.0 μm in diameter.

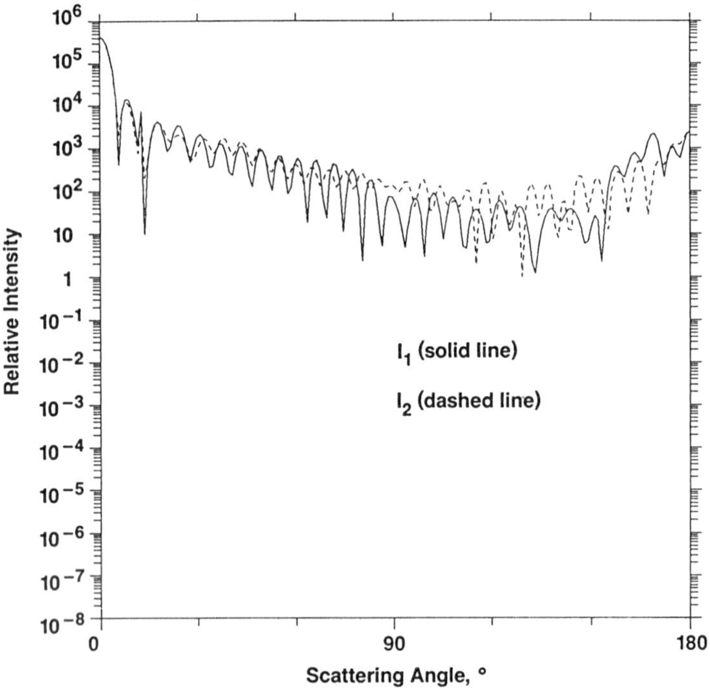

Figure 4-9. Relative intensity vs. scattering angle for a sphere 7.0 μm in diameter.

which the dip is a minimum is now more than 90°. Values of I_1 and I_2 are two to three orders of magnitude higher for the 0.3-μm particles than for the 0.1-μm particles of Figure 4-4.

In Figure 4-6, for 0.7-μm particles, the scattering pattern is substantially different from that for the smaller particles. Now I_2 is greater than I_1 for most angles. Instead of showing a dip for an angle close to 90°, I_2 has a relatively flat scattering profile, (although it still varies by nearly two orders of magnitude). I_1 has developed a scalloped pattern, and varies over three orders of magnitude with changes in scattering angle. The values of I_1 and I_2 for these 0.7-μm particles are about two orders of magnitude larger than those for the 0.3-μm particles shown in Figure 4-5.

In Figures 4-7, 4-8, and 4-9, for spheres of 1, 3, and 7 μm, the trends first shown in Figure 4-6 for 0.7-μm spheres continue. I_2 remains relatively flat for the 1-μm particles but develops a scalloped pattern for the larger particle sizes. The scalloped pattern for I_1 becomes more pronounced as particle size increases. For the larger particles, 3 and 7 μm in diameter, scattering in the forward and backward directions becomes more pronounced. Comparison of the scattering patterns shown in Figure 4-4 through 4-9 demonstrates that light scattering from a sphere becomes increasingly complex as particle diameter increases.

Nevertheless, some generalizations can be drawn from these plots. Although values of I_1 and I_2 differ at most scattering angles, for scattering at 0° or 180°, I_1 and I_2 always coincide. As particle diameter increases, scattering in the forward and backward directions becomes more pronounced. For particles larger than several micrometers in diameter, scattering in the forward direction is especially important.

Figure 4-10 is a plot of the sum of $I = I_1 + I_2$ against scattering angle with particle diameter as parameter. It allows comparison of the total intensity of light that spheres scatter at any angle. This figure again shows that as particle diameter increases, the complexity of the scattering pattern increases. Note that scattering intensity is strongly dependent on particle diameter; as size increases by a factor of 70, from 0.1 to 7 μm, scattering intensity increases by over eight orders of magnitude from 10^{-3} to nearly 10^{6}! Again, note that for larger particles, scattering intensity is highest in the forward and backward directions. At some angles in the near forward direction, scattering intensities coincide. For example, Figure 4-10 shows that at about 15°, the total scattering intensity for particles 1, 3, and 7 μm in diameter is about the same, about 700 on the scale used in this figure. An instrument that reads this value of intensity cannot distinguish among particles 1, 3, and 7 μm in diameter. For particles substantially submicrometer in size, these overlap regions do not occur.

Figure 4-11 is a plot of relative intensity against size parameter α and particle diameter for spheres with refractive indices corresponding to glass and carbon in air at a scattering angle of 0°. For both types of particles, intensity increases strongly with increasing particle size for particles smaller than about 0.5 μm in diameter. Differences in refractive index cause

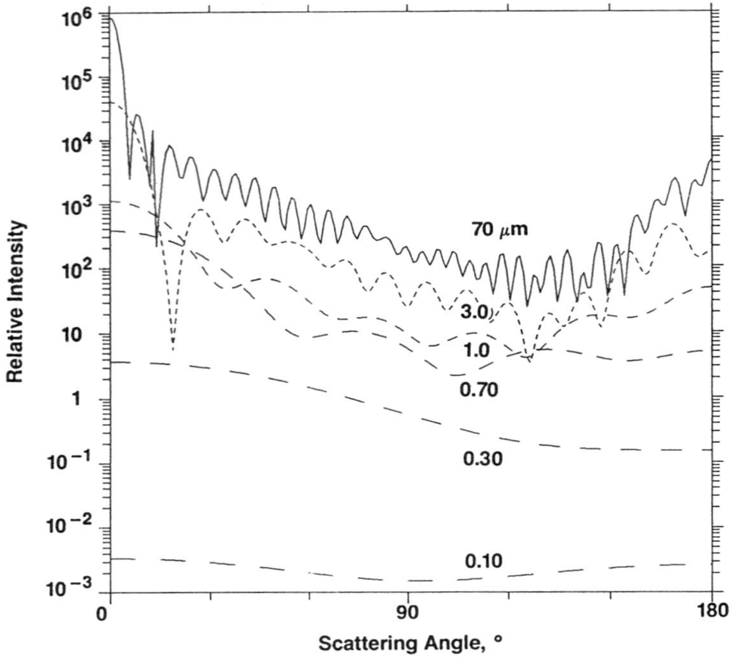

Figure 4-10. Relative intensity vs. scattering angle for spheres from 0.10 to 7 μm in diameter.

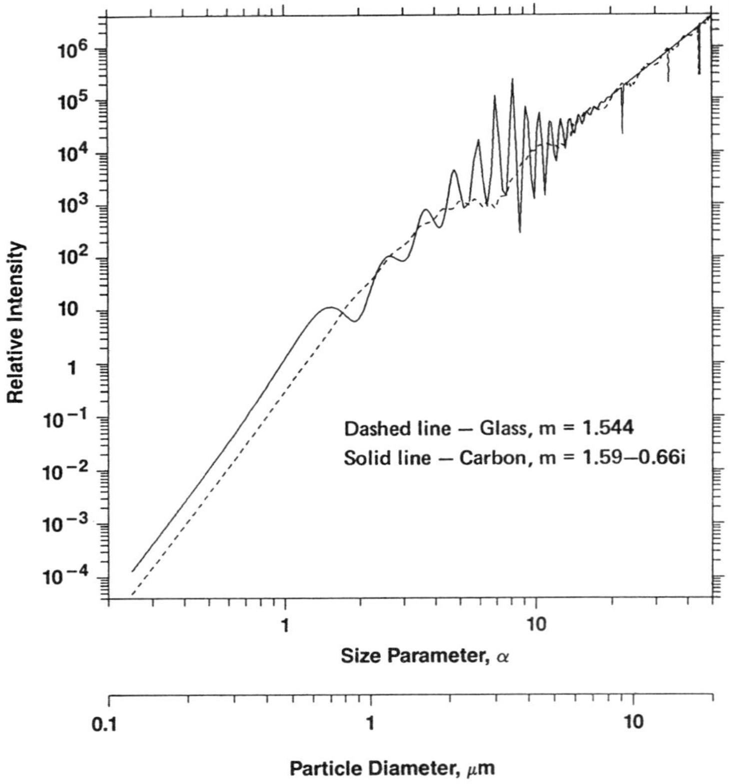

Figure 4-11. Relative intensity for $I_1 + I_2$ vs. size parameter and sphere diameter for illumination by a HeNe laser for particles with two different refractive indices, m.

differences in scattering intensity of three or four, corresponding to an uncertainty in "measured" particle diameter of 20 to 30 percent. For carbon particles larger than 0.5 μm, the scattering intensity oscillates with increasing frequency and amplitude until the oscillations damp out for particles larger than about 5 μm in diameter. This oscillation region leads to ambiguity in an instrument that attempts to size particles by measuring the intensity of the light they scatter. Carbon particles between 1 and 3 μm in diameter cannot be distinguished. Glass (or sodium chloride) particles are better behaved, and display less oscillation in this region.

Instruments that size particles by light scattering combine the intensities from a range of scattering angles. Under these circumstances, the differences in intensities for particles of different refractive indices can be less than those shown in Figure 4-11.

In summary, the intensity of scattered light depends strongly on particle diameter for particles between 0.01 and 10 μm in diameter. Instruments take advantage of this strong dependence by measuring intensity and inferring particle size from that measurement. Because the relationship between intensity and particle diameter can have oscillations and also depends on particle refractive index, more than one kind and size of particle can scatter light with the same intensity so that optical instruments are necessarily somewhat inexact.

Light Scattered by Particles on Surfaces

Scanners used to detect or size particles on a silicon wafer rely on detecting the light scattered by these particles. Although analytical solutions are available to describe light scattered by isolated spheres in space, no such solutions have been developed to describe light scattered by spheres on a surface. As a result, these instruments must be calibrated experimentally. Questions reasonably arise regarding the accuracy with which these calibrations can be done. If refractive index, scattering angle, and particle size affect the amount of light scattered by a sphere in space, these same factors should affect the amount of light that the particle scatters when it rests on a surface. In addition, the amount of light the particle on the surface scatters probably depends as well on the optical properties of that surface. If the surface has coatings with optical properties that vary with depth below the surface, the problem becomes even more complex.

Recently, Wojcik, Vaughan, and Galbraith (1987) have demonstrated that Maxwell's equations can be solved numerically to describe the light scattered by particles on surfaces. Each solution requires considerable time on a Cray 2 supercomputer, so computer simulation and solution of all problems relevant to the microelectronics industry is unlikely to follow soon. Nevertheless, the techniques described in this article are encouraging, particularly because the simulations match well the results from experimental measurements for light scattering by polystyrene latex spheres on a silicon wafer. Figure 4-12, taken from their article, shows that the numerical simulations they present match their experimental data in qualitative terms for particles from 0.54 to 1.6 μm in diameter.

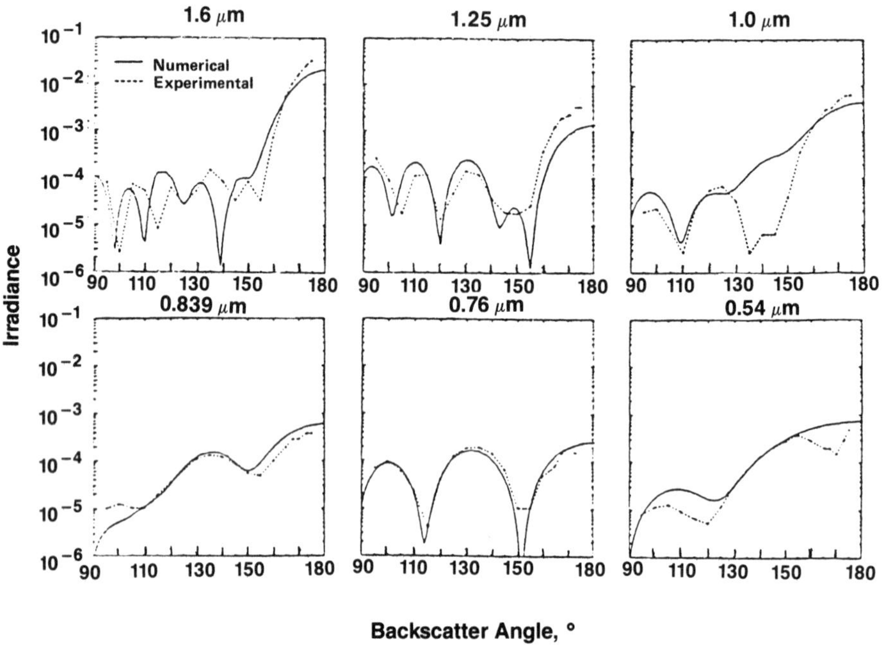

Source: Wojcik, Vaughan, and Galbraith, 1987.

Figure 4-12. Comparison of numerical and experimental values for relative intensity for polystyrene latex particles on silicon.

SUMMARY

Light scattered by particles depends strongly on particle diameter, but also upon the scattering angle, wavelength and polarization of the illuminating light, and on the refractive index of the particle. The scattering pattern becomes more complex as particle diameter increases. Instruments that measure particle diameter in air or in water, or on a surface, cannot be exact unless the particle sizes measured are in a region where no ambiguities in the response curve exist, and unless the instrument is calibrated with particles of the same refractive index as that of the particles measured.

Theory for light scattering has been developed for spheres. Nonspherical particles will scatter light as well. The size of a nonspherical particle can be expressed as that of a sphere that scatters light in the same way as the particle of interest; however, the physical dimensions of such a nonspherical particle and the equal-scattering sphere will not necessarily be close. The principles discussed in this chapter apply to particles in air as well as to particles in liquids.

REFERENCE

Wojcik, Gregory L., David K. Vaughan, and Lee K. Galbraith. 1987. Calculation of Light Scatter from Structures on Silicon Surfaces. SPIE Vol. 774, *Lasers in Microlithography*, pp. 21-31.

HOMEWORK PROBLEM

Consider the design limitations of instruments that use light scattering principles to measure the concentration of particles with various sizes in a clean room.

- Why are these instruments sometimes more accurate at measuring particles about 0.5 μm in diameter than at measuring particles 3 μm in diameter? (Answer: See, for example, Figure 4-11.)
- Why are light scattering instruments not available to measure particles smaller than 0.05 μm in diameter? (Answer: Light scattering falls off very quickly with particle diameter; insufficient light is scattered by particles this small.)

5. DEVELOPING TRUST IN PARTICLE COUNTERS

B. TULLIS Hewlett-Packard Company, Palo Alto, California

INTRODUCTION: HIDDEN VARIABLES (WHAT CAN GO WRONG)

Particle counters are necessary in semiconductor manufacturing to control contamination and ultimately to achieve high functional yields. Because small particles are so prevalent and their sources so numerous in integrated circuit (IC) processing, special instrumentation and procedures are needed to permit accurate diagnoses of particle-caused impacts on yield.

Particle-counting instruments are sophisticated in design and sometimes appear to be temperamental in response. They therefore need to be understood by their users. Significant and disappointing losses in person-hours, materials, and business can result from placing unwarranted trust in the particle count numbers reported by particle-counting instruments. It is essential that the instruments' responses to known particle sizes be well calibrated and characterized and that the instrument output be interpreted in conjunction with all other available information. To illustrate the importance of complete and detailed analyses consider the following scenario.

If you were an engineer or manager attempting to increase your company's IC yield by reducing particle counts in the various steps of the manufacturing process, would you believe technicians' or engineers' reports if they stated that:

- The wafer-start area was Class 150.
- Most particle problems were in metal deposition.
- The new counter was much more sensitive than the old one.
- Our masking isle has always contributed the fewest defects.
- Our supplier of sulfuric acid meets tighter particle specifications than our hydrofluoric acid (HF) supplier.

Perhaps you want to know a little more about what is going on. Additional clues follow:

- An aerosol particle counter sampled at 0.1 cfm (2.8 Lpm) for 1 min and obtained a total count of 15. The probe was placed 4 ft (1.2 m) above the floor next to the operator.
- The highest counts our surface particle counter ever counted on test wafers was on those immediately after metalization.
- The new counter always gives us higher counts than our older counter, no matter where it is used in the facility.
- We have aerosol count data from a continuous, facilities monitoring system that shows that the lowest counts always come from our masking isle probes.
- The counts from both our counters were always more than twice as great in HF as in sulfuric acid.

Now you should be really suspicious of the correctness of the reports. Here is the real story:

- The operator was probably the major source of particles, and the noise count was 10/min. One 0.1 ft^3 (2.8 L) sample is inadequate.
- Because particles on metal produce stronger signals than those on other films or on silicon, counts on metal include smaller particles, which results in higher counts.
- The new counter was used to count TiO_2 particles in a filter test, and now it counts 500/min anywhere (dirty optics).
- The facilities system has long sampling tubes. Tubes to the masking area are the longest, therefore sampling efficiency is poorest for the masking isle.
- Because the index of refraction of HF is less than that for sulfuric acid, the counter can detect smaller particles in the HF.

A quote from the 19th century humorist, Artemus Ward: "It's what we think we know, that ain't so, that causes all our problems."

Trust is too easily bestowed on particle counts and on the calibration accuracy and precision of the instruments used to obtain them. The first objective of this chapter is to give the reader a healthy dose of distrust in particle counters and in the particle counts quoted by one's staff or by others inside or outside one's own company, especially in published reports and equipment manufacturers' quoted specifications. A second objective is to equip the readers with an intelligent set of questions and concerns to guide them in systematically addressing the important issues of instrument calibration, use, and data interpretation.

PARTICLE COUNTER EVALUATION

The primary measurement made by particle counters is the number of signal responses found greater than a threshold value within a sample. A sample may be defined by volume, area, or area and time interval. This count per sample is then interpreted as the number of particles counted in a sample that are greater in "size" than some reference standard. The meaningfulness of such a measure depends on many instrument, particle, and environmental variables. Without going into why an instrument behaves as it does, one can generally characterize what it does. To this end, there are a number of measures with which one can describe the performance of a particle-counting instrument. These characterizing measures follow:

- Accuracy—how close to the truth a measurement comes
- Precision—the smallness of meaningful increments
- Repeatability—how closely repeated measurements agree
- Linearity—how any of these other measures change with the magnitude measured

- Dynamic range—the range of magnitudes measurable
- Sensitivity—how small a measurement can be made
- Resolution—how closely particles can be positioned near one another and still be distinguished
- Uniformity—how any of the above seven measures change with position of particles in the sample space.

The three principal measurements involved in a measurement of particle concentration are:

- Count
- Size threshold
- Sample size (volume, area, or time interval and area).

The first six of the characterizing measures can be applied to each of these three principal measurements, whereas the last two (resolution and uniformity) apply only to count and size threshold.

These eight characterizing parameters are the subject of a previously published article (Berger and Tullis, 1987) and will not be elaborated upon here. Their determination, however, requires certain practices in calibration.

CALIBRATION

Calibration means "to determine, rectify, or mark the graduations of a measuring device," or "to standardize by determining the deviation from a standard so as to ascertain the proper correction factors." One important point is that calibration requires standards of comparison that do not change.

Two kinds of standards generally exist in practice: (1) absolute standards and (2) relative standards. Absolute standards are those traceable to National Bureau of Standards standards (Raloff, 1985) or to some other standard that is accepted within an industry to be stable and usable by all from an operational point of view. Relative standards are any standards that produce measurement results like those of the absolute standard without actually being identical to the absolute standard.

Of the three quantities requiring calibration (count, size threshold, and sample size), the first two (count and size threshold) are particularly troublesome. The most fundamental issue in particle counting of micrometer and submicrometer particles is how to define an acceptable and universal reference for "particle size." Once a reference for "size" is defined, defining a reference for count in a sample of particles equal to or larger than such a reference "size" is, in theory, a straightforward concept, but in practice is difficult to achieve. The calibration of sample size (e.g., volume or area, not count) is the least difficult, but unfortunately is often overlooked.

Calibration for Particle Size (Nuances)

One difficulty with calibrating the detection threshold(s) of particle counters to particles of standard sizes is that the best known and practiced primary measurements of small particle sizes require the employment of energy waves to sense the bulk and/or edges of the particles. These signal-amplitude-based measurements, whether made with light, electron beams (E-beams), or acoustics (Tolliver, Davenport, and Abts, 1982), interact with the particles in such a manner as to leave the final outcome dependent not only on the particle's general "size," but also on other particle characteristics and on the measurement system. These other characteristics include the physical shape (morphology) of the particle, the optical constant(s) of the material(s) comprising the particle's surface and interior, the wavelength(s) of the energy used to sense the particle, the nature of any physical objects near the particle (e.g., a supporting wafer or a fluid surrounding the particle), and the optics or geometry of the energy collection and sensing hardware, not to exclude the electronics involved in the processing of the particle's signal.

Many of the above factors can vary significantly from measurement to measurement and from instrument to instrument. To define a truly usable universal standard all must be accurately specified, matched, and controlled. The trouble only begins here, however. Once one has calibrated a particle counter using agreed-upon reference particle(s), it is a fact of life in the IC application area that the calibration no longer strictly applies when the "calibrated" instrument is used to measure "real" contamination-causing particles in the manufacturing environment. The reason is simply that the counter has been calibrated only for the response it has to the reference particles. Because the "real" particles are nothing like the reference particles in shape or material, they more often than not produce responses in the measurement instrument that may have no known relationship to their physical sizes.

Figure 5-1 depicts this "size" confounding situation. Whereas a reference particle of rather ideal spherical geometry of known and homogeneous material composition produces a

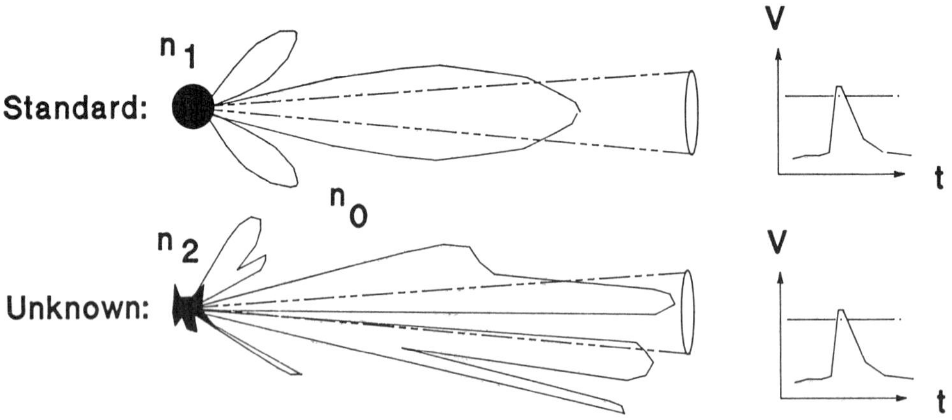

Figure 5-1. Concept of "equivalent-response" standard.

calibratable response, another particle of maverick shape, orientation, or physical composition may produce an identical response when its size differs markedly from that of the standard.

The best that can be done is to define the concept of standards of "equivalent-response." That is, one can use reference particles to produce a standard response. Like responses from otherwise unknown particles can then be referred to as being equivalent in "apparent size" to the standard.

Understanding this limitation, one can proceed to calibrate a particle counter to detect particles producing signal responses equal to or greater than the response produced in the counter by a standard particle. (With appropriate signal processing and calibrations using standard particles of more than a single size, counts of particles producing signal responses between two different "apparent sizes" are obtained.) The particles most favored for use as standards in this manner are latex spheres. These are favored because they are widely available, relatively inexpensive, stable, spherical, homogeneous, and available in a range of monodisperse size distributions (i.e., all the particles in a sample are of nearly the same size). The difficulties in first establishing what the sizes and refractive index of these particles are in order to be able to use them as standards are discussed well by Robillard and Patitsas (1973, 1974a,b,c).

One limitation that cannot be overcome is that no matter what the chosen particle, if two instruments of different optical design are each calibrated against the same standard particle sizes, there is no guarantee that counts will be comparable between instruments if the particles being counted are anything other than identical to those used for the calibrations. The explanation has to do with three variables:

- Particles scatter differing amounts of incident energy in different directions (see Chapter 4).
- The relative amounts scattered in different directions change as the particle size or other physical and optical parameters change (see Chapter 4).
- Instruments of different design may collect different solid angles of scattered energy from the particle (see Liu, Szymanski, and Ahn, 1985).

Figure 5-2 describes some of the art of threshold calibration. Six plotted lines appear in this figure. Understanding their meaning is necessary for a full appreciation of what sizing-thresholds mean for a particle counter. The abscissa on the left-hand plot represents voltage increasing to the right. This represents the voltages that may be generated in the instrument detection circuitry when a particle scatters energy from the beam, incident from the instrument, back into the detection optics and photosensor. The corresponding ordinate represents the fraction of a particle population (comprising a sample) that has been detected as having generated signal response amplitudes greater than a specified threshold voltage. Each plot was generated by obtaining counts as a function of threshold settings. Plot generation required that an identical sample be measured at each of a series of different threshold settings.

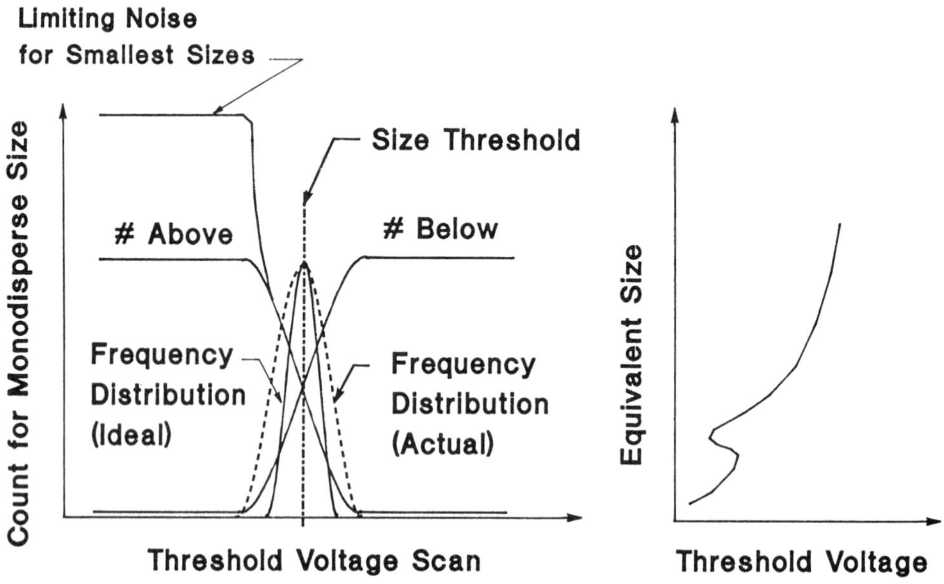

Figure 5-2. Size calibration.

What do the plots mean? The plot labeled "# Above" is a record of measurements when the instrument used has sufficient sensitivity to detect the lower voltage signals from smaller particles as well as the higher voltages from larger particles. The plot labeled "Limiting Noise for Smallest Sizes" was generated in an identical manner except that noise pulses on the voltage signal added false counts, which obscured the relatively less numerous particle counts at small response amplitudes (smaller particle sizes). The plot labeled "# Below" is the result of subtracting the ordinate values of plot "# Above" from the maximum value of "# Above." Note that "# Below" would be much different if one had subtracted the ordinate value of plot "# Above" from the maximum value of the plot "Limiting Noise for Smallest Sizes," where the maximum count may be as high as the saturated maximum count of the counting circuitry. The plot labeled "Frequency Distribution (ideal)" represents the frequency distribution obtainable with a monodisperse size distribution of latex spheres whose median size produces a signal voltage equal to the particular threshold value labeled "Size Threshold." This symmetrical, peaked and narrow distribution has breadth because there is a small spread in actual particle sizes about the average value. The dotted distribution labeled "Frequency Distribution (actual)" is the distribution of the signal response amplitudes actually produced by the distribution of particle sizes described in the preceding sentence. This latter distribution is broader for two reasons. First, small fluctuations in particle size can effect large fluctuations in signal response amplitudes because of the "ripples" in the particle scattering function (see Chapter 4). Second, a particle of any one given size can effect different signal response amplitudes depending on its various possible positions in the region of space from which the instrument collects scattered energy. Note that this broader distribution is the derivative of the plot labeled "# Below." Also note that if the threshold value positioned at the centerline of these distributions becomes the calibrated value, half the distribution of standard particles used in obtaining it will be counted as having exceeded this threshold. Thus, in the absence of limiting noise, the crossing point of the two plots "# Above" and "# Below" serves in this case as a good marker of the threshold value whose voltage represents the voltage produced by the size value at the median of the particle population being used as a calibrating standard.

The right-hand plot in Figure 5-2 represents a typical relationship between signal voltage (or its analog: Equivalent Size) and Threshold Voltage. Note the incidence of triple values at some threshold voltages, which is to say that some of the signal voltages are indeterminately related to any of three possible Equivalent Sizes of particles. This phenomenon demonstrates what can result from light-scattering physics as explained in Chapter 4, and has differing consequences in particle sizing, depending upon detail of the particle counter design.

Calibration of particle counters for particle size thresholds has an added complexity when using instruments designed for counting particles on wafer surfaces. The added complexity (Tullis, 1986) derives from the fact that known solutions of light scattering, as discussed in Chapter 4, no longer apply because the amount of light a particle scatters now depends on the interaction of the light's electromagnetic energy with both the geometry and optics of the particle and with the geometry and optics of the surface supporting the particle. This interaction is depicted in Figure 5-3, and although illustrated using rays from geometrical optics, really is a physical optics problem requiring boundary value solutions to partial

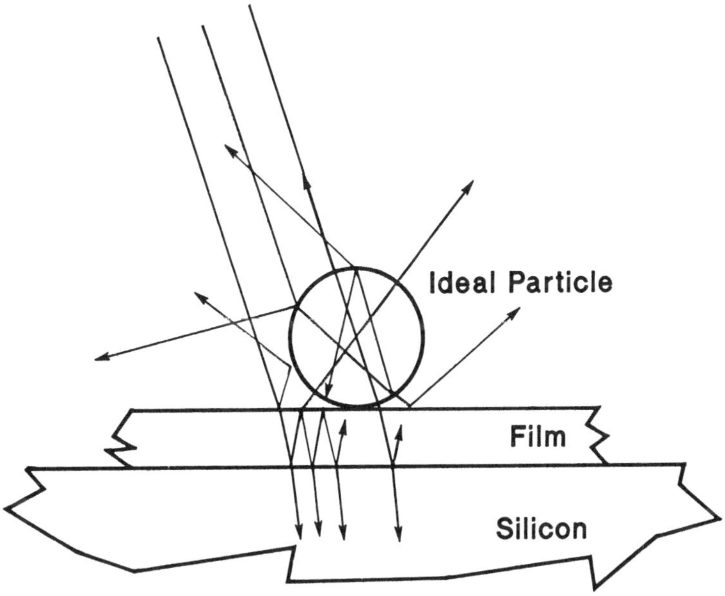

Figure 5-3. Geometric optics, though suggestive, is not sufficient.

differential equations. This interaction is not one that can be treated by superposition. There is no known mathematical solution with which we can accurately model the interaction. We must instead resort to extremely expensive numerical solutions on supercomputers (see Chapter 4) or determine the relationships between light scattering and the combinations of particles with supporting structures empirically.

Although inadequate and overly simplified, Figure 5-4 helps illustrate the kind of phenomena that can occur when light interacts with more than one interface between material media, as when accounting for both (a) the interface(s) between a particle boundary and the medium surrounding it (e.g., air) and (b) the interfaces between the medium and a surface supporting the particle. For illustration purposes, think of the particle as being supplanted by a film of comparable thickness to the particle. This film may be of infinite extent, whereas a true particle is very finite in extent; that is, the shape of the boundary of a particle is more complicated and three-dimensional, unlike the one-dimensionality of the film. We readily agree that light scattering from a particle will be much more complicated than from a film. But a particle presents both a front and rear surface segment to an incident beam or wave of light. The reflections from these two surfaces join in a strong way in determining the interference patterns created about both the forward and backscattered directions. In this sense the first and second surfaces of a film may similarly represent the type of interactions that can take place in scattering from the front and back surfaces of a particle. The advantage of working with a film instead of a particle is that there is a complete and exact mathematical

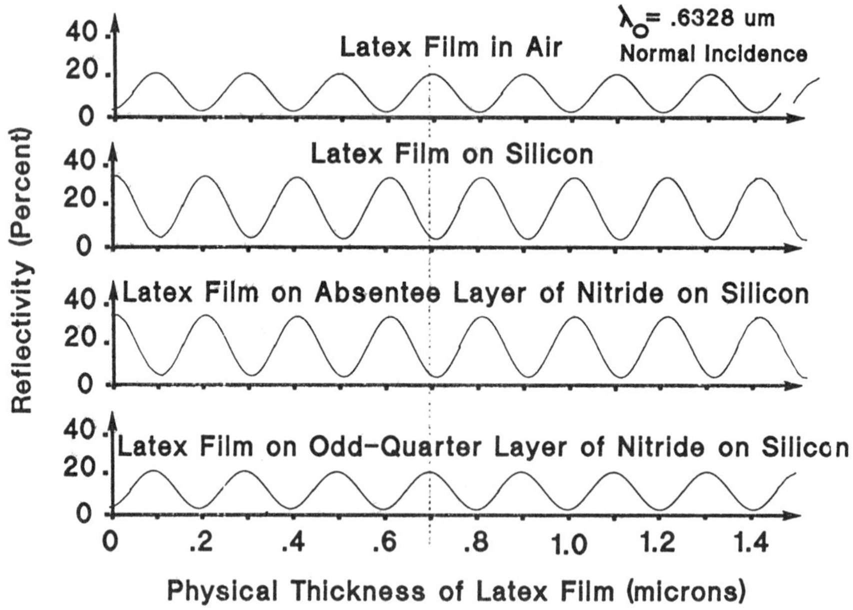

Figure 5-4. Suggestion of highest order interaction when a latex "object" overlays a process film on silicon.

solution to light interaction with a film on a supporting surface. Figure 5-4 illustrates what this solution can tell us.

If the film is suspended in free space and grows in thickness, the reflectivity of light incident upon it will rise and fall as the film thickens. A similar oscillation occurs if the film is grown on a silicon surface, only then the maxima in reflectivity almost double. If the film is grown on an absentee layer of silicon nitride (i.e., a silicon nitride film of thickness equal to an even quarter-wavelength multiple of the light wavelength measured within the silicon nitride, which is in turn on the silicon substrate), the reflectivity of the overall structure is no different than if the silicon nitride film were absent and the latex film were directly on the silicon. But finally, if the film of silicon nitride between the latex film and the silicon substrate is of a thickness that is an odd-multiple of the quarter-wavelength (measured in the silicon nitride), the reflection maxima are reduced to that which occurred when the latex film was in free space. And note, for example, at a thickness ("particle diameter") of 0.7 μm, that with a supporting film like the silicon nitride film pictured, the reflectivity of the combination of a latex film ("particle") on another film on a silicon substrate is less than the reflectivity from the latex film of the same thickness in free space. This is an attempt to communicate that an apparent particle size will be different when the particle is on a bare silicon wafer having only a very thin or native oxide surface than when the particle is on a silicon wafer that has a film of thickness not near to an even-multiple of a quarter-wavelength of the incident light, where the wavelength is measured in the material of the film (where zero is considered an even number).

It would be foolish to think that this illustrates exactly, or even approximately, all that can occur in the interaction of light with a particle on a substrate covered by a film. When a particle is supported by a substrate of any kind, the substrate will most likely interact with any light scattering from particles it supports, and in a way that confounds any theoretical attempt to ascertain its physical size from the amount of light that is scattered from it and its neighborhood, unless the instrument used is calibrated with standard particles on a wafer having a similar film material and thickness. Even then, however, the added complexity of an underlying film may make the dissimilarities between standard calibration particles and "real" contamination particles even more glaring.

All this aside, for empirical purposes with surface particle counters, one must deposit particles of known shape and material onto standard surfaces like silicon wafers and calibrate the response of particle counters using these combinations of particles and substrates to produce standards of apparent particle size. Aerosol dispersers and monodisperse latex particles are commercially available for this purpose. It is helpful to use a static-electricity-dissipation device at the output of these dispersers to ensure a uniform deposition of the particles onto wafers. The deposition can be simply achieved under a clean laminar downflow workbench holding the outlet of the disperser about 1 in. above the wafer and moving both the outlet and the wafer in figure-eight patterns that will bring the outlet over all the area of the top of the wafer. It helps also to tip the wafer, otherwise held horizontally, in a manner to allow the downdraft of particles to have a chance to "spill" over all portions of the perimeter of the wafer. Checks with optical and E-beam microscopes are helpful in ascertaining whether or not any particle agglomeration is occurring, which could spoil the results. Because it is so difficult to focus these microscopes onto polished silicon wafers, it is desirable to use wafers that have an etched reference grid on which to focus for evaluation purposes. With optical microscopes having both fluorescence and darkfield capability, excellent evaluation results can be obtained without reference features on the wafer by using fluorescent latex spheres (also commercially available). What makes the fluorescent beads so useful with optical microscopes is that they overcome the short depth of focus that otherwise makes it nearly impossible to find latex spheres against the background of a polished silicon wafer. Once the deposition process is proven to produce uniform distributions over the surface of a wafer and at a density of around $10/cm^2$, standard calibration wafers can be made that will last for years if kept in a clean environment, such as an individual wafer shipping container made of polypropylene. They will even endure being dropped if the dropping does not dirty them.

Before leaving the subject of size calibration, it should be mentioned that TSI, Inc., manufactures an aerosol counter that does not depend upon light-scattering amplitudes for particle sizing. The technique used is called aerodynamic particle sizing. The aerosol under test is accelerated through a nozzle. The accelerating gas drags the particles with it, but because drag-force-to-mass ratio on a particle decreases with increasing particle size, small particles will travel through the measurement section faster than large particles. Particles are detected and timed as they cross through two closely adjacent regions of focused light mutually separated along the particle path by a known distance, permitting their velocities to be determined. Finally, the particle velocities obtained can be somewhat reasonably related by aerodynamic principles to the aerodynamic cross-sectional area of the particle. Somewhat analogous to the uncertain relationship between particle sizes and detected light scattering from particles under test with other products, the relationship between particle drag coefficients and the shapes, masses, and surfaces of the particles under test with aerodynamic particle sizers is imperfect, which leads to difficulties in interpreting calibrations in terms of actual rather than apparent physical sizes. Information on the calibration of this product can be obtained from the manufacturer.

Calibration of Particle Concentration

Once signal thresholds have been successfully related to apparent sizes of the standard particles, the job of a particle counter remains to detect and count particle number within a selected sample space (volume of fluid or gas or spatial area of a surface) to derive a measure of particle concentration (or flux through an area). Again, calibration requires a reference standard against which to compare the numbers reported by the counting instrument. This is a difficult task and has often been overlooked by users. Some experimental work has been done to compare the counting results of numerous instruments (Dahneke and Johnson, 1986; Gebhart, 1985; Peacock, Accomazzo, and Grant, 1986; and American Society for Testing and Materials, 1985b,d). Some of these comparisons show that counts from various instruments can differ from one another by factors as large as 2 and even in some instances 6. Such discrepancies are in many cases a sign of poor calibration practices, the use of incompatible standards, and to some extent inadequacies of instrument designs to permit standardizing responses.

There are distinctly different ways of calibrating numbers, depending on whether the counter being calibrated is for counting particles on surfaces or in sampled volumes of gas or liquid.

Counting particle concentrations in gas and fluids requires the generation of a sample of known volumetric concentration. An aerosol cloud or seeded liquid stream entraining standard-sized particles must be created, mixed to ensure uniform distribution throughout the volume to be sampled, and then sampled by both the counter in question and by a reference counter (Peacock, Accomazzo, and Grant, 1986; American Society for Testing and Materials, 1985a,d). In the case of seeding fluid flows, a concentrate of higher concentration must be injected into the flow, and the flow must in some cases be pressurized to suppress bubbles from forming and being counted as particles. The reference counter can either be another counter that has previously been calibrated or a method that can give an accurate count. The method of choice for obtaining a count of the particle number in the referee sample is to collect the particles onto a filter or impaction surface on which particles can be found and counted under a microscope. Traditionally such counting has been visual (American Society for Testing and Materials, 1985c). With particle sizes of interest in the semiconductor business being much less than the traditional visual limit of 5 μm in a microscope, the industry is turning to automated surface counters for such inspections of referee counts. This brings us to the question of how to calibrate particle counts on surfaces.

Using wafers with latex particles deposited upon them as described at the end of the previous section, the job of calibrating count is that of comparing the counts obtained by the automatic counter in question with those visually obtained microscopically. Because microscopic counting of particles on these standard wafers requires hours of painfully careful counting by an alert microscopist, it is not something to be done more often than is absolutely necessary. The technique is illustrated in Figure 5-5 and involves overlapping raster scans across the wafer. It is important before commencing with such scans to define a minimum "neighborhood distance," that is, a distance from one particle such that if another one or more particles is found, these other particles will be ignored for counting purposes. It is necessary to define this distance because all automatic surface particle counters have a resolution limit that causes them to count in a similar way. For laser scanning instruments that divide the image field into pixels and only count, at most, one particle per pixel, the "neighborhood distance" should be defined as the interpixel distance. Some counters may have one interpixel distance along the scanning direction and yet another across the scanning direction; in this case, two "neighborhood distances" must be dealt with.

For the count to be statistically significant, one must scan a sufficient amount of a wafer surface to obtain a count of at least 500 particles (American Society for Testing and Materials, 1985c). For these same statistically based reasons, the calibration procedure should be designed to ensure that there are 500 or more particles in the area from which counts are to be obtained. On the other hand, densities greater than about 70/cm^2 (Tullis, 1986) start to lead to pixeling errors in some surface counting instruments. It is therefore desirable on 100-mm-diameter wafers, for example, to limit the total number of standard particles to less than 2,000.

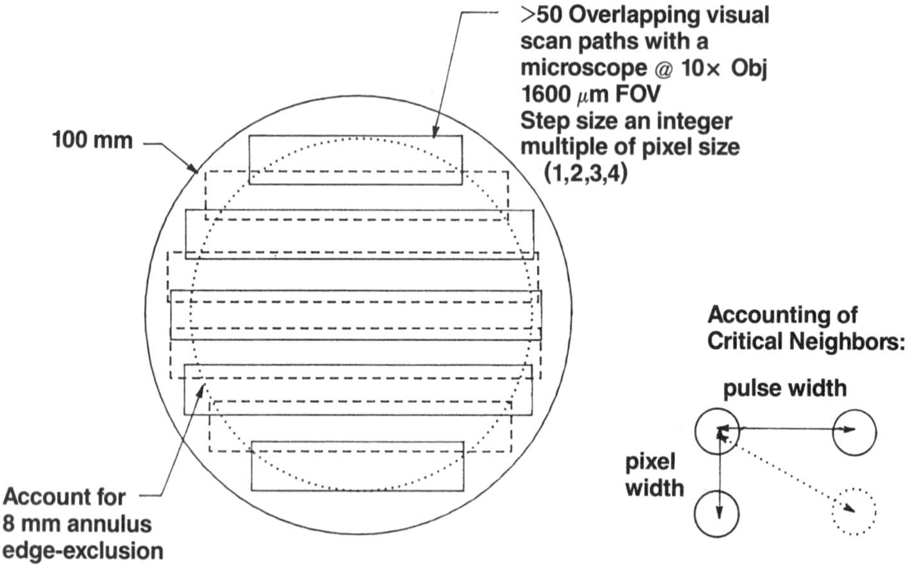

Figure 5-5. Absolute count of monodisperse latex fluorescent spheres.

In this way one ends up having to sample from 25 to 100 percent of the wafer surface to authenticate a standard calibration wafer. As discussed in the preceding section, the use of fluorescent latex spheres can be used to vastly ease the difficulty of counting standard particles across the surface of a polished silicon wafer.

The number 500 in the above paragraph came from an assumption that a 95-percent confidence level is needed when sampling less than the full wafer so that the resulting count per sampled area will represent, within an accuracy of ±10 percent, the true average surface density of particles on the wafer. Obviously, though, if it is a calibration wafer in question and one is counting the whole wafer, no such sampling statistics apply. They do apply when one is not inspecting the entire wafer, when one is attempting to estimate the average count over more than the one wafer, or when the particle count on a wafer is itself a sample for deriving the concentration in an aerosol or fluid from which the particles were obtained. The fundamental difference is the one between a count on a calibration wafer that is to be an end itself and a count on a test wafer (or filter surface) that is only a sample estimate.

If one is dealing with a sample to obtain an estimate and is interested in accuracies other than ±10 percent or in confidence levels other than 95 percent, the following formula can be used to calculate the minimum required count for a meaningful single sample estimate, where it is assumed no counting errors are made: $N = (100^*Z/X\%)^2$. In this formula, X% is the plus-or-minus tolerance desired on the count, and Z is the Z-score in a Gaussian distribution. For this formula to be correct, the number N must be greater than approximately 20 to permit a Gaussian approximation of the assumed Poisson distribution for the particle density over the wafer surface. The assumed Poisson distribution is in turn correct only under the assumption that the distribution of particles over the wafer surface is uniform, random, and independent. Then, to achieve a confidence level of 95 percent, Z has the value 2.0 (i.e., 95 percent of repeated count results on the same wafer would result in a count within ±X percent of the sample result); this is to say that any given sample count has a 95% chance of being within ±X% of the sample estimate. Thus, according to this formula and the assumptions of uniformity, randomness, independence, and error-free counting, for an accuracy of ±10 percent, a count of at least 400 (500 is a good round number approximation) should be obtained for the aforementioned 95-percent confidence level.

Once a surface particle counting instrument has been calibrated using one of these calibrated wafers, other techniques exist for more easily and routinely verifying the stability of such primary calibrations. The American Society for Testing and Materials (1985a) describes the accepted technique of secondary calibrations for aerosol counters that have the same size settings. The Society (1985a) also describes the accepted technique of secondary calibrations for aerosol counters that do not have the same size settings. Berger and Tullis (1987) describe a technique using etched silicon wafers to verify the stability of count calibration of surface particle counters.

Billat (1987) describes a new holographic surface particle counter. This instrument has two significant features that may allow it to play an important role in the production of primary calibration standards for surface particle counters: (1) it counts particles in a real image of the wafer under test and achieves a very deep depth of focus, which aids accurate counting across the full wafer surface, and (2) the holograms created of each wafer under test represent an archivable and robust image-record of the wafer tested. Together these are a means to calibrate the holographic counter once against a visually authenticated standard wafer as described earlier. Thereafter, recalibrations require only the use of the robust and reusable original hologram made during the initial calibration. This calibrated holographic counter can then be used to certify the particle counts of new calibration wafers, thus eliminating the human visual labor of further authentications.

ERROR SOURCES

The calibration of a particle counter can be meaningless if errors occur in the calibration process or in the use of the equipment afterward. This section categorizes these potential error sources. The four most general categories are those that are (1) particle related (due to the particles themselves or the medium that surrounds them), (2) sample related (due to the means of collecting the particles to be counted), (3) optics related (due to the optics of the instrument), and (4) processing related (due to the electronics circuitry and internal data processing). For more information on some of these errors and their effects, see Dahneke and Johnson, 1986; Gebhart, 1985; Peacock, Accomazzo, and Grant, 1986; and American Society for Testing and Materials, 1985a. Also see the additional references used in the lists below.

Particle Related Errors

- Particle shape and structure
- Particle index of refraction
- Particle size
- Particle position
- Particle orientation
- Particle motion
- Other objects in the neighborhood of the particle
- The medium surrounding the particle, its index and homogeneity (Knollenberg, 1987; Grant and Schmidt, 1987)

Sample Probing Errors (Fissan and Schwientek, 1987; Cooper, 1986)

- Impact of sampling on environment being measured
- Anisokinetic sampling
- Transport losses
- Errors in flow rates
- Contamination of the sample

Optics Errors

- Misaligned, dirty, or damaged optics
- Noncoincidence of illumination and detection focii
- Poor focus
- Nonuniform illumination of focal volume
- Sweep instabilities and misalignments
- Incomplete sensor coverage of the entire sample volume or area
- Irregular scattering distributions
- Changes in collection angles
- Polarization anisotropies
- Background and pattern noise

Processing Errors

- Source and detector instabilities, nonuniformities, and aging
- Threshold instabilities
- Unmatched offset voltages
- Transit time saturation
- Electrical noise
- Pixel overcounting (Pecen et al., 1987)
- Pixel saturation and undercounting (Tullis, 1986)

STATISTICS

Users of particle counters must be well versed in the statistics that apply to their situations. The following distributions and charts should be understood:

Frequency Distributions

- Poisson (applicable to frequencies of rare events, e.g., particles/sample)
- Binomial (applicable to proportions, e.g., percent yield)
- Student (applicable to sample averages)
- Chi-square (applicable to variances)
- F (applicable to ratios of variances)

Confidence Calculations, Charts, or Tables (for predicting repeatabilities)

Control Charts

- C charts (for monitoring stability of frequencies when sample size remains constant)
- U charts (for monitoring stability of frequencies when sample sizes do not remain constant)
- P charts (for monitoring stability of proportions)
- X-bar and R charts (for monitoring measures that are continuous-variables like flow rates and threshold voltage settings)
- Moving range charts (for monitoring continuous variables as with X-bar and R charts but where significant time elapses between measurements to cause a shortage in data)

The user should become competent in making statistical inferences using these statistics tools. The tools for confidence calculations and frequency distributions are covered completely in Gibra (1973); Box, Hunter, and Hunter (1978); Bzik (1986a,b); and Slooten (1986). AT&T Technologies (1956) and Duncan (1986) offer a full coverage of control charts. These are all reviewed in Chapter 13.

MAJOR CURRENT CHALLENGES IN PARTICLE COUNTER CALIBRATION

In conclusion, we recognize that major challenges remain in the calibration and use of automated particle counters. The most important of these are the following:

- Number certification of latex standards on surfaces and on surfaces with films
- Creation of known concentrations of standard particles in aerosols and fluids
- Cross-calibrations between instruments, not only within facilities but between facilities, between users, and between vendor companies

- Standardizing calibration techniques and making them routine procedures
- Demonstrating consistent, appropriate, and reliable use of statistics in designing measurement procedures and in data interpretation and reporting.

REFERENCES

American Society for Testing and Materials. 1985a. Determining Counting and Sizing Accuracy of an Airborne Particle Counter Using Near-Monodisperse Spherical Particulate Materials (Standard Practice Designation 328-80). *Annual Book of ASTM Standards*. Vol. 15.03, pp. 676-680.

____. 1985b. Determining Relative Bias of Two Particle Counters Using Comparison Procedures (Standard Practice Designation F-650-80). *Annual Book of ASTM Standards*. Vol. 15.03, pp. 786-791.

____. 1985c. Particle Size Analysis of Particulate Substances in the Range of 0.2 to 75 Micrometers by Optical Microscopy (Standard Practice Designation E-20-85). *Annual Book of ASTM Standards*. Vol. 14.02, pp. 18-28.

____. 1985d. Secondary Calibration of Airborne Particle Counter Using Comparison Procedures (Standard Practice Designation 649-80). *Annual Book of ASTM Standards*, Vol. 15.03, pp. 781-785.

AT&T Technologies. 1956. *Statistical Quality Control Handbook* (Western Electric Co.), Select Code 700-444 (1-800-432-6600).

Berger, J., and B. Tullis. 1987. Calibration of Surface-Particle Detectors. *Microcontamination*, (July):24, 26-29.

Billat, S. 1987. Automated Defect Detection on Patterned Wafers. *Semiconductor International*, (May):116-119.

Box, G., W. Hunter, and J. Hunter. 1978. *Statistics for Experimenters*. John Wiley & Sons, NY.

Bzik, T. 1986a. Statistical Management and Analysis of Particle Count Data in Ultraclean Environments: Part I. *Microcontamination*, (May):58-63, 98-99.

Bzik, T. 1986b. Statistical Management and Analysis of Particle Count Data in Ultraclean Environments: Part II. *Microcontamination*, (June):34-41.

Cooper, D. 1986. Minimizing Bias in Sampling Contaminant Particles. *Solid State Technology*. (January):73-79.

Dahneke, B., and B. Johnson. 1986. A Method for Cross Calibrating Particle Counters. *The Journal of Environmental Sciences*, (September/October):31-36.

Duncan, A. 1986. *Quality Control and Industrial Statistics*. Richard D. Irwin Press, Homewood, IL.

Fissan, H., and G. Schwientek. 1987. Sampling and Transport of Aerosols. *TSI Journal of Particle Instrumentation*, 2(2):3-10.

Gebhart, J. 1985. Counting Efficiency and Sizing Characteristics of Optical Particle Counters. *Proceedings of the Institute of Environmental Sciences*, pp. 102-105.

Gibra, I. 1973. *Probability and Statistical Inference for Scientists and Engineers*. Prentice-Hall, NJ.

Grant, D., and W. Schmidt. 1987. Improved Methodology for Determination of Submicron Particle Concentrations in Semiconductor Process Chemicals. *The Journal of Environmental Sciences*, (May/June):28-33.

Knollenberg, R. 1987. The Importance of Media Refractive Index in Evaluating Liquid and Surface Microcontamination Measurements. *The Journal of Environmental Sciences*, (March/April):50-58.

Liu, B., W. Szymanski, and K. Ahn. 1985. On Aerosol Size Distribution Measurement by Laser and White Light Optical Particle Counters. *The Journal of Environmental Sciences*, (May/June):19-24.

Peacock, S., M. Accomazzo, and D. Grant. 1986. Quantitative Count Calibration of Light Scattering Particle Counters. *The Journal of Environmental Sciences*, (August):23-27.

Pecen, J., A. Neukermans, G. Kren, and L. Galbraith. 1987. Counting Errors in Particulate Detection on Unpatterned Wafers. *Solid State Technology*, (May):149-154.

Raloff, J. 1985. Yardsticks from Space. *Science News*, 128(August 10):92-93.

Robillard, F., and A. Patitsas. 1973. Diameter Measurement of Dow Latexes EP-1358-38 by Three Optical Methods. *Canadian Journal of Physics*, 51:2395-2402.

Robillard, F. and A. Patitsas. 1974a. Determination of the Particle Diameter of Dow Latex 642-6 by Four Independent Techniques. *Powder Technology*, 9:247-255.

____. 1974b. Determination of Size, Size Distribution, and Refractive Index of Dow Latexes EP-1358-38 by the Mie Scattering Method. *Canadian Journal of Physics*, 52:1571-1576.

____. 1974c. A Study of Diameter, Polydispersity and Refractive Index of Dow Latex 642-6 by Light Scattering Method. *Powder Technology*, 10:307-315.

Slooten, R. 1986. Statistical Treatment of Particle Counts in Clean Gases. *Microcontamination*, (February):32-38.

Tolliver, D., N. Davenport, and L. Abts. 1982. The Detection of Microcontaminants in Semiconductor Process Fluids Using an Acoustic Technology. *Solid State Technology*, (September):116-123.

Tullis, B. 1986. Measuring and Specifying Particle Contamination by Process Equipment: Part III. Calibration. *Microcontamination*, (January):50-55, 86.

6. ULTRAFINE (<0.1-μm DIAMETER) PARTICLES

GILMORE J. SEM TSI, Incorporated, St. Paul, Minnesota

CHARACTERISTICS OF ULTRAFINE PARTICLES

Ultrafine particles are defined in this chapter as particles with dimensions smaller than 0.1 μm. The smallest airborne particles are clusters of a few molecules with dimensions of about 2 nm. Smaller clusters are unstable and quickly disperse into individual molecules. Therefore, the limits of particle size for this discussion are 0.002 to 0.1 μm or 2 to 100 nm.

Whether inside a clean room, processing equipment, or elsewhere, ultrafine particles are generated primarily by combustion or evaporation at elevated temperatures, followed by homogeneous nucleation and condensation while the gas cools. Examples inside a clean room include:

- Furnaces
- Overheated components on circuit boards
- Reactors
- Undersized electrical wiring
- Electric motors
- Worn bearings
- Flames
- Hot lights
- Microscope lamps.

HEPA and ULPA filters collect nearly all particles smaller than 10 nm in a single pass by the mechanism of Brownian diffusion. Most nuclei particles in a clean room probably come from internal sources. Nearly all sources have profuse rates of production, millions or even billions per second. In limited measurements in steady-state, at-rest clean rooms, particles larger than 10 nm outnumber those larger than 0.3 μm by 10 to 1,000 times (Donovan et al., 1984, 1985b; Fisher, 1987). Figure 6-1 from Donovan et al. (1984) and Figures 6-2 and 6-3 from Fisher (1987) illustrate the point. As seen in Figure 6-1, the optical particle counter (OPC), a PMS LPC-555, measured particles larger than 0.3 μm in an at-rest clean room. The condensation particle counter (CPC), a TSI 3020, measured 100 to 1,000 times as many particles per cubic foot as the OPC. In Figure 6-2, using the same equipment from Figure 6-1, the CPC measured in an at-rest horizontal-flow clean room about 100 times as many particles per cubic foot as the OPC. The same equipment from Figures 6-1 and 6-2 was used for Figure 6-3, but for measurements in an at-rest vertical-flow clean room. The CPC measured about 100 times as many particles per cubic foot as the OPC.

Liu et al. (1986) performed an interesting experiment involving a university clean room used for teaching and research. They measured particle concentration and size distributions both inside the clean room and at the entrance to the makeup air inlet on the roof of the four-storied building in an urban area. The outside aerosol consisted of a constant level of background aerosol with clouds of higher concentration aerosol passing by with the wind. A 27-hour plot of particle concentration measured by a CPC on the roof, Figure 6-4, shows that each high-concentration cloud that passed by resulted in a concentration peak lasting 1 to 20 minutes. The same 27-hour plot shows corresponding peaks of concentration inside the clean

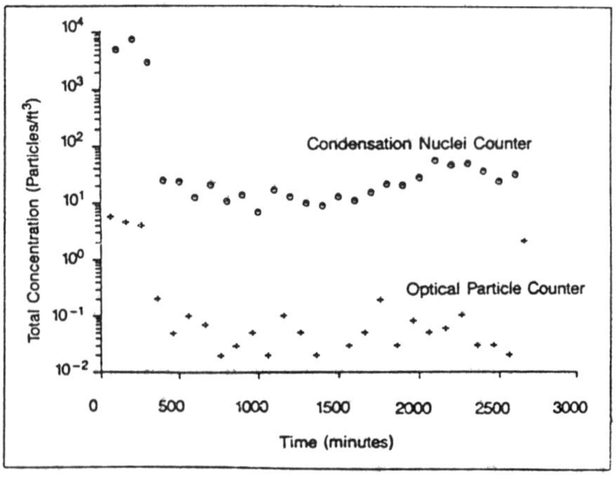

Source: Donovan et al., 1984

Figure 6-1. Comparison of concurrent CPC and OPC measurements in an at-rest clean room.

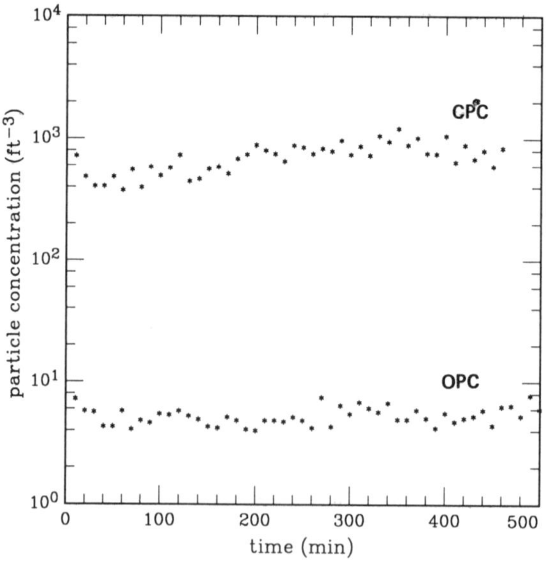

Source: Fisher, 1987

Figure 6-2. Comparison of concurrent CPC and OPC measurements in an at-rest, horizontal-flow clean room.

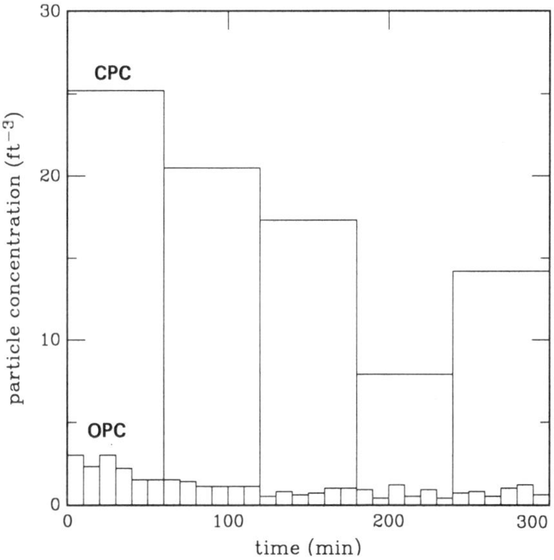

Source: Fisher, 1987

Figure 6-3. Comparison of concurrent CPC and OPC measurements in an at-rest, vertical-flow clean room.

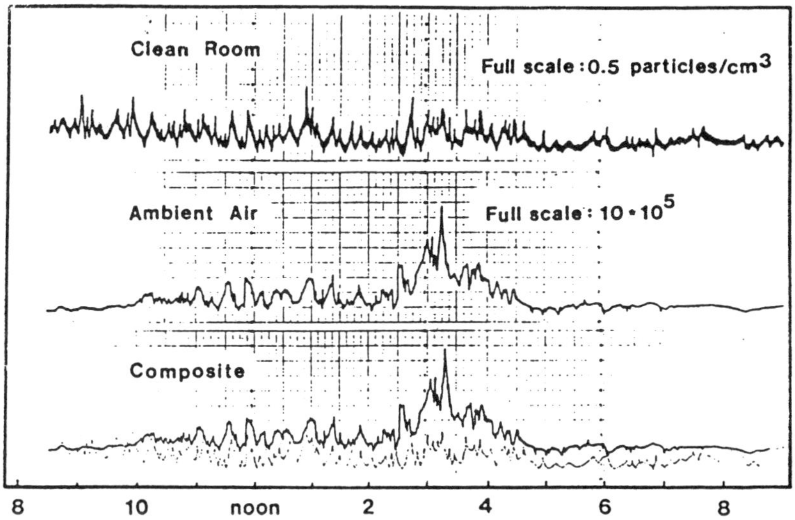

Source: Liu et al., 1986

Figure 6-4. Comparison of CPC measurements inside a clean room and on the roof of the four-storied building that contains the clean room.

room. In that clean room during that particular time, particles < 0.15 μm appeared to come mostly from makeup air, while particles > 0.15 μm originated partially within the room. In Figure 6-4, the outdoor CPC is near the inlet for the makeup air to the clean room. Both CPCs are TSI 3020s. The upper graph is clean room concentration, the middle graph is outdoor concentration, and the bottom graph is an overlay of one on top of the other. Notice the close correlation between most indoor and outdoor spikes. In parallel tests, little correlation was found between indoor and outdoor particles larger than 150 nm.

Donovan et al. (1984) and Liu et al. (1986) found that steady-state size distributions of particles inside clean rooms at rest do not contain very many particles smaller than 50 nm. Figures 6-5 and 6-6 illustrate those results. Notice that the measurement in Figure 6-5, in an at-rest clean room with no known internal nuclei mode sources, indicates that only about half the particles are smaller than 100 nm. The data in Figure 6-6, again taken under at-rest conditions, indicate that particles under 100 nm outnumber those over 100 nm by a factor of about 5. Much more data are needed, especially during normal nonsteady-state operations, before we can draw any general conclusions. These data do not include any data from internal particle events that provide the majority of particles to many critical clean areas.

Ultrafine particles do not settle significantly in gravity because they weigh so little. They migrate toward deposition surfaces by Brownian diffusion resulting from collisions with gas molecules (Chapter 2). Brownian diffusion makes HEPA and ULPA filters very efficient for 10-nm particles; it is a less effective capture mechanism for 100-nm particles. Brownian diffusion also determines the rate of particle loss inside aerosol sampling tubes for ultrafine particles.

Coagulation is generally insignificant in clean rooms because of rapid dilution. Once the particles have dispersed sufficiently from the source, they are not close enough to neighboring particles for much coagulation to occur. Coagulation is important during the first second or two after particle formation by condensation. The size of the particle is determined during this short time period.

Once ultrafine particles touch a surface, they are nearly impossible to dislodge without solvents and/or detergents (Bowling, 1985; Fisher, 1987). Because particles within solvents and detergents are so difficult to get rid of, it is usually best to minimize the deposition of ultrafine particles onto clean surfaces.

Ultrafine particles are more chemically reactive than larger particles or bulk quantities of a material because their surface area per unit mass is very high. For a constant mass of spherical particles, 10-nm particles contain 1,000 times as much surface as 10-μm particles.

Summarizing, particles as small as 10 nm have been measured in large numbers in steady-state, at-rest clean rooms. Even larger numbers are expected in fully operational clean rooms.

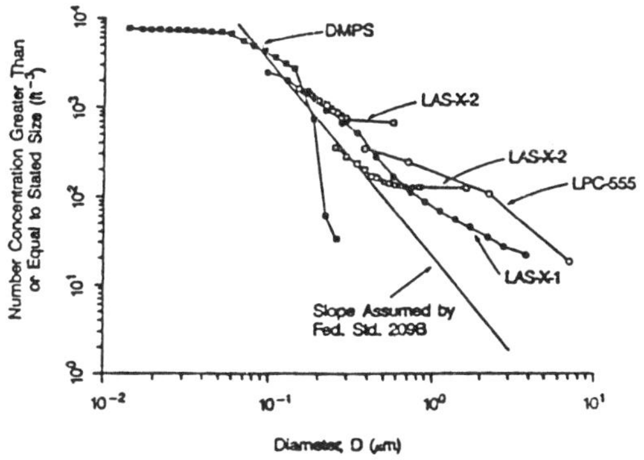

Source: Donovan et al., 1984

Figure 6-5. Grand average cumulative number distribution of airborne particles in a university semiconductor fabrication laboratory as measured by the indicated PMS OPCs and the TSI DMPS.

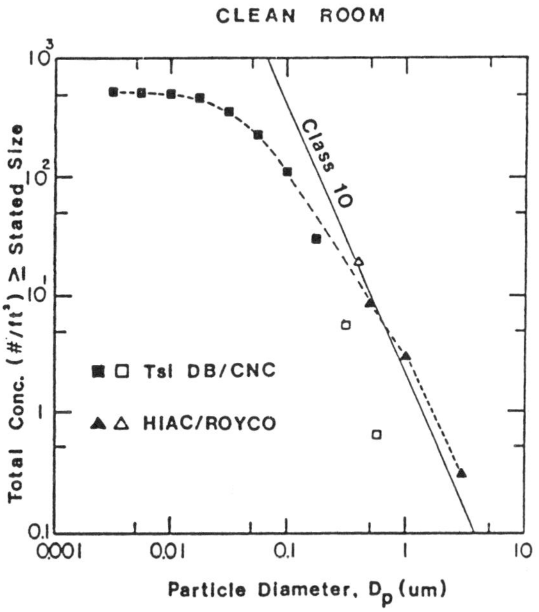

Source: Liu et al., 1986a

Figure 6-6. Cumulative number distribution of airborne particles in a university semiconductor fabrication laboratory as measured by the indicated instruments.

WHY MEASURE ULTRAFINE PARTICLES?

Rule of 10

Traditionally, the "rule of 10" has been used for determining the smallest particle that can cause damage to a semiconductor. If the minimum dimensions on the semiconductor are 1 μm, the "rule of 10" says that particles larger than 0.1 μm may cause failure of the integrated circuit. This dimension, one-tenth of the minimum dimensions on the chip, has determined the type of particle counter purchased to measure airborne particles in many clean rooms.

The Third Dimension: Layer Thickness

The "rule of 10" does not consider the equally important third dimension of the semiconductor. Many of today's chips have layer thicknesses that are much thinner than 1 μm, even as thin as 20 nm on some CMOS oxide layers. During certain parts of the manufacturing process for these components, particles of 10 nm may cause failures (Fisher, 1987). Because some of these particles may contain NaCl, they may degrade areas on the semiconductor that are much larger than the particle itself.

Evidence That Ultrafine Particles Cause Semiconductor Defects

Fisher (1987) presents detailed evidence that ultrafine particles cause defects (see Chapter 1). He says that there are no particles so small that their presence in the manufacturing environment can be safely ignored. He identifies three ways that ultrafine particles can cause failures:

- Particle interactions with thin layers
- Chemical interaction
- Particle-induced defects that are much larger than the originating particle.

The thicknesses of junctions and of oxide layers in the gates of MOS transistors of DRAMs are now in the range of 20 to 40 nm. For these films, it is the size of a particle relative to the film thickness that determines whether the particle causes a killing defect. A particle that is the same size as the film thickness may clearly cause a short through an insulating film.

To illustrate the second way that ultrafine particles can cause defects, Fisher cites a documented case of chemical reaction between particles and silicon dioxide films during growth of the films by thermal oxidation of silicon. The resulting particles were 2 to 25 times their original size. The size of the defect depends on the chemical composition and physical size of the original particle, the oxidation time, and the temperature of the oxidation process.

The third way that ultrafine particles can cause defects is by the physical growth of a defect or particle that is much larger than the original particle. Fisher illustrates this method by an etching process. Masking by an ultrafine particle during a critical point of a photolithography process can grow a cone- or column-shaped defect that is as long as the thickness of the layer. This defect may cause semiconductor failure where it is formed, or it may break loose and cause failure elsewhere.

MEASUREMENTS OF AIRBORNE ULTRAFINE PARTICLES

Condensation Particle Counters

Condensation particle counters (CPCs), traditionally called condensation nucleus counters (CNCs), are the primary detector to sense and count particles smaller than 100 nm. We will cover them here in more detail than any other instrument.

Definitions

A *condensation nucleus* is a *small particle* upon which a vapor will condense. Many vapors will condense on any particle, regardless of the composition or other characteristics of the particle.

Therefore, we use the terms *nucleus* and *particle* interchangeably.

Ultrafine particle = condensation nucleus.

Nucleus is singular. *Nuclei* is plural.

The term *condensation nucleus* comes from atmospheric cloud physics where condensation nuclei are the core upon which water condenses to form raindrops. However, water does not condense as readily on hydrophobic particles as it does on hydrophilic particles.

Instruments

Condensation particle counters are simple enhancements of optical particle counters (OPCs). The optical section is simply an optical particle counter that can count relatively large particles. Upstream of the OPC section is a device that grows the particles until they are easy for the OPC to detect and count. The particle-growing device is also simple. It condenses a convenient vapor onto the particles until they become micrometer-sized droplets.

The above description applies only to continuous-flow, single-particle-counting CPCs. Older models were not capable of detecting single particles. They required sufficient concentration to produce a cloud of particles. The older photometric CPCs (CNCs) are not useful in areas cleaner than Class 1,000,000.

Figure 6-7, from Keady, Quant, and Sem (1986), shows how gas passes through a CPC. Aerosol, drawn into the CPC by a vacuum pump, first becomes saturated with the working fluid at a temperature a few degrees above ambient. The saturated aerosol enters a cooled condenser where the vapor becomes supersaturated and condenses onto the particles. The particles grow until the vapor is no longer supersaturated. They then pass through the sensing volume of an optical detector, producing a pulse from the photodetector. A computer counts pulses for a specified period of time, corresponding to a specified aerosol sample volume. The result is simply counts (or particles) per unit volume of aerosol.

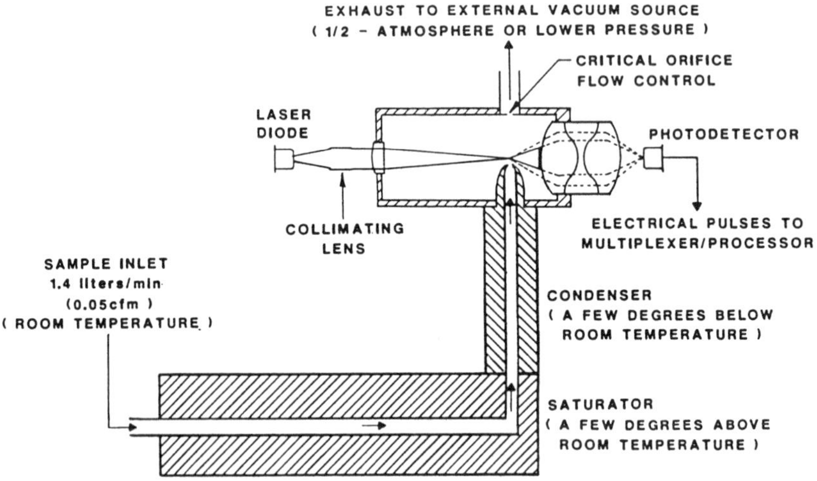

Source: Keady et al., 1986

Figure 6-7. Schematic diagram of a TSI 3760 clean room CPC illustrating how aerosol passes through the instrument.

Another type of continuous-flow, single-particle-counting CPC uses the mixing of two gas streams to create supersaturated conditions (Kousaka et al., 1982; Fukushima et al., 1986; Sekigawa and Ohno, 1987). One of the gas streams is saturated with vapor at an elevated temperature; the other gas stream is the aerosol stream.

Primary Standard

Single-particle-counting CPCs are primary measurement standards. CPCs count droplets as they pass one-at-a-time through the sensing volume. Particle concentration is simply:

$$\text{Conc} = \frac{\text{Counts/second}}{\text{Sample flow rate}} \quad .$$

Sample flow rate is expressed in ft^3/s or cm^3/s. Therefore,

$$\text{Conc} = \frac{\text{Counts}}{\text{Unit volume}} \quad .$$

Therefore, the measurement is a primary standard that does not require independent calibration if:

- The electronic counter operates properly (a simple function to check).
- The sample flowmeter is calibrated and operates properly.
- All particles of interest are activated by the condensation process.

There is another primary standard for concentration measurement for ultrafine aerosol particles, developed by Liu and Pui (1974). The comparison in Figure 6-8 shows that the two primary standard methods agree with each other almost perfectly.

Particle Size Dependence

Figure 6-9 shows the lower detection limit of the CPCs for which we have data. The lower detection limit is specified at the size where 50 percent of the particles of that size are detected. The TSI 3020's 50 percent detection point is 10 nm; the TSI 3760's 50 percent detection point is 14 nm. Notice that all of them have a lower limit between 2 nm and 20 nm. All detect some smaller particles and do not detect some larger particles, a characteristic shared with all aerosol particle counters.

The upper limit of detection for most CPCs is 3 to 5 μm. Thus, CPCs detect the particles of primary importance for today's semiconductor manufacturing.

CPC Working Fluids

There are many possible working fluids for CPCs:

- ***Butanol***: Used extensively, best-known performance, disagreeable odor, somewhat toxic.

- ***Isopropanol***: Absorbs water from air changing the amount of vapor available for saturation; evaporates quickly requiring frequent replenishment; somewhat toxic.

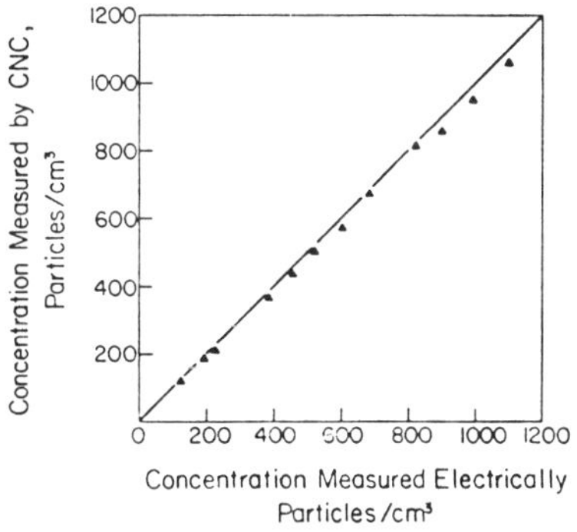

Source: Agarwal and Sem, 1980

Figure 6-8. Comparisons between electrically measured and CPC-measured concentrations (TSI 3020 CPC in the count mode) with 50-nm monodisperse NaCl aerosol.

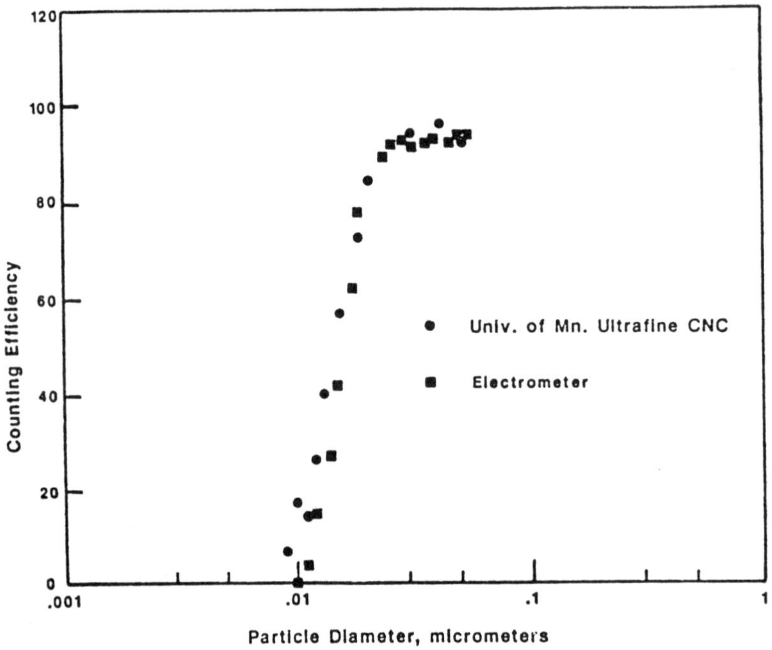

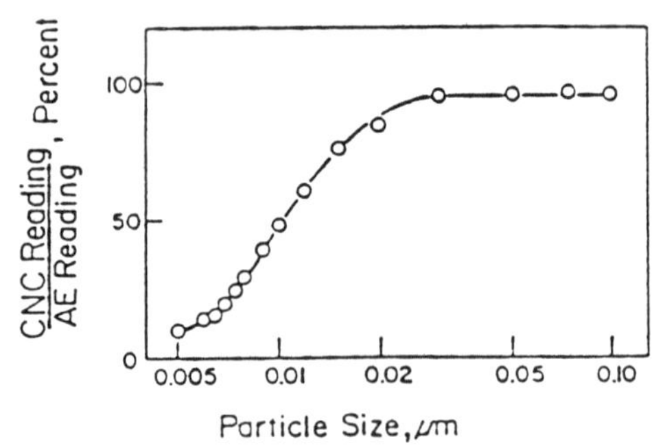

Source: Smaller graph from Agarwal and Sem, 1980

Figure 6-9. Detection efficiencies for two CPCs: TSI 3020 (smaller graph) and TSI 3760 (larger graph).

- ***Ethanol***: Difficult to obtain, absorbs water, requires frequent replenishment.
- ***Diethylene glycol, dipropylene glycol, dibutyl phthalate, and hexanol***: Used in mixing-type CPCs with success (Kousaka et al., 1982; Fukushima et al., 1986; Sekigawa and Ohno, 1987).
- ***Inert fluids***: Several organizations are currently testing several inert fluids, including perfluorinated hydrocarbons, for measurements of particles in air, nitrogen, argon, helium, oxygen, and hydrogen. Fluorocarbons are not harmful to the ozone layer as are chlorofluorocarbons.
- ***Water***: May be a viable fluid for some CPC designs, but water may not condense on hydrophobic nuclei, such as nearly pure mineral oils.

Commercial CPCs

Table 6-1 summarizes properties of contemporary CPCs.

Diffusion Particle Size Selectors

Brownian diffusion is the only practical method we have to classify low concentration ultrafine aerosol particles by size so we can measure their size distribution. Brownian diffusion is the more-or-less random motion of airborne particles caused by collisions with gas molecules. Thus, a particle that is carried along with a laminar airstream has motion perpendicular to the bulk air motion as well as parallel to the air. Clearly, the magnitude of diffusive motion is greater for small particles that have masses closer to the mass of air molecules than it is for large particles.

Ultrafine particles passing laminarly through a tube have a finite chance of colliding with the tube wall because of diffusive motion. The smaller the particle, the greater the chance the particle will not complete the trip through the tube. The same can be said about the fraction of ultrafine airborne particles that will successfully pass through a screen. The smallest ones have a greater chance of ending up on the screen wires.

The collection of particles on a screen by Brownian diffusion is not intuitively obvious. This mechanism operates opposite to the normal particle collection by a screen. For macrosized particles, the smallest particles have the greatest chance of penetrating the screen.

The traditional method of using Brownian diffusion to measure aerosol particles is called a diffusion battery, illustrated in Figure 6-10. Smaller particles deposit preferentially on earlier screens; larger particles penetrate the screens more easily. The commercially available diffusion battery (TSI 3040) consists of fine-mesh (250 wires/cm) stainless steel screens (shown in Figure 6-11) arranged in series. After the aerosol has passed through each set of screens, a CPC measures the concentration. The data must then be inverted to obtain an estimate of the particle size distribution.

Table 6-1. Single-Particle-Counting CPCs That Measure Ultrafine Particles

Manufacturer:[a]	TSI	TSI	TSI	Kanomax	Kanomax	Rion	Dan Sci
Model No.:	3766	3022	3025	3851	3861	KC-60	?
Introduced:	1988	1988	1988	1986	1987	1988	1988?
Country:	USA	USA	USA	Japan	Japan	Japan	Japan
Mininum part nm:	14	10	2.5	10	10	10	?
Flow cfm:	0.10	0.01	0.001	0.10	0.20	0.10	?
Readout:	Options[b]	Panel +	Panel +	Options[c]	Panel	Panel	?
Computer comp:	Yes	Yes	Yes	Yes	Yes	?	?
False ct/ft^3:	< 0.2	?	?	?	?	?	?
Maximum conc/ft^3:	10^7	10^{11}	10^9	10^7	10^6	10^6	?
Light source:	LaserDi	LaserDi	LaserDi	LaserDi	LaserDi	LaserDi	?
Laser power mW:	3	3	3	3	3	?	?
Recom fluid:	Inert	Inert	Inert	EGlycol	EGlycol	PGlycol	?
Days/fill:	7	10 +	30 +	14	14	?	?
Ext. vac. req'd:	Yes	No	No	Yes	No	No	?
Noncontam. des:[d]	Yes	No	No	Yes	No	No	?
Power req'd. W:	30	200	200	?	?	230	?
Volume cm^3:	4,774	18,144	22,464	16,800	39,680	37,400	?
Weight Kg:	4	14	16	9.5	~15	20	?
US Price $:	5,800	17,500	19,500	~14,000	~20,000	> 20,000	?

[a]TSI Inc., 500 Cardigan Road, PO Box 64394, St. Paul, MN 55164 USA.
Kanomax Japan, Inc., 2-1 Shimizu, Suita, Osaka, Japan. The Model 3851 is sold, in somewhat modified form with a readout panel, in North America by Atcor, 2350 Charleston Road, Mountain View, CA 94043.
Rion Company Limited, Kokubunji, Tokyo, Japan. We do not know if Rion markets this model in North America.
Dan Science, Japan. A rumor says that Dan will introduce a CPC in 1988. Specs unavailable. Dan Science does not market instruments in North America.

[b]TSI Model 3766 is designed to fit into the TSI multipoint continuous monitoring system. Concentration readout will normally be available on the personal computer. Separate readout modules are available. This model does not contain a pump.

[c]The Kanomax 3851 does not contain a pump.

[d]Noncontaminating design means no pump, no cooling fan, and no printer.

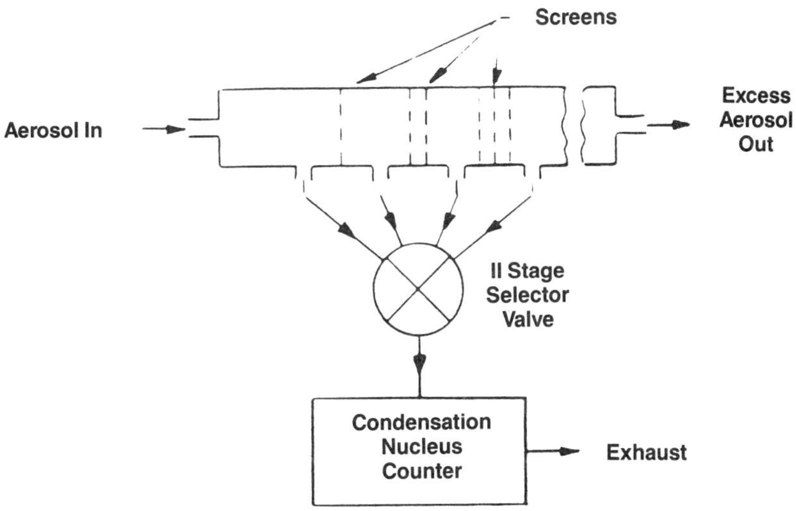

Figure 6-10. Schematic diagram of a screen battery assembly.

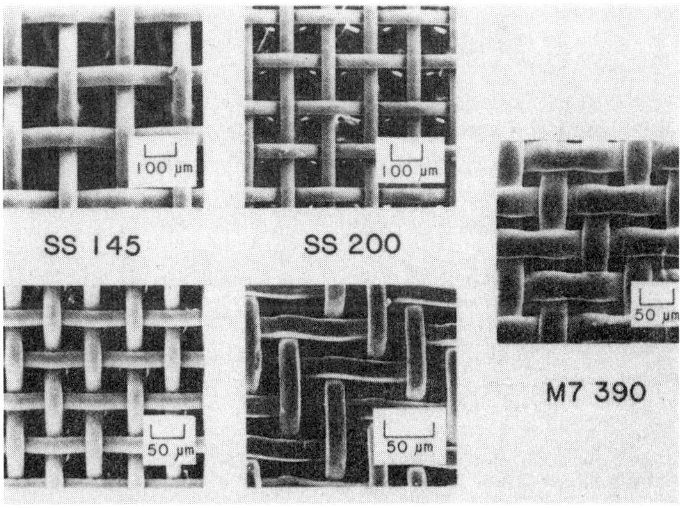

Source: Cheng and Yeh, 1983

Figure 6-11. Photographs of screens used in diffusion batteries.

The inversion process substantially magnifies any experimental error. Because diffusion is a random process and does not have very sharp size resolution, experimental errors are already quite large. Therefore, the repeatability and the accuracy of diffusion measurements is not high.

In spite of the problems with the diffusion measurement, it remains the best available method of measuring ultrafine particle size distributions in clean areas. It shares the same lower concentration limits as its CPC concentration detector. It requires measurements at each of several stages, depending on how many data points you need over the entire size range. To minimize random errors, each stage should have at least 100 particle counts. Therefore, it is not a fast measurement.

TSI and Research Triangle Institute are currently evaluating the diffusion measurement technique for clean areas. One of the options we are looking at is the use of four CPCs, each with a different number of diffusion screens mounted ahead of it, to measure four "size" channels simultaneously. Some preliminary results are reported in Chapter 9 (see p. 9-12).

Perhaps the most useful way of utilizing Brownian diffusion to measure airborne particles in clean areas is as a simple way to remove particles smaller than a specified size from the sample entering a CPC. Thus, a CPC can measure particles larger than 10, 20, 50, 100, or 200 nm. The number of diffusion screen determines the size limit of the CPC.

The CPC with diffusion conditioner has two substantial advantages over 100-nm laser particle counters:

- Much lower false count rate, regardless of the size selected
- Lower hardware cost.

The CPC with diffusion conditioner also has a disadvantage compared to 100-nm LPCs. The LPCs have a substantially sharper particle size cutoff at 100 nm because of the 6th-power dropoff of light scattered from ultrafine particles. Figure 6-12 shows the lower size limit characteristic of the CPC with diffusion conditioner. The 50 percent efficiency point for this particular combination of screens and flow rate is 100 nm.

Useful references on diffusion include works by Remiarz et al. (1980); Yeh, Cheng, and Orman (1982); Soderholm (1983); Cheng and Yeh (1983); Brown, Beyer, and Gentry (1984); Locke, Donovan, and Ensor (1985); and Sinclair (1986).

Light-Scattering Particle Counters

The purpose of this section is not to describe laser particle counters, but to discuss the usefulness of such counters for the measurement of ultrafine particles.

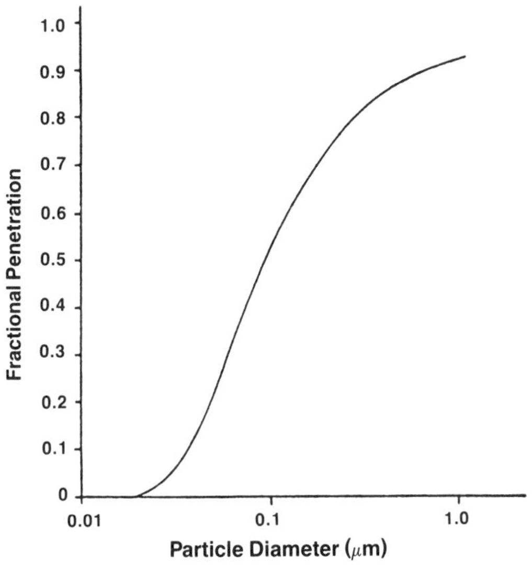

Figure 6-12. Lower detection efficiency for a CPC with a diffusion conditioner attached to its inlet.

Light-scattering particle counters, both the active cavity and passive cavity types, face two formidable barriers at about 0.1 μm:

- The light scattered by an airborne particle is proportional to the 6th power of particle diameter in this size range.
- The gas molecules in the sensing zone scatter as much light as a particle in this size range.

To illustrate the significance of this 6th power relationship, let us assume that we want to design a light-scattering particle counter that detects 0.05 μm particles rather than the current 0.1 μm diameter limit. This "simple" factor of 2 requires a detectability improvement of $2^6 = 64$ from our instrument. We can try to gain this factor of 64 by a combination of these design factors:

- Increase laser intensity by 64 times. But we are already inside the laser cavity itself, where the intensity is highest, and we are using large and expensive lasers.
- Improve signal-to-noise by 64 times. But we are already pushing the state of the art for detection, and our current signal-to-noise is about 2:1 for 0.1-μm particles.
- Decrease sensing volume by a factor of 64 to increase laser intensity. But if we decrease sensing volume any more, we will not be able to sample 0.1 cfm, or even 0.01 cfm.

It appears that we have reached a basic lower limit for the particle size that we can detect by active or passive cavity light scattering. Then we discover the second basic factor that makes the push to lower minimum detectable size even more difficult:

- The light scattered by gas molecules in a reasonably sized sensing volume (large enough for reasonable aerosol flow rate) is comparable to the light scattered by 0.1-μm particles.

Instrument designers have done an excellent job of optimizing the lower detection limit of laser particle counters. There will certainly be improvements such as less expensive lasers, perhaps laser diodes, that will replace the expensive, fragile lasers of today, but there will not be much improvement in the minimum detectable particle size of particle counters because of these formidable and basic laws of physics.

Laser particle counters that can detect and count particles as small as 0.1 μm are available from Particle Measuring Systems (PMS), Climet, Met One, Pacific Scientific, Hitachi (Japan), Rion (Japan), and Dan Science (Japan). PMS and Hitachi offer models with detectability down to 0.05 μm at 0.1 cfm and for high price. The lower limits are almost always specified for polystyrene latex spheres.

The false count when no particles are present is specified at less than 1 particle (> 0.05 μm) per cubic foot for the PMS 0.05 μm model. The false count rate is a critical specification for particle counters. With signal-to-noise limits approaching 2:1, expect to find significant false counts in the lowest size channels of many models. With false counts of 10 to 20 per cubic foot, many particle counter models of the mid-1980s are not useful in today's Class 1 and 10 environments.

Electrical Mobility Particle Size Analyzers

Two electrical mobility methods can measure the size distribution of aerosol particles. The oldest, named the electrical aerosol analyzer (EAA), was used extensively during the 1970s to measure atmospheric, smog chamber, and stack pollution aerosols. It is the instrument most responsible for discovering the trimodal nature of air pollutant aerosol. Unfortunately, the EAA requires concentrations greater than 1,000,000/ft^3. Contrary to a suggestion by Donovan, Locke, and Ensor (1985a), replacement of the electrometer detector with a CPC does not solve the lower-concentration-limit problem. Therefore, the EAA will not be discussed further here.

The newer, more precise electrical mobility sizing method is called the differential mobility particle sizer (DMPS). The core of the instrument is often called a differential mobility analyzer (DMA). Figure 6-13 shows schematically how the DMPS works. The DMPS is a combination of a DMA to select a narrow size-band of particles and a CPC to detect the concentration of those monodispersed particles. The DMPS is not practical for clean rooms cleaner than Class 300. In a DMPS, aerosol enters the DMA where large particles and small particles are electrically stripped away, with only a narrow size-band of particles remaining in the airborne state. The size of the particles in the narrow size-band is selectable from 0.01 to 0.5 μm. These selected particles then pass out of the DMA and into a CPC that measures their concentration. By scanning the DMA over the entire size range, we obtain the submicrometer size distribution.

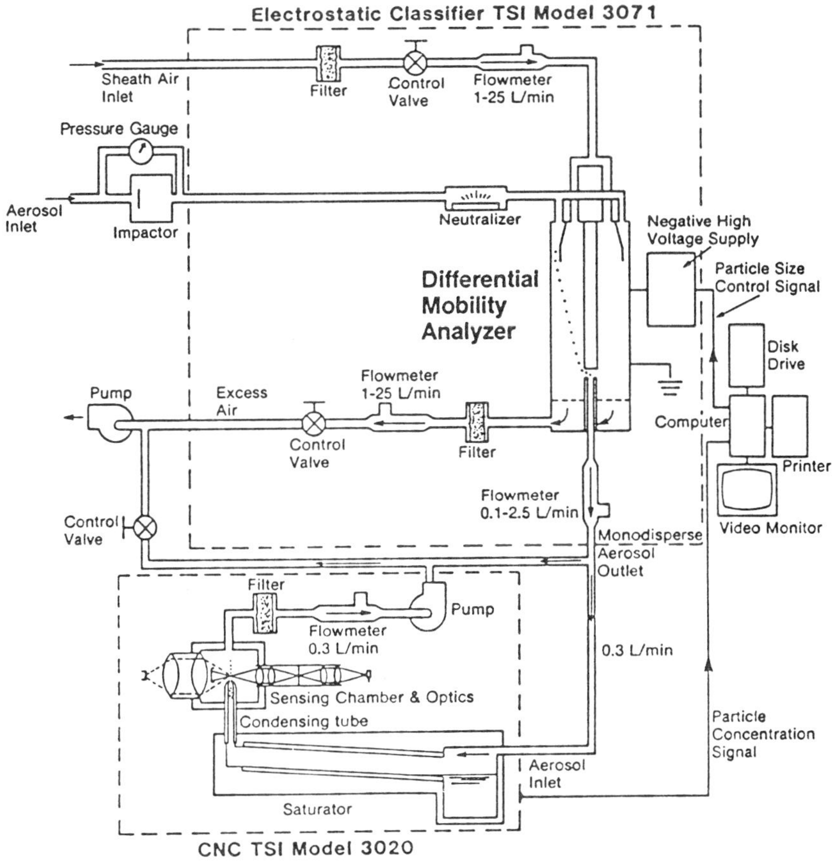

Source: Keady, Quant, and Sem, 1983

Figure 6-13. Schematic diagram of a differential mobility particle sizer (DMPS).

The DMPS sizing method is able to measure lower concentrations than the EAA. However, Donovan et al. (1985b) found that Class 300 environments were the lowest concentration level practically measurable by the DMPS. Figure 6-14 shows the distribution measured by Donovan et al. in a clean room. Figure 6-15 shows the lower concentration limit for the DMPS measured in the same experiment. A limitation pointed out by Donovan et al. is the 20+ minutes required by the DMPS to scan the entire size distribution. A problem with some DMPSs is that the high voltages (up to 10,000 V) required to classify the particles generates particles by corona discharge. The result is that the DMPS is not very useful for measuring airborne particles in most semiconductor clean rooms. It is a highly useful tool in filter testing and other higher concentration environments.

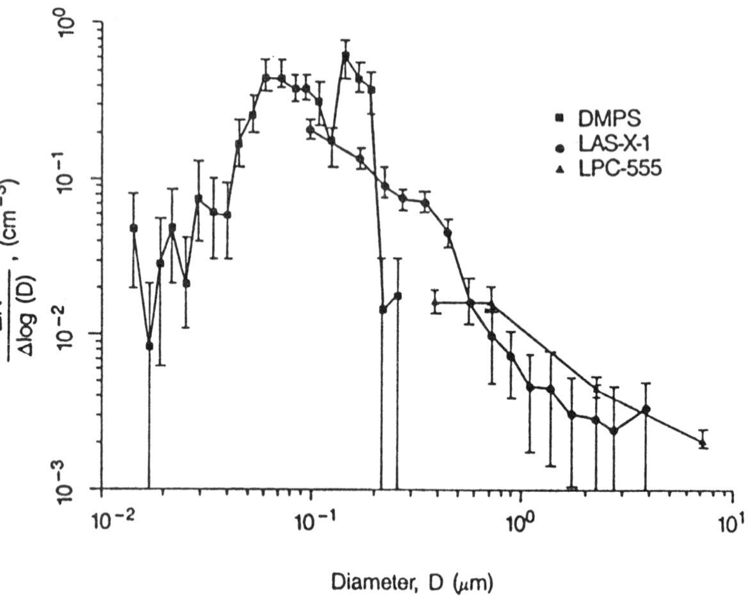

Source: Donovan et al., 1985b

Figure 6-14. Grand average differential number size distributions in an at-rest university semiconductor fabrication laboratory.

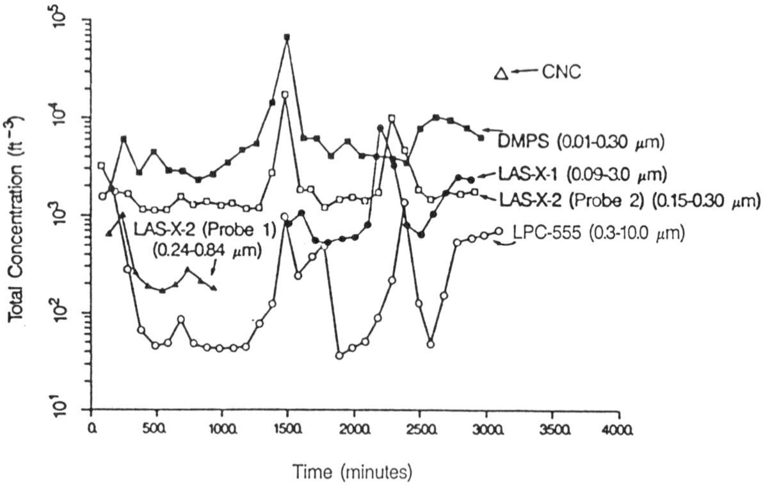

Source: Donovan et al., 1985b

Figure 6-15. Total aerosol concentrations in the tests shown in Figure 6-14.

MEASUREMENTS OF ULTRAFINE PARTICLES IN PRESSURIZED GAS LINES

No instrument is available to measure or count ultrafine particles in pressurized gas lines at pressures greater than about 1.5 atmospheres. Light-scattering methods cannot detect particles in this size range because:

- The light scattered from a particle drops off proportional to the 6th power of particle diameter.
- The gas molecules in the sensing volume surrounding the particle scatter light comparable to the particle itself.

It will probably be possible to extend condensation particle counting to pressurized gases. However, instruments do not now exist for this purpose.

Fortunately, many uses for pressurized gas in semiconductor manufacturing are at atmospheric pressure. Therefore, point-of-use measurements using condensation particle counters are feasible. Current instruments can measure air, nitrogen, argon, and helium. Work is progressing to design instruments that will extend the list to hydrogen, oxygen, and other hazardous gases.

MEASUREMENTS OF ULTRAFINE PARTICLES IN VACUUM

No known instrument can detect ultrafine particles in a vacuum. Particle flux instruments, such as the one marketed by High Yield Technology, are probably not extendable to this size range because:

- Light scattered by particles drops off proportional to the 6th power of particle diameter.
- Any surrounding light will provide noise to the signal.

Because several methods of manipulating molecules use high-vacuum technology, it would seem feasible to use particle beams, charged particles, electric fields, and charge detection to measure particles in this size range. However, no known instruments yet exist using this technology.

MEASUREMENTS OF ULTRAFINE PARTICLES IN DIFFUSION FURNACES

Current particle counters cannot fit into the environment of a diffusion furnace, nor can they operate in such hot gases. However, Cheung and Roberge (1987) have described and demonstrated a sampling system that can allow both CPCs and laser particle counters to sample the gas inside a diffusion furnace and measure the particles in that gas. They used an isokinetic quartz sampling probe, stainless steel sampling tube, and a flow splitter to remove a sample from the furnace and deliver it to both instruments simultaneously. The length and tortuousness of all flow paths were minimized. The wandlike design of the quartz probe allows it to be put in close proximity to suspected particle sources within the furnace. The probe and sampling system were certified experimentally to produce no particles. The sampling efficiency was also measured experimentally and was found to be 85 percent for particles smaller than 100 nm and 90 percent for particles larger than 100 nm.

Cheung and Roberge (1987) used the system to measure many diffusion furnaces. In one case with nitrogen gas, they found that valves, joints, fittings, regulators, and mass flow controllers generated large quantities of particles if there was no point-of-use filter in place. The CPC measured about 700,000 particles/ft^3 above 15 nm; the LPC measured about 100 particles/ft^3 above 100 nm. In another case, they identified large quantities of sub-100-nm particles generated during and for several hours after heat-up of a diffusion furnace from 700 to 1,000 °C. It is clear that this method can provide useful data about particles present inside operating tube furnaces.

MEASUREMENTS OF LIQUIDBORNE ULTRAFINE PARTICLES

Traditional Liquidborne Particle Counters

There are currently no known methods for measuring or counting single particles smaller than 0.1 μm in liquids. Light-scattering methods do not work because the surrounding liquid scatters an amount of light comparable to the light scattered by the particle itself. This is especially true for particles that have an index of refraction near that of the liquid. The Coulter method sees too much noise in this range. No other method is known for detecting such small single particles in liquid.

It is important that any device for this purpose detect single particles, rather than clouds of particles. The concentration of particles in ultraclean liquids needed for semiconductor manufacturing is so low that only single particles will likely be in any sensing volume at any one time.

Nonvolatile Residue Monitors: An Alternative

The measurement of nonvolatile residue may be an alternative to ultrafine particle measurements in liquids. Nonvolatile residue is what is left when the volatile part of a liquid evaporates. Ultrafine particles are only one part of nonvolatile residue. Dissolved salts are another major part of nonvolatile residue.

When removing liquid from the surface of a wafer, we are usually interested in the total contamination that remains. It often does not matter much whether the remaining contaminating foreign matter is in the form of particles or films. Therefore, a convenient monitoring method that senses nonvolatile residue may serve as an alternative to one that senses only ultrafine particles.

Wen, Kasper, and Chesters (1986) and Kousaka et al. (1987) have recently introduced two new closely related methods for monitoring the nonvolatile residue of certain liquids.

Wen's method requires that a sample of the liquid be aerosolized into a fine mist of monosized droplets. The volatile portion of the mist droplets then evaporates, leaving only an aerosol of nonvolatile residue. The size of the nonvolatile residue particles is directly proportional to the amount of nonvolatile residue in the original mist. Their original tests used a vibrating orifice aerosol generator to produce the monosized mist and laser aerosol spectrometer to measure the size of nonvolatile residue particles. Beginning with monosized droplets of 40 μm, 1 ppm of nonvolatile residue results in nonvolatile particles of 0.4 μm. Beginning with monosized droplets of 100 μm, 10 ppb of nonvolatile residue results in nonvolatile particles of 0.215 μm. The only assumptions are that the original liquid is not recontaminated during introduction into the vibrating orifice generator and that the nonvolatile solid particles are nearly spherical and dry. Both are reasonable with the correct design of the equipment. Figure 6-16 shows schematically the components of Wen's method. As shown in Figure 6-16,

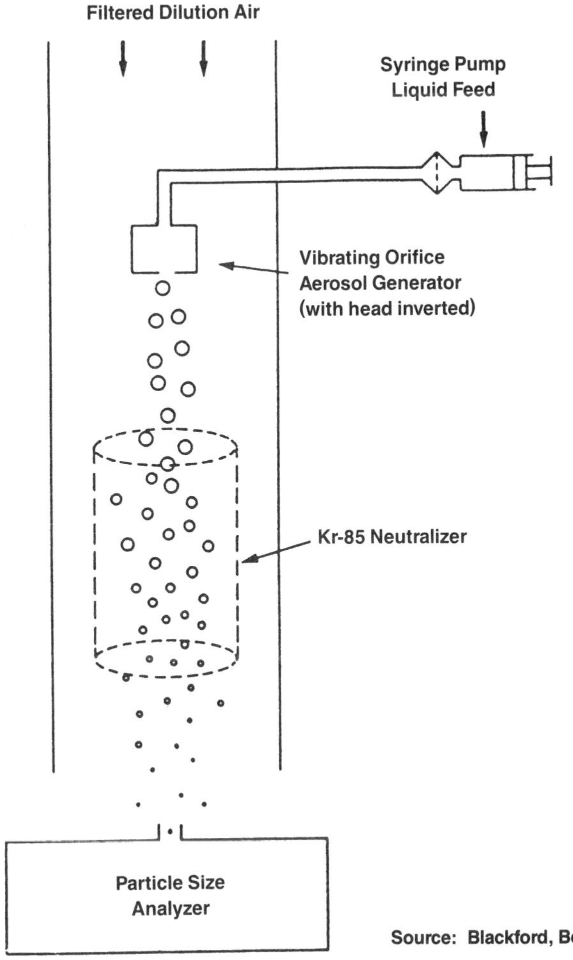

Source: Blackford, Belling, and Sem, 1987

Figure 6-16. Schematic diagram of the experimental setup for the Wen, Kasper, and Chesters (1986) method of precisely measuring the nonvolatile residue of a volatile ultrapure liquid.

the vibrating orifice aerosol generator produces uniformly sized droplets of the ultrapure liquid. The volatile fraction of the droplet then evaporates, leaving a nonvolatile residue particle to be measured by a particle size analyzer.

Kousaka's method, shown schematically by Blackford, Belling, and Sem (1987) in Figure 6-17, is simpler from a hardware point of view, but requires calibration. This method uses an atomizer to generate an aerosol. The resulting polysized droplets with a wide range of diameters around 1 μm pass through a diffusion dryer so that only nonvolatile residue remains in the particles (the volatile fraction of the droplets evaporates). A CPC then measures the concentration of the aerosol that is within the detectable size range of the CPC. If the nonvolatile residue concentration in the original liquid is high, the CPC will measure nearly all of the dry particles. If the nonvolatile residue concentration is very low, most of the dry particles will be below the detection limit of the CPC, resulting in a low detected concentration. Thus, CPC concentration is directly related to nonvolatile residue content of the liquid by calibration. Calibration, using a technique such as that of Wen, Kasper, and Chesters (1986), can result in a fast, accurate, ultrasensitive measure of nonvolatile residue content.

The lower detection limit of this technique appears to be below 10 ppb. The Kousaka method appears easily adaptable for continuous, on-line monitoring of DI water and other volatile liquids. The Wen method is necessary to calibrate the Kousaka method. Nomura Micro Science Company Limited, Kanagawa, Japan, offers an instrument based on the Kousaka design. The components of the Wen method are available from TSI Incorporated, Box 64394, St. Paul, MN 55164, and Particle Measuring Systems Incorporated, 1855 South 57th Court, Boulder, CO 80301.

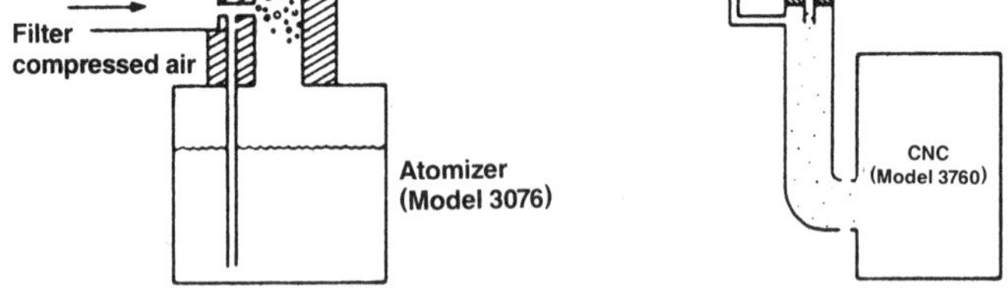

Source: Blackford, Belling, and Sem, 1987

Figure 6-17. Schematic diagram of the experimental setup for the Kousaka et al. (1987) method of measuring nonvolatile residue of a volatile ultrapure liquid.

MEASUREMENTS OF SURFACEBORNE ULTRAFINE PARTICLES

There are currently no known methods for measuring or counting particles smaller than 100 nm on a surface. Light-scattering methods cannot be extended to this range because:

- The amount of light scattered from a particle drops off proportional to the 6th power of particle diameter.
- The surface the particle is resting on also scatters light, providing a relatively large noise signal.
- The gas molecules in which the particle is immersed scatter light comparable to the light scattered by the particle.
- Patterns on wafer surfaces scatter light much like particles.

Lilienfeld (1986) describes the complexities of using light-scattering to detect ultrafine particles on surfaces.

Several people, including Lilienfeld, have proposed condensation growth, analogous to the condensation nucleus counting method for airborne particles, to grow ultrafine particles on surfaces until they can be counted by ordinary surface scanners. The proposed technique would cool the wafer surface until a selected vapor surrounding the wafer condenses onto the particles. There is no assurance that this technique would work, for several reasons:

- Vapor may prefer to condense onto the wafer surface because that surface has a preferable radius of curvature.
- Finding a vapor that is phobic to the surface, but philic to the particles, may not be easy.
- Patterns on wafer surfaces may react much like particles.

It may be many years before someone develops practical ways to detect ultrafine particles on surfaces, especially patterned surfaces.

REFERENCES

Agarwal, J. K., and G. J. Sem. 1980. Continuous Flow, Single-Particle-Counting Condensation Nucleus Counter. *Journal of Aerosol Science*, 11:343-357.

Blackford, D. B., K. J. Belling, and G. J. Sem. 1987. A New Method for Measuring Nonvolatile Residue for Ultrapure Solvents. *Journal of Environmental Science*, 30(4):43-47.

Bowling, R. A. 1985. An Analysis of Particle Adhesion on Semiconductor Surfaces. *Journal of the Electrochemical Society*, 132(9):2208-2214.

Brown, K. E., J. Beyer, and J. W. Gentry. 1984. Calibration and Design of Diffusion Batteries for Ultrafine Particles. *Journal of Aerosol Science*, 15:133-146.

Cheng, Y. S., and H. C. Yeh. 1983. Performance of a Screen-Type Diffusion Battery. *Aerosols in the Mining and Industrial Work Environments, Vol. 3: Instrumentation*, V. A. Marple and B.Y.H. Liu, eds. Ann Arbor Science Publishers, Ann Arbor, Michigan, pp. 1077-1094.

Cheung, S. D., and R. P. Roberge. 1987. An Inside Look at Measuring Particles in Process Equipment. *Microcontamination*, 5(5):45-50,94,96.

Donovan, R. P., B. R. Locke, and D. S. Ensor. 1985a. Real-time Measurements of Single Submicron Aerosol Particles in Clean Rooms. *Solid State Technology*, 28(9):139-148.

Donovan, R. P., B. R. Locke, D. S. Ensor, and C. M. Osburn. 1984. The Case for Incorporating Condensation Nuclei Counters into a Standard for Air Quality. *Microcontamination*, 2(6):39-44.

Donovan, R. P., B. R. Locke, C. M. Osburn, and A. L. Caviness. 1985b. Ultrafine Aerosol Particles in Semiconductor Clean Rooms. *Journal of the Electrochemical Society*, 132(11):2730-2738.

Fisher, W. G. 1987. Particle Monitoring in Clean Room Air with the TSI 3020 Condensation Nucleus Counter. *TSI Journal of Particle Instrumentation*, 2(1):3-19.

Fukushima, N., K. Hironaga, K. Sugimoto, and T. Takahei. 1986. CNC Sensor for Clean Room Monitoring. *1986 Proceedings, 32nd Annual Technical Meeting, Institute of Environmental Sciences, Dallas/Ft Worth, Texas, 6-8 May*. pp. 439-444.

Keady, P. B., F. R. Quant, and G. J. Sem. 1983. Differential Mobility Particle Sizer: A New Instrument for High-Resolution Aerosol Size Distribution Measurement Below 1 μm. *TSI Quarterly*, 9(2):3-11.

Keady, P. B., F. R. Quant, and G. J. Sem. 1986. A Condensation Nucleus Counter for Clean Rooms. *1986 Proceedings, 32nd Annual Technical Meeting, Institute of Environmental Sciences, Dallas/Ft. Worth, Texas, 6-8 May*. pp. 445-451.

Kousaka, Y., T. Niida, K. Okuyama, and H. Tanaka. 1982. Development of a Mixing Type Condensation Nucleus Counter. *Journal of Aerosol Science*, 13(3):231-240.

Kousaka, Y., T. Niida, Y. Tanaka, Y. Sato, H. Kano, N. Fukushima, and H. Sato. 1987. Development of a New Continuous Monitor for Nonvolatile Solute in Ultrapure Water by Atomization. *Journal of Environmental Science*, 30(4):39-42.

Lilienfeld, P. 1986. Optical Detection of Particle Contamination on Surfaces: A Review. *Aerosol Science Technology*, 5(2):145-165.

Liu, B.Y.H., and D.Y.H. Pui. 1974. A Submicron Aerosol Standard and the Primary Absolute Calibration of the Condensation Nuclei Counter. *Journal of Colloidal Interface Science*, 47:155-171.

Liu, B.Y.H., J. W. Lee, D.Y.H. Pui, K. Ahn, and S. L. Gilbert. 1986. Performance of a Model Clean Room. *Proceedings, 8th International Symposium on Contamination Control, Milan, Italy, 9-12 Sept.* pp. 810-821.

Locke, B. R., R. P. Donovan, and D. S. Ensor. 1985. Assessment of the Diffusion Battery for Determining Low Concentration Submicron Aerosol Distributions in Microelectronic Clean Rooms. *1985 Proceedings, 31st Annual Technical Meeting, Institute of Environmental Sciences, Las Vegas, Nevada, 30 April-2 May.* pp. 13-17.

Remiarz, R. J., G. J. Sem, J. K. Agarwal, and C. E. McManus. 1980. An Automated Diffusion Battery/ Condensation Nucleus Counter System for Sizing Submicron Aerosols. *TSI Quarterly*, 6(2):5-12.

Sekigawa, K., and T. Ohno. 1987. Dependency of Mixing-Type Condensation Nucleus Counter on Material of Aerosol Particles. *Proceedings, 5th Symposium on Aerosol Science & Technology*, Tsukuba, Japan, 19-21 August. pp. 244-246. In Japanese.

Sinclair, D. 1986. Measurement of Nanometer Aerosols. *Aerosol Science Technology*, 5(2):187-204.

Soderholm, S. C. 1983. Analyzing Screen Diffusion Battery Data. *Aerosols in the Mining and Industrial Work Environments, Vol. 3: Instrumentation.* V. A. Marple and B.Y.H Liu, eds. Ann Arbor Science Publishers, Ann Arbor, Michigan, pp. 1095-1115.

Wen, H. Y., G. Kasper, and S. Chesters. 1986. A New Method of Residue Analysis for Ultrapure Liquids. *Microcontamination*, 4(3):33-39,68.

Yeh, H. C., Y. S. Cheng, and M. M. Orman. 1982. Evaluation of Various Types of Wire Screens as Diffusion Battery Cells. *Journal of Colloidal Interface Science*, 86:12-16.

7. PARTICLE CONTROL METHODS: I

DAVID S. ENSOR Research Triangle Institute, Research Triangle Park, North Carolina

INTRODUCTION

The protection of critical components from contamination by particles is the core of clean high-technology manufacturing. Filters are used to purify the air in clean rooms and in process liquids and gases. This chapter is an overview of filtration technology. Only the general performance of filters will be discussed; for detailed specifications, the manufacturers' information should be consulted. For further information on filters, see Pich (1987), Hinds (1982), Davies (1973), Kirsch and Stechkina (1978), and Ensor and Donovan (1988).

The types of particulate filters developed over the years include:

- Fibrous
- Membrane
- Fabric
- Granular bed
- Electrostatic
  -- Active: applied fields and corona discharge
  -- Passive: Rosin-wool, electrets.

Of these filters, only the first two (fibrous and membrane) are used to control particulate contamination in clean rooms. Fibrous filters are characterized by a high gas flow rate, low pressure drop, and the ability to retain collected particles without clogging. Fibrous fillers are used mainly in air filtration to take advantage of low pressure drop inherent in the media. Another application of fibrous filters in contamination control is as a prefilter in high-efficiency process gas filtration systems. Membrane filters, on the other hand, are used in the purification of process gases and liquids and to obtain samples of particles for analysis. Membrane filters have very high efficiency, but they may also have a high pressure drop. Membrane filters are used as final filters on gas delivery systems to take advantage of the nearly absolute retention of particles and infrequent particle shedding. Rosin-wool filters used for respirators are made by beating rosin with wool to fix a long-lasting charge in the filter to improve collection efficiency. Filters made from electrets have a charge fixed in a polymer used to make the filter. The other filters mentioned in the above list have been applied to air pollution control and industrial hygiene; they will not be discussed further.

High-efficiency fibrous filters are formed from bulky loose fibers on paper-making machines. The fiber stock is slurred in water with additives such as binders and then allowed to flow on a moving screen. The water drains from the slurry, leaving the fibrous material in a layer. The formed paper is then dried and stored in rolls. The result is a porous fibrous matrix with a solids fraction of about 10 percent. The particles are collected within the structure of the filter relying on the tortuous pathways through the filter to remove particles by impaction, interception, and diffusion. An example of a fibrous filter is shown in Figure 7-1.

Membrane filters can be considered to be porous sheets of polymeric material. The primary collection of particles is on the surface by direct interception and impaction. Collection by diffusion is of less importance in membrane filters as opposed to fibrous filters

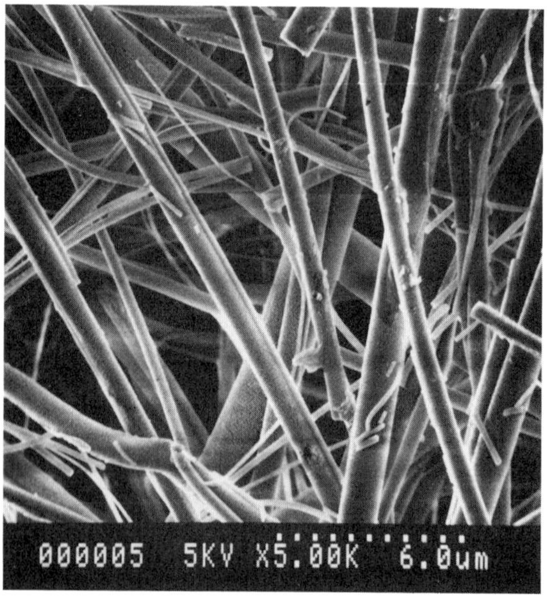

Micrograph by John Posthill, RTI.

Figure 7-1. Fibrous filter media.

because of the very thin media. Three general types have different processes of manufacture: cast membrane, Nuclepore, and Goretex. The cast filters are formed by dissolving a polymer in a solvent and then casting the liquid on a flat surface. As the solvent evaporates, a porous structure is formed by the polymer, as shown in Figure 7-2. The Nuclepore filter is formed by irradiating a thin sheet of polycarbonate film in a nuclear reactor and then etching the radiation-damaged sites. The result is a field of round holes "drilled" through the sheet. Nuclepore filters are very useful for obtaining samples for analysis because the holes can be differentiated from the contaminant particles when viewed by a microscope. An example of a clean filter is shown in Figure 7-3. Goretex filters are made by stretching Teflon^{TM} films and causing the formation of pores and a fiberlike structure. A photomicrograph is shown in Figure 7-4.

FILTRATION MECHANISMS

Filter Properties

The performance of a fibrous filter depends on the physical construction or projected fiber area of the filter and single particle collection mechanisms. Membrane filters in principle have similar features.

The collection efficiency for a filter is given by:

$$\nu = 1 - \exp(-\nu_s S) \quad , \tag{7-1}$$

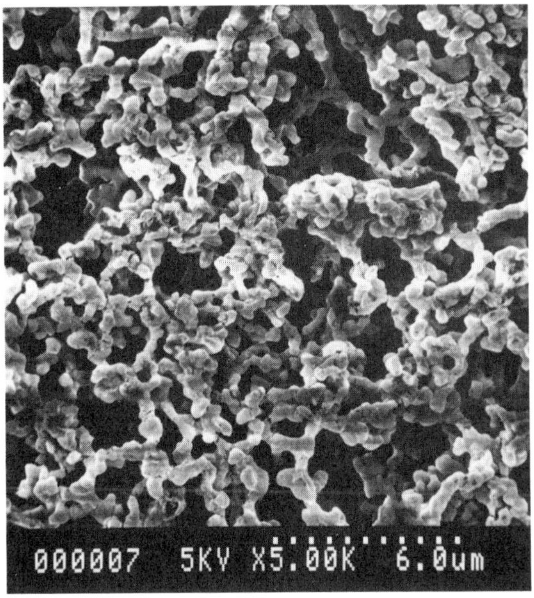

Micrograph by John Posthill, RTI.

Figure 7-2. Cast membrane filter.

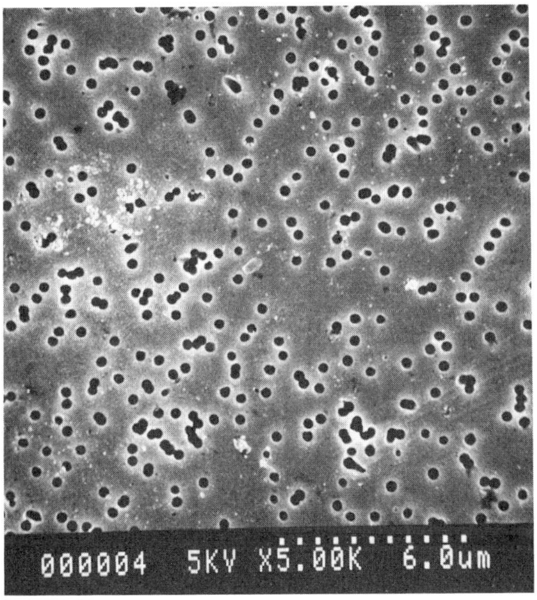

Micrograph by John Posthill, RTI.

Figure 7-3. Standard Nuclepore filter.

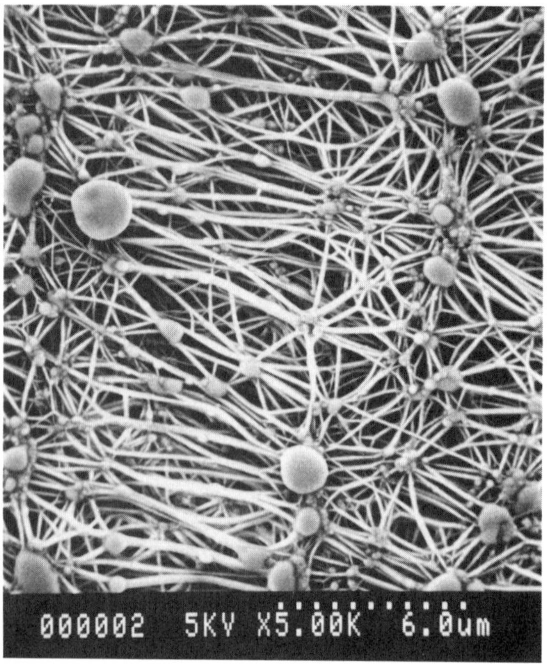

Figure 7-4. Teflon membrane filter.

where:

ν_s = single fiber collection efficiency
S = projected fiber area within the filter per unit area.

As the gas passes through the filters layer by layer, the particles are removed sequentially. This results in an exponential decay of the particles as a function of depth.

The single fiber efficiency is given by:

$$\nu_s = \frac{\text{Particles collected by fiber}}{\text{Particles in volume of air swept by fiber}} \quad . \tag{7-2}$$

The projected area of fiber per unit area is given by:

$$S = D_f \, l_f = 4 \, m_f/\pi \, d_f \, \rho_f = 4 \, \alpha L/\pi \, d_f \quad , \tag{7-3}$$

where:

d_f = fiber diameter
m_f = specific filter mass, mass of filter per unit filter area
ρ_f = fiber density
α = fiber volume fraction, filter solidity or packing density

$\alpha = m_f/\rho_f L$
L = filter depth
l_f = specific fiber length (length of filter fiber per unit area of filter)
$l_f = 4\, m_f/\pi\, d_f^2\, \rho_f$.

Therefore, the performance of the filter depends on two aspects: the projected area of the fibers in the filter and the single fiber particle collection efficiency. One of the difficulties of using fundamental filter characteristics to predict filtration performance is the ability to adequately measure fiber length and diameter, as well as media density in the filter. The single fiber efficiency will be discussed at length beginning on p. 7-7.

Pressure Drop

The pressure drop through a filter depends on the specific fiber length and fiber drag:

$$\Delta p = F_d\, l_f \quad , \qquad (7\text{-}4)$$

where:

F_d = dimensionless fiber drag parameter
l_f = fiber length.

$$F_d = F_*\, \mu\, U_o \qquad (7\text{-}5)$$

where:

μ = gas viscosity
U_o = face velocity, volumetric flow/filter area.

F* depends on the assumed flow model. For the case of Kuwabara flow, F* is given by:

$$F_* = 4\,\pi/K \quad , \qquad (7\text{-}6)$$

where:

K = hydrodynamic factor, given by:

$$K = -\,1/2 \ln \alpha + \alpha - 1/4\,\alpha^2 - 3/4 \ .$$

The Kuwabara hydrodynamic factor takes into account the fiber interaction in the filter. Kuwabara (1959) modeled the system with randomly placed and homogeneously distributed fibers. The spaces surrounding the filters are assumed to be cells adjacent to one another. At the cell boundary, the vorticity and radial velocity are assumed to be zero from the model. Kuwabara developed analytical expressions for the stream functions. The hydrodynamic factor is used in both pressure drop and fiber efficiency equations.

Particle Collection

The mechanisms responsible for collecting particles are illustrated in Figure 7-5. The following is a summary of each mechanism.

Gravity. Particles greater than about 5 μm will settle to the bottom of the filter. This mechanism is independent of the filter construction and is not an important mechanism for ultrahigh efficiency filters.

Interception. Particles follow the fluid streamlines that intercept the fiber; the particles are collected upon contact.

Sieving. This is a special case of interception in which the particle is too large to penetrate the pores or pathways in the filter. A reentrained particle captured by interception can penetrate the filter; a reentrained particle captured by sieving cannot.

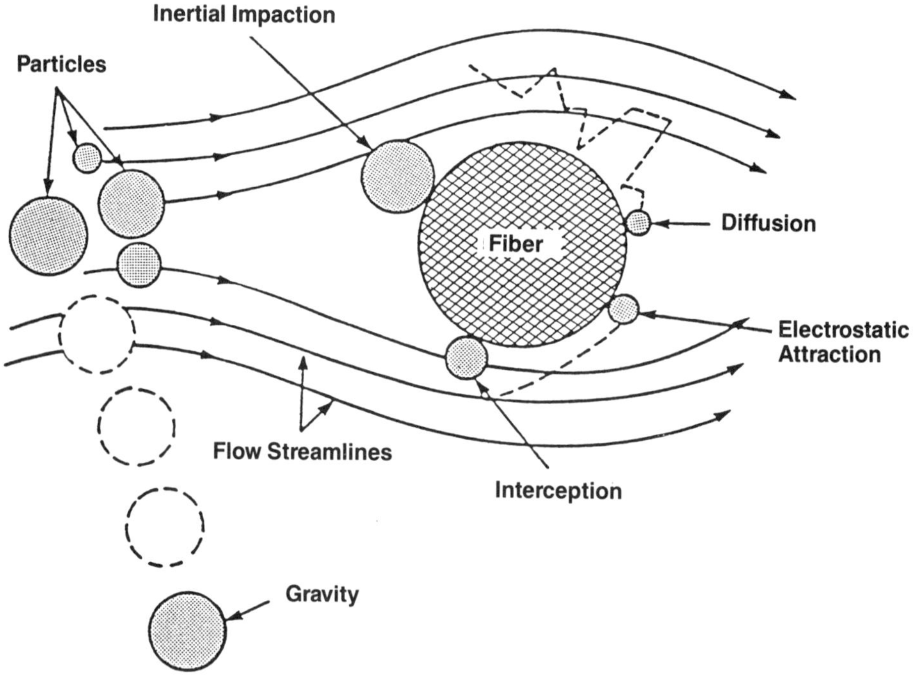

Figure 7-5. Classical mechanisms: aerosol capture by a fiber.

Inertial Impaction. Large particles tend to follow straight lines from inertia even though the fluid is flowing around the fibers. As a result, the particles collide with the fibers and are collected. Impaction is usually important for roughing filters because of the large fibers in the media and the high face velocities. This mechanism is usually not important for high-efficiency filters because of low face velocity and the small diameter fibers used in the media. For this reason, impaction will not be explored in detail. Impaction depends on the particle diameter and density, and the gas viscosity and velocity. Impaction is an important mechanism for high-velocity flows and particles greater than 0.5 μm.

Diffusion. The random movement of small particles from the bombardment of surrounding gas molecules is called Brownian motion. This produces a diffusion process that will cause a deflection of small particles from the streamlines. Thus, the probability is enhanced that the particle will be collected by the fiber.

Electrostatic Mechanisms. Such mechanisms depend on electric fields, created by charges on the fiber and/or particles or introduced by external fields applied to the filter. There are three major electrical forces between fibers and particles:

- Image force (charged particles only)
- Dielectrophoretic force (charged fibers only)
- Coulombic force (charged particles and fibers).

These mechanisms are assumed to act independently of one another.

For the submicrometer-sized particles of interest, gravity is usually not a significant mechanism. Electrical effects are considered to be of secondary importance except in special filters designed to take advantage of charge. The combination of interception and diffusion and, to a lesser extent, impaction appears to provide an adequate description of high-efficiency filters.

FILTER THEORY

Single Fiber Efficiencies

The single fiber efficiencies described earlier are a function of mechanisms that are assumed to be additive. For the present application, only interception and diffusional collection are significant for high-efficiency filtration of submicrometer particles.

Interception

Interception is simply the collection of a particle contacting a fiber when following a flow streamline. Strictly speaking, pure interception occurs only when the particles have size but no mass. Thus, the idea of collection resulting only from interception is conceptually useful but physically unobtainable. The interception parameter, R, is given by:

$$R = d_p/d_f \quad , \tag{7-7}$$

where:

d_p = particle diameter
d_f = fiber diameter.

If potential flow is assumed, collection by interception is a function of R only. For viscous flow, the Reynolds number becomes important. Also, for a system of fibers, the packing density may affect interception through fluid-dynamic interaction of the fibers. For a fixed fiber size, the interception mechanism becomes more significant as the particle size increases.

Diffusion

The Brownian motion of the particles produces a diffusion phenomena. The particles exhibit a random motion caused by collisions with surrounding air molecules. This mechanism is a function of the diffusion coefficient for the particles, gas velocity, and fiber diameter. Zig-zag deviations from the flow streamlines result in a net statistical movement. As the particle size decreases, the importance of Brownian motion relative to the fluid flow increases until the distinction between aerosol particle and gas molecule becomes insignificant. Diffusional particle collection is significant only for submicrometer particles with large diffusion coefficients. For collection by filters, the residence time in the filter matrix is important. The Peclet number, Pe, relates fluid motion to the diffusion coefficient and hence is useful in prediction of collection. The Peclet number is defined as:

$$Pe = \frac{d_f \, U}{D} \; , \qquad (7\text{-}8)$$

where:

d_f = fiber diameter
D = particle diffusion coefficient = $k\,T\,C/(3\,\pi\mu\,d_p)$
k = Boltzmann's constant
T = absolute temperature
d_p = particle diameter
μ = gas viscosity
C = Cunningham correction for slip
U = gas velocity.

In terms of filter and aerosol properties, the collection efficiency from diffusion increases with decreasing particle diameter, gas velocity, and fiber diameter.

Combined Diffusion and Interception Collection

For high-efficiency filters used for air and gas, the two most important mechanisms are diffusion and interception. An empirical expression for single particle efficiency, νd, r, is given by (Lee and Lin, 1982a,b):

$$\nu_{d,r} = 1.6 \left(\frac{K}{1-\alpha}\right)^{-0.333} Pe^{-0.667} + 0.6 \left(\frac{1-\alpha}{K}\right) \left(\frac{R^2}{1+R}\right) \quad . \tag{7-9}$$

The second term on the right-hand side of Equation (7-9) represents the contribution of the particle collection by interception; the first term represents particle collection by diffusion. This expression was obtained by dimensional analysis and by obtaining the coefficients by statistical analysis of laboratory-derived filter data.

Most Penetrating Particle Diameter

The cumulative effect of these collection mechanisms is shown in Figure 7-6. A region of minimum efficiency is found for particles from 0.05 to 0.5 μm. In this region, both mechanisms are the least effective.

Lee and Liu (1980) differentiated the above equation for the combined diffusion and interception mechanism with appropriate corrections for slip. They obtained:

$$d_{p\,min} = 0.885 \left[\left(\frac{K}{1-\alpha}\right) \left(\frac{l^{1/2}\, k\, T}{\mu}\right) \left(\frac{d_f^{\,2}}{U}\right)\right]^{2/9} \tag{7-10}$$

for $0.075 < l/d_p < 1.3$

where:

$d_{p\,min}$ = most penetrating particle diameter
d_p = particle diameter
K = hydrodynamic factor
l = mean free path
k = Boltzmann's constant
μ = viscosity of gas
d_f = fiber diameter
U = gas velocity
α = fiber volume fraction (packing density).

The results predicted from this equation are plotted in Figure 7-7. The implication of this analysis is that, as smaller diameter fibers are used in filters to improve efficiency, the most penetrating particle diameter will likely be reduced from 0.3 μm.

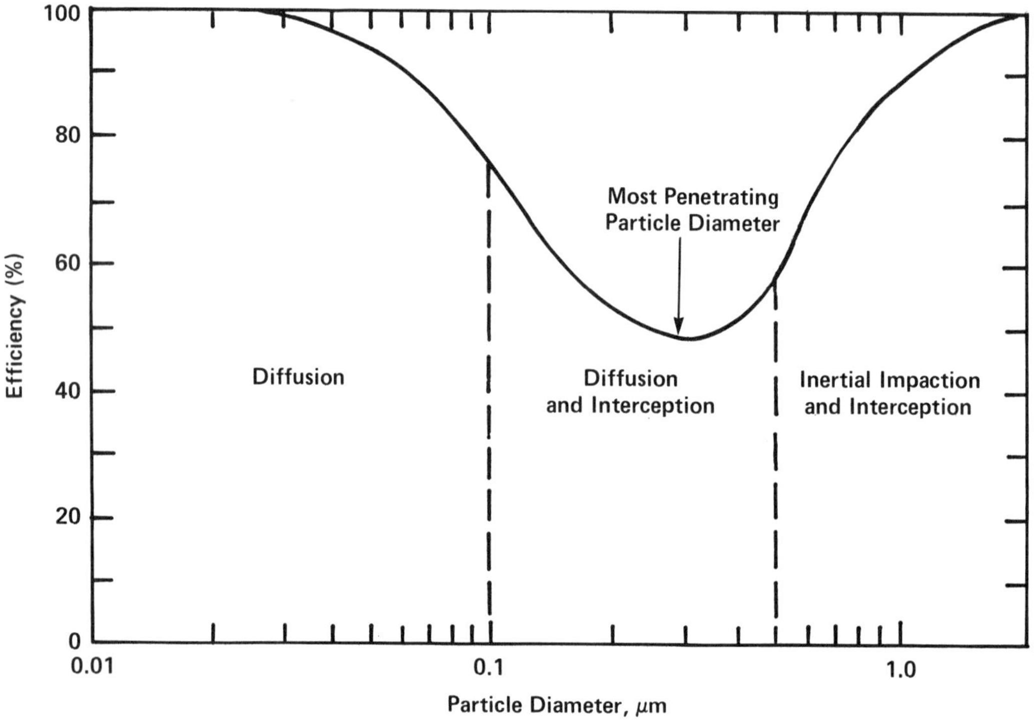

Figure 7-6. Interrelationship of collection mechanisms.

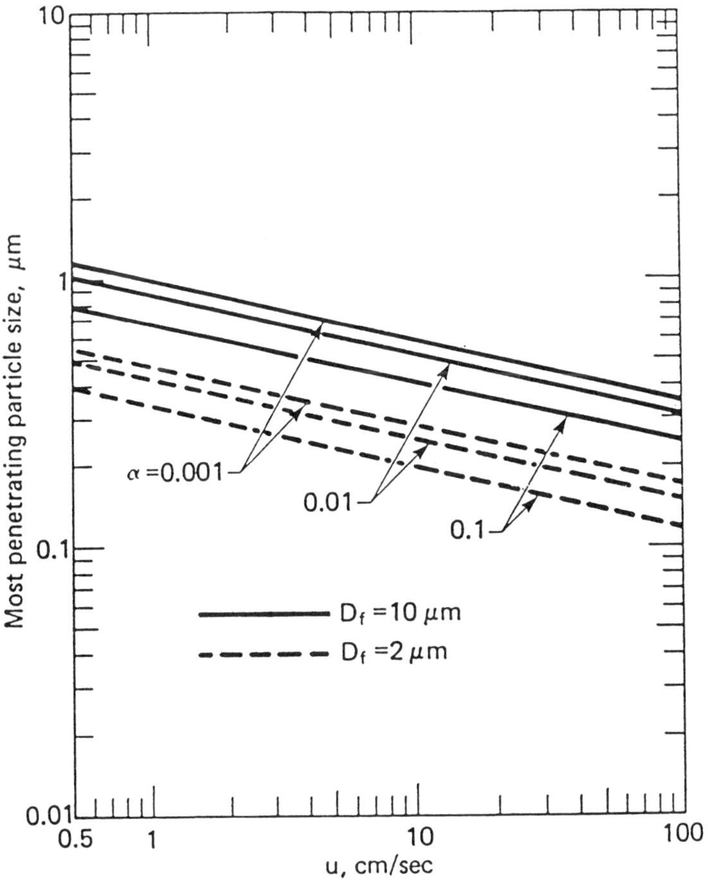

Source: Lee and Liu, 1980.

Figure 7-7. Most penetrating particle size as a function of velocity, fiber volume fraction, and fiber predicted by theory.

Membrane Filters

The prediction of the efficiency of a cast membrane filter, according to Rubow (1981), can be accomplished best by using the fibrous filter model described in the previous section. Rubow and Liu (1986) found that, by using the diffusion and interception mechanisms, the particle-size-dependent penetration for Millipore AA filters, with a pore diameter of 0.8 μm, could be matched with an assumed fiber diameter of 0.555 μm and actual values of filter thickness and porosity. This work demonstrates that applying the fibrous filter model to membranes is not ideal, yet it certainly can develop an understanding of the effect of the filter parameters on performance. This is especially useful because the complex structure of cast membrane filters is not well described by simple pore models.

The Nuclepore-type filters can be described by a pore model. Mathematical modeling of this medium is greatly aided by the nearly round straight pores. Also, the filter collects particles primarily on the surface with some collection of smaller particles in the pores. For more information, see a standard reference such as Pich (1987) or Spurny et al. (1969).

TESTING AND RATING

Air filters are available in a wide range of efficiencies and pressure drops (see Table 7-1). The rating is accomplished by challenging the filter under controlled conditions of flow, temperature, and pressure with a known aerosol. The concentration of the aerosol is measured upstream and downstream of the filter to determine the efficiency. Membrane filters are rated by the pore size required for application to liquid filtration and by the bubble point, i.e., the air pressure required to produce a visible bubble on a solvent-wetted filter (Ensor and Donovan, 1988).

Table 7-1. Typical Characteristics of Air Filters

Filter	U, cm/s	S	α	Δp in. H_2O	Efficiency
Roughing	150	4	0.005	0.1	50% ASHRAE
Medium	25	15	0.005	0.25	80% ASHRAE
High	10	30	0.01	0.5	97% ASHRAE
HEPA	2.5	100	0.1	1.0	>99.97% DOP

American Society of Heating, Refrigerating, and Air-Conditioning Engineers (ASHRAE) Test Methods

ASHRAE has developed three test methods for low-efficiency filters used in ventilation systems. These are the (1) dust holding capacity, (2) weight arrestance, and (3) atmospheric aerosol challenge. The dust holding test and arrestance test use a mixture of 72 percent test dust, 23 percent carbon black, and 5 percent cotton lint to challenge the filter in a square test duct under rated airflow rates. The efficiency is determined by weighing the filter and the quantity of dust suspended. The atmospheric aerosol method uses ambient air as the challenge. The concentration of aerosol is measured by using the soiling or opacity of small test filters used to sample upstream and downstream of the ventilation filter under test. The efficiency of the filter is obtained by the difference in the opacity of the small test filters. For more information, see the ASHRAE Standard 52-76 (1976).

DOP Testing

Filters of higher efficiency are tested with DOP challenge aerosol generated by vaporizing and recondensing the fluid to produce a nearly monodisperse aerosol at 0.3 μm. The test aerosol was selected to match the most penetrating particle diameter for the filters available during the 1940s. The efficiency of an acceptable high-efficiency particulate air (HEPA) filter is >99.97 percent. Light-scattering photometers are used to measure the upstream and downstream concentration (MIL-STD-282, 1956). The recently developed ultralow penetration air (ULPA) filters are rated to >99.997 percent at 0.12 μm. In many test fixtures, the light-scattering photometers have been replaced with laser aerosol spectrometers (Cadwell, 1985; Scripsick, 1986).

Testing of State-of-the-Art Filters

In the last few years, clean-room filters with 99.9999999 percent efficiency have been manufactured. The standard test methods are inadequate to test these filters because of the need to measure accurately low concentration aerosols downstream of the filter. In Figure 7-8, a test facility developed by Ensor et al. (1988) is shown. This equipment is operated at slightly positive pressure in a Class 100 clean room to prevent artifacts from inleakage. The back of the filter can be scanned with a probe connected to a condensation nucleus counter (CNC) by a technician working through an access door. A CNC is used to measure the concentration of the challenge and penetrating aerosol. About a 10 to 1 dilution is used on the inlet stream. A wide range of challenge aerosols has been used, such as ambient air, KCl, and oils. With the use of polydisperse aerosols and multichannel instruments, the particle-size-dependent efficiency can be determined for the filter under test.

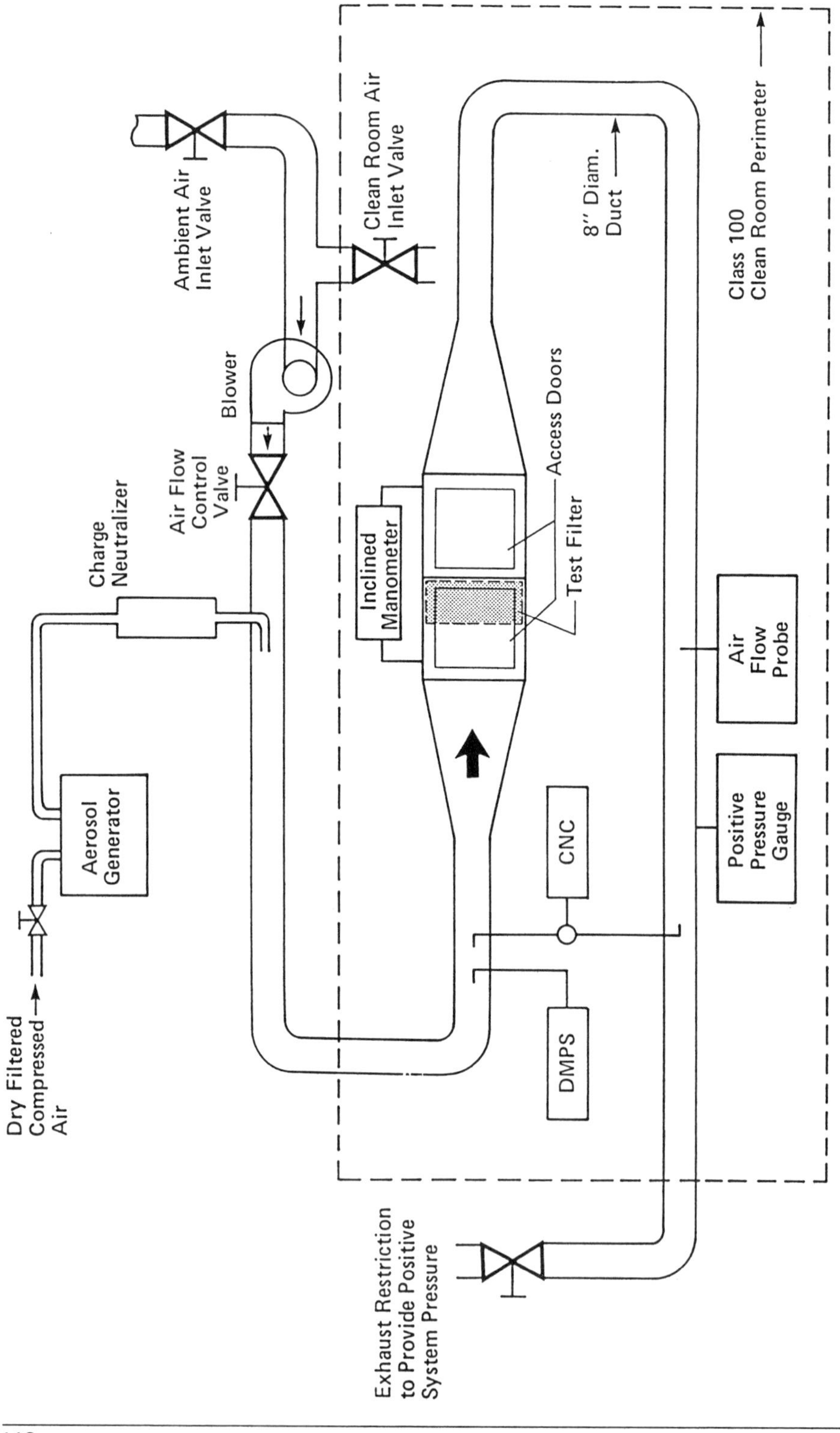

Figure 7-8. Schematic diagram of ultrahigh efficiency filter testing facility.

PARTICLE-SIZE-DEPENDENT FILTER TEST DATA

Fibrous Filters

The particle-size collection efficiency determines the protection provided by the filter for submicrometer particles. Figure 7-9 shows the particle penetration for ventilation and clean-room filters. Penetration is the outlet concentration divided by inlet concentration. It is related to efficiency by the following:

$$E_t = (1 - P_t)\ 100 \quad . \tag{7-11}$$

As indicated by the analysis of the collection mechanisms, the filter will exhibit a most penetrating particle diameter. The minimum depends somewhat on the face velocity of air passing through the filter.

The data were obtained by a variety of test methods, and the original papers should be consulted for details.

Membrane Filters

Only a limited set of data is available for membrane data. An example of a cast membrane performance is given in Figure 7-10. The size-dependent efficiency of a Nuclepore filter is shown in Figure 7-11.

LIQUID FILTRATION

Introduction

The filtration of particles from liquids is much more complex than filtration of particles from gases. Additional forces come into play, such as charge effects, adhesion, absorption, and the hydrophobic or hydrophilic properties of the filter. The theoretical aspects were recently reviewed by Grant et al. (1988). When applying the theoretical development for gas filters, it would be expected that a most penetrating particle size would also be identified.

Mechanisms

The following is a brief overview of the mechanisms expected in liquid filtration.

- **Interception.** Expected to be of the same magnitude as in liquid filtration.
- **Impaction.** Probably not important because the high viscosity of the fluid reduces impaction efficiency.
- **Diffusion.** Diffusion coefficient is smaller, but residence time is longer within the filter.

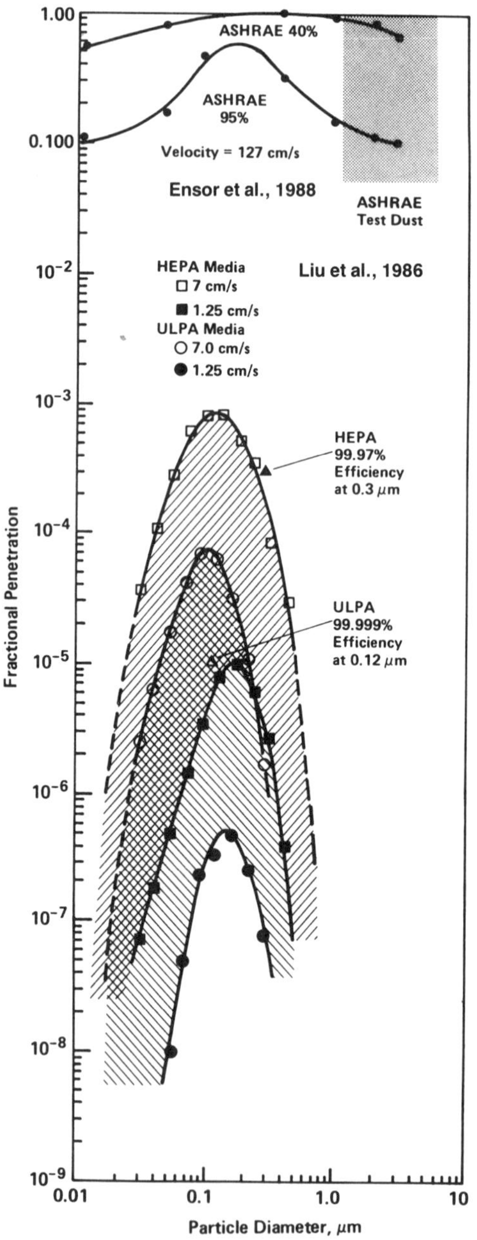

Figure 7-9. Particle-size-dependent fraction penetration for ventilation filters and clean-room filters.

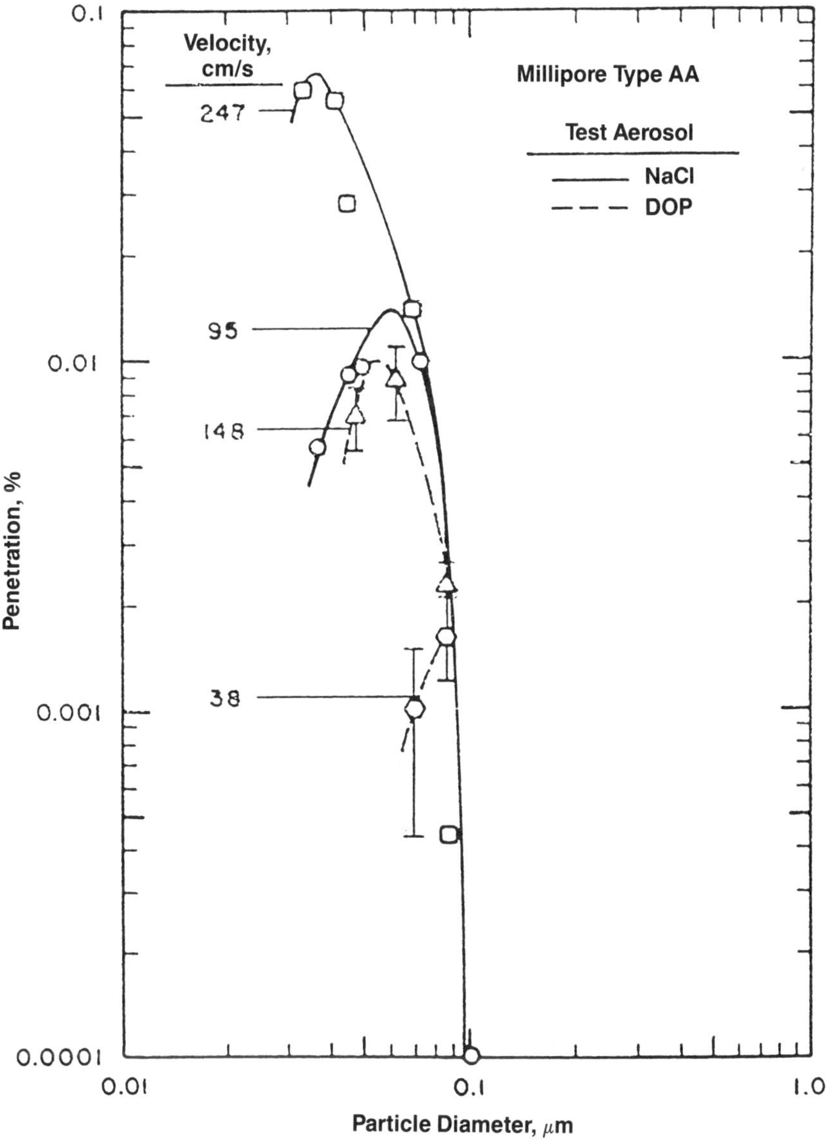

Reported on by Rubow and Liu, 1984.

Figure 7-10. Experimentally determined aerosol penetration data for the AA filter.

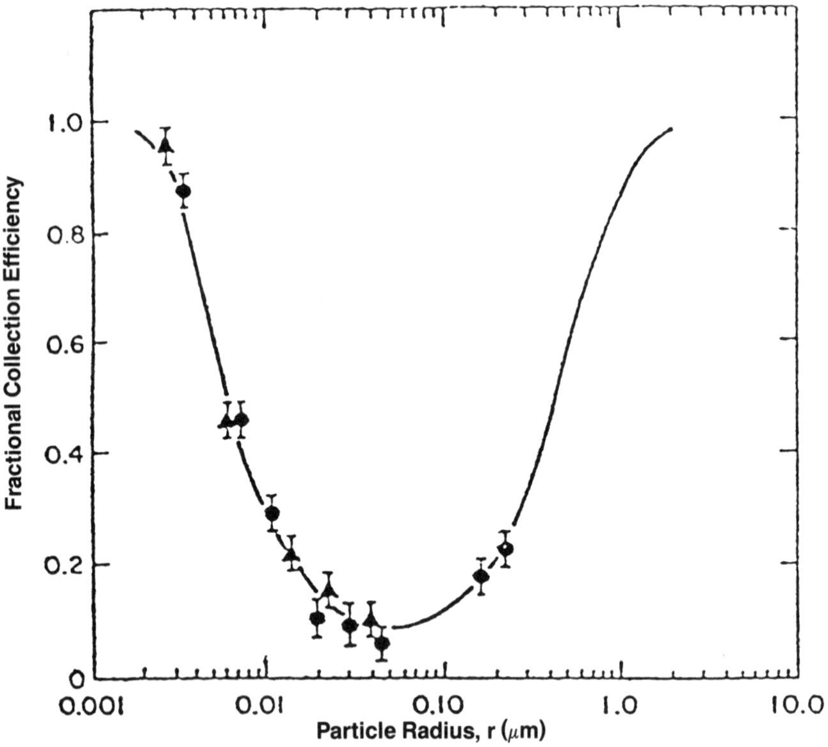

Source: Rubow and Liu, 1986.

Figure 7-11. Comparison of the collection efficiency as theoretically predicted and experimentally determined for a 5 μm pore diameter Nuclepore filter at a face velocity of 5 cm/s.

- **Electrostatic.** Liquids have double-layer effects, which can repel particles depending on the ionic balance in the fluid.

- **Sieving.** Sieving may prove to be the most important because of the strong dependence of collection efficiency on surface, fluid, and particle charge.

- **Adhesion.** Particles are more easily reentrained from the filter because of the thinner boundary layer and the higher drag.

Double-Layer Effects and Filter Efficiency

The major difference between air and liquid filters is the ability to capture and retain particles on the filter structure. In air filters, if a submicrometer particle touches the filter surface, it is usually retained. However, the ability to collect particles in a liquid filter depends on the molecular dispersion force or Van der Waal's force, which will always attract, and double ionic layer interactive forces. The net surface interaction is the sum of the two forces and can either attract or repel particles depending on the signs of the surface potential. The double-layer force is affected by the ionic concentration of the solution and the properties of the surfaces (Tien and Payatakes, 1979). The case for an unfavorable condition for collection of particles is shown in Figure 7-12.

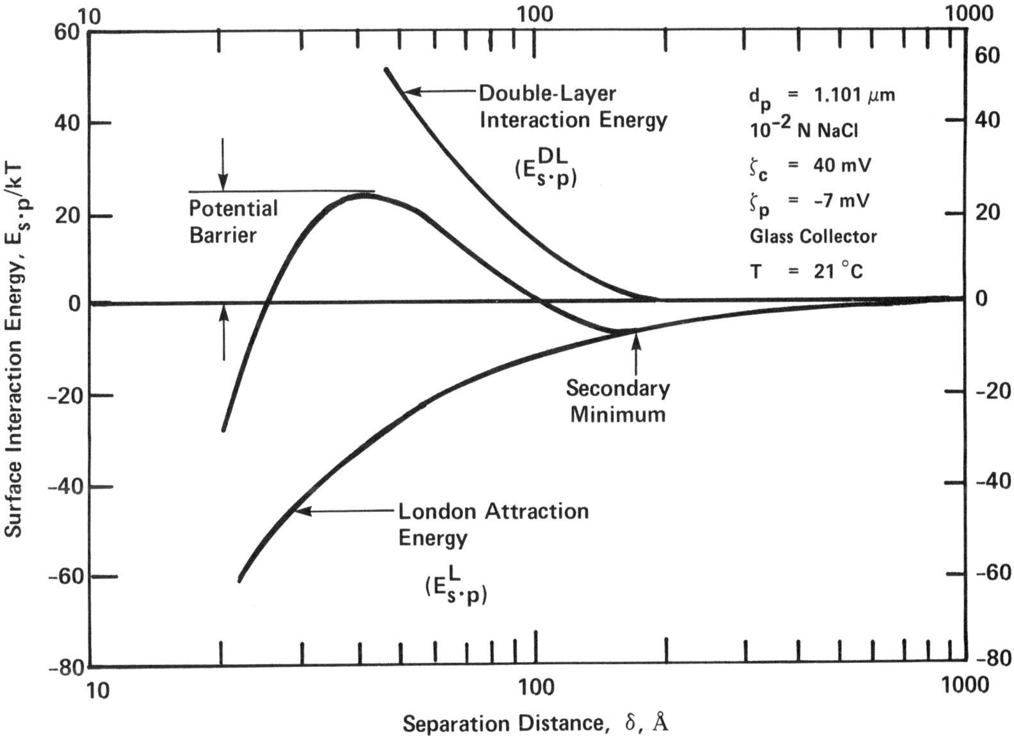

Source: Tien and Payatakes, 1979.

Figure 7-12. Surface interaction energy (for the sphere-plane system) corresponding to an "unfavorable" condition.

The results of the theoretical analysis and experimental work reported by Grant et al. (1988) is shown in Figure 7-13. The fractional penetration of particles under conditions favorable to collection is indicated by the combined mechanisms of interception and diffusional collection. The most penetrating particle size (MPPS) was computed from theory to be 0.065 μm. The experimental data are shown for a system of 0.1 percent Triton X-100 surfactant in water. This represents an unfavorable collection condition. The only collection mechanism effective for this fluid is sieving by the pores of the filter. The penetration under these conditions is orders of magnitude greater than predicted without an ionic double-layer repulsive effect.

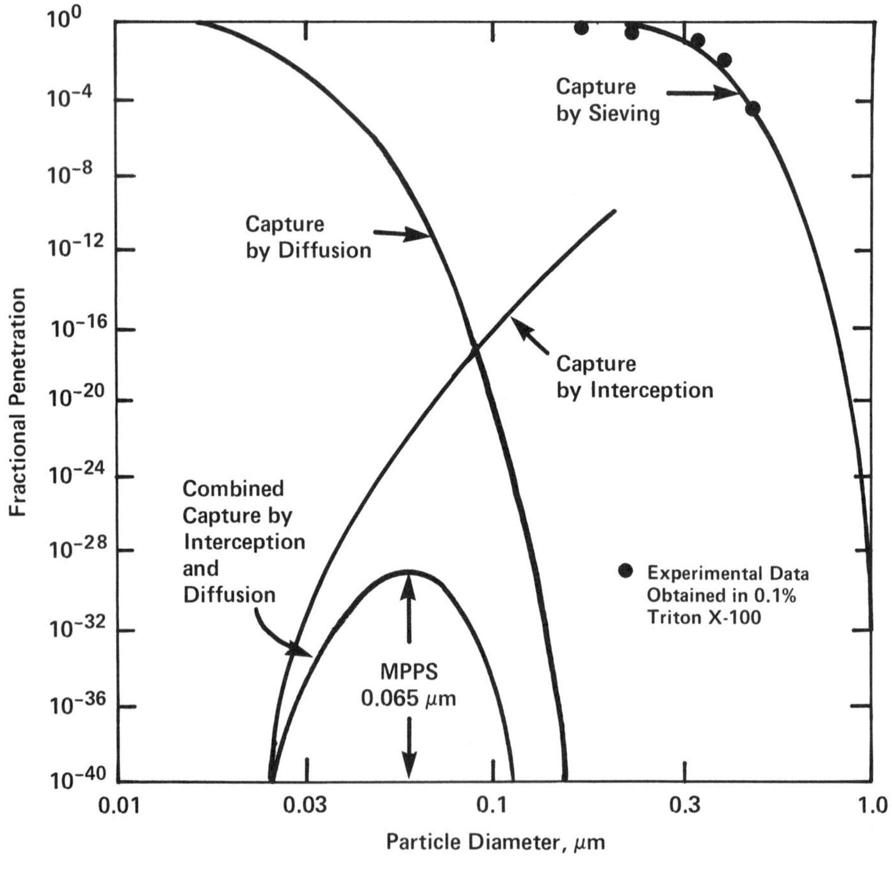

Source: Grant et al., 1988.

Figure 7-13. Comparison of liquidborne particle capture by sieving to capture by diffusion and interception and experimental data.

SUMMARY

Two general types of filters are used in contamination control. The fibrous filters have a low pressure drop and a high gas flow rate, as well as an ability to hold particles within the filter. For these reasons, they are used mainly for clean-room filters and as prefilters in point-of-use filters. The membrane filters have a very high efficiency, but they also have a high pressure drop. Membrane filters are used as final filters in point-of-use filters.

For gas filters, the mechanisms of capture that are important under most conditions are interception and diffusion. Other mechanisms are sieving, impaction, and electrostatics.

The dominant mechanisms of interception and diffusion have a region of particle size where both are relatively ineffective. Thus, a most penetrating particle size is usually observed in the 0.05 to 0.5 μm range.

A number of test methods are used depending on the application of the filter. ASHRAE test methods are used to rate ventilation filters, which generally are up to 95 percent efficient. DOP tests are used to test HEPA filters with particles at 0.3 μm and efficiencies of 99.97 percent. Recently, filters with very high efficiency have required custom test apparatus to obtain reliable data.

When filters are applied to liquids, additional forces come into play. Double-layer charges can build up on the surfaces and cause repulsion of particles. The safest approach is to specify pore sizes smaller than the particles of concern and depend on sieving.

REFERENCES

ASHRAE. 1976. *Method of Testing Air-Cleaning Devices Used in General Ventilation for Removing Particulate Matter.* ASHRAE Std. 52-76. American Society of Heating, Refrigerating, and Air-Conditioning Engineers, 1791 Tullie Circle, NE, Atlanta GA 30329.

Cadwell, G. H. 1985. A Brief History of the Modern ULPA Filter. In: *Proceedings of the Institute of Environmental Sciences, Las Vegas, NV, April 30 - May 2, 1985.*

Davies, C. N. 1973. *Air Filtration.* Academic Press, New York, NY.

Ensor, D. S., et al. 1988. Air Cleaner Technologies for Indoor Air Pollution. In: *Engineering Solutions to Indoor Air Problems.* ASHRAE, Atlanta, GA. pp. 111-129.

Ensor, D. S., and R. P. Donovan. 1988. Aerosol Filtration Technology. In: *Handbook of Contamination Control in Microelectronics*, D. L. Tolliver (ed.). Noyes Publishing Co. pp. 1-67.

Grant, D. C., B.Y.H. Liu, W. G. Fisher, and R. A. Bowling. 1988. Particle Capture Mechanisms in Gases and Liquids: An Analysis of Operative Mechanisms. In: *Proceedings of the Annual Technical Meeting of the Institute of Environmental Sciences.*

Hinds, W. C. 1982. *Aerosol Technology*. John Wiley & Sons, Inc., New York, NY.

Kuwabara, S. 1959. The Forces Experienced by Randomly Distributed Parallel Circular Cylinders or Spheres in a Viscous Flow at Small Reynold's Numbers. *Journal of the Physics Society of Japan*, 14(4):527-532.

Kirsch, A. A., and I. B. Stechkina. 1978. The Theory of Aerosol Filtration with Fibrous Filters. In: *Fundamentals of Aerosol Science*, Ch. 4, D. T. Shaw (ed.). Wiley-Interscience, New York, NY.

Lee, K. W., and B.Y.H. Liu. 1980. On the Minimum Efficiency and the Most Penetrating Particle Size for Fibrous Filters. *JAPCA*, 30(4):377-381.

Lee, K. W., and B.Y.H. Liu. 1982a. Experimental Study of Aerosol Filtration by Fibrous Filters. *Aerosol Science & Technology*, 1:35-46.

Lee, K. W., and B.Y.H. Liu. 1982b. Theoretical Study of Aerosol Filtration by Fibrous Filters. *Aerosol Science & Technology*, 1:147-161.

Liu, B.Y.H., K. L. Rubow, and D.Y.H. Pui. 1986. Performance of HEPA and ULPA Filters. In: *Proceedings of the Institute of Environmental Sciences, Las Vegas, NV, April 30 - May 2, 1986.*

Micro Filtration Systems. 1984. Catalog 85: A User and Purchasing Guide. Micro Filtration Systems, 6800 Sierra Court, Dublin, CA 94568.

Nuclepore Corporation. 1988. Products for Laboratory and Process Filtration (Catalog). Nuclepore Corporation, 7035 Commerce Circle, Pleasanton, CA 94566-7711.

Pich, J. 1987. Gas Filtration Theory. In: *Filtration Principles and Practices*, 2nd Edition, M. J. Matteson and C. Orr (eds.). Marcel Dekker, New York and Basel. pp. 1-117.

Remiarz, R. J., B. R. Johnson, and J. A. Agarwal. 1986. Automated Systems for Filter Efficiency Measurement. In: *Fluid Filtration: Gas, Volume I, ASTM STP 975*, R. R. Raber (ed.). American Society for Testing and Materials, Philadelphia, PA.

Rubow, K. L., and B.Y.H. Liu. 1984. Evaluation of Ultra-High Efficiency Membrane Filters. In: *Proceedings of the Institute of Environmental Sciences, Las Vegas, NV.*

Rubow, K. L. 1981. Submicron Aerosol Filtration Characteristics of Membrane Filters. PhD Dissertation, University of Minnesota.

Rubow, K. L., and B.Y.H. Liu. 1984. Evaluation of Ultra-High Efficiency Membrane Filters. In: *Proceedings of the Institute of Environmental Sciences, Las Vegas, NV.*

Rubow, K. L., and B.Y.H. Liu. 1986. Characteristics of Membrane Filters for Particle Collection. In: *Fluid Filtration: Gas, Volume I, ASTM STP 975*, R. R. Raber (ed.). American Society for Testing and Materials, Philadelphia, PA.

Scripsick, R. C. 1986. New Filter Efficiency Tests Being Developed for the Department of Energy. In: *Fluid Filtration: Gas Volume I, ASTM STP 1975*, R. R. Raber (ed.). American Society for Testing and Materials, Philadelphia, PA.

Spurny, K. R., et al. 1969. Aerosol Filtration by Means of Nuclepore Filters: Structural and Filtration Properties. *Environmental Science & Technology*, 3:453-464.

Tien, C., and A. C. Payatakes. 1979. Advances in Deep Bed Filtration. *AICHE Journal*, 25:737-759.

MIL-STD-282. 1956. Filter Units, Protective Clothing, Gas-Mask Components and Related Products: Performance-Test Methods. United States Government Printing Office, Washington, DC.

8. PREDICTED AND MEASURED CLEAN-ROOM CONTAMINATION

ANDREW S. VINER 3M Company, St. Paul, Minnesota

INTRODUCTION

A clean room is an enclosed space bounded by a leak-free structure such that the only air entering the environment passes through banks of HEPA (or ULPA) filters. The purpose of a clean room is to provide as particle-free an environment as possible for manufacturing. Thus, a clean room is a key component of the semiconductor manufacturing process, and like any other piece of equipment, it must be properly designed, built, and maintained to function effectively. Unfortunately, the evaluation of clean-room performance is an inexact science. Consequently, it is not easy to predict how a clean room will perform under any given set of circumstances. The purpose of this chapter is to provide a simple means for obtaining an approximate estimate of contamination levels in a clean room under various conditions. In addition, representative data will be presented to show contamination levels in an operating clean room from some typical (and some atypical) particle sources.

MODELING CLEAN-ROOM PERFORMANCE

A great deal of effort has been invested in modeling efforts to understand and predict the performance characteristics of clean rooms (see Chapter 18). The bulk of these efforts begin with a two- or three-dimensional description of a facility and obtain numerical solutions to the Navier-Stokes equations describing airflow patterns in the clean room (Matsuhashi et al., 1984; Jinno, 1986; Busnaina, 1987; Shanmugavelu, Kuehn, and Liu, 1987; Yamamoto, Donovan, and Ensor, 1987 and 1988; Busnaina, Abuzeid, and Sharif, 1988). Once the flow field is determined, the equations describing particle transport can be solved (Jinno, 1986; Busnaina, 1987; Shanmugavelu, Kuehn, and Liu, 1987; Busnaina, Abuzeid, and Sharif, 1988; Fujii and Hayakawa, 1985). The results of these efforts are impressive. In particular, these models can be used to consider the effect of equipment placement on airflow patterns in a clean room, predict the location of air recirculation zones where particles can become trapped, and predict when harmful vapors may be drawn out of a fume hood into the general clean-room atmosphere. Two factors limit the utility of these modeling programs. First, there is a tradeoff between computational speed and modeling flexibility. The programs either require a significant amount of computation time or they are limited to steady-state conditions (and hence are not well-suited to computing the effects of a time-dependent particle source). Second, the amount of effort required to compute the concentration profiles can be prohibitive because it is more than simply a matter of computing streamlines. The problem is compounded by the fact that particle size can range over three orders of magnitude (0.01 to 10.0 μm). Consequently, while these models are useful and important in the understanding and improvement of clean-room performance, they are not yet able to predict clean-room performance under any but the simplest conditions.

A simpler model of clean-room contamination has been developed based on a particle-mass balance in the clean room (Ensor, Donovan, and Locke, 1987). This model is quite simple in that it assumes that all gas streams are perfectly mixed and flow rates and particle source terms are constants. Clearly, these assumptions are naive; however, as shown below, the insight gained is well worth the sacrifice of modeling rigor. The geometry of this model is illustrated in Figure 8-1. In this model, a given flowrate of prefiltered makeup air (Q_a) with a

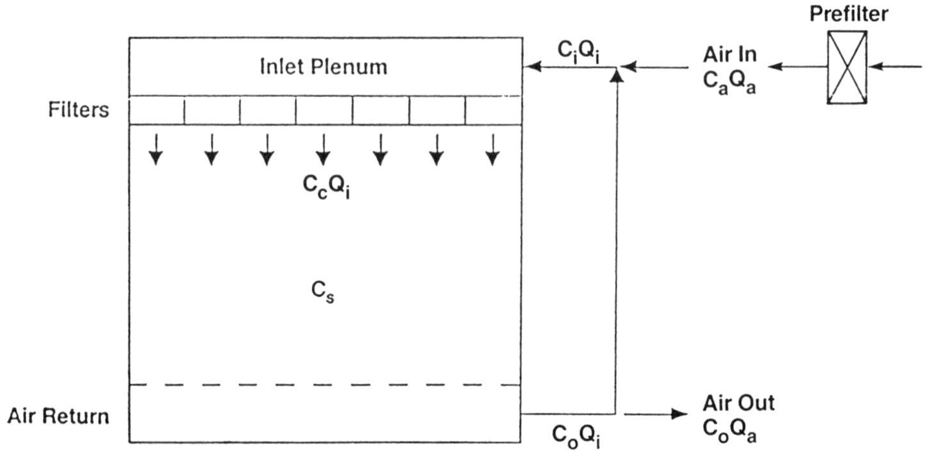

Source: Adapted with permission from Ensor, Donovan, and Locke, 1987.

C_a = prefiltered atmospheric concentration
C_c = particle concentration entering the room
C_s = sources in the room
C_o = particle concentration in exhaust air
Q_a = volumetric flow rate of make-up air
Q_i = volumetric flow rate of air in the clean room

Figure 8-1. A clean room filtration system.

particle concentration C_a combines with air recirculated within the ventilation system at the inlet plenum of the filter system. The resulting flow rate Q_i passes through a bank of HEPA (or ULPA) filters, thereby reducing the particle concentration from C_i to C_c. As the air flows down over the people and processing equipment in the room, it encounters particle sources that produce a contamination concentration C_s. The outlet concentration C_o is the sum of the "clean" concentration C_c and the source concentration C_s. A fraction of the air exits the clean room through exhaust hoods which discharge to the outside. The remainder is recycled to combine with new makeup air, which replaces the exhausted air.

The total particle mass at any point in the clean room is governed by mass and flow balance equations and the filtration efficiency of the roughing and prefilters as well as that of the HEPA (or ULPA) filters. By inspection, the particle mass balance equations can be written as:

$$C_o\, Q_i = C_c\, Q_i + C_s\, Q_i \ , \tag{8-1}$$

$$C_i\, Q_i = C_o\, Q_i - C_o\, Q_a + P_R\, C_a\, Q_a \ , \tag{8-2}$$

$$C_c\, Q_i = P_H\, C_i\, Q_i \ , \tag{8-3}$$

$$Q_a = Q_i\, F \ , \tag{8-4}$$

where:

C_o = concentration of particles in the air leaving the clean room
C_c = concentration of particles leaving the high-efficiency filters
C_s = concentration of particles generated in the clean room
C_i = concentration of particles upstream of the filter
C_a = concentration of incoming atmospheric aerosols
Q_i = volumetric airflow in the clean room
Q_a = volumetric flow of incoming and exhaust air
P_H = fractional filter penetration
P_R = fractional penetration of the fresh air filters
F = fraction of makeup air.

Given an estimate of the particle source concentration in the clean room and a knowledge of the air flow rates, the lowest particle concentration (C_c) can be calculated by manipulating Equations (8-1) to (8-4). The resulting expression is

$$C_c = P_H \frac{C_a \; F \; P_R + C_s \; (1-F)}{1 - P_H \; (1-F)} \quad . \qquad (8\text{-}5)$$

Note: There is a slight error in the denominator of Equation (8-5) as it appears in the paper by Ensor, Donovan, and Locke (1987); the equation given above is the correct one. Equation (8-5) gives the relation between the clean air exiting the high-efficiency filter bank and the various flow rates (and the balance of flows), particle source terms, and filter efficiencies. To a certain extent, most of the terms in Equation (8-5) are constants. That is, neither the roughing filter efficiency P_R nor the HEPA (or ULPA) filtration efficiency P_H change significantly over time, and the balance of recirculating and makeup air is relatively constant for long periods. Similarly, although the particle concentration in the ambient air C_a follows a diurnal cycle, typical daytime concentrations can be defined. The greatest uncertainty in Equation (8-5) arises from the uncertainty in the term C_s. In general, the concentration of particles emitted by people and processes in the clean room is a highly variable function of time; however, there are two special cases when the source term can be considered constant. In a new clean room (i.e., the "as-built" condition), there are no particle sources (assuming that no equipment has been placed in the room), so that $C_s = 0$. Similarly, in a clean room "at rest" (i.e., when there are no people in the clean room and no wafers are being processed), then the source terms (e.g., idling equipment) can be considered constant over time. Under these circumstances, Equation (8-5) can be used to estimate the particle concentration downstream of the filters (C_c). Alternatively, given a measured value of C_c and a known value of C_a, one can estimate the constant source strength, C_s.

Although not explicitly shown in Equation (8-5), the concentration and penetration terms are particle size-dependent. That is, the concentration of particles depends on the size under consideration (see Chapters 2 and 3). In addition, the filtration efficiency of roughing filters, HEPA and ULPA filters depends on particle size (see Chapter 7 and Liu, Rubow, and Pui [1985]). Therefore, in applying Equation (8-5) to calculate particle concentrations in a clean room, one must consider the size-dependence of each term.

As an illustration of the application of their model. Ensor, Donovan, and Locke (1987) calculated expected particle size distributions in a clean room as built (i.e., when $C_s = 0$), with no roughing filters in place (i.e., $P_R = 0$) and 20 percent makeup air ($F = 0.2$). The four cases were: Case 1, HEPA filtration system in an urban setting; Case 2, HEPA filtration system in a rural setting; Case 3, an ULPA filtration system in an urban setting; and Case 4, ULPA filtration in a rural setting. The results of their calculations are shown in Figure 8-2, along with the size/concentration limits for Class 100 and Class 1000 clean rooms (FED-STD 209D). Several conclusions can be drawn from the results. First, although the model is simple, it can predict particle concentrations that are typical of those found in clean rooms. Second, the concentration of particles in a clean room can be strongly affected by the concentration of particles in the outdoor ambient air. The difference in concentration for Cases 1 and 2 (i.e., a rural versus an urban setting) is more than an order of magnitude. A similar difference is seen for Cases 3 and 4. Of course, the differences between these two cases would be reduced if prefilters were used. Finally, the model points out an important aspect of clean-room design and operation: none of the large particles (i.e., those larger than 1.0 μm) found in a clean room is introduced from the outside air, even when there is no prefilter. The important point of these results is that a simple mass-balance model can be used to estimate expected particle concentrations for a variety of clean-room designs. If particle concentrations in a new clean room (i.e., as-built) are significantly greater than the numbers predicted by the model, then it may be an indication of improperly installed or damaged filters, especially if a great number of large particles are found.

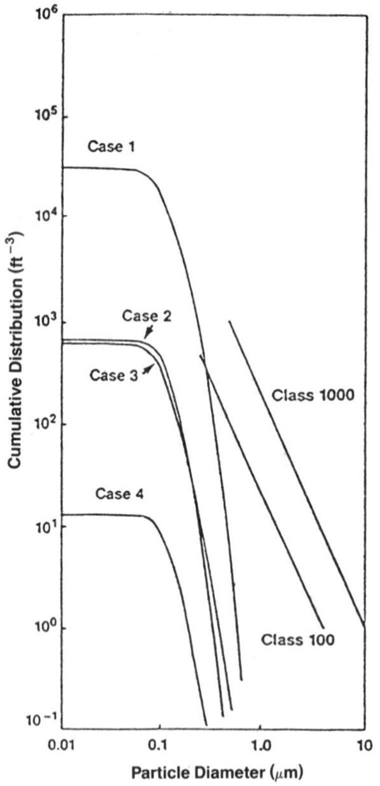

Source: Ensor, Donovan, and Locke, 1987 (with permission).

Figure 8-2. Predicted clean room contamination for HEPA and ULPA filters in urban and rural settings.

The relationship between clean-room and outdoor ambient particle concentrations has been verified by at least two experimental studies (Liu et al.. 1987; Viner and Donovan, 1988). In the first study by Liu et al. (1987), condensation nucleus counters (CNCs) were used to simultaneously measure particle concentration as a function of time within a Class 100 clean room and at the makeup air intake vent. A comparison of the measured concentrations showed that the concentration of particles within the clean room followed the same trends as the outdoor particle concentration. That is, concentration peaks in the outdoor air were mirrored by similar peaks in the clean-room air. In a similar study (Viner and Donovan, 1988), outdoor ambient concentrations (as measured by a CNC) were compared with concentrations measured at two locations in a Class 100 clean room (also with CNCs). Cross-correlation analysis provided a means to quantify the extent of correlation between the indoor and outdoor counts and also yielded information on the time-lag between the two signals (i.e., the time required for propagation of a particle spike in the ambient air to be detected within the clean room). The results of these two studies further illustrate the relationship between the outdoor environment and the wafer processing environment.

MEASURED CLEAN-ROOM PERFORMANCE

Mathematical models can provide a great deal of guidance for minimizing clean-room contamination; however, they cannot predict which processing steps or equipment (or people) will generate particles. That is, all of the available models require that the user specify the sources of particles and their locations within the clean room. The Research Triangle Institute (RTI), under the sponsorship of the Semiconductor Research Corporation (SRC), is currently conducting a study to identify various particle sources and measure the time-dependent particle concentrations emitted by various pieces of processing equipment.

The measurement program consists of an array of CNCs, laser particle counters (LPCs), and environmental sensors placed at various locations in a Class 100 clean room. The clean room consists of two rows of five bays separated by a central aisle. Adjacent bays are separated by a chase that houses utilities and serves as the return air duct for the sidewall air return. The monitoring equipment used in the study is listed in Table 8-1. In the first phase of the study the instruments were all located in the photolithography bay, as illustrated in Figure 8-3. For the purposes of orientation, the safety shower in the figure is adjacent to the central corridor of the clean room. A CNC/LPC pair was used at each particle monitoring location to provide particle size as well as source strength information. One instrument pair (numbers 3 and 4) was placed on a shelf mounted on the wall of the chase approximately 1 ft above the photoresist spinner. Another pair (numbers 5 and 6) was placed inside the environmental chamber housing the photomask stepper. The third pair of counters (numbers 7 and 8) was placed atop the cabinet that houses the electronics for the mask stepper, approximately 3 ft below the HEPA filter in the ceiling. The last pair of counters was placed on top of a table adjacent to a reticle cleaner, in the back corner of the bay. The environmental sensors (temperature/humidity, air velocity, and overpressure) were mounted on the chase wall above the photoresist spinner. The data from each instrument were transmitted to a multiplexer that accumulated the count data for 15 s intervals and then transmitted it to a microcomputer for storage on disk. A video camera was also installed in the clean room to monitor activity in the

Table 8-1. Clean-Room Monitoring Hardware

Equipment	Particle detection limit
TSI Clean Room CNC Model 3760	>0.014 μm
TSI CNC Model 3020	>0.005 μm
TSI Laser Particle Counter Model 3755	>0.5 > 5 μm
TSI Temperature/Humidity Monitor System Model 3721	--
TSI Velocity System Model 3722	--
TSI Room Pressure System Model 3723	--

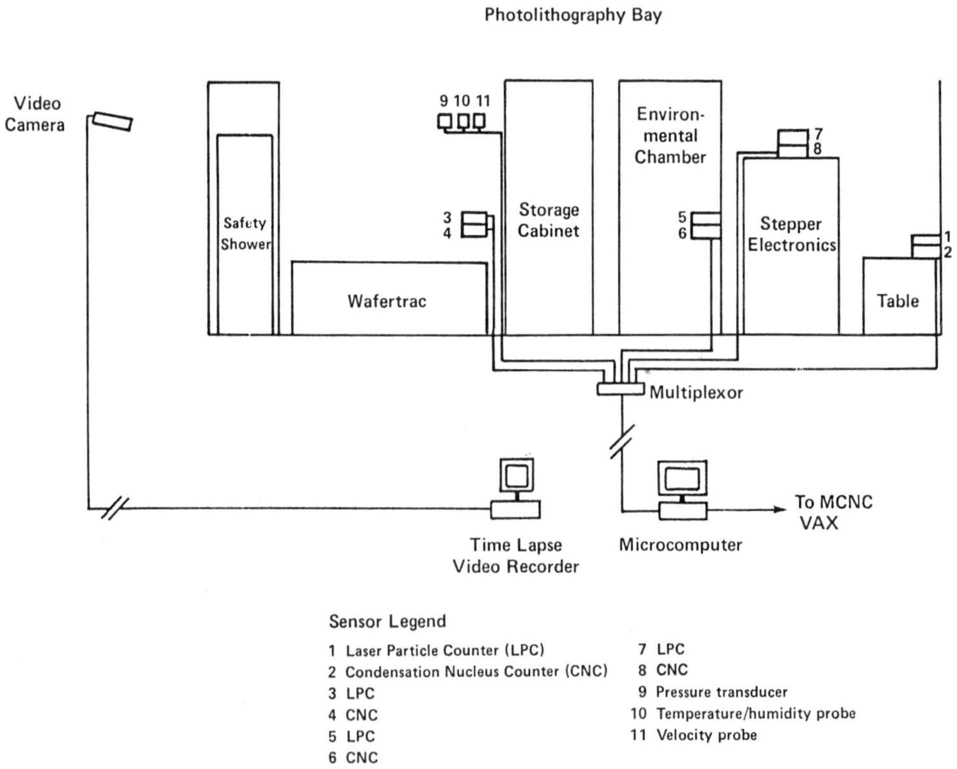

Source: Viner and Donovan, 1988 (with permission).

Figure 8-3. Schematic of instrument placement for the continuous monitoring program.

vicinity of the processing equipment. The video signal was transmitted to a time-lapse recorder that recorded approximately 1 frame per second. Synchronization of the video and computer clocks aided in identifying the sources of high particle counts.

A typical workday particle concentration profile is illustrated in Figure 8-4, which presents a 24-h trace of particle contamination in the vicinity of the Wafertrac photoresist spinner for the CNC and both size channels of the LPC. Each vertical bar in the graphs is a 3-min average concentration computed from 12 adjacent 15-s points. The uppermost profile is the concentration of large particles (> 5.0 μm). None of these large particles is detected in the clean room during the early morning and late evening hours. During the workday, a few large particles are detected, with peaks as high as 100 particles/ft^3 for brief periods. The middle profile in the figure is from the smaller size LPC channel (> 0.5 μm). A few particles are detected during the early morning and late evening hours; however, it appears these are random events and do not indicate any problems with the HEPA filters. During the workday, concentration peaks over 1,000 particles/ft^3 are observed for brief periods. Thus, for short times the clean room is out of compliance with Class 100 requirements (as defined by FED STD 209D). The bottom profile in the figure was obtained with the CNC. Unlike the LPC data, the CNC shows a persistent background concentration on the order of 100 particles/ft^3. This is not surprising though because the CNC detects particles that fall in the most penetrating size range of HEPA filters. The data shown in the figure are consistent with the discussion of HEPA filter performance above: For a clean room at rest, there should not be any large (> 0.5 μm) particles because they are removed by the HEPA filters; however, a large concentration of submicrometer particles can be expected because HEPA filters are not very efficient at removing them.

It is reasonable to expect that the particle counts would be highest when equipment is in use; however, it is not correct to assume that the bulk of emissions is the result of people-related activities. For example, when the CNC data of Figure 8-4 is plotted with an expanded time scale, the pattern shown in Figure 8-5 emerges. In this figure, each vertical bar represents the concentration in one 15-s period. What appears as a constant high concentration from noon to 1 p.m. in Figure 8-4 takes on a very different appearance when viewed at higher resolution. The concentration changes from the overnight background level of approximately 100 particles/ft^3 to a high of over 10^5 particles/ft^3 in a matter of seconds and then drops just as rapidly. This pattern repeats with a period of approximately 90 s. The regular pattern of emissions strongly suggests a mechanical (rather than human) source. Upon review of the video record, it was found that the high particle counts correspond to the application of a photoresist primer during operation of the Wafertrac. The high particle counts have disappeared since a new system was installed for applying the primer.

Another important point arises from the consideration of Figure 8-5: Particle emissions can be highly variable, changing by several orders of magnitude in brief periods. Sem (1986) has reported that particle generation events can be as short as 10 s long, and the RTI program has documented bursts of fewer than 5 s duration. Therefore, it can be misleading to consider long-time averages of particle count data. For example, consider an average count of 100 particles/ft^3 over a 1 h period. Given only that information, it is impossible to know if the

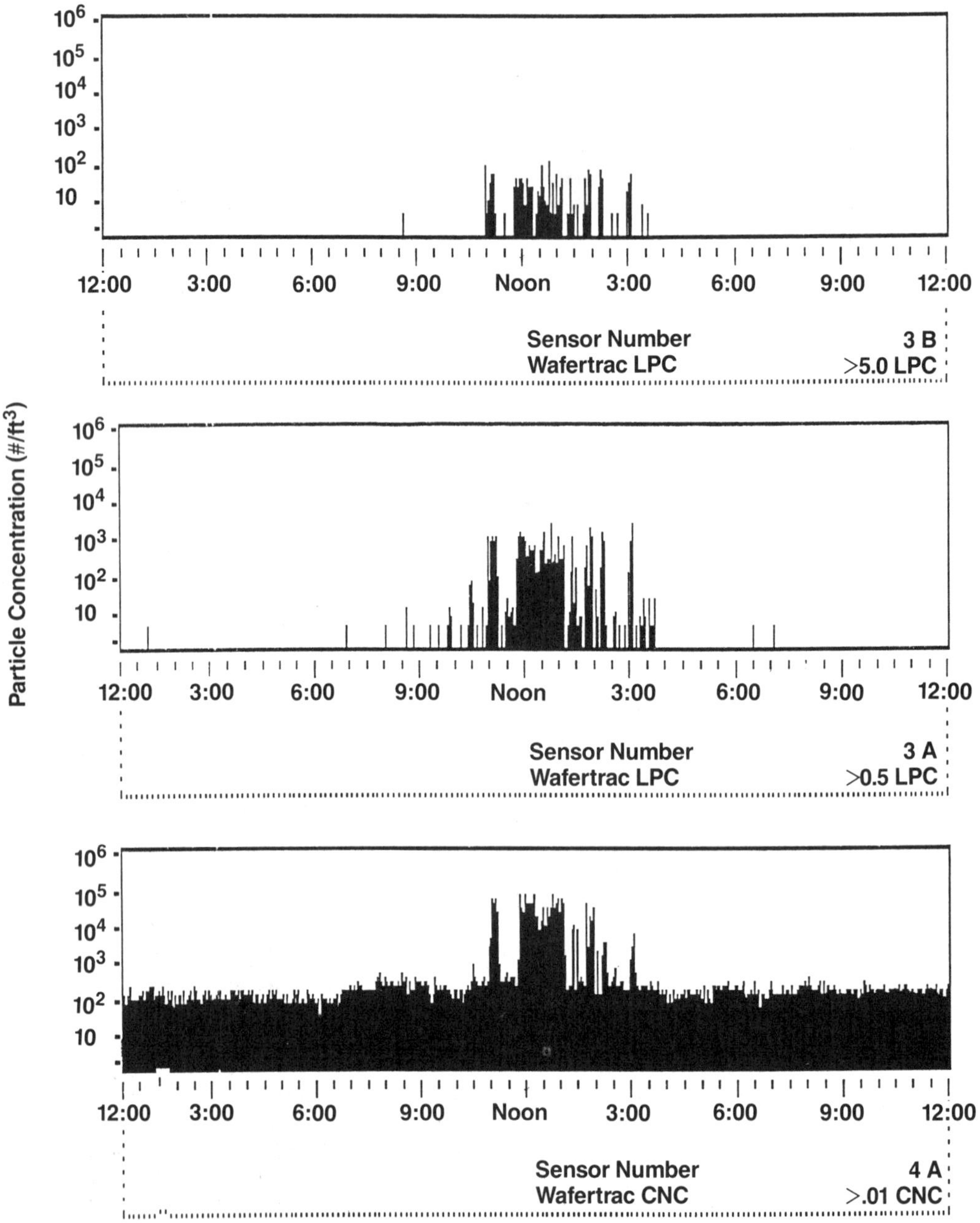

Figure 8-4. Particle size and concentration profiles for a typical workday (midnight to midnight).

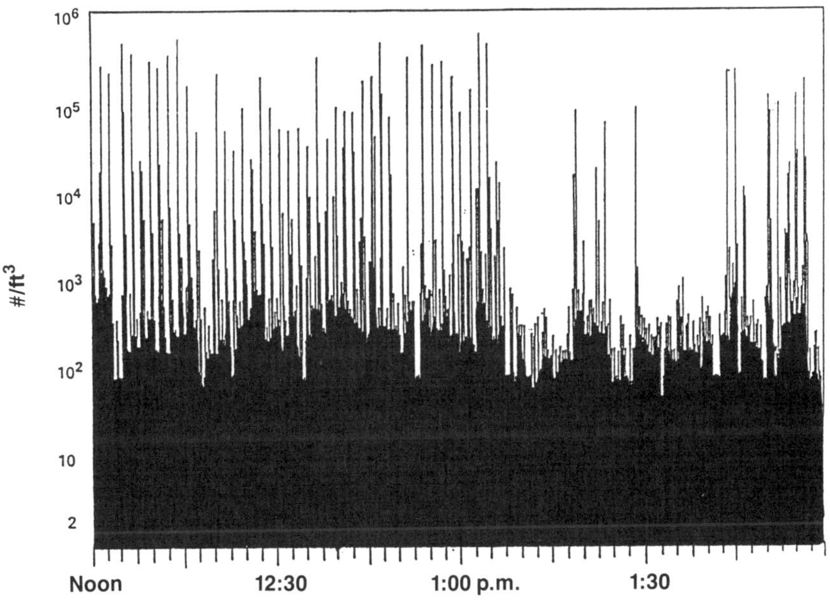

Figure 8-5. A high-resolution plot of the condensation nucleus counter data from Figure 8-4. Each bar is a 15-s average value.

concentration was constant or if it consisted of 6,000 particles/ft^3 over a 1-min period accompanied by 59 min of no particles. Consequently, in measuring particle emissions, one must collect data over short intervals.

Particle sources are not always obvious. For example, the LPC labelled #7 in Figure 8-3 was placed on top of the cabinet that houses the electronics for the photomask stepper. The original intent was to obtain a measure of the particle concentration in the air as it left the HEPA filters. The location on top of the cabinet was thought to be sufficiently far from any potential contaminating sources to provide an accurate measure of the quality of the HEPA-filtered air. It was quite surprising when the results shown in Figure 8-6 were obtained. The LPC measured a high steady background of approximately 100 particles/ft^3 for particles greater than 0.5 μm. This result was not consistent with the results shown in Figure 8-4, where the concentration of 0.5 μm and larger particles near the photoresist spinner is seen to drop when there is no activity in the clean room. At first, the reading from the LPC was not trusted, and some simple tests were run to verify that the LPC was functioning properly. Upon further investigation, it was found that the stepper electronics cabinet contains several potential particle sources, such as printed circuit boards, a video monitor, and a power supply. Cooling fans tended to disperse the particles within the cabinet. Because the cabinet is not sealed and exhausted, the particles escape through the open grillwork on top of the cabinet. Given the potential for high emissions from unexpected sources, it is best to establish a policy of testing every piece of equipment for emissions, a sort of "guilty until proven innocent" policy.

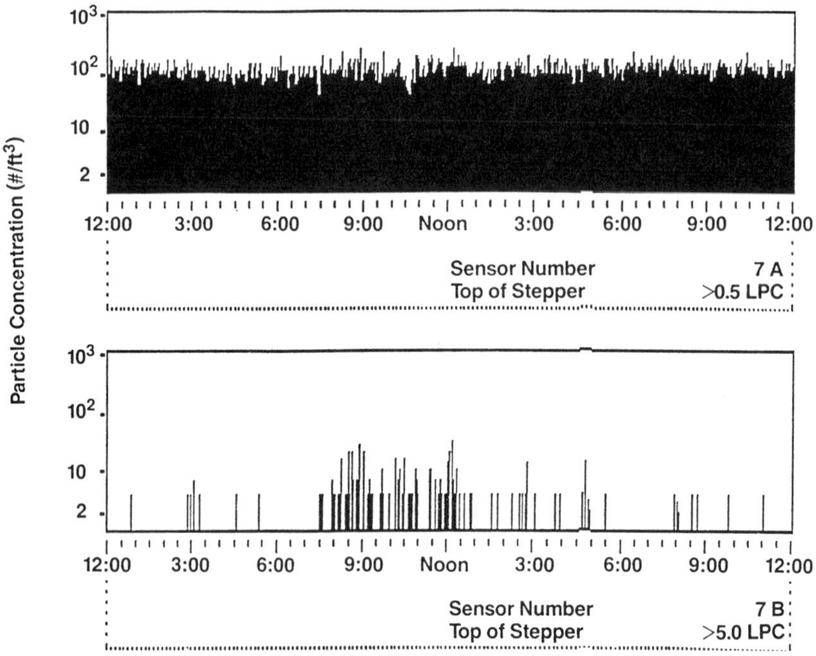

Figure 8-6. Particle concentration as measured by a laser particle counter atop the stepper electronics cabinet.

The problem of unforeseen particle sources becomes more complicated when the sources are intermittent. An example of intermittent high particle concentrations is shown in Figure 8-7a. This particle record was measured by the LPC/CNC pair located in the corner of the photo bay adjacent to the reticle cleaner. The most striking feature of the data is that the concentration exhibits large step changes at arbitrary times. In the figure, the concentration is low from midnight until approximately 6:30 a.m., at which time it increases by four orders of magnitude and remains high, although there is some variability. Suddenly at about 2:15 p.m., the concentration drops dramatically for 1/2 h and then rises again, finally dropping again at 4:30 p.m. While the high particle concentrations in the figure correspond fairly closely with normal workday hours, there was no correlation of particle counts with activity of the reticle cleaner. Also, the data records for other days show similar high particle concentrations but at random times, such as in the middle of the night. The mystery was solved when the particle concentration records were compared with the clean-room overpressure sensor (Figure 8-7b). The overpressure sensor (located above the Wafertrac spinner) measures the pressure differential between the clean room and the surrounding laboratory corridor. Under normal circumstances, the clean room is maintained at a positive pressure with respect to the building shell, so that any air leakage between the two is in the direction out of the clean room, thereby preventing inleakage of particles. An intermittently functioning control circuit in the ventilation system was upsetting the normal flow balance of the clean room, thereby lowering the pressure differential. While the pressure data in the figure show that the pressure differential remained positive above the wafer spinner, it may well have been negative near the reticle cleaner (approximately 20 ft away). At that point, unfiltered air began leaking into the clean

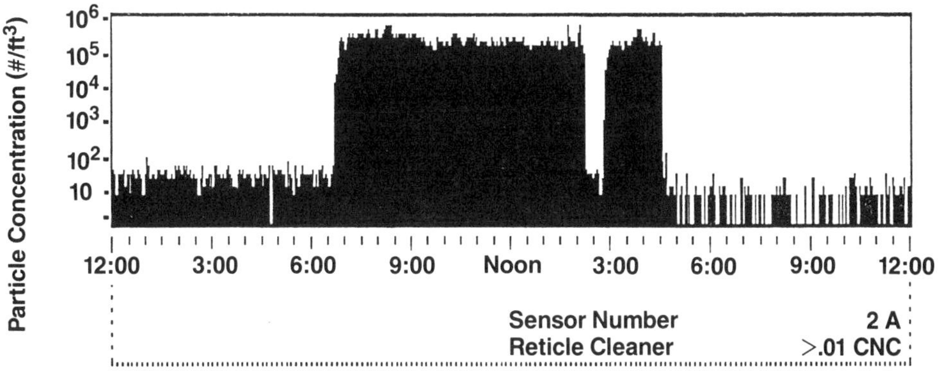

Figure 8-7a. A 24-h particle concentration profile measured near the reticle cleaner.

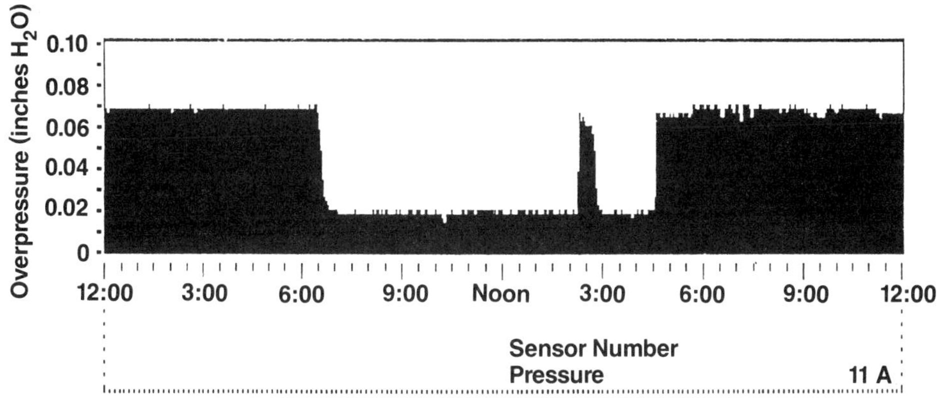

Figure 8-7b. A 24-h trace of clean-room overpressure.

room from a poorly sealed filter joint in the ceiling and from around the door jamb of an emergency exit near the particle counter. The striking correlation between the concentration and overpressure data leaves little doubt as to the source of the high particle counts.

The data of Figure 8-7 serve to emphasize the importance of proper construction and operation of a clean room. Although small air leaks are harmless under normal (positive pressure) situations, they can be disastrous when the ventilation system malfunctions. The clean room then no longer serves as a clean room, i.e., a room that isolates the work area from the outside ambient air. Identifying such unlikely sources requires a painstaking effort that can be greatly aided by auxiliary information (such as clean room overpressure; temperature/ humidity; and airflow velocities). As before, one must adopt a "guilty until proven innocent" attitude when pursuing sources of clean-room contamination.

SUMMARY

The ultimate goal of any contamination control program is to minimize the number of chip defects caused by particle contamination. Several approaches have been adopted by researchers to achieve this goal, including computer modeling and clean-room monitoring. The results of clean-room monitoring programs such as the one described here will help by providing information on source strength and time-dependent behavior. As computer models become more sophisticated, they will be able to use these time/concentration signatures and predict how particles from these sources will disperse within the clean room. Additional experimental and modeling studies of particle deposition on wafers will provide the link between contamination sources, transport within the clean room, and chip defect densities. Once this goal is achieved, progress will be made from understanding contamination processes to predicting when contamination will occur, and thereby being able to prevent it.

REFERENCES

Busnaina, A.A. 1987. Modeling of Clean Rooms on the IBM Personal Computer. In: *Proceedings of the 33rd Annual Technical Meeting, IES, San Jose, CA*, pp. 292-297.

Busnaina, A.A., S. Abuzeid, and M.A.R. Sharif. 1988. Three-Dimensional Numerical Simulation of Flow Field and Particle Transport in a Clean Room. In: *Proceedings of the 34th Annual Technical Meeting, IES, King of Prussia, PA*, pp. 326-330.

Ensor, D.S., R.P. Donovan, and B.R. Locke. 1987. Particle Size Distributions in Clean Rooms. *Journal of Environmental Science*, 30(6):44-49.

Fujii, S., and I. Hayakawa. 1985. Theoretical Studies of Particulate Behavior in Laminar Airflow Clean Room. In: *Proceedings of the 31st Annual Technical Meeting, IES, Las Vegas NV*, pp. 70-73.

Jinno, N. 1986. Study of Airflow Pattern in the Clean Room for Manufacturing Semiconductors. In: *Proceedings of the 32nd Annual Technical Meeting, IES, Dallas TX*, pp. 556-564.

Liu, B.Y.H., K.L. Rubow, and D.Y.H. Pui. 1985. Performance of HEPA and ULPA Filters. In: *Proceedings of the 31st Annual Technical Meeting, IES, Las Vegas, NV*, pp. 25-28.

Liu, B.Y.H., J.W. Lee, D.Y.H. Pui, K. Ahn, and S.L. Gilbert. 1987. Performance of a Laboratory Clean Room. *Journal of Environmental Science*, 30(5):22-25.

Matsuhashi H., T. Izumoto, R. Nakayama, and Y. Yamaji. 1984. Airflow Visualization in a Clean Room, *Journal of Flow Visualization Society of Japan*, 4(July):145-148.

Sem, G.J. 1986. A Case for Continuous Multipoint Particle Monitoring in Semiconductor Clean Rooms. In: *Proceedings of the 32nd Annual Technical Meeting, IES, Dallas TX*, pp. 432-438.

Shanmugavelu, I., T.H. Kuehn, and B.Y.H. Liu. 1987. Numerical Simulation of Flow Fields in Clean Rooms. In: *Proceedings of the 33rd Annual Technical Meeting, IES, San Jose, CA*, pp. 298-303.

Viner, A.S., and R.P. Donovan. 1988. A Comparison of Clean Room Air Quality with Outdoor Ambient Air Quality. In: *Proceedings of the 34th Annual Technical Meeting, IES, King of Prussia, PA*, pp. 342-349.

Yamamoto, T., R.P. Donovan, and D.S. Ensor. 1987. Clean Room Air Flow Modeling. In: *Proceedings of the 33rd Annual Technical Meeting, IES, San Jose, CA*, pp. 304-309.

Yamamoto, T., R.P. Donovan, and D.S. Ensor. 1988. Modeling of Flow in Clean Room Hoods. In: *Proceedings of the 34th Annual Technical Meeting, IES, King of Prussia, PA*, pp. 321-325.

9. THE DYNAMICS OF AEROSOLS IN CLEAN-ROOM ENVIRONMENTS: IMPLICATIONS FOR MONITORING AND CONTROL OF AIRBORNE PARTICLES AND THEIR SOURCES

GILMORE J. SEM TSI, Incorporated, St. Paul, Minnesota

INTRODUCTION

At least one major semiconductor manufacturer has recently declared that airborne particles are a process variable that will be continuously monitored and controlled. The company believes that airborne particles reduce yield. The direct correlation of a specific incidence of particle contamination with yield reduction has been done in just a few isolated cases. The difficulty with such correlations is that there are hundreds of steps in the semiconductor manufacturing process and many of those steps are complex, with large numbers of possible yield-reducing factors.

Few people associated with semiconductor manufacturing believe that yield is unaffected by airborne particle contamination. We can make a very convincing case for the connection of low yield with particles.

This chapter develops the scientific basis for understanding the dynamics of aerosols in clean room environments. Such understanding plays a major role in the monitoring and control of airborne particle contamination, and the detection and elimination of particle sources.

RELEVANT CHARACTERISTICS OF OUTDOOR AIRBORNE PARTICLES AND THEIR SOURCES

During the 1970s, Whitby (1978) and Whitby and Sverdrup (1980) discovered and verified that outdoor ambient aerosol nearly everywhere in the world fits the form of a trimodal particle size distribution. Figure 9-1, from Whitby (1978), shows what a typical trimodal distribution looks like. The mass of particles in the nuclei mode is usually not so pronounced, even though this mode dominates the number concentration of nearly all outdoor aerosols. The trimodal distribution is almost universally applicable. Some areas may have one or two dominant modes of the distribution so that the nondominant mode(s) is(are) barely noticeable. Urban areas usually have all three modes with one or another rising or falling depending on the local sources and weather conditions.

It is important to note that the vertical axis of Figure 9-1 is in mass concentration units. When converted to number concentration, the particles in the smallest-diameter mode are almost always dominant.

Figure 9-2 (Whitby and Sverdrup, 1980) shows why there are three modes in the distribution of outdoor aerosols. The schematic shows the main source of mass for each mode, the primary processes for inserting mass into each mode, and the primary mechanisms for removing mass from each mode. The mode associated with the smallest particles, those smaller than 0.1 μm, is generated by the condensation of vapor to form particles. This may

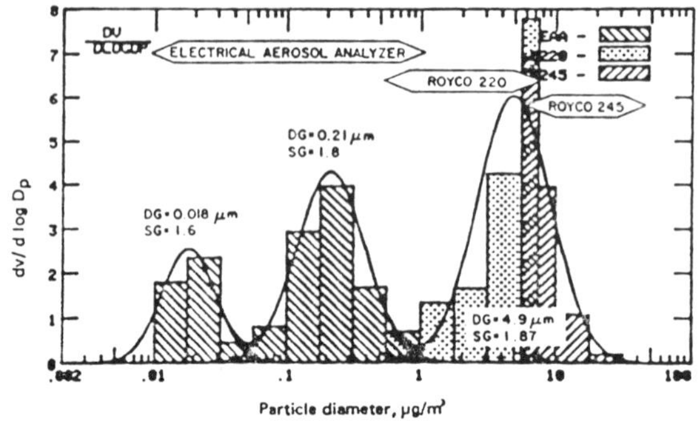

Source: Whitby, 1978

Figure 9-1. A trimodal atmospheric aerosol volume size distribution measured downwind of an automobile test track.

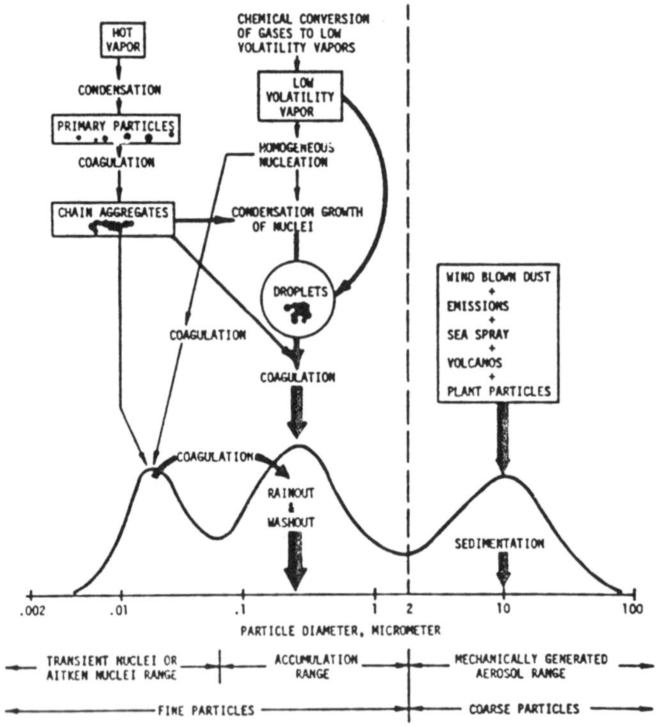

Source: Whitby and Sverdrup, 1980

Figure 9-2. Schematic of an atmospheric aerosol distribution showing the three modes.

occur during combustion, or it may occur during condensation of a vapor at ambient temperature. These nuclei mode particles are usually generated in very dense clouds, resulting in rapid coagulation during the first few seconds, until the concentration is dilute enough to slow the coagulation. Thus, a very high concentration of nuclei mode particles ends up, after a few minutes, as a much lower number concentration of larger particles. Assuming no loss of particles to surfaces or by evaporation, mass is conserved.

The middle range of particle sizes, between 0.1 and 1 to 2 μm, is called the accumulation mode, for reasons that will become apparent. As nuclei mode particles grow by condensation and coagulation, the particles grow into the accumulation range of sizes. By this time, the concentration is usually so dilute that coagulation no longer occurs. The same dilution reduces vapor concentration, resulting in reduced condensation as well. In addition, these particles are not heavy enough to settle out of the atmosphere by gravitational means. In fact, there is no efficient means of removing these particles from the atmosphere. They have airborne lifetimes of weeks or even months, travelling across continents and oceans during that time. This is the size range that is responsible for the transport of acid rain from central Europe to Scandinavia and from the Ohio River Valley to New England and Canada; and for the transport of radioactive particles from Chernoble to other parts of Europe,

The third mode is the coarse mode because it contains the large particles, generally larger than 1 to 2 μm. These particles are formed by completely different mechanisms. They result from the breakup of larger bulk materials into particles of this size. It is difficult, but not impossible, to break 1 μm particles into smaller particles. The energy required for the production of submicrometer particles by this mechanism is very high. In the outdoor atmosphere, this size range consists of windblown dust, particles thrown into the air by car tires, by weather systems, and by volcanoes, and ash left over from coal and wood combustion (such combustion also produces some nuclei particles by evaporation followed by condensation). Particles larger than about 5 to 10 μm settle rapidly by gravity, remaining air-borne for only seconds or a very few minutes.

Two other characteristics of outdoor aerosols are important to discuss here. One is that the concentration of particles in any location changes rapidly, depending on the location and strenth of local- and medium-range particle sources and the wind conditions. The second is that the concentration of particles at any time varies greatly from place to place, depending on the same factors. A single continuous measurement cannot characterize an entire city. On the other hand, a single "snapshot" of concentrations at many points in a city cannot characterize what happens during a 24-hour period in that city.

CHARACTERISTICS OF CLEAN ROOM AEROSOLS

The characteristics of outdoor aerosols apply to clean room aerosols in several ways. First, the same basic mechanisms of airborne particle formation and deposition apply inside clean rooms as well as outside them. Second, outdoor makeup air challenges the HEPA filters that clean the air entering the clean room.

Coarse Mode, Particles >0.5 μm

Coarse mode particles are generated primarily by mechanical abrasion. Examples include:

- Skin flakes, dandruff, hair pieces
- Cosmetics, powders
- Sneezes, coughs
- Clothing pieces, paper pieces
- Sand and soil from shoes and floor
- Mechanical wear and abrasion of bearings
- Droplets shed by rotating machinery
- Particles shaken from surfaces
- Any breakup of large objects into small particles.

Coarse particles cause semiconductor failure by bridging line spacings, breaking conductive paths, and obliterating devices. Control of coarse particles is quite simple because HEPA and ULPA filters remove nearly all of them in one pass. Sources of coarse particles are found both inside and outside clean rooms. They settle rapidly in gravity (a 10-μm particle settles 6 mm/s) so they do not transport well horizontally. Particle sensors must be directly below the point of interest. Ordinary optical or laser particle counters detect coarse particles from 0.5 to 5 or 10 μm quite easily.

In one Class 100 clean room, Larrabee (1987) attributes airborne particles primarily to coarse mode particle sources:

Source	Percent
• Human	46
• Dust-free garments	8
• Cleaning materials	4
• Photoresist pieces	12
• Silicon, quartz	22
• Other	8
	100

Nuclei Mode, Particles <0.1 μm

Nuclei mode particles are generated primarily by combustion or evaporation followed by condensation. Examples include:

- Furnaces and heaters
- Hot components on circuit boards
- Electric motors
- Hot bearings
- Flames
- Hot lights
- Microscope lamps.

Nuclei particles cause semiconductor failures by bridging through layers, chemically degrading much larger spots, and physically causing contamination that is much larger than the original particle (Fisher, 1987). HEPA and ULPA filters collect nearly all nuclei particles in a single pass (Liu, Rubow, and Pui, 1985). Nuclei particle sources are located both inside and outside the clean room. Sources of nuclei particles usually have profuse production rates, millions or even billions of particles generated per second. Nuclei particles are usually present in at-rest clean rooms with concentrations 10 to 1,000 times the level of particles >0.3 μm (Donovan et al., 1985). In a recent study of an operational semiconductor clean room by Donovan et al. (1988), similar ratios of nuclei to coarse particles were detected.

Nuclei mode particles do not settle significantly by gravity; the settling speed of a 10-nm particle is just 150 nm/s. They migrate rapidly toward deposition surfaces by Brownian diffusion. Therefore, their sensor should be located just downstream of the point of interest, with a short sampling tube. Coagulation may be high during the first second after generation, but it is generally insignificant at concentrations found in clean rooms because of rapid dilution. Nuclei mode particles are easily counted with condensation particle (nucleus) counters (Keady, Quant, and Sem, 1986).

Accumulation Mode, Particles from 0.1 to 0.5 μm

Accumulation mode particles may come from three sources:

- Nuclei mode particles that have grown into this size range; probably not many are able to do this.
- Coarse mode particles that have been broken up into smaller sizes; probably not many are able to do this either.
- Accumulation mode aerosols in the makeup air from the outdoor atmosphere that have penetrated the HEPA filters.

Ensor, Donovan, and Locke (1986) developed a model to describe the size distribution of airborne particles in an at-rest clean room (Chapter 8). Figure 9-3 shows the particle source terms used in the model. The combination of a peak in the outdoor aerosol size distribution (Figure 9-1) and a sharp peak in the penetration of HEPA and ULPA filters (Figure 9-4, from Liu, 1985), both in the 0.1 to 0.5-μm range, leads to many, if not most, of the particles in the accumulation size range.

Accumulation mode particles cause failures of semiconductors by bridging across line spacings, penetrating through layers, breaking conductive paths, and degrading larger areas. These particles can be transported through a few meters of tube with little wall deposition because they are nearly unaffected by gravity, coagulation, and Brownian diffusion. Some laser cavity particle counters count particles larger than 0.1 μm. Condensation particle counters count particles larger than 0.01 μm.

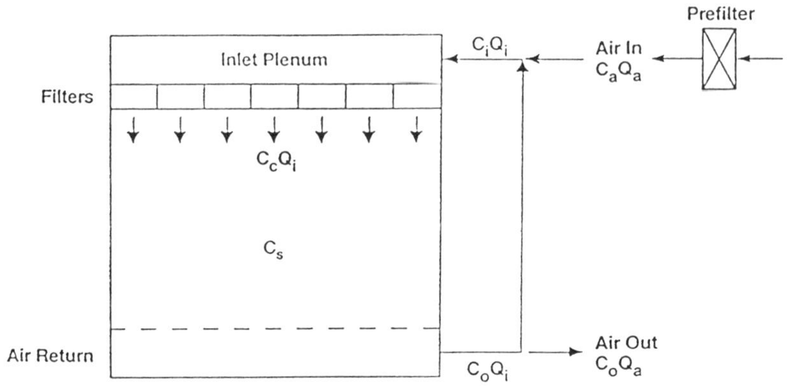

Source: Adapted with permission from Ensor, Donovan, and Locke, 1987.

C_a = prefiltered atmospheric concentration
C_c = particle concentration entering the room
C_s = sources in the room
C_o = particle concentration in exhaust air
Q_a = volumetric flow rate of make-up air
Q_i = volumetric flow rate of air in the clean room

Figure 9-3. Schematic of a clean room illustrating source and sink terms for a model describing particle dynamics for the room.

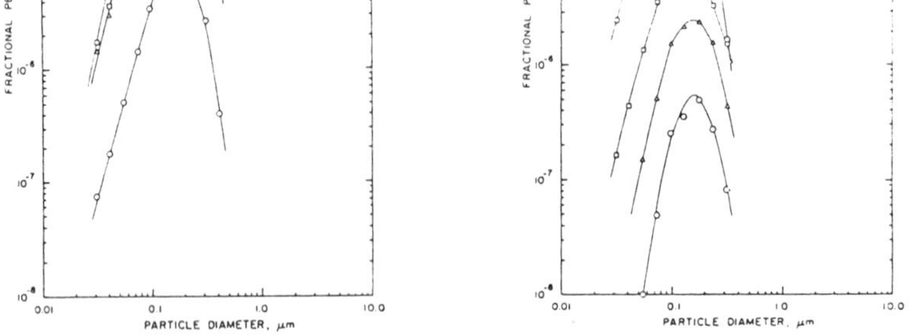

Source: Liu, Rubow, and Pui, 1985

Figure 9-4. Penetration of particles through HEPA and ULPA filter media as a function of particle diameter. Notice the relatively high penetration of particles just above 100 nm.

Summary of Clean Room Size Distributions

Figure 9-5 is for clean rooms the equivalent of Figure 9-2 for ambient outdoor aerosol. It illustrates the sources of particles in clean rooms for each of the three size modes. The smallest particles are generated by condensation of vapors. The midsized particles penetrate through HEPA filters. The largest particles are generated by mechanical abrasion of larger materials. These source characteristics play an important role in the particle detection, monitoring, and control strategies for any clean room. Each of the three size ranges of particles has a different generation mechanism and a different method of becoming suspended in clean room air. Also illustrated in Figure 9-5 are ranges of particle counters and relative physical sizes of particles.

Particle Events (Temporal Variation)

Particles penetrating through HEPA or ULPA filters produce relatively constant levels of particle concentration, consisting of isolated single particles appearing more-or-less randomly throughout the clean room. The randomness is certainly disturbed by the nonuniformity of the filter media, filter mounting frame, and the nonrandomness of the sources of particles inside or outside the clean room.

Superimposed on the constant background of particles penetrating the filters are short bursts of particles from internal sources of many kinds. A "typical" event may produce tens, thousands, or millions of particles in a few milliseconds or seconds. In most clean areas, event particles dominate the number of particles seen by a silicon wafer at any chosen location. Events are highly localized, affecting only small areas directly downwind of the source. An event is likely to do its greatest harm before it enters the recirculating system for the first time. By the time the particles that define the event have been swept into the recirculating air system, the evidence of the event is lost by dilution and by HEPA filtration. Measurement and control of an event require continuous monitoring within very few centimeters of the critical point.

Figure 9-6, from Sem (1986), shows particle concentration (particles per cubic foot) in a cleanbench averaged over 6-minute periods during a 24-hour day. Sensor 11 was near the rear wall of the bench. Sensor 13 was just inside the bench's window. The window was shut except for three times during the day. The bench easily met Class 10 requirements while the window was shut. The three bursts occurred when a technician opened the window, reached over Sensor 13 and then Sensor 11 to check an electrical data cable behind Sensor 11, and shut the window again. The operation took less than 6 seconds/check. Each particle burst lasted only about 10 seconds, not the 6 minutes averaged on the figure. During the third check, the operator first opened the window for 90 minutes before reaching over Sensor 13 to check behind Sensor 11.

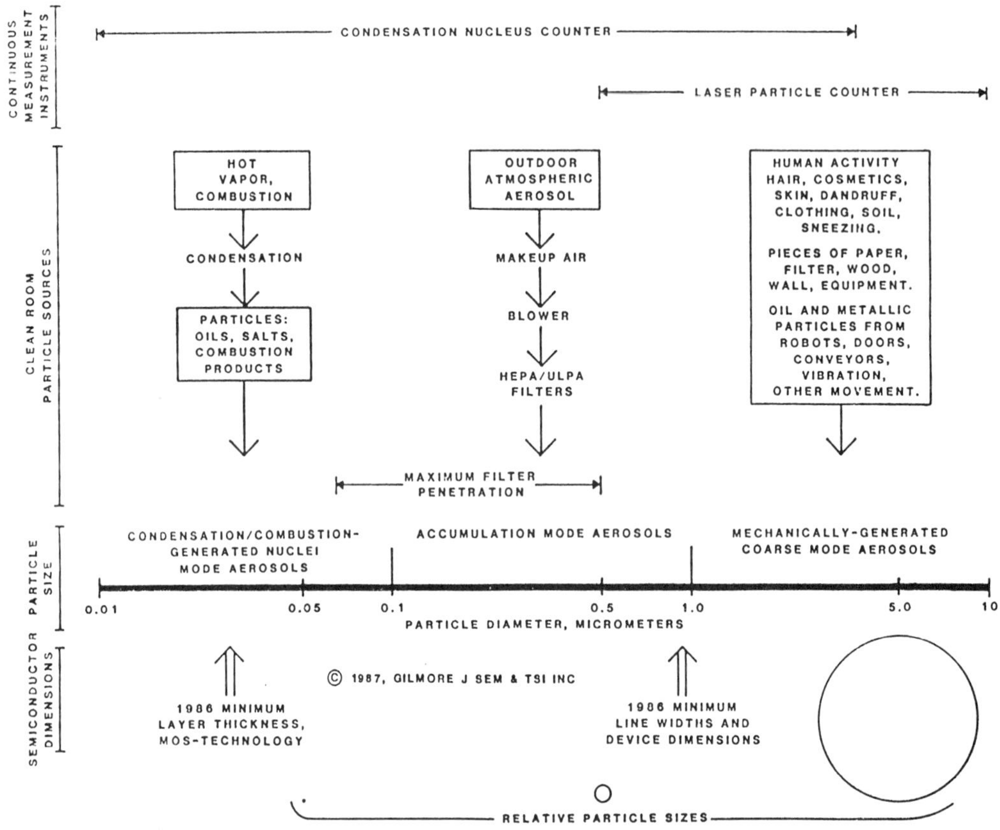

Figure 9-5. Illustration of airborne particle sources within clean rooms, emphasizing the trimodal nature of the particles.

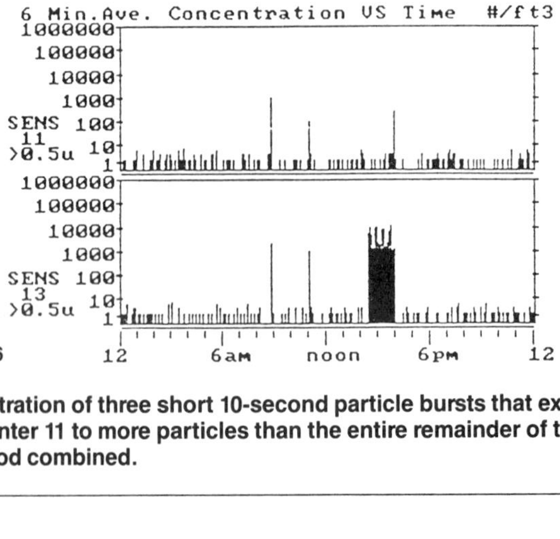

Source: Sem, 1986

Figure 9-6. Illustration of three short 10-second particle bursts that exposed Counter 11 to more particles than the entire remainder of the 24-hour period combined.

Sensor 11, in the cleanest location, counted more particles during the 30 seconds when the three bursts took place than during the entire remainder of the 24-hour period. Sensor 13 indicated Class 10,000 conditions during the 90-minute period; Sensor 11, just 30 cm away, indicated Class 1 during the same time. We see both the spatial and temporal nature of particle bursts.

Donovan et al. (1988) present several examples of particle burst data. Figure 9-7 illustrates particle bursts in the vicinity of a photoresist spinner; the bursts correlate precisely with the operation of the spinner. The entire graph in Figure 9-7 spans 2 hours, and each particle entry represents particle concentration averaged over 15 seconds. Whenever the spinner was activated, the CPC detected a burst. The concentration during the bursts is usually over 10,000 particles/ft^3. Figure 9-8 illustrates the particle concentration in the loading area of a low-pressure chemical vapor deposition furnace. Sensor 1 is a CPC, and Sensor 2 is a laser particle counter. Many of the particle bursts are attributable to specific steps in the process, but not all bursts are detectable by all sensors. In this case, some CPC-detected bursts can be attributed to specific activities of the process. Detectors for particles greater than 0.5 μm and greater than 5 μm did not see all of the bursts as clearly as the CPC. However, the combination of the three sensors yielded extra information about the types of particles that the wafers were exposed to. For example, when the quartz cradle lid was removed, many of the particles were greater than 5 μm.

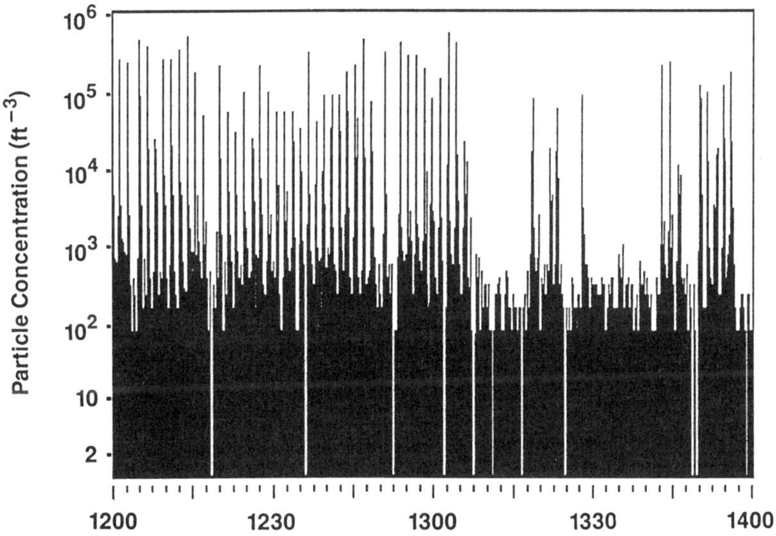

Source: Donovan et al., 1988

Figure 9-7. Periodic bursts of particles larger than 10 nm in the vicinity of a photoresist spinner.

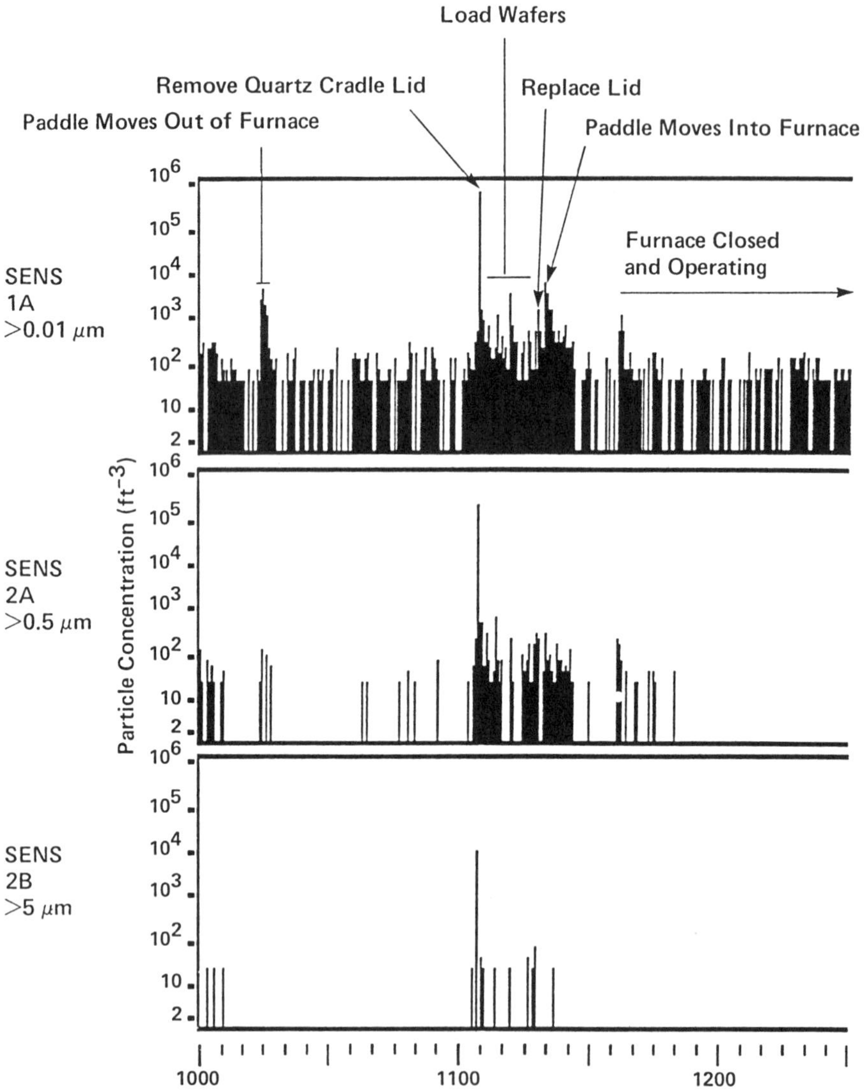

Source: Donovan et al., 1988

Figure 9-8. Particle concentration from the loading area of a low pressure chemical vapor deposition furnace.

Donovan et al. (1988) point out that, in their limited experience, the origins of particle bursts are not always easy to identify. They found surprisingly weak correlation between bursts and human activity. They found good correlation between bursts and equipment operation, as shown in Figures 9-7 and 9-8.

Donovan et al. (1988) also measured particle size distributions during three periods of distinctly different activity within a semiconductor clean room. They used four CPCs with four different sets of diffusion battery screens upstream of the CPCs to measure the size distributions below 200 nm. They used two types of LPCs to measure four more size ranges above 200 nm. They measured the clean room (1) at night while no human activity took place in the clean room (at rest), (2) during normal daytime activity, and (3) during particle bursts. The size distributions for the three activities are radically different, as shown in Figure 9-9. At night, while the room is at rest, particles in the air are dominated by those penetrating the filters. During normal daytime activity, concentrations of both large and small particles increase. During bursts, distributions are highly variable, depending on the source of the particles.

Spatial Variation of Clean Room Particles

Figure 9-10 illustrates the spatial distribution of airborne particles in a clean room. The rapid continuous replacement of clean room air with filtered air leads to great variation of particle concentration from one area to other nearby areas. If a particle source is located just upstream of a critical area, contamination occurs. If the particle source is displaced slightly to the right or left, contamination does not occur. The distance between contaminated and noncontaminated areas may be just 10 cm. Thus, unless the particle counter is located within the same airstream as the particle source, it will not detect the correct particles. Also, unless the counter is located within the same airstream as the critical process, it will not detect particles that may cause yield loss.

The source may be external, such as a hole in a filter, or internal. Airflow that is optimized to avoid recirculation and contamination leads to greater spacial variation of particles throughout the room. People or equipment that move horizontally in the airflow disturb the air, spreading the contaminating particles over a wider area. That makes particles easier to detect, but harder to control.

Airflow in a clean room is seldom laminar in a scientific sense. The Reynold's Number defines whether the airflow is turbulent or laminar. Clean room flow is not laminar, but is highly turbulent. A properly designed clean room has unidirectional airflow so that localized areas quickly flush clean.

Top: At night while no operators were present
Middle: During normal workday activity
Bottom: During five particle bursts

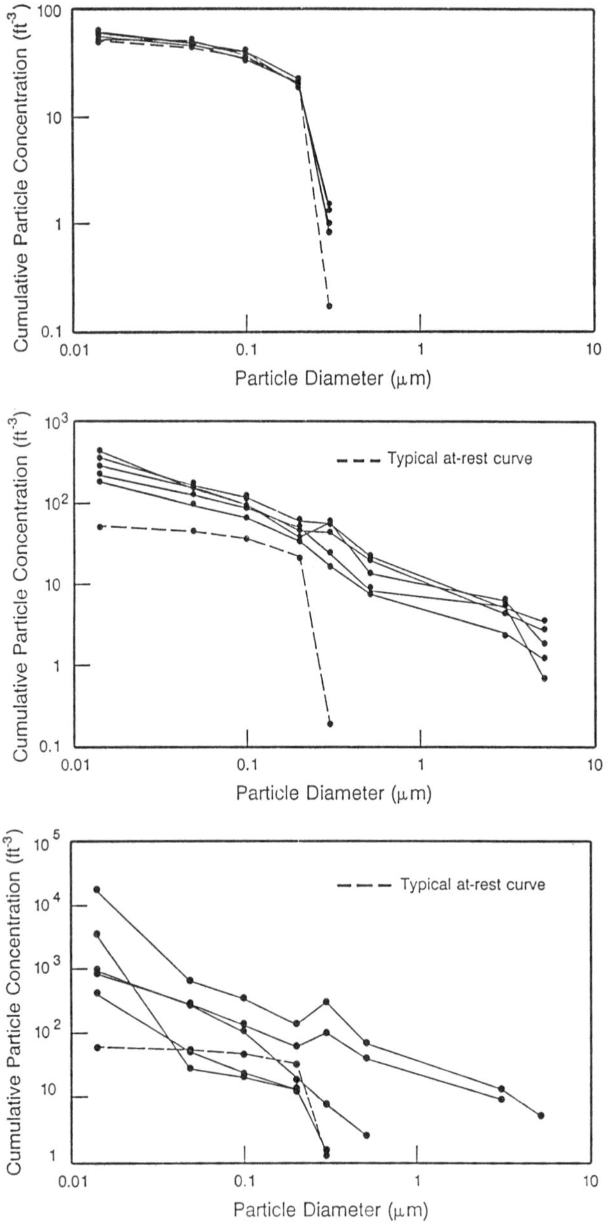

Source: Donovan et al., 1988

Figure 9-9. Particle size distributions of clean room air measured under three different conditions.

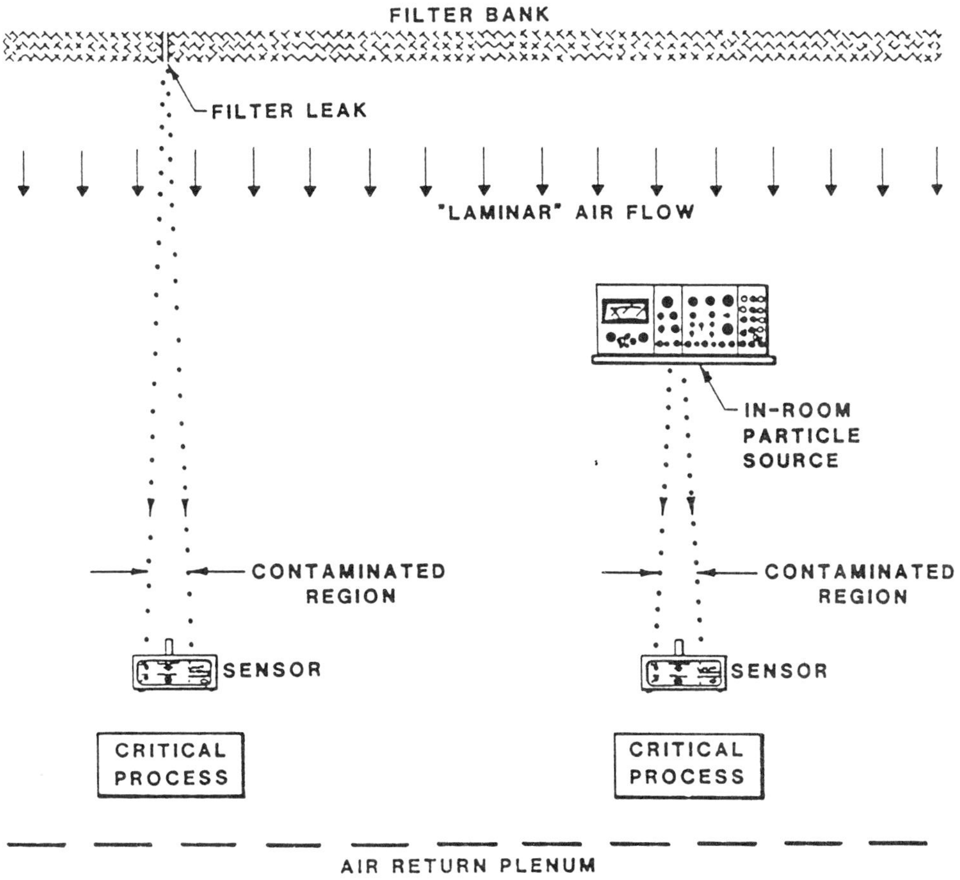

Source: Sem, 1987

Figure 9-10. Schematic of a clean room illustrating the variations of spatial distributions of airborne particles.

WHAT IS THE PURPOSE OF PARTICLE MEASUREMENT IN CLEAN ROOMS?

Burnett (1986) described two reasons for particle measurement in clean rooms:

- Certification
- Yield improvement.

Both reasons are appropriate, but for differing circumstances. Because the goals and focus of the two reasons are so different, it is important not to confuse the two when designing a measurement system.

Certification

Certification requires a specified standard method such as Federal Standard 209D. The purpose is to certify that a clean room or clean area meet all the requirements of the standard method. The procedure is used primarily to satisfy a contractual agreement between two or more parties. Examples include certification of a newly constructed clean room so that the contractor can get paid for the project, and certification of a new filter installation by leak detection and efficiency measurements so that the filter manufacturer and installer can be paid for the job.

During certification measurements, the emphasis is on stable conditions. "Normal operations" are often suspended so as not to disturb the certification process. If a short-term particle burst is measured, methods are available to eliminate the "statistical outlier" by taking more data points. The assumption is that the burst data indicated an instrument malfunction or an unusual process upset that is not typical of the process and will not be repeated.

Yield Improvement

Financial results are the driving force behind yield improvement. Industry-wide standards play no role in yield improvement. If the presence of particles at a critical point of a manufacturing process will lead to yield reduction, the operator of the process wants to detect, locate, and eliminate any possible particle sources in the area. This is a detective process, usually involving the continuous monitoring and detection of particle events, followed by the hunt for the particle source.

During yield-improvement measurements, the emphasis is on the unusual events that create particles. The operator wants to highlight particle bursts so that he/she can track down their source. Stable conditions are welcomed, but are not the item of interest. The operator spends little time on stable, low-level particle measurements, and may even program the data computer to throw away those portions of the data.

Continuous particle detection at each of many critical process locations is a key to particle burst detection. Control of particle bursts is a key to control of airborne particles in clean rooms. Particle control is a key to yield improvement.

REFERENCES

Burnett, J. 1986. Particle Counting Practices for Yield Improvement. *Microcontamination,* 4(3):20-22.

Donovan. R. P.. B. R. Locke. C. M. Osburn. and A. L. Caviness. 1985. Ultrafine Aerosol Particles in Semiconductor Cleanrooms. *Journal of the Electrochemical Society,* 132(11):2730-2738.

Donovan. R. P., A. S. Viner, D. S. Ensor, and E. W. Johnson. 1988. Aerosol Particle Metrology for Semiconductor Manufacturing. Presented at TECHCON '88, Dallas, TX, 12-14 October.

Ensor, D. S., R. P. Donovan. and B. R. Locke. 1986. Particle Size Distributions in Clean Rooms. *Proceedings, 8th International Symposium on Contamination Control, Milan, Italy, 9-12 September.*

Fisher, W. G. 1987. Particle Monitoring in Clean Room Air With the TSI 3020 Condensation Nucleus Counter. *TSI Journal of Particle Instrumentation,* 2(1):3-19.

Keady, P. B., F. R. Quant, and G. J. Sem. 1986. A Condensation Nucleus Counter for Clean Rooms. *1986 Proceedings, 32nd Annual Technical Meeting, Institute of Environmental Sciences, Dallas, Texas, 5-9 May.*

Larrabee, G. B. 1987. Contamination Control in Semiconductor Device Manufacturing. Notebook for Microcontamination Control, The 21st One-Day Videoconference. The Institute of Electrical and Electronics Engineers and The Learning Channel, 2 December.

Liu, B.Y.H., K. L. Rubow, and D.Y.H. Pui. 1985. Performance of HEPA and ULPA Filters. *1985 Proceedings, 31st Annual Technical Meeting, Institute of Environmental Sciences, Las Vegas, NV, 30 April - 2 May.*

Sem, G. J. 1986. A Case for Continuous Multipoint Particle Monitoring in Semiconductor Clean Rooms. *1986 Proceedings, 32nd Annual Technical Meeting, Institute of Environmental Sciences, Dallas, Texas, 5-9 May.*

Sem, G. J. 1987. Detecting Sources of Airborne Particles in Clean Rooms. Presented at the Conference of the Integrated Circuit Manufacturers' Consortium, Conventry, England, February.

Whitby, K. T. 1978. The Physical Characteristics of Sulfur Aerosols. *Atmospheric Environment,* 12(1-3):135-159.

Whitby, K. T., and G. M. Sverdrup. 1980. California Aerosols: Their Physical and Chemical Characteristics. *The Character and Origins of Smog Aerosols,* G. M. Hidy et al., eds. John Wiley & Sons, New York, NY, pp. 477-517.

10. MEASUREMENT AND MINIMIZATION OF PARTICLES IN PROCESS GASES

JEFFREY M. DAVIDSON The BOC Group, Inc., Murray Hill, New Jersey

INTRODUCTION

Among the many particle sources in the semiconductor fabrication environment, process gases must be considered as one possibility (Duffalo and Monkowski, 1984; Osburn et al., 1988; Tolliver, 1984). Although gases are generally regarded as the least significant contamination source, they must nonetheless be carefully scrutinized because of their contact with the semiconductor throughout processing. As detailed in Table 10-1, a variety of high-purity gases are used in virtually every process step in semiconductor fabrication. The gases are often exposed directly to the wafer during fabrication; occasionally, the gases become part of the semiconductor chip itself. In contrast, semiconductor contamination during processing from other sources, such as air, people and equipment, is continually being minimized by increased isolation of the wafer from the fabrication environment.

In the past, the semiconductor technologist was required to ensure only that the concentration of particles in a process gas met specifications. Increasingly, the technologist is faced with the additional task of improving gas cleanliness beyond industry specification levels, especially as more stringent specifications are being applied to processes for higher scale integration of semiconductor chips. Therefore, advanced technologies for both particle measurement/identification and particle minimization are required.

This chapter highlights the important issues related to particle control in semiconductor gases and provides references in which these topics are discussed in greater detail. For background information, procedures involving the production and distribution of semiconductor gases are initially reviewed in the next section. The important features of particle measurement/identification and particle minimization are summarized in subsequent sections. Chapter 20 discusses particle identification techniques in greater detail. Future directions of these technologies are noted in this chapter's last section.

Table 10-1. Gases Used in Semiconductor Manufacturing

Epitaxial Silicon	SiH_4, $SiCl_2H_2$, AsH_3, PH_3, H_2, N_2
Oxidation	O_2, H_2, Ar, HCl
SiO_2 Deposition	SiH_4, $SiCl_2H_2$, CO_2, H_2, O_2, N_2O
SiN Deposition	SiH_2, $SiCl_2H_2$, NH_3, Ar, N_2
Doping	AsH_3, AsF_3, PH_3, BCl_3, BF_3, Ar, He, H_2, N_2
Etching	CF_4, Cl_2, HCL, BCL_3, CCl_3F, CCl_2F_2, $CClF_3$, NF_3, SF_6, Ar, O_2
Annealing	N_2, Ar, H_2
Blanketing	N_2, Ar, He

PRODUCTION AND DISTRIBUTION OF HIGH-PURITY GASES

To appreciate the complexities of particle analysis and control in process gases, it is useful to first understand the gas production and delivery route to a semiconductor facility point-of-use. It is especially important to be cognizant of potential particle-generating mechanisms and sites.

Gases used in semiconductor fabrication are generally divided into two broad categories—atmospheric gases and specialty gases (Lattimer, 1967; Ruheman, 1949). The industrial atmospheric gases, i.e., nitrogen, oxygen, and argon (as well as the rare gases, i.e., neon, krypton, and xenon), are produced by air separation. Other industrial gases also found in air, such as carbon dioxide, helium, and hydrogen, are more economically produced by alternative processes. In air separation units (ASU), atmospheric air is drawn through a filter and compressed to the required pressure by one or more compressors. The air temperature is reduced in heat exchangers during and following compression. The air is brought to near liquefaction temperature (-170°C at 5.5 x 10^5 Pa [5.5 atm]) in reversing heat exchangers, where carbon dioxide and water impurities are removed with waste gas. In some plants, pre-purification units (PPU) consisting of molecular sieve adsorption beds are used to remove these impurities. The cooled and purified air then enters a single or double distillation column where, through the processes of boiling, evaporation, condensation, and expansion, the desired elements are separated from the air. Both liquid and gaseous products are possible, and purities can be increased through process and equipment design modifications.

On a volume basis, nitrogen is the single most used gas in semiconductor production. Nitrogen may be provided to a facility as gas from an on-site nitrogen generator, with a backup liquid supply, or through a pipeline from an ASU. Alternatively, nitrogen may be shipped as liquid in road tankers to an on-site liquid storage tank. Upon demand, the liquid is turned into gas in an ambient heat exchanger or vaporizer. In all cases, the nitrogen gas should be filtered before entering the facility, again at all points-of-use, and possibly at intermediate points as well. Distribution to a point-of-use requires the nitrogen to also pass through a series of pressure regulators, pressure gauges, valves, and possibly flow monitors. Pressures within the distribution piping are generally maintained at approximately 4 to 8 x 10^5 Pa (60 to 120 psig).

As shown in Table 10-1, semiconductor processing also requires the use of many specialty gases, such as silane, arsine, phosphine, hydrogen chloride and other halogen gases, and halocarbon gases. Specialty gases are produced by a variety of techniques, but most of these are either through specifically designed chemical reaction processes or from side streams of industrial processes. These gases are subsequently purified, usually through adsorption in molecular sieves, and loaded into cylinders as pure gases or as mixtures. Pressures can range anywhere from just above atmospheric to approximately 1.4 x 10^7 Pa (2,000 psig). At a semiconductor production facility, the cylinder is stored within a gas cabinet, which provides safety and contains the gas distribution hardware. From the cylinder, the gas passes through a cylinder valve, pressure regulators, tubing, and numerous other valves for flow control, mixing, and purging. Although filtration is increasingly being used during cylinder loading and in gas cabinets, it is not yet a universally accepted practice for all specialty gases.

For either atmospheric or specialty gases, particle evaluations are generally not performed until the final stage in the gas life cycle, before entering the specific semiconductor fabrication process.

PARTICLE MEASUREMENT AND IDENTIFICATION IN HIGH-PURITY GASES

Particle measurement in high-purity gases is most frequently performed to ensure that gas cleanliness is within specification, which generally is defined by an upper concentration limit of particles larger than a specified size. For example, the Semiconductor Equipment and Materials Institute (SEMI) has recently adopted a specification for pipeline nitrogen or argon after filtration to contain no more than 20 particles larger than 0.2 μm per standard cubic foot (SCF) of gas (700 particles/m^3) (SEMI, n.d.). More stringent specifications are often agreed upon between the gas supplier and customer. Indeed, Ohmi's group at Tohoku University claims no particles greater than 0.1 μm in the gas supplied to their "super cleanroom" (Kanno and Ohmi, 1988). In any event, before a newly installed gas delivery system is commissioned, particle concentrations are measured at points-of-use and throughout the system with dynamic particle counters to ensure that particle-level specifications are being met. Increasingly, these counters are also incorporated into continuous on-line monitoring systems to ensure maintenance of gas cleanliness. These specifications and practices for bulk gases are currently being extended by gas suppliers and users to high-purity cylinder gases as well.

Unfortunately, these cleanliness certifications provide little insight regarding particle origins and minimization. This information is essential during upset conditions when gas cleanliness is not within specification. Furthermore, as semiconductor fabrication facilities are upgraded to produce higher density chips, gas cleanliness levels must also be improved to meet the new specifications. In either case, the particle source(s) must be identified and eliminated.

In some cases, dynamic particle counters can be used to isolate particle-generating sources (Davidson, 1987; Davidson and Ruane, 1987; Malczewski, Vorkman, and Vardian, 1986; Borkman and Malczewski, 1986; Malczewski and Goddard, 1986; Thorogood et al., 1986; Cheung and Roberge, 1987; Thorogood and Schwartz, 1986). Particle counters are especially effective in indicating particle sources that are intermittent or time-dependent. For example, by measuring particle concentrations in real-time, bursts of particles in nitrogen gas have been time-correlated with liquid nitrogen deliveries to storage tanks, valve cycling, and mechanical vibrations (Davidson, 1987; Borkman and Malczewski, 1986; Malczewski and Goddard, 1986; Thorogood et al., 1986). Particle counters can also be used to indicate sources that are not time-dependent through comparisons of particle concentrations upstream and downstream of particular components or processes in operation. For example, particle measurements at several locations in a nitrogen distribution system (e.g., ASU, liquid delivery truck, storage tank, vaporizer, filters, piping) may reveal where in a system particles are being introduced (or removed). In laboratory evaluations of gas-handling hardware, highly filtered gas can be passed through a component of interest and particle concentrations measured at the component outlet.

Generally, compositional information on particles is more useful in determining particle origins (Davidson, 1987; Davidson and Ruane, 1987). If the compositions of the materials used in gas production and handling and the compositions of other contamination sources are known, then compositional match-ups can help pinpoint a particle source among several possibilities. For example, in a study on a low-purity nitrogen line, the predominance of Fe - (5 to 10 percent) Ni particles at a point-of-use revealed the source of these particles to be the 9 percent Ni-steel tank used for liquid nitrogen storage (Davidson, 1987).

In the following sections, dynamic particle counting techniques will be reviewed briefly. Particle analysis techniques are reviewed in Chapter 20. The salient features of these techniques that are especially relevant to semiconductor process gas evaluations are highlighted.

Dynamic Particle Counting

Optical particle counters (OPC) are the most commonly used instruments for counting and sizing particles in process gases. The lower detection limit of laser-based OPCs is generally 0.1 to 0.2 μm. Condensation nucleus counters (CNC), capable of counting particles as small as 0.01 μm, are used to a lesser extent for high-purity gas measurement, but their use is steadily increasing. The theory and operation of both OPCs and CNCs are discussed in detail in Chapters 4 and 6. A number of additional concerns related specifically to particle measurement in process gases also need to be considered and are discussed in the following sections.

Sampling Techniques

A major challenge in particle measurement in high-purity process gases is to interface the particle counter with the gas line. Because most particle counters operate at atmospheric pressure with relatively low flow rates, a pressure-reducing sampling system is generally required, and the sample provided is often a small portion of the total flow in the process gas line. The sampling system therefore must withdraw an aerosol that is representative of the gas at the sampling location, without either contributing or trapping particles. Furthermore, the system should be rugged and reliable for field evaluations and must be designed for safe gas handling, especially when used with reactive and toxic gases.

The sampling of aerosols is the subject of continuing intensive research (Cooper, 1986; Fuchs, 1975; Belyaev and Levin, 1974; Liu and Pui, 1985; Tufto and Willeke, 1982). Withdrawing a representative sample and delivering it to a particle counter requires that a number of particle loss mechanisms be minimized:

- **Anisokinetic Losses**. Gas samples are generally specified to be withdrawn with a pitot probe tube parallel to and facing the gas flow, with matched gas velocities in the probe and the pipeline being sampled. These isokinetic vector (direction and velocity) conditions ensure that the sample particle concentrations and size distributions are representative of the gas. Biases due to anisokinetic conditions are a complex function of the gas and sample velocities, the gas density and viscosity, the particle size, and the probe diameter. Sampling probe diameters of 1 cm or more are desirable.

- **Gravitational Losses.** The loss of sampled particles due to settling in the sampling tube is dependent on the nature of gas flow, i.e., laminar or turbulent.

- **Inertial Losses.** Particles with Stokes numbers larger than 0.5 to 1.0 deviate significantly from gas streamlines when the gas direction is changed. Therefore, larger particles tend to impact and adhere to sampling tube walls at bends or constrictions in the tube.

- **Diffusional Losses.** Brownian motion, which becomes more significant for smaller particles (< 0.1 μm), causes particles to randomly deviate from gas streamlines. As these particles approach the sampling system surfaces, they are likely to adhere.

Based on these loss mechanisms, the tubing that connects the probe to the counter should be short, vertical (if possible), and contain as few bends as possible.

Pressure reduction is most commonly achieved through the use of a critical orifice, often in conjunction with an expansion chamber (Pui et al., 1987; MacDermott and Ockovic, 1987; Wen and Kasper, 1987). Pressure reduction with regulators or valves is less desirable because, as will be discussed later, these can introduce additional contamination or trap particles in dead spaces. When sampling atmospheric gases at pressures up to 1 x 10^7 Pa (150 psig), orifices are generally noncontaminating if they are made of suitable materials and are properly designed and cleaned. For sampling cylinder gases at much higher pressures, typically 0.7 to 1.4 x 10^7 Pa (1,000 to 2,000 psig), the development of noncontaminating orifices is not trivial. Selection of the proper orifice material is essential because some materials will shed particles more than others due to the erosive force of the gas. Also, orifices must be carefully cleaned; contamination from a single water drop was reported in one study to result in hours of sample contamination (Wen and Kasper, 1987).

Instrument Selection

The designs of dynamic particle counters are optimized for specific applications or capabilities. Generally, each commercially available counter represents a tradeoff between a number of different features, as shown in Table 10-2. The most striking tradeoff is between sample flow rate and lower particle-size detectability. Recently, instrument manufacturers have developed new designs that enable sampling both at high flows, 4.7 x 10^{-4} m^3/s (1 ft^3/min or 1 cfm), and with small particle size, 0.1 μm, detectability (Knollenberg, 1987).

Instruments are also being developed to measure particles in gases at pressures up to 1 x 10^6 Pa (150 psig). This can be achieved either through the use of closed cells or through internal pressure reduction (Knollenberg, 1987; Nowakowski and Martinez de Pinillos, 1985). Lower particle-size detectability in closed-cell systems is approximately 0.3 μm. For bulk gases, these at-pressure instruments preclude the use of pressure-reducing samplers and may therefore provide a more direct and reliable measurement of the gas. These instruments also make feasible particle measurement in hazardous or reactive gases that could not suitably be sampled with atmospheric pressure counters.

Table 10-2. Characteristics of Laser Optical Particle Counters Used in High-Purity Gas Evaluation

Manufacturer	Model	Flow rate $(ft^3/min)^2$	Pressure $(atm)^b$	Size range (μm)
Particle Measuring Systems, Inc.				
	LPC-HS	0.1	1	0.05 - 3.0
	LAS-X	0.01	1	0.09 - 3.0
	LPC-101	0.1	1	0.1 - 1.0
	LPC-525	1.0	1	0.2 - 5.0
	LPC-110	1.0	1	0.1 - 5.0
	HPLAS	6.7(max)	3-11	0.3 - 5.0
	LPC-101-HP	0.1	3-7	0.1 - 1.0
Climet Instruments				
	CI-6300	1.0	1	0.19 - 5.0
	CI-7500	0.1	1	0.1 - 0.5

[a] 1 ft^3/min (CFM) = 4.719×10^{-4} m^3/s.
[b] 1 atm = 1.013×10^5 Pa.

Data Analysis

The measurement of clean gases is especially difficult with conventional particle counters because particle concentrations are often at, or below, the detection limits of counting instruments. Therefore, appropriate background analysis of the particle counters and sampling systems, and a statistical analysis of the data, are required (Malczewski, Vorkman. and Vardian, 1986; Borkman and Malczewski, 1986; Malczewski and Goddard, 1986; Thorogood and Schwartz, 1986; Kasper and Wen, 1985; Lieberman, 1985; Van Slooten, 1986; Bzik, 1986). Instrument background is generally determined by running the instrument with a high-efficiency filter at the inlet or, as a less desirable alternative, without any sample flow to the instrument. Van Slooten (1986) has detailed the statistical procedure for specifying real particle counts based on subtracting background particle levels from measured levels. A direct result of these analyses is the desirability of higher sampling flow rates for a given instrument's background level.

Time Resolution

Although most commercially available dynamic particle counters are capable of outputting particle concentrations at short time intervals (on the order of seconds), data output is often set by the user at much longer time intervals of several minutes to as much as 1 or 2 hours. The use of long time intervals is a virtual necessity if data are acquired on paper tape. To observe short-lived time-dependent phenomena, e.g., effects of vibrations, counters should be integrated with computers for rapid data accumulation. Data obtained in this fashion are also more suitably formatted for real-time statistical analysis.

Compositional Analysis of Particles

An ideal particle analyzer for process gases would provide not only particle counts and sizes, but also particle compositions in real-time. Some experimental systems are being developed for this purpose (Sneddon, 1986), but their commercialization and application to gas evaluations are not imminent. Consequently, most existing techniques for analyzing particle compositions provide only time-averaged information. With these techniques a large number of particles must be collected over a period of time from the gas stream by filtration, impaction, or liquid scrubbing, and these particles are subsequently analyzed.

The chemical identification techniques that have been or could be applied to particles collected from gas streams are borrowed mostly from environmental pollutant and surface contaminant identification techniques. For the specific case of clean gases, some of these techniques might be unsuitable because a sufficient sample cannot be obtained within a reasonable length of time, or the particles are too small for analysis. Two general types of analysis are considered and are discussed further in Chapter 20: (1) single particle analysis, and (2) bulk analysis, which averages the compositions of all particles in a sample. Single particle analysis provides more detailed information regarding particle origins because its composition is uniquely determined without interference from other particles with different compositions. Also, single particle analysis techniques are often combined with microscopy techniques, which can yield additional information regarding particle physical characteristics and concentrations. Single particle analysis is especially advantageous for clean gas studies where less sample is available. On the other hand, bulk analysis can be less tedious and time-consuming, especially if information is required for large particle populations.

PARTICLE MINIMIZATION

Particle measurement and identification enable the semiconductor technologist to determine particle origins and consequently upgrade the cleanliness of the process gases. Before discussing the conclusions of such studies, it is useful to first consider the gas production and distribution schemes reviewed earlier, and the several possible particle sources or generating mechanisms that require evaluation:

- **Raw Materials.** The raw material for the atmospheric gases—air—contains approximately 10^7 to 10^8 particles > 0.1 μm/m^3 (~ 10^6 to 10^7/ft^3) (Malczewski and Goddard, 1986; Thorogood et al., 1986; Dennis, 1976). For specialty gases, raw material particle concentrations are not well documented, but they are likely to be very high as well.

- **Erosion.** The force of gases flowing at high velocities, in some cases even sonic velocities, may be sufficient to remove loosely adhering debris from the surface of gas-handling hardware; the force may even erode the hardware material itself. Such a mechanism would depend largely on the gas flow patterns (Cooper, 1986).

- **Corrosion.** The reaction of various gases with gas-handling hardware surfaces can yield corrosion products that subsequently can be detached from the surfaces as large flakes and smaller particles. Such particles can be removed from surfaces through gas erosion, mechanical abrasion, or through internal and thermal stresses generated in the corrosion product film or scales (Tien and Davidson, 1975).

- **Wear.** The equipment for gas production and delivery includes many components with moving parts, such as compressors, regulators, and valves. In any such equipment, abrasive or adhesive wear debris can be expected when moving surfaces are in contact (Yust, 1985). Somewhat related to this phenomenon would be the formation of particles through attrition as gases pass through granular powder beds, e.g., adsorbent or catalyst beds (Fenz and Kelleher, 1980).
- **Condensation.** Condensation of particles from the vapor or liquid state is the primary mechanism for the production of natural aerosols. For certain gases, trace hydrocarbons and other impurities present in a gas or removed from hardware surfaces can condense out as particles (Periasamy, 1988; Wen, Kaspar, and Montgomery, 1988).
- **Foreign Particles.** Particles from ambient contamination sources, such as air or fabrication processes, can adhere to the surfaces of gas-handling hardware and subsequently be released into the gas stream.

For atmospheric gases, particles originating from the first source, raw materials, must be considered unlikely to be present at a process gas point-of-use. First of all, a large percentage of inlet airborne particles is removed during gas production. Airborne particles are first removed by the inlet air filters. Further reductions are caused by water sprays that precede air cooling; cryogenic distillation; and the condensation of water, ice, and carbon dioxide crystals on particle nuclei that are subsequently swept into the waste gas in reversing heat exchangers. Consequently, particle concentrations tend to decrease throughout the production route, yielding product gas from ASUs with particle concentrations on the order of 10^3 to 10^4 particles > 0.1 $\mu m/m^3$ (~ 10^2 to $10^3/ft^3$) (Malczewski and Goddard, 1986; Thorogood et al., 1986). Design and process modifications can reduce these levels even further. Through redundant filtration, virtually all of these particles are ultimately removed from the gas before reaching the point-of-use. Although specialty gases have not been similarly studied, we can expect a reduction in raw materials' particles through the purification and filtration processes.

Consequently, the most likely sources of particles at a point-of-use are in some way related to, or affected by, gas-handling hardware. Gas cleanliness must therefore be viewed from a total systems approach, which includes all hardware from production to point-of-use. To test these hypotheses and the hardware-related parameters, particle measurement and analysis studies are performed on individual components under controlled conditions in the laboratory and on complete gas production, storage, and distribution systems in the field. Additional information on hardware surfaces, obtained with SEM and XPS, provides clues as to what surface species can be eroded or the propensity for these surfaces to trap foreign particles. By combining these various surface, component, and systems analyses, complete systems can be designed or upgraded for particle minimization. Some of these design considerations are reviewed in the following sections.

Material Composition

Increasingly, austenitic stainless steels, especially AISI types 316L and 304L, are being specified for gas distribution hardware and, in some cases, for liquefied atmospheric gas delivery trucks and storage tanks (Davidson and Ruane, 1986; Davidson, Sherman, and Irion, 1987; Thomas, 1985; Zawierucha, 1986; Jensen, 1987; Kumar and Dyer, 1987; Anewalt, Drummer, and Martinez de Pinillos, 1984; Boyd, 1986; Lowles, 1985; Kroll, 1984; Kroll and Duran, 1988; Ohmi et al., 1987; Lowles and Tom, 1986). Compositions of these alloys are detailed in Table 10-3. This selection is based primarily on the excellent corrosion resistance, or passivity, of stainless steels, which results from a chromium oxide/oxyhydroxide film that forms on the surface of stainless steels (Sedricks, 1979; Adams, 1983). In mildly oxidizing environments, this stable and tightly adherent film, only 4 nm or less thick, inhibits further reactions between the steel and the environment.

In constantly flowing inert gas systems, such corrosion resistance may seem unnecessary. Tests on straight tubing lengths of electropolished 316L and electropolished Cu under pure nitrogen flow revealed no corrosion-related particle generation for either tube, as shown in Table 10-4 (Davidson and Ruane, 1986; Davidson, Sherman, and Irion, 1987). However, in the real world, systems can be exposed to moisture and other gaseous contaminants during system fabrication, or due to system "trauma" such as leaks, pressure loss, back diffusion, improper purging, and gas supply contamination (Thomas, 1985). In these cases, conventional gas supply systems materials, such as copper, aluminum, and low alloy steels, may readily corrode. Corrosion products on these materials are less adherent than on stainless steels, and they can be released into the gas stream as particles. Table 10-4 illustrates this phenomenon for electropolished 316L and Cu following exposure to a simulated air environment.

Table 10-3. Typical Stainless Steel Alloy Compositions (wt. %)

AISI type	Cr	Ni	Mo	Mn[a]	Si[a]	P[a]	S[a]	C[a]	Fe
304L	18-20	8-12	--	2.0	1.0	0.045	0.030	0.03	Bal
316L	16-18	10-14	2.0-3.0	2.0	1.0	0.045	0.030	0.03	Bal

[a]Maximum levels.

Table 10-4. Concentrations of Particles Larger than 0.1 μm Measured at the Outlet of 20 ft Sections of 3/8 in. Tubing that Have Compositions Matching Those of the Tubing or Corrosion Products

	Particles/SCF at tubing outlet	
Tubing	Before exposure	After exposure
Electropolished 316L	0	0
Electropolished Cu	0	3,000
Mill Finished 316L	7	ND[a]
Mill Finished Cu	7	ND[a]

[a]ND = Not determined.

Note: Filtered nitrogen was the inlet gas. Concentrations on the electropolished tubes were measured before and after exposure to an accelerated atmospheric corrosion test environment.

For more aggressive gases, the alloy composition of gas-handling hardware is becoming of increasing concern, and in some cases stainless steels may not be the most desirable materials (Martinez De Pinillos, Maroulis, and Drummer, 1986). For halogen-bearing gases, nickel and nickel-containing alloys are being considered because of their superior corrosion resistance to these gases in the presence of moisture (Friend, 1976).

High-pressure gas cylinders pose a particularly difficult challenge to the materials engineer (Lowles, 1985; Kroll, 1984; Kroll and Duran, 1988; Ohmi et al., 1987). The cylinder material should not only minimize reactions with the contained gas, but it should also possess the necessary mechanical and physical properties for pressure containment. Aluminum cylinders with specially passivated surfaces are used for many electronic-grade gases except halogens. For these corrosive gases, internally coated or composite cylinders appear to be the most promising containment vessels for contamination minimization.

Surface Finish

In addition to specifying 316L or 304L for gas-handling hardware, the internal surfaces of these stainless steel components are frequently specified to be electrolytically or chemically polished, with a very low surface roughness. The surface roughness usually is defined by an arithmetic mean (R_a) or a root-mean-square (R_{RMS}) of surface displacement from a mean line (Whitehouse, 1974). A number of benefits toward particle minimization can be provided by these polishing processes. First of all, polishing removes machining debris, lubricants, surface oxide scale, and other residual contaminants, all of which might be eroded and entrained into the gas stream, as shown in Table 10-4. Any residual free-metallic (nonpassivated) iron particles are especially deleterious because they may not only corrode, but also lead to accelerated corrosion of the stainless steel upon which they rest. Although polishing does remove these extraneous contamination sources, the electropolishing process tends to bring stringers of nonmetallic inclusions within the steel into surface relief (Kumar and Dyer, 1987). Because even these inclusions are potential particle sources, high-purity steel with a low inclusion rating may be specified. Chempolishing tends to etch out these inclusions.

Finely polished surfaces have also been shown to minimize the trapping of foreign particles, but only under certain conditions. High magnification scanning electron microscopy of surfaces reveals mill-finished surfaces to be highly irregular and porous, with numerous trapping sites for particles, and gaseous/liquid contaminants as well. Chempolishing tends to smooth out this surface, although grain boundary etching is not uncommon. Electropolishing yields the microscopically smoothest surface although, as noted previously, surface depressions appear around the nonmetallic inclusions at the surface. Laboratory tests on air-contaminated mill-finished and electropolished 316L tubes have shown that the mill-finished tube trapped a greater amount of particles, and subsequently released them into a high-purity gas stream over an order of magnitude longer time period (Davidson, Sherman, and Irion, 1987). This effect is illustrated in Figure 10-1. This phenomenon may explain why seemingly clean gas systems may display sudden bursts of particles during pressure or flow surges, or mechanical vibration, and why these bursts are more prevalent in systems with nonpolished surfaces. In contrast, in ongoing work being performed by Stenhouse and Arnold (in preparation), in which tubes were contaminated by polystyrene latex spheres or metal particles carried in dry air, no differences were observed between mill-finished and electropolished tubes.

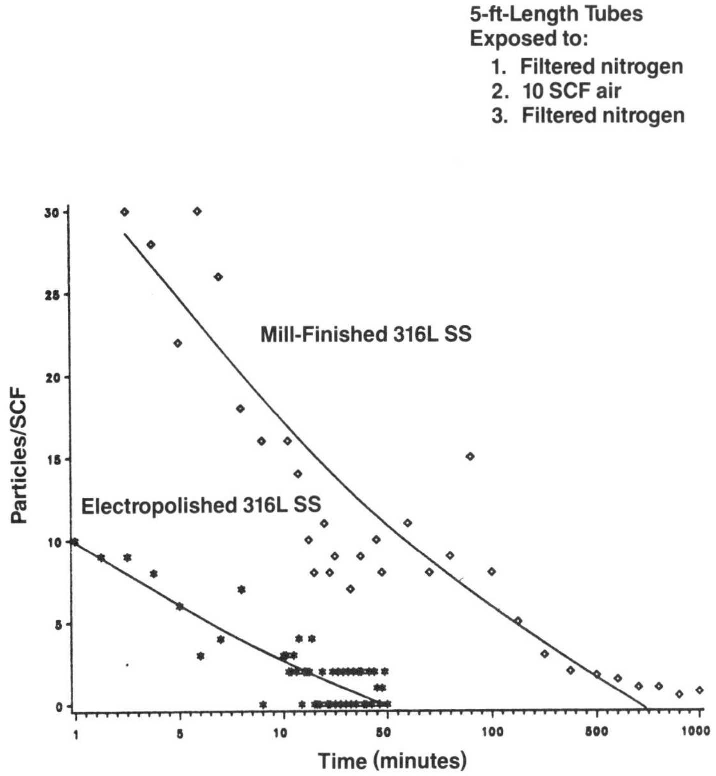

Figure 10-1. Concentrations of particles larger than 0.2 μm measured at the outlet of 5 ft sections of 3/8 in. diameter 316L SS tubes, using filtered nitrogen as the inlet gas. Tubes were previously exposed to 10 SCF of flowing air containing approximately 200,000 particles/SCF.

Finally, polishing processes, especially electropolishing, enhance surface passive film formation (Jensen, 1987; Kumar and Dyer, 1987; Sherman, Goto, and Davidson, in preparation). As detailed in Table 10-5, this enhancement is most accurately revealed through XPS analyses as surface Cr enrichment and higher Cr/Fe ratios. Consequently, chemically or electrolytically polished 316L surfaces exhibit corrosion resistance superior to mill-finished or bright-annealed surfaces (Seo and Sato, 1980).

Despite this advantage, polishing solutions also can impart impurities into the surface film. Of particular concern are S and P impurities that can be slowly released into the process gas or, if bound with water molecules, can cause localized corrosion within surface crevices or grain boundaries. Frictional cleaning may remove some of these impurities, but specially designed cleaning treatments have been shown to be more effective (Kumar and Dyer, 1987; Sherman, Goto, and Davidson, in preparation). Figure 10-2 illustrates the removal of S and P impurities from an electropolished 316L surface by a subsequent surface treatment.

The above discussion applies to all gas-handling hardware surfaces exposed to process gas streams. Although most attention to these materials and surface parameters initially focused on tubing and piping, which account for the major portion of gas-exposed surface area, such criteria are also important for components such as valves, regulators, filter housings, fittings, etc. In addition to these effects, many components have additional characteristics that can affect particulate concentrations.

Table 10-5. Elemental Analyses of Inner Diameter Surfaces of Type 316L Tubes with Different Processing Histories and from Different Vendors

Tubing	Cr	Fe	Ni	Mo	O	C	ΣP_i[a]	Cr/Fe
Mill-Finished Vendor 1	4.6	4.3	0.4	0.4	38.1	47.9	4.3	1.1
Mill-Finished Vendor 2	2.4	1.7	0.1	1.0	31.6	49.0	14.4	1.4
Chempolished Vendor 2	5.2	3.3	0.2	0.4	48.5	36.5	7.6	1.6
Chempolished Vendor 3	7.5	5.5	0.5	0.5	44.9	37.5	4.7	1.4
Electropolished Vendor 1	4.8	2.4	0.5	0.5	35.1	47.1	9.5	2.0
Electropolished Vendor 2	8.2	3.6	0.3	0.5	48.7	31.1	7.6	2.2
Electropolished Vendor 4	6.8	3.9	0.5	0.4	43.1	41.1	3.9	1.7

[a]Sum of process-related impurities.

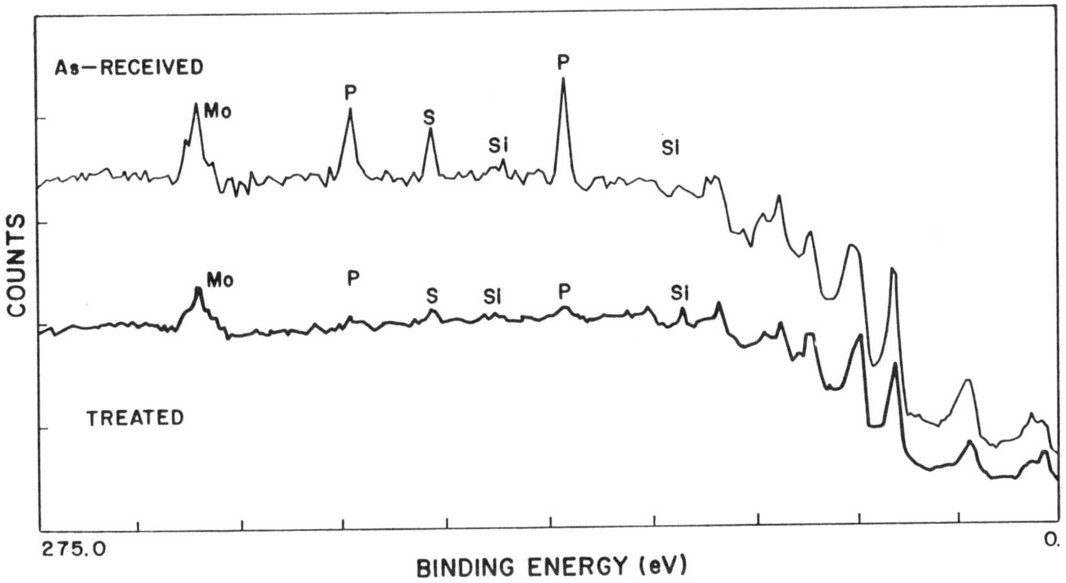

Figure 10-2. XPS spectra obtained from 316L SS in the (a) as-electropolished condition and (b) electropolished and surface-treated condition. S and P contaminants are reduced by subsequent surface treatment.

Valves

The particle-generating characteristics of valves depend on the valve design, manufacturing and packaging processes, and on the service applications (Davidson, Sherman, and Irion, 1987; Govers et al., 1986). Table 10-6 shows particle concentrations measured downstream of several valves cycled between the fully open and half-open, or throttling, positions. Filtered nitrogen was used as the inlet gas. Most valves exhibit maximum particle generation when the valve is between the fully open and closed positions. When fully open, valves exhibit either no (Davidson, Sherman, and Irion, 1987) or minimum (Govers et al., 1986) particle generation. Valve evaluations have indicated that these particles can result from frictional wear in the moving parts or sliding mechanisms, or from lubricants or other foreign contaminants on the gas-exposed surfaces. These sources can be minimized through proper engineering design and careful cleaning and packaging procedures. The dead-zones in valves also can be efficient trapping sites for particles, which can later be re-released into the gas stream (MacDermott and Ockovic, 1987; Ohmi et al., 1987). Recently, valves designed specifically for high-purity gas service have been shown to be virtually contamination-free (Davidson, Sherman, and Irion, 1987; Ohmi et al., 1987).

Table 10-6. Concentrations of Particles > 0.1 μm/SCF at Outlet of Valves in Full Open Position, During Initial Throttling, and Over 16 h of Throttling

Valve	Particles > 0.1 μm/SCF		
	Fully open	Initial throttle	16 h throttle
Ball			
Vendor 1	2.2	362.2	8,139.1
Bellows			
Vendor 2	1.1	887.4	166.2
Bellows			
Vendor 3			
"Off the shelf"	6.8	24.0	100.5
Cleaned	2.1	26.3	14.8
EP & Cleaned	1.2	29.3	2.6
Diaphragm			
Vendor 1			
With purge ports	1.2	11.6	4.0
Without purge ports	1.2	2.0	1.8
Diaphragm			
Vendor 2	4.1	266.0	31.7
Diaphragm			
Vendor 3	1.9	2.9	2.3
Diaphragm			
Vendor 4	1.5	1.5	1.4

Filters

Filters are certainly the most critical components in gas delivery systems and the most effective means for minimizing particle concentrations at points-of-use. As noted earlier, they are used not only at points-of-use, but also at other locations within the gas storage and distribution route. Filtration mechanisms and filter characteristics are discussed in Chapter 7 and Chapter 16. Some additional concerns should be considered, and possibly evaluated, for high-purity semiconductor gases:

- The manufacturing and packaging processes can leave residual contaminants on filters. When new, filters can shed these particles over periods of several hours or even days, depending on gas face velocity (Davidson, Sherman, and Irion, 1987). Precleaning treatments can eliminate this particle source.
- Particles may also be released from the filter housing and associated downstream tubing (Kasper and Wen, 1987; Grant, 1987). Such particles can be erroneously attributed to filter shedding.
- Although filters may remove particles, they also may outgas intolerable levels of chemical impurities into the high-purity gas stream. For this reason, bakeable ceramic filters are being considered for some demanding applications (Ohmi et al., 1987).
- Filters are often tested in inert gases or air at pressures up to approximately 7×10^5 Pa (100 psig). Questions remain regarding the suitability and efficiency of present filters for reactive gases, or at high pressures and pressure drops. The effects of mechanical shock likewise need to be carefully evaluated.

Tube Bending

The processes used in gas system fabrication, although frequently overlooked, are a common source of contamination. For example, construction of gas-handling systems requires an extensive number of bends. However, the bending of finely polished tubing can significantly increase the surface roughness through the introduction of slip steps, deformation twins, and microtearing. Again, these surface features become effective particle-trapping sites. Although bending also has been suggested to destroy the passive Cr-rich film on 316L surfaces (Jensen, 1987; Kumar and Dyer, 1987), recent studies have indicated no major changes in passivity after bending (Sherman, Goto, and Davidson, in preparation).

Large quantities of particles also can be generated during the bending process. Compositional analyses of these particles indicate that they originate from tubing surface species, such as oxide scale or nonmetallic inclusions (Gerristead, Sherman, and Davidson, 1987). As detailed in Table 10-7, this particle generation is greatest for mill-finished tubes, less for chem-polished tubes, and least for electropolished tubes (Davidson, Sherman, and Irion, 1987). Because of these characteristics of bending, the use of pre-polished elbows, or the in situ polishing of systems after fabrication are often considered (Kumar and Dyer, 1987).

Joining

As with tube bending, the primary concerns with joining of high-purity gas-handling components are the effects of the process on surface morphology and chemistry, and particle generation during the process. Based on both of these considerations, stainless steels are

Table 10-7. Concentration of Particles > 0.1 μm/SCF in Gas Flowing at 180 SCF Through 3/8 in. 316L Tubes During Bending, and Time Required After Bending to Return to Original Background

Tubing	Bend number	Particles > 0.1 μm/SCF during bending	Time after bending to return to original background
Electropolished 316L	1	33.3	60 s
	2	33.3	0 s
	3	0	0 s
	4	100	0 s
Chempolished 316L	1	300	0 s
	2	50	0 s
	3	450	30 s
	4	400	6 s
Mill-finished 316L	1	432,733	>48 h
	2	119,800	>48 h
	3	136,300	>48 h
	4	409,033	>48 h

again desirable, in that they are compatible with clean gas-tungsten arc welding (GTAW) procedures. GTAW is capable of producing relatively smooth, autogenous joints on stainless steel components, without requirements for potentially contaminating filler metals or fluxes. In contrast, copper systems must be brazed or soldered, and the fluxes and low melting point metals are a primary source of contamination in these systems (Davidson, Sherman, and Irion, 1987; Thomas, 1985). Furthermore, GTAW of tubular components can be performed with automatic orbital welding machines, which provide, on a macroscopic scale, consistently smooth welds without pockets or spatter that can occur with other welding systems (Henon and Overton, 1988). On a microscopic scale, however, even the smoothest GTAW surface will have a greater surface roughness than the unwelded polished surface, and welding parameters need to be adjusted to minimize this roughness. Finally, the inert gas shielding used in GTAW limits oxidation of the weld, although XPS studies have shown some loss of passivity at the weld and within the heat affected zone (Jensen, 1987). Passivity is impaired to a lesser extent if higher purity shielding gases are used (Davidson, Sherman, and Irion, 1987).

GTAW is also considered an ultraclean joining process relative to other welding processes because it limits the generation of welding fumes, which are essentially aerosols of micrometer and submicrometer particles (Speight and Campbell, 1979). Most of these comparisons were made on the basis of bulk collection techniques. However, with the use of OPCs, even GTAW of 316L is found to generate large numbers of particles (Davidson, Sherman, and Irion, 1987). SEM/microprobe analyses of these particles indicate that they are predominantly manganese-rich oxides, likely resulting from the reaction of metallic Mn fumes with residual oxygen in the piping system and the purge gas. As illustrated in Figure 10-3, the use of ultrahigh-purity shielding gases reduces this particle generation, whereas any leaks of air into the shielding gas significantly increase particle levels (Davidson, Sherman, and Irion, 1987).

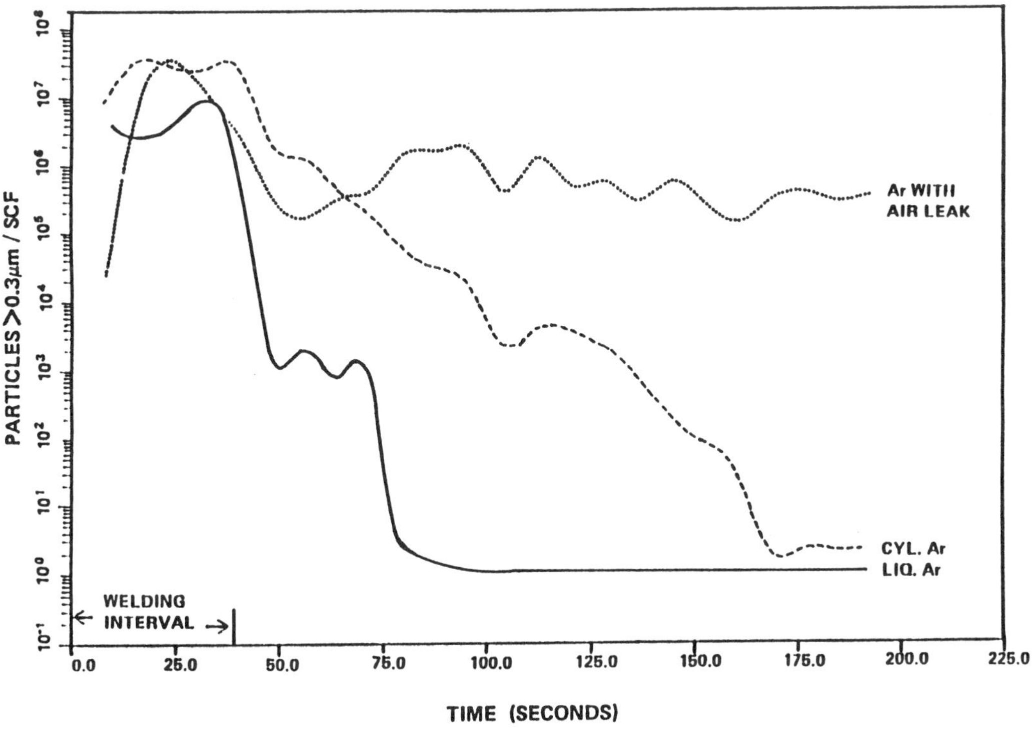

Figure 10-3. Concentrations of particles larger than 0.3 μm in argon shielding gas during gas-tungsten arc welding.

An additional consideration in the welding of stainless steels is the minimization of grain boundary sensitization, or chromium depletion, in the heat-affected zone of the weld. In this condition, the steel becomes very susceptible to intergranular corrosion attack (Sedricks, 1979). The most convenient method for preventing welding sensitization is to use the "L" (low carbon) grades of stainless steels, in which carbon levels are kept to less than 0.03 wt.%.

Cleaning

The post-fabrication cleanup of gas delivery systems warrants much further research and development. Although sophisticated techniques are used to ensure the cleanliness and purity of component surfaces, fabrication procedures can degrade and contaminate these systems. In the past, a commonly accepted technique for cleaning and/or final inspection was to force a "lint-free" cloth plug, soaked with solvent, through tubing and fittings. These components were considered clean if the plug appeared clean to the naked eye. Not only is this technique incapable of detecting even substantial amounts of submicrometer particles, but it also may contaminate the system with chemical residues from the solvent (Thomas, 1985). High-purity gas purging also may be performed before commissioning a system. With on-line particle detection systems, this purging can be performed until residual particles are removed. Figure 10-4 shows a typical example of particle concentrations measured at an unfiltered point-of-use on a nitrogen distribution system during initial startup and after 1 week of purging. On conventional nonstainless steel systems, purging techniques may not be feasible if continuously generating particle sources exist within the system.

Total Systems

Clearly, particle concentrations in gases can be affected by numerous parameters related to the gas-handling system, such as materials, surface finish, fabrication techniques, and component design. Gas cleanliness therefore is defined by the most contaminating portion of the system. By approaching gas cleanliness from this total systems point of view, each of these parameters can be rationally selected or designed for total particle minimization.

FUTURE DIRECTIONS

Particle control in semiconductor process gases relies on two distinct but interactive technologies of particle measurement/identification and particle minimization. State-of-the-art capabilities in both areas enable current particle specifications to be easily measured and achieved. However, in the future, the need to reach even lower particle concentrations is anticipated, and with specifications set for particles much smaller than the present 0.1 to 0.2 μm limits. Furthermore, current particle specifications for bulk gas systems are being extended to gases that are at higher pressure and that are reactive or toxic. Such requirements demand continued improvements in particle analysis techniques and in gas-handling materials and processes evaluated with these techniques. Finally, although this chapter has focused on particle contamination in gases, chemical impurities must be recognized as an equally important form of contamination that also must be analyzed and minimized to far lower levels than in the past. Also, both forms of contamination can be interrelated; filters used to remove particles can add chemical impurities, whereas purifiers use to remove trace impurities can add particles to the gas. Therefore, gas purity must be considered from a thoroughly integrated approach in both analyses and gas system design.

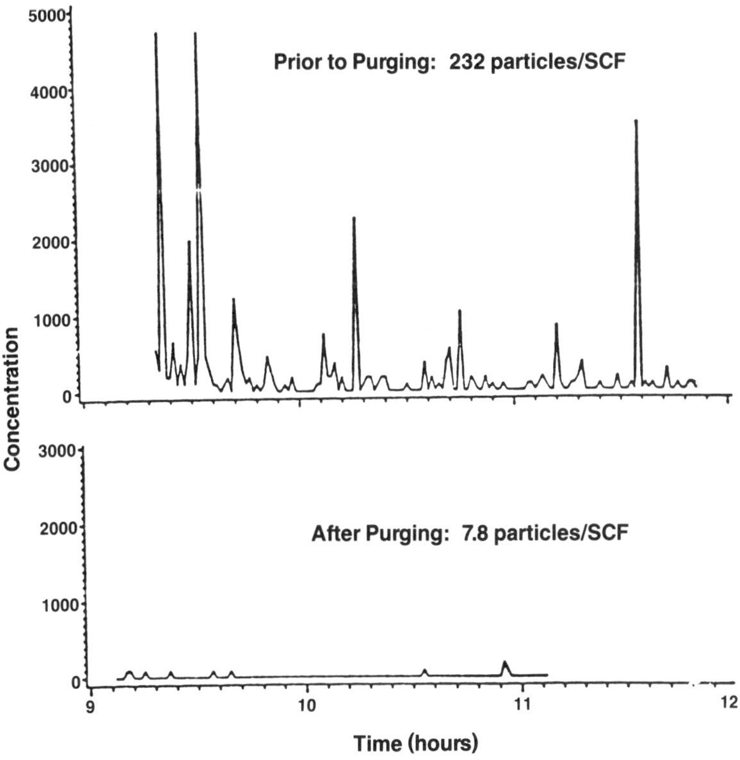

Figure 10-4. **Concentrations of particles >0.1 μm/SCF at an unfiltered point-of-use (a) during initial purging and (b) after 1 week of purging of a new nitrogen storage and distribution system.**

REFERENCES

Adams, R.O. 1983. A Review of the Stainless Steel Surface. *J. Vac. Sci. Tech.*, A1(1):12-18.

Anewalt, M.K., D.M. Drummer, and J.V. Martinez de Pinillos. 1984. Surface Studies of Semiconductor Processing Piping: 316L Stainless Steel. In: *Proceedings - Inst. Env. Sci. 1984*, pp. 32-43.

Belyaev, S.P., and L. M. Levin. 1974. Techniques for Collection of Representative Aerosol Samples. *Aerosol Science*, 5:325-338.

Borkman, J.D., and M.L. Malczewski. 1986. High Purity Nitrogen Pipeline Certification. In: *Proceedings of Semicon/West '86*, pp. 194-203.

Boyd, H. 1986. Achieving Future Requirements in VLSI Gases. *Semiconductor International*, Aug., pp. 72-75.

Bzik, T.J. 1986. Statistical Analysis of Data for Ultraclean Systems. In: *Proceedings - Microcontamination Conf. 1986*, pp. 49-61.

Cheung, S.D., and R.P. Roberge. 1987. An Inside Look at Measuring Particles in Process Equipment. *Microcontamination*, 5(5):45-50, 90.

Cooper, D.W. 1986. Minimizing Bias in Sampling Contaminant Particles. *Solid State Tech.*, 29(1):73-79.

Davidson, J.M. 1987. Determination of Particle Origins in High Purity Gases. *Solid State Tech.*, 30(7):63-67.

Davidson, J.M., and T.P. Ruane. 1986. Gas Handling Hardware: Considerations for Ensuring Gas Purity. In: *Proceedings - Semicon/West '86*, pp. 185-193.

Davidson, J.M., and T.P. Ruane. 1987. Particle Control in VLSI Gases. In: *Emerging Semiconductor Technology*, ASTM-STP 960, D.C. Gupta and P.H. Langer, eds. American Society for Testing and Materials, Philadelphia, PA.

Davidson, J.M., R. Sherman, and P. Irion. 1987. Particle Minimization in Gases Used in IC Fabrication. Proceedings of the 3rd symposium on Automated Integrated Circuits Manufacturing, Vol. 88-3, pp. 136-149, the Electrochemical Society, 10 South Main St., Pennington, NJ 08534-2896.

Dennis, R., ed. 1976. *Handbook on Aerosols*. Technical Information Center, U.S. Energy Research and Development Administration.

Duffalo, J.M., and J.R. Monkowski. 1984. Particle Contamination and Device Performance. *Solid State Tech.*, 27(3):109-114.

Fenz, F.A., and E.G. Kelleher. 1980. Studies of Attrition Rates in Fluid-Particle Systems via Free Fall, Grid Jets, and Cyclone Impact. *J. Powder & Bulk Tech.*, 4(2/3):13-20.

Friend, R. 1976. *Corrosion of Nickel and Nickel-Base Alloys*. John Wiley & Sons, New York, NY.

Fuchs, N.A. 1975. Sampling of Aerosols. *Atm. Env.*, 19:697-707.

Gerristead, W.R., Jr., R. Sherman, and J.M. Davidson. 1987. Automated SEM/Microprobe Finds Gas Contaminants, Determines Sources. *Research & Development*, 29(6):93-100.

Govers, T., G. Torelli, D. Gary, and C. diGiulio. 1986. Particle Generation by 1/4" and 3/8" Manual Valves for High Purity Gas Service; A Comparative Evaluation. In: *Proceedings – Microcontamination Conf. 1986*, pp. 105-110.

Grant, D.C. 1987. Particle Shedding from Semiconductor Process Gas Filters. In: *Proceedings – Inst. Env. Sci. 1987*, pp. 370-374.

Henon, B.K., and J.S. Overton. 1988. Design Planning for Class 1 UHP Stainless Steel Process Gas Piping System. *Microcontamination*, 6(2):43-48.

Jensen, D. 1987. Reducing Microcontamination Generation and Entrapment Within High Purity Gas Distribution Systems. *Microcontamination*, 5(5):52-62, 97.

Kanno, Y., and T. Ohmi. 1988. Components Key to Developing Contamination Free Gas Supply. *Microcontamination*, 6(12):23-30, 56.

Kasper, G., and H.Y. Wen. 1985. Particle Technology for High Purity Gases. In: *Proceedings - Inst. Env. Sci. 1985*, p. 106.

Kasper, G., and H.Y. Wen. 1987. On the Origin of Residual Particles After HP Gas Filtration. In: *Proceedings – Inst. Env. Sci. 1987*, pp. 375-377.

Knollenberg, R.G.. 1987. Sizing Particles at High Sensitivity in High Molecular Scattering Environments. In: *Proceedings - Inst. Env. Sci. 1987*, pp. 428-435.

Kroll, W. 1984. Contamination Prevention and Protection for Process Gases. *Solid State Tech.*, 27(5):220-227.

Kroll, W.J., and J.A. Duran. 1988. Semiconductor Cylinder Gas Purity. *Microelectronic Man. and Test.*, Feb., pp. 28-30.

Kumar, A., and C.E. Dyer. 1987. Characterization of Surface Treated 316L Stainless Steel Tubings to Prevent Contamination in Gas Distribution Systems. *Solid State Tech.*, 30(2):89-94.

Latimer, R.E. 1967. Distillation of Air. *Chem. Eng. Prog.*, 63(2):35-59.

Lieberman, A. 1985. Perspectives on Particle Counting Technology. *Microcontamination*, 3(3):31-40.

Liu, B.Y.H., and D.Y.H. Pui. 1985. A Universal Aerosol Sampling Probe for High Efficiency Sampling from Flowing Gas Streams. *Part. Charact.*, 2:125-132.

Lowles, D.C. 1985. Introduction to Improved Gas Quality for High Technology Industries. In: *Proceedings - Microcircuit Pure Materials Conf. 1985*, pp. 214-225.

Lowles, D.C., and G.M. Tom. 1986. Improving Process Gas Purity Through System Design and Testing. *Microcontamination*, 4(12):35-39.

MacDermott, W.T., and R.C. Ockovic. 1987. Particle Deposition in Pressurized Gas Sampling Systems. In: *Proceedings - Inst. Env. Sci. 1987*, pp. 392-399.

Malczewski, M.L., and J.B. Goddard. 1986. Particle Analysis in Nitrogen: A Case Study. In: *Proceedings Microcontamination Conf. 1986*, pp. 81-88.

Malczewski, M.L., J.D. Vorkman, and G.T. Vardian. 1986. Measurement of Particulates in Filtered Process Gas Streams. *Solid State Tech.*, 29(4):151-157.

Martinez de Pinillos, J.V., P.J. Maroulis, and D.M. Drummer. 1986. Metals in Corrosive Chemicals: An Overview. In: *Proceedings - Microcontamination Conf. 1986*, pp. 80-104.

Nowakowski, E.E., and J.V. Martinez de Pinillos. 1985. In-Situ Monitoring of Particles in Gases. In: *Proceedings - Microcontamination Conf. 1985*, pp. 160-176.

Ohmi, T., J. Murota, Y. Kanno, Y. Mitsui, K. Sugiama, T. Kawasaki, and H. Kawano. 1987. Ultra Clean Gas Delivery System for VLSI Fabrication and Its Evaluation. Presented at Electrochemical Society Spring Meeting, Philadelphia and to be published in the *Proceedings*.

Osburn, C.M., H. Berger, R. Donovan, and G. Jones. 1988. The Effects of Contamination on Semiconductor Manufacturing Yield. *J. Inst. Env. Sci.*, 31(2):45-57.

Periasamy, R., F.L. Chen, D.S. Ensor, R.P. Donovan, and R. Denyszyn. 1988. Particles in High Pressure Cylinder Gases: Gas to Particle Conversion During Expansion. *J. Env. Sci.*, 31(4):20-25.

Pui, D.Y.H., F. Romay-Novas, S.Z. Wang, and B.Y.H. Liu. 1987. Evaluation of Pressure Reducing Devices for High Purity Gas Sampling. In: *Proceedings - Inst. Env. Sci. 1987*, pp. 388-391.

Ruheman, M. 1949. *Separation of Gases*. Oxford University Press, Cambridge, England.

Sedricks, A.J. 1979. *Corrosion of Stainless Steels*, John Wiley & Sons, New York, NY.

SEMI. n.d. SEMI Particle Specification for 20/0.2 Nitrogen or Argon Delivered as Pipeline Gas. Semiconductor Equipment and Materials Institute, Mountain View, CA.

Seo, M., and N. Sato. 1980. Surface Characterization of Stainless Steel Prepared with Various Surface Treatments. *Trans. Japan Inst. Met.*, 21(12):805-810.

Sherman, R., M. Goto, and J.M. Davidson. (In preparation). Surface Evaluations of Tubing for High-Purity Gas Handling.

Sneddon, J. 1986. Direct and Near Real Time Determination of Metallic Compounds in the Atmosphere by AA. *Amer. Lab.*, March 1, pp. 43-50.

Speight, F.Y., and H.C. Campbell, eds. 1979. *Fumes and Gases in the Welding Environment*. American Welding Society, Miami, FL.

Stenhouse, J.I.T., and L. Arnold. (In preparation). Particle Contamination of Process Gases from Pipe Internal Surfaces.

Thomas, R.C., 1985. Non-Contaminating Gas Distribution Systems. *Solid State Tech.*, 28(9):153-158.

Thorogood, R.M., and A. Schwartz. 1986. Performance Measurement of Gas Ultrafiltration Cartridges. In: *Proceedings - Inst. Env. Sci. 1986*, pp. 460-467.

Thorogood, R.M., A. Schwartz, W.T. McDermott, and C.D. Holcomb. 1986. Production of Ultrapure Nitrogen for the Electronics Industry. *Microcontamination*, 4(8):29-34.

Tien, J.K., and J.M. Davidson. 1975. Oxide Spallation Mechanisms. In: *Stress Effects and the Oxidation of Metals*, pp. 200-219. AIME, New York, NY.

Tolliver, D.L. 1984. Contamination Control: New Dimensions in VLSI Manufacturing. *Solid State Tech.*, 27(3):129-140.

Tufto, P.A., and K. Willeke. 1982. Dependence of Particulate Sampling Efficiency on Inlet Orientation and Flow Velocities. *Am. Ind. Hyg. Assoc. J.*, 43:436-443.

Van Slooten, R.A. 1986. Statistical Treatment of Particle Counts in Clean Gases. *Microcontamination*, 4:32-38.

Wen, H.Y., and G. Kasper. 1987. Particle Analysis in Cylinder Gases. In: *Proceedings - Inst. Env. Sci. 1987*, pp. 400-402.

Wen, H.Y., G. Kasper, and D. Montgomery. 1988. Nucleation of Trace Amounts of Condensible Vapors in an Expanding Gas Jet. *J. Aerosol Sci.* 19(1), pp. 153-156.

Whitehouse, D.J. 1974. The Measurement and Analysis of Surfaces. *Trib. Int.*, Dec., pp. 249-259.

Yust, C.S. 1985. Tribology and Wear. *Int. Met. Rev.*, 30(3):141-154.

Zawierucha, R. 1986. High Purity Materials for Gas Distribution Systems. *Microelectronic Man. and Test.*, Feb., pp. 16-19.

11. PARTICLES IN PROCESS LIQUIDS

GILBERT DRAB, JIM TICHICH, and LES NICHOLS Ashland Chemical, Dallas, Texas

INTRODUCTION

Particle control in high-purity chemicals has become a major concern in the past 3 to 5 years as the electronic industry progressed toward 256K to 4M memory devices with smaller line widths and larger chip sizes.

Particles and dissolved metals in chemicals cause special concern because their diffusion into the silicon substrate can cause changes in electrical properties. Maximum concentrations for a wide range of metallic impurities have been incorporated by some of the major chemical suppliers in their low particle-grade chemicals. This product line provides the electronic industry with a more consistent chemical to use in their processes, which results in higher yields.

New manufacturing techniques and process controls incorporated by major chemical suppliers and processors in the past 4 years have made major improvements in the quality of process chemicals used in the semiconductor industry. These improvements include reduction of metallic ions, major reductions in particulate levels in the 1.0 and 0.5 μm range, and more consistent products.

Packaging used for process chemicals has also made improvements in the past 4 years. Resin used for bottles and drums have tighter control through statistical process control (SPC) programs that ensure chemical purity. Moreover, specialty-designed bottles and dip tube drum systems have enhanced cleaning and improved safe handling of process chemicals.

Quality control and SPC programs are being incorporated in chemical manufacturing and packaging processes to ensure that product quality meets new rigid specifications incorporated by aggressive chemical suppliers. These programs are being initiated to provide semiconductor manufacturers with chemicals to meet their needs today and in the near future so that they can be competitive in the worldwide marketplace.

CHEMICAL MANUFACTURING

All of the inorganic acids used in semiconductor manufacturing are either water solutions or result from the reaction of products with water. The use of various sources of water, ranging from city water to filtered 18 megohm water, gives rise to not only different degrees of chemical purity but also different levels of particles in the final product.

The material of construction for reactors, pumps, transfer lines, absorbers, and storage tanks affects the particle levels for all chemicals produced. The former practice of merely isolating a side stream from a reactor for use as semiconductor-grade material is no longer acceptable as demands for chemical purity increase and particle levels must decrease. All-glass or all-Teflon™ systems may replace the steel and composite polymer systems now used to manufacture these chemicals.

Impurities in the basic raw material used to manufacture semiconductor chemicals can contribute or lead to the formation of insoluble or high molecular weight components in the final purified product.

PACKAGING

Selection of a package for semiconductor chemicals presents several choices for material of construction (glass, polyethylene, and PFA, TEF, or FEP fluoropolymers). The material used to make a package is chosen to optimize the parameters of cost, chemical compatibility, and (only recently) the particle-generating characteristics of the package's surface. Fluoropolymers have been recognized for chemical stability, but we are finding that their surfaces are sources of particles. Selecting a container size ranging from 1 pint to a tankwagon presents us with the dilemma of choosing between a small container having a large surface-to-volume ratio but good manufacturing control and a large container with a small surface-to-volume ratio but less manufacturing control of the surface.

Manufacturing Specifications

Manufacturing specifications have been found to be inadequate for characterizing what constitutes a good package for high-purity semiconductor-grade chemicals. Different suppliers can make materials having significantly different characteristics but all meeting a common "specification."

The process of shaping the container contributes to different particle-shedding characteristics. For example, blow molding and rotary molding processes result in different surface characteristics. Blow molding produces a much smoother and uniform surface over rotary molding, which in turn is much easier to clean for packaging high-purity chemicals.

CHEMICAL SYSTEMS MATERIAL SELECTION

Selecting the proper material of construction in the packaging systems for pipelines, surge tanks, and fill manifolds is critical in maintaining the chemical purity and reducing particle shedding during processing. Table 11-1 lists acceptable materials of construction for some of the most widely used chemicals in manufacturing semiconductor devices.

CONTAINER CLEANING

High-purity filtered demineralized water must be used to clean containers for packaging. (See Figure 11-1 for a photograph of an automatic bottle washer.) Air used to remove dust and extraneous material from a bottle or other containers must also be filtered. The environment in which the cleaning process takes place must also be controlled to ensure that a clean bottle or container remains clean until it is filled with the product and sealed. Because of the difficulty of handling larger containers such as drums, different techniques and equipment are used to clean these.

FILLING

The environment in which a clean package is filled with a clean chemical is critical. These environments include open air, clean hood, and clean-room types and can be manual or automated. Automated clean-room facilities give the best control of a packaging environment (Figure 11-2).

Table 11-1. Suggested Materials for Construction of Selected Chemical Process Systems

	Elastomers					Piping Materials				
Chemicals	Viton A	Ethylene propylene	Teflon™	304 L Stainless Steel	316 Stainless Steel	Aluminum[a]	PVDF[b]	Poly-propylene	Teflon™	ABS[c]
Acids										
Sulfuric Acid (Conc.)	•		•				•		•	
Phosphoric Acid (Conc.)	•	•	•	•	•					
Nitric Acid (Conc.)	•		•				•		•	
Acetic Acid (Glacial)		•	•		•					
Hydrochloric Acid (Conc.)	•		•				•			
Hydrofluoric Acid (49%)		•	•				•	•	•	
Solvents										
Acetone		•	•	•						
n-Butyl Acetate		•	•	•						
Isopropyl Alcohol	•	•	•	•						
Methanol Alcohol		•	•	•						
Xylene	•		•	•						
Deionized Water	•		•				•		•	•
Hydrogen Peroxide	•		•		•	•		•	•	

• = Acceptable material.
[a] Aluminum = Type 1060 for hydrogen peroxide service.
[b] ABS = Dura Plus ABS Industrial; acrylonitrile butadiene styrene.
[c] PVDF = Polyvinlidene fluoride.

Figure 11-1. Automatic bottle washer.

Figure 11-2. Automatic fill station.

Once the filling process has begun, the batch size becomes a factor. The shorter the time that the equipment runs, the higher the percentage of material that is likely to contain large amounts of particles. That is, filter efficiency tends to increase as packaging runs proceed.

Filling lines are critical in maintaining low particle levels in chemicals being transferred from filters to packages. The length of these lines, as well as the material of construction, affects the particle levels of the product at the fill nozzle. Recirculation of the product can reduce the particles in the chemical to a desired level before packaging.

Caps for packages can be a source of particles due to the way they are applied and the type of liner used. Cap liners must be compatible with chemicals packaged and must provide a tight sealing surface. Cap liners used by most packaging suppliers have a very high cold flow characteristic that causes the cap to become loose in storage in a relatively short period.

Process control of particles in packaged chemicals is also critical. Three approaches can be used. The first and most ideal is using a continuous monitoring system, but current technology cannot do this. Second, grab samples of packaged product can be taken and analyzed within minutes of being packaged. In a third approach, material can be packaged and samples can be taken according to a quality assurance plan.

HOLDING/SHIPPING CONTAINERS

Bulk storage tanks are the most extensively used systems to store or stage process chemicals by the chemical industry (Figure 11-3). These systems are dedicated by chemical and are constructed of special materials to maintain the purity of these chemicals. Selection of material of construction, valve types, and elastomer gaskets is very important to maintain cation and anion stability of stored chemicals. High-purity chemicals that are hydroscopic require special purge systems such as nitrogen for bulk storage.

Trailer Tankers/Rail Cars

Trailer tankers/rail cars varying in capacity from 10 to 80 m^3 (3,000 to 20,000 gal) are the most common method for transporting chemicals from a manufacturer to a chemical processor (Figure 11-4). Chemical contamination during this movement is the most common cause of quality deterioration and could result from:

- Tankers or rail cars manufactured from incorrect material
- Tankers not properly cleaned between loads of different products
- Tankers or rail cars contaminated by dust or dirt
- Valves made out of materials incompatible with chemicals
- Gaskets and valve seals made out of improper elastomers

Figure 11-3. Bulk storage tank farm.

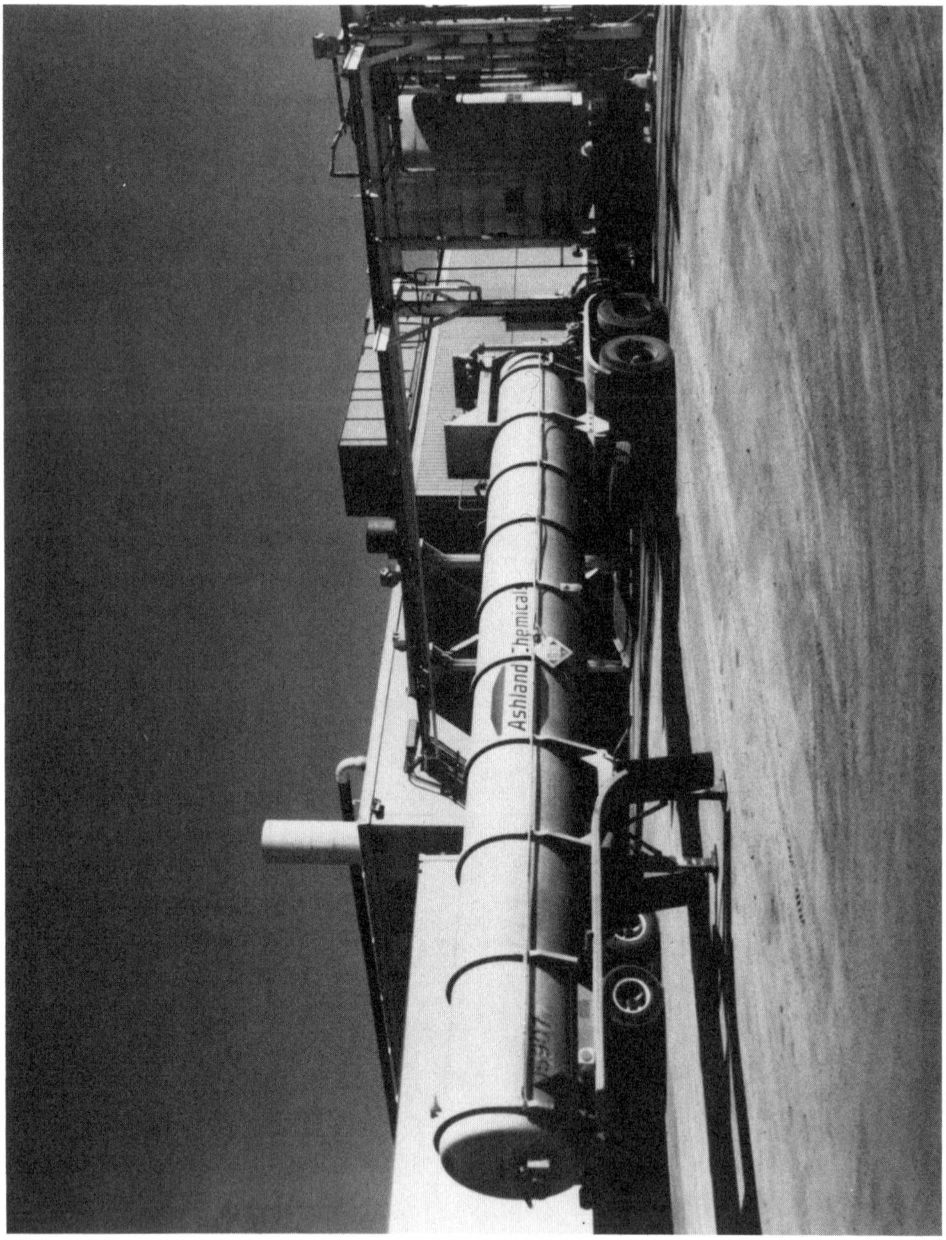

Figure 11-4. 12 m^3 (3,200 gal) sulfuric tanker.

- Transfer hoses that are not dedicated, are made out of wrong material, or are dirty
- Improper design of loading and unloading transfer systems.

Bulk Transfer Systems

Bulk transfer systems used to load or unload bulk chemicals are subject to the same factors affecting chemical purity in trailer tankers and rail cars.

Personnel training is very important in chemical transfer and transportation operation. Maintaining the cleanliness of equipment and using proper techniques in transferring chemical to tankers or rail cars minimize contamination from the outside environment.

Moreover, bulk transfer stations should be dedicated by product to ensure there is no contamination from using multipurpose transfer systems. It takes a very small amount of contaminant chemical to produce off spec product and usually takes a large volume of chemical to flush out contamination from an unlikely source.

FILTRATION

Selecting the most economical filter type to obtain the desired results is a challenge faced by every filter user.

There are basically two types of filters from which to choose. The first is a depth-type filter that is constructed of randomly oriented fiber or beads pressed or wound together to form a tortuous maze of flow channels. The advantage of this filter type is low cost and high particle load capacity. Disadvantages of this filter type include the tendency of media to migrate into filtered product, no well defined pore size, and the absorbtion and retention of a large volume of the filtered fluid by the filter.

The second filter type is a screen or membrane made of rigid, uniform continuous material having a specific pore size determined during manufacturing. The advantage of this filter type is that there is little or no media migration because of the continuous structure of the filter material. Also, it has a well-defined pore size, and filter media are thin. Disadvantages of this filter type are that it has a very low particle-loading capacity, the media are easily fractured in pulse-mode operations, and in some cases it is difficult to manufacture.

Operations using the screen type or membrane style of filter should seriously consider the disadvantages of this filter type.

The surface area of the filter media is the most important design parameter to consider in sizing a filtration system using screen media. A large filter surface area is required to maintain low delta pressure across the filter media, operate at acceptable loading, and minimize hydraulic shock across filter media.

Filter systems can be staged to obtain desirable filtration levels. Pre-filtration reduces particle load on the final filter, which in turn improves the efficiency or end results of the filtration system. This is accomplished by maintaining a lower delta pressure for a longer period of time across the final filter media. This design concept also reduces system operation cost in that pre-filters are usually less expensive than the final filters.

CHEMICALS FILTERED

Low viscosity chemicals are easier to filter than are chemicals that have high viscosity. Higher viscosity chemicals create a much higher delta pressure across the filter media, which degrades filter efficiency. This phenomenon drives up the cost of obtaining the desired level of filtration by requiring either increased surface area of the filter media or thermally induced change in the viscosity of the chemical.

SYSTEM DESIGN CONCEPT

System design parameters required to maintain high chemical purity levels and low particle levels after filtration require consideration of the following:

- System material of construction
- Piping connections
- Velocity of chemicals in a system
- Surface area of filtration media and housing types
- Hydraulic stability
- Locations of filters
- Pump selection
- Valve selection.

The above parameters must all be considered in designing a system that will give desired filtration levels in a given product or process.

STATISTICAL PROCESS CONTROL (SPC)

An important feature of a filter cartridge is its ability to be integrity-tested before and after each filtration. To prevent waste of effort, time, and valuable product, a filter integrity test should be performed prior to putting a filter in use. The test will detect a damaged membrane, ineffective seals, or other system leak. After the filtration run, an integrity test confirms that the filter is still intact and that the system remained leak free throughout the run. Two kinds of acceptable integrity tests, the bubble point test and the diffusion test, are described in this section.

Bubble Point Testing

An important advantage of a membrane filter is its ability to be tested for integrity just prior to filtration. This is a simple, nondestructive check on performance known as "bubble point" testing. Membrane filters have passages penetrating from one side to the other that can be thought of as fine capillaries. The bubble point test is based on the fact that liquid is held in these capillary tubes by surface tension and that the minimum pressure required to force liquid out of the tubes is a measure of the tube diameter.

A bubble point test is performed by pre-wetting the filter, increasing the pressure of air upstream of the filter, and watching for air bubbles—downstream—to indicate the passage of air through the filter capillaries. The pressure at which a steady continuous stream of bubbles appears is the bubble point pressure. An observed bubble point pressure significantly lower than the bubble point specification for that particular filter indicates a damaged membrane, ineffective seals, or a system leak. A bubble point that meets specifications ensures the system is integral.

Diffusion Testing

In high-volume systems, where a large volume of downstream water must be displaced before bubbles can be detected, a diffusion test can be performed instead of the bubble point test. The diffusion test is based on the fact that, in a wetted membrane filter, air under pressure flows through the water-filled pores at differential pressures below the bubble point pressure of the filter by a diffusion process following Fick's Law. In small surface area filters, this flow of air is very slow, but in large surface area filters used in high-volume systems, it is significant and can be measured to perform a sensitive filter integrity test.

In performing the diffusion test, pressure is applied at a lower pressure than the bubble point pressure established for the particular filter system. The volume of airflow is determined by measuring the rate of flow of displaced water.

Applying pressure at a lower pressure than the filter bubble point validates filter integrity because there would be a dramatic increase in air (and water) flow at this lower pressure if there were damaged membranes, wrong pore size filter, ineffective seals, or system leaks.

Membrane filters are usually hydrophobic which requires that the filter media be: (1) wetted with a nonaqueous chemical such as isopropyl alcohol to initiate aqueous fluid flow, (2) rinsed with D.I. water under pressure flow, (3) then installed in a filter housing. All new filters should be tested by diffusion flow to check for filter integrity before using. A percentage of new filters will not pass this test due to leaking end caps, membrane fractures, or poor bonding of membranes to end seals. This test requires special equipment and trained personnel.

During the operation of a filtering system, monitoring of delta pressure across each filter housing is required to determine the status of the filter system. These data will provide a baseline to show when the desired filtration is being obtained and when it is time to change the filters. Maximum delta pressure across filter membrane media should not exceed 300 kPa (40 psig) and in most cases should stay below 140 kPa (20 psig). A correlation between delta pressure and particle level achieved in a given chemical can be projected after enough data have been obtained and correlated.

INCOMING QUALITY CONTROL

An important aspect of any particle or impurity reduction process is quality control. Continuous monitoring of the raw material and incoming packaging is essential to alleviate unexpected contamination problems in the final packaged product. Bottles and drums used to package high-purity chemicals must be treated with the same quality control measures as the chemical itself to maintain the consistent particulate level demanded by the industry. Three points of particular interest are described in this section: (1) gross contamination, (2) surface defects, and (3) chemical compatibility.

Gross Contamination

A large contributor to total particle contamination of high-purity chemicals is dust or dirt present in the bottles or drums arriving at the packaging plant. Quality control measures of container materials must be exercised on the incoming packages to ensure low particulate levels in the final product. Polyethylene bottles are checked for any obvious flaws in the mold or excess mold shavings that may remain in the bottles. These shavings are not always completely removed in the bottle washing step and can show up as floaters once the chemical is packaged.

Surface Defects

Surface defects in the bottle material are of particular concern. Bubbles in the glass bottles or thinly sealed seams in polyethylene bottles may contribute to product contamination or even failure of the bottle. Many times a few residual mold release agents may remain on the bottle, making it very difficult to clean and contaminate the product. Random samples of each bottle lot can be pulled and visually checked for surface contamination or defects prior to packaging.

Chemical Compatibility

Polyethylene bottles and drums are subject to chemical compatibility testing to ensure that the resin is consistent with previous lots. Bottles are pulled periodically from the line, filled with approximately 0.5 L (1 pint) of sulfuric acid, and placed in a 49°C (120°F) oven overnight. The color of the acid is measured before and after heating. Any substantial darkening of the acid after testing indicates an incompatible lot of resin. Similarly, poly drums are tested by cutting one drum into small pieces and placing these test chips into different chemicals. Any color change of the chemicals indicates that the drum resin is incompatible with that chemical.

PARTICULATE LEVEL VARIATIONS DUE TO STORAGE AND SHIPPING

Some questions exist concerning the changes in particle levels that occur during storage and shipping of high-purity chemicals. This section presents data collected on different chemicals in different types of containers. See Figure 11-5 for a photograph of a particle-counting test station.

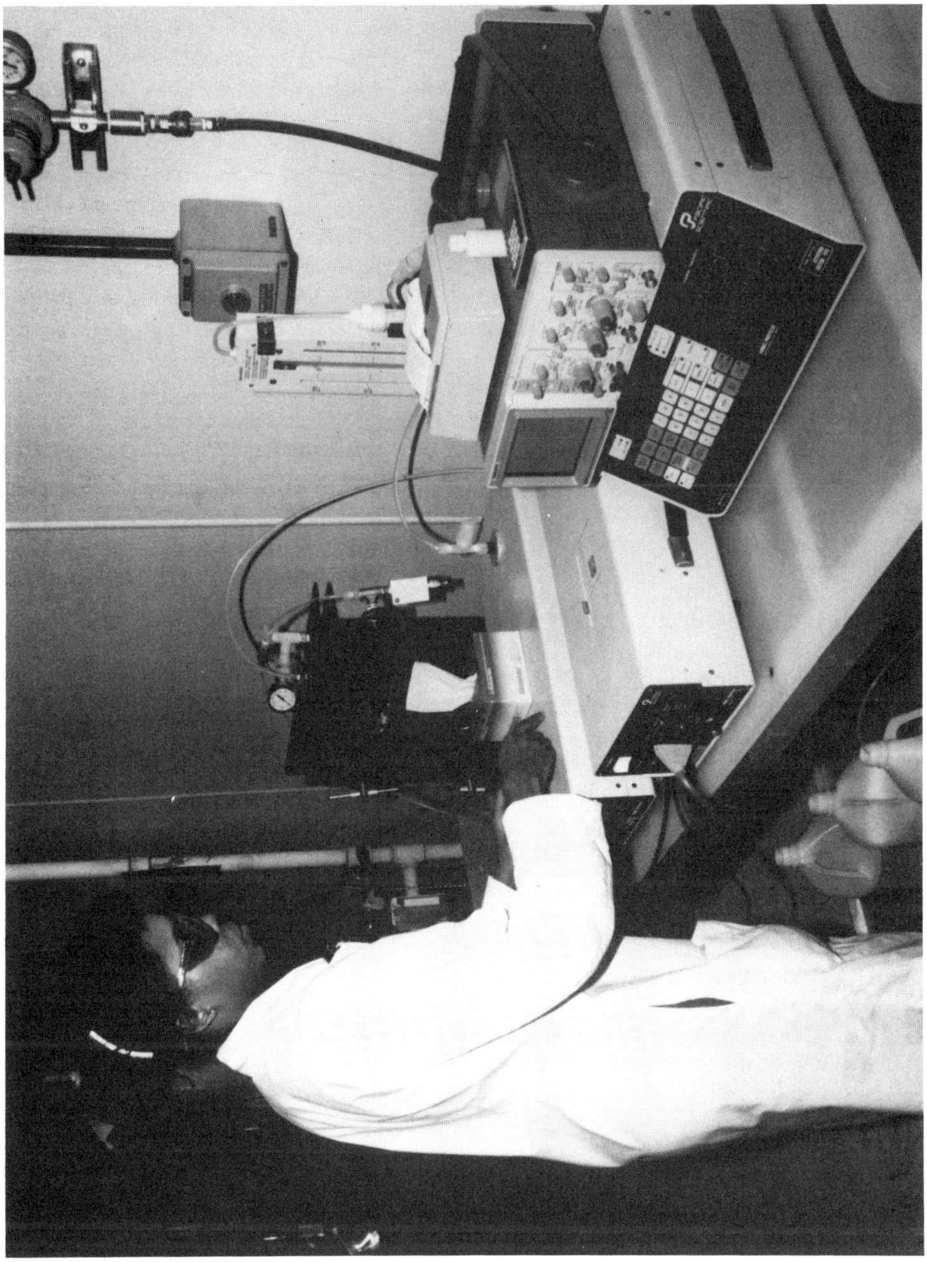

Figure 11-5. Particle-counting test station.

Acids

Particle counts were measured on samples of the same lot of 49 percent hydrofluoric acid (see Figures 11-6 and 11-7). As the data indicate, particle levels tend to decrease quite dramatically and rather rapidly from bottle filling to only 24 hours later. The reasons for this are not entirely clear; however, it is assumed that the highly corrosive nature of 49 percent hydrofluoric acid is responsible for the reduction of particles with time.

In another recent study, eight 1 gal (3.8 L) polyethylene bottles of sulfuric acid were taken from the same lot of product. The bottles were counted by the manufacturer and taken the next day to customer "A." After customer "A" counted the same samples, they were returned to the manufacturer for a recount. The results of these tests are presented in Table 11-2 and show good correlation between the particle counters at the two sites and only a slight increase in counts over 9 days. The increase is in fact so slight that it can be easily attributed to reopening and retesting of the same bottles. Each time a bottle is opened outside a clean-room environment, particles are added to the sample. It has been shown through recounts of numerous retained samples that sulfuric acid particulate levels remain constant after being packaged.

Etches

Extensive studies have been done on the effects of storage and transportation on particulate levels in 10:1 etch solutions. As Table 11-3 shows, particulate levels tended to increase between a shipment of a product from Newark to Dallas. This tendency is attributed to temperature fluctuation experienced during transport and storage. As 10:1 etch is exposed to decreasing temperatures approaching the freezing point of the solution (3.3°C [38°F]), crystals are formed on a submicrometer level that tend to resist dissolving back into solution. This accounts for part of the increase in submicrometer particles in these solutions.

Solvents

Particulate levels in solvents packaged in like containers run very low relative to acids. The fact that they are less dense with lower viscosity makes them flow more freely through the particle counter, and any entrapped air bubbles are quickly dispersed. Historically, solvents have not shown any increase in particulate levels due to storage or transportation.

PARTICLE CONTAMINATION FROM CONTAINERS

Bottles

Three types of bottles are used for packaging high-purity chemicals: polyethylene, glass, and amber glass. All three are potential particulate contamination sources.

Polyethylene Bottles

Most of the data that has been compiled on particulate contamination of high-purity chemicals concerns chemicals packaged in polyethylene bottles. The preceding data show, therefore, that polyethylene bottles do not add to the particle levels of the products packaged in them.

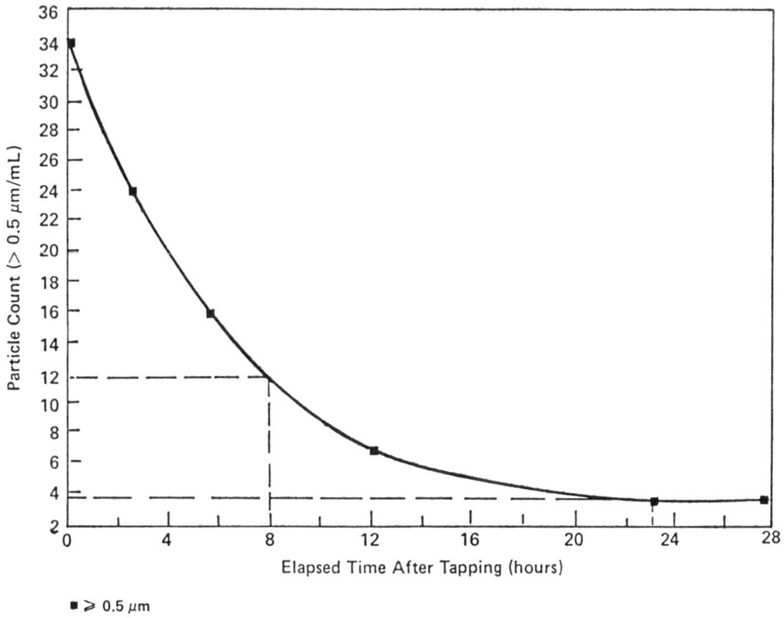

Figure 11-6. Particle count drop after tapping (49% HF #080086N Sample A).

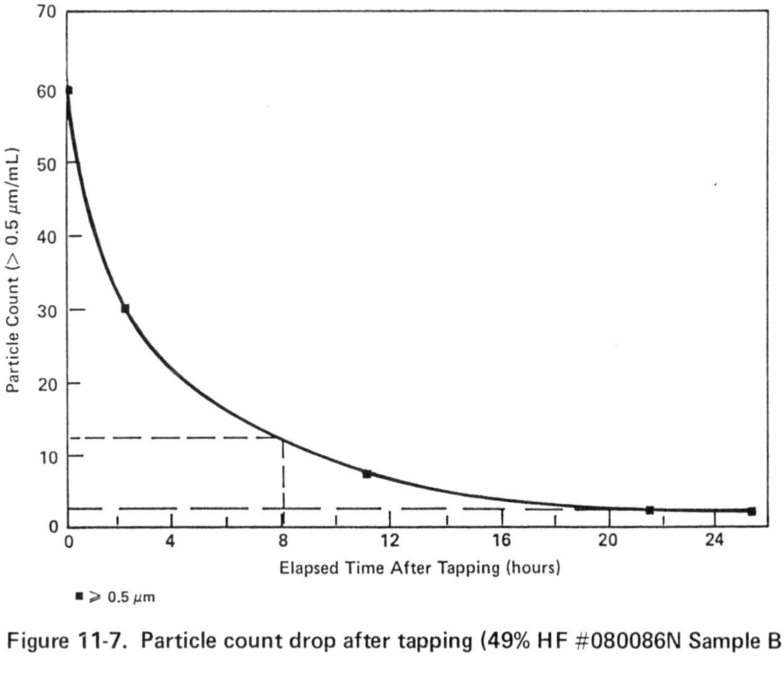

Figure 11-7. Particle count drop after tapping (49% HF #080086N Sample B).

Table 11-2. Correlation Between Particle Counters at Two Sites

Lot Number	Bottle Number	Particle Count (> 0.5 μm/mL) Manufacturer	Customer "A"	Manufacturer
102687101	A	25.9	26.6	50.2
102687101	B	28.9	26.0	58.8
102687101	C	17.2	18.6	27.6
102687101	D	23.5	22.0	36.3
102687101	E	13.3	25.6	11.4
102687101	F	12.2	10.8	22.6
102687101	G	13.9	33.7	16.8
102687101	H	9.5	25.4	34.7
Date Counted		12-8-87	12-9-87	12-17-87

Table 11-3. Particle Data (10:1 BOE)

	Shipper Newark Counts		Receiver Dallas Counts	
Bottle #	0.5 μm/mL	1.0 μm/mL	0.5 μm/mL	1.0 μm/mL
Lot #010828N				
1	12.32	0.40	83.04	1.80
2	9.56	0.44	71.08	1.64
3	6.24	0.32	73.12	0.12
4	4.84	0.28	57.00	1.16
Lot average	8.24	0.36	71.06	1.18
Lot #010548N				
1	10.32	0.64	75.48	1.40
2	8.76	0.60	95.40	3.68
3	3.36	0.44	126.60	3.52
4	9.48	0.32	185.64	5.44
Lot average	7.98	0.50	120.78	3.51

Glass Bottles

Glass bottles, both clear and amber, are used to package many chemicals, including both acids and solvents. Glass is not used instead of polyethylene due to any contamination advantage, but because of customer preference. Only certain organics require clear or amber glass because they are incompatible with polyethylene and are light sensitive, thus requiring storage in a dark bottle. Glass is typically more difficult to handle, harder to clean, and tougher to apply quality control methods to than are polyethylene bottles. Also, only one glass producer in the United States produces glass bottles for high-purity chemicals. Because glass constitutes a very small share of their total bottle production, there is very little control that can be exercised by the chemical industry over the end product bottles they receive. Any slight process change on the part of a bottle manufacturer can contribute a higher level of particles to the packaged product. Thus, process alterations on the packaging line may be required to offset these contributions and keep the end product within specification. Table 11-4 shows a comparison of particle levels of sulfuric acid packaged simultaneously in glass and poly bottles. The bottles went through identical cleaning and preparation steps prior to being filled with the same sulfuric acid at the same time. Particulate levels in poly bottles can be driven to a lower level more consistently than those in glass. Numerous counting and rechecking of acids and solvents packaged in glass have shown no significant increase in particles over time.

Table 11-4. Direct Correlation Poly vs. Glass Sulfuric-Particle Levels

Sample	Poly Bottles (≥0.5 μm/mL)	Glass Bottles (≥0.5 μm/mL)	% Difference
1	4.92	22.04	448
2	54.32	99.64	183
3	57.96	80.04	138
4	13.82	42.76	309
5	50.72	93.24	184
6	23.00	45.48	198

Drums

Particle data on drummed product are in the early stages of development. However, it appears that the drumming of high-purity chemicals will contribute significantly to total particle levels in the product. Table 11-5 shows particle counts taken from five types of drums containing four different products. In most cases, especially concerning solvents, particle counts are 10 to 20 times higher than can be expected from bottled product. Drum cleaning is performed manually, which may explain some excess particles. But the sampling system of pulling a sample from the drum into a bottle, then counting under pressure, is probably a large contributing factor. When the particle count is taken directly from the drum, there is too much entrapped air, which causes interference. As sampling procedures are further developed and refined, a better count of the true particle levels will be available.

Table 11-5. Particle Counts from Various Drums

Product	Drum Type	Drum Number	> =0.5 μm/mL	> =1.0 μm/mL
IPA	Steel	1	847.16	9.48
		2	1,400.00	60.04
	Vendor:	3	1,005.72	12.20
	C	4	793.64	7.60
		5	496.36	6.08
		6	1,139.72	8.28
		7	894.92	8.88
IPA	PFA	1	246.88	18.28
	Refills	2	203.04	8.60
		3	524.04	42.28
	Vendor:	4	250.84	15.80
	F	5	522.92	15.88
		6	1,075.44	34.08
		7	370.88	22.80
		8	185.12	9.80
		9	570.48	32.28
		10	83.64	6.08
IPA	Poly/	1	132.00	0.68
	Steel	2	31.84	0.60
		3	50.80	1.00
	Vendor:	4	29.64	0.80
	C	5	37.32	1.16
		6	130.04	0.64
Hydrogen	Poly	1	1,894.28	52.84
Peroxide		2	187.20	8.76
	Vendor:	3	156.28	8.32
	C	4	169.08	10.12
		5	139.48	8.56
		6	261.16	13.08
		7	180.96	10.20
		8	86.08	4.16
Sulfuric	Poly	1	75.60	0.48
		2	66.12	0.20
	Vendor:	3	50.32	0.32
	G	4	69.16	0.80
		5	33.20	0.24
		6	45.48	0.72
		7	33.24	0.36
		8	43.16	0.12
		9	63.04	1.36
		10	96.64	1.04

Drums (210 L [55 gal]) used in centralized distribution systems have an advantage over smaller-sized containers in that these systems can be dedicated to a chemical family and additional filtration can be achieved at the discharge of the drums and at point-of-use locations in the distribution system. Drums used in bulk or distribution systems also provide a safer way to distribute large volumes of chemicals to front end processes. Special dip tube systems that are color-coded and keyed are available to prevent cross-contamination of chemical family by personnel hooking up wrong chemical drums to a distribution system. A drum distribution system is the most popular method selected by the electronic industry for new front ends and retrofitting existing systems due to the following reasons:

- Improved safety of plant personnel by reducing movement of a large volume of chemicals in small containers
- Interlock key systems available to prevent cross-contamination
- Additional filtration to reduce particle level at point-of-use stations
- Relatively small size with large volume of chemicals that can be moved throughout the plant
- Reduction of manpower in chemical distribution department
- Quality control of incoming material guaranteed by the chemical supplier. In-place large bulk systems are very expensive and present a high potential of total plant shutdown due to gross contamination of all distribution systems plus the added personnel and cost needed for incoming quality control and maintenance.

REFERENCES

Dillenbeck, Keith. 1984. "Characterization of Particle Levels in Incoming Chemicals." *Microcontamination* (December).

Johnson, Donald W. 1985. "Evaluating Low-Particulate Chemicals." *Semiconductor International* (April).

Jones, Donadee. 1986. "Submicron Particle Levels In Process Chemicals." *Microelectronics* (May).

Peacock, S. L., M. A. Accomazzo, D. C. Grant, and P. M. Meier. 1985. "A Comparison of Particle Shedding from Different Chemical Filtration Products." Presented at the 1st *Microcontamination* Conference, San Jose, CA (November).

Przybytek, James T., and Karen L. Calabrese. 1985. "Particle Counts in Solvent." *Microcontamination* (June).

PROBLEMS

Q11.1 Fab 6 is going to incorporate a drum distribution system for sulfuric acid with the following specifications:

1. System materials to be compatible with sulfuric acid to maintain chemical purity.
2. System to incorporate filtration at drum and point of use. Point-of-use filtration level is to be 0.2 μm.
3. Drum connection to be designed where other chemicals cannot be connected to sulfuric system.

QUESTIONS

1. What materials of construction should be used for the following: bulk distribution pipeline system, and elastomers for valve seats and sealing surfaces.
2. What type of disconnect system should be used to connect drums to distribution system?
3. What level of filtration would you use at the outlet of the drum?
4. What other recommendation would you suggest for this drum distribution system?

Q11.2 What factors contribute to particle levels and the degrading of chemical purity during a packaging operation?

12. THE IMMERSION WET PROCESS SYSTEM

KEITH DILLENBECK Ashland Chemical Company, Columbus, Ohio

INTRODUCTION

Within an integrated circuit manufacturing process, there are several wet chemical etching or stripping operations that use a single chemical immersion bath. Examples of single chemical immersion process operations include:

- Photoresist develop
- Silicon dioxide etch, using dilute hydrofluoric acid (HF) or buffered oxide etch
- Photoresist strip, using a hot sulfuric acid/hydrogen peroxide mixture (piranha)
- Silicon nitride etch, using hot phosphoric acid.

These single-chemical process operations comprise three integrated steps: (1) the etch step, (2) the ultrapure water rinse step, and (3) the wafer-drying step. Each of these steps, if not properly designed and maintained, can add a significant amount of contaminants to the wafer surface.

CHEMICAL IMMERSION BATH

The most apparent feature of the chemical immersion bath is that many wafers are processed through the bath before fresh chemical is substituted for used chemical. It is a fact that the wafers and the cassettes that they are contained in add a significant amount of particles or dissolved contaminants (Jones et al., 1987). In an effort to provide a consistently clean chemical process environment, a continuous recirculating filtration bath is often used. As shown in Figure 12-1, the chemical is continuously recirculated and filtered to remove

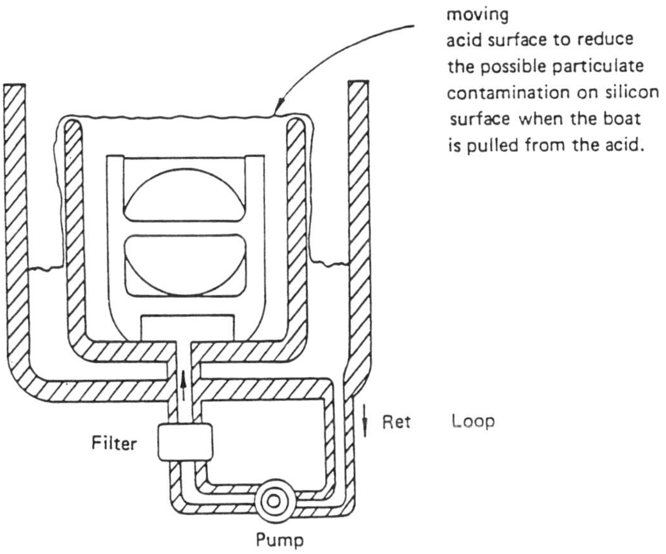

Figure 12-1. Schematic of recirculation immersion tank.

particles as they are generated within the process bath. The design of a recirculation bath must optimize the fluid flow characteristics and the filtration rate to quickly remove particles in all regions within the process bath environment. The specific filter to be used must be carefully chosen to ensure that it is compatible with the chemical to be filtered. In addition, the filter must not cause an excessive pressure drop that would overly restrict fluid flow. There is a tradeoff here between filter pore size (i.e., particle removal efficiency) and flow conditions within the process bath.

Figure 12-2 gives an example of the particle removal efficiency within an **empty** recirculating bath of HF. The number of particles per volume within the bath (as measured by a liquid particle counter) declines quickly as a function of recirculation time. These data can be misleading, as they show how initial particles from the chemical supplier's container are removed, but do not reflect the time necessary for the bath to return to a low particle level after a cassette of wafers has been processed. It is the particle recovery time that determines whether the bath will be clean enough for subsequent cassettes of wafers to be processed.

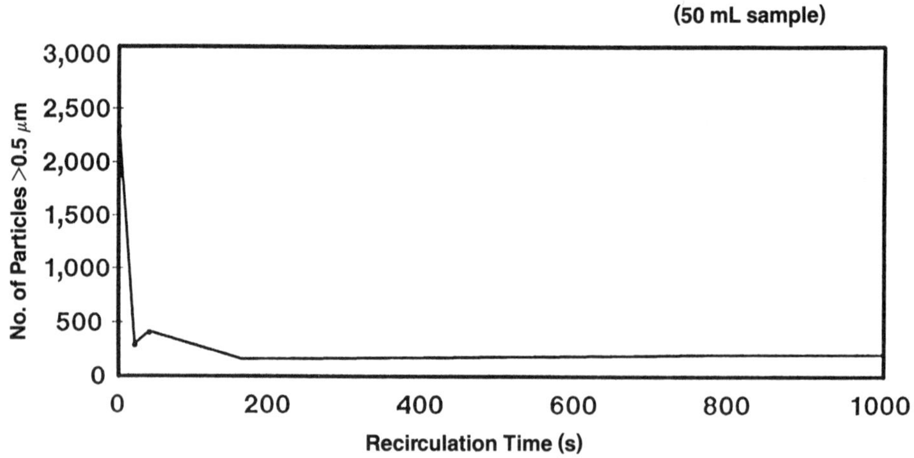

Figure 12-2. Hydrofluoric acid particle levels in a recirculation bath.

Figure 12-3 shows an example of bath recovery rates for a recirculation bath used for developing positive photoresist; the numbers of particles of diameter greater than 0.5 and 1.0 μm are included in the graph. When the bath is initially filled with new developer solution, low particle levels in the bath are achieved after 60 min of recirculation. When the first batch of 25 wafers is processed, the number of particles larger than 0.5 μm in the bath quickly rises to over 600/25 mL. Only 10 min were allowed for recirculation before the second batch of wafers

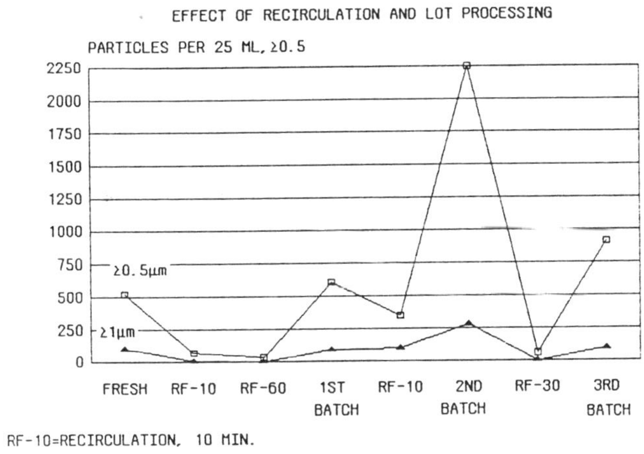

Figure 12-3. Particle levels within a developer bath.

was processed, after which the level of particles larger than 0.5 μm rose to over 2,200; it rose to 280 for particles larger than 1.0 μm. After 30 min of recirculation, the bath particle level declined to a level below 50 particles per 25 mL. From Figure 12-3, the following conclusions can be drawn:

- A large number of particles are generated within the bath during the photoresist developing process. A significant number of large particles (i.e., greater than 1.0 μm) are generated. These particles are most likely fragments of partially dissolved photoresist.
- Approximately 30 min of recirculation time is required between batches of wafers before the particle concentration in the bath decays to an acceptable level.
- Higher recirculation rates (e.g., by use of a larger pump) would be required to improve the particle removal efficiency and recovery time within the recirculation bath.

Recirculating chemical baths for use at room and slightly elevated temperatures are common (Dillenbeck, 1987). However, the ability to recirculate hot chemicals (e.g., phosphoric acid, at 140 °C) is significantly more difficult to achieve due to the higher probability of degradation of the materials that come into contact with the very aggressive fluids. Chemical purity and hardware reliability problems have been encountered because of past limitations in the various components of a hot acid recirculation system. Turnkey systems that can reliably and safely recirculate and filter hot acids, up to 150 °C, have only recently been introduced.

It is important to note that continuous acid-recirculation systems do not remove metallic, organic, or other dissolved impurities that are introduced into a bath during wafer processing. These dissolved impurities accumulate in the bath as wafers are processed, at a rate that depends on the specific process operation. The dissolved impurities are transported into the bath by the wafers themselves or by the wafer cassettes (Jones et al., 1987). New technology is being introduced to continually recirculate and distill the chemical during processing. The ability to provide a consistently pure chemical environment relative to particles and dissolved impurities is becoming increasingly important as Ultra Large Scale Integration (ULSI) process technology continues to evolve.

ULTRAPURE WATER RINSE TANKS

After a cassette of wafers is etched in an acid immersion bath, it is moved to an ultrapure water rinse tank. Rinse tanks can take on a variety of configurations, but two are most common. In the first type, the spray/dump rinse tank, as depicted in Figure 12-4, uses a water spray to quickly remove chemical from a wafer surface. Following a short spray period, water is rapidly dumped through large ports in the bottom of the tank. Several spray/dump cycles are repeated until all portions of the wafer are adequately rinsed. In the second type, an

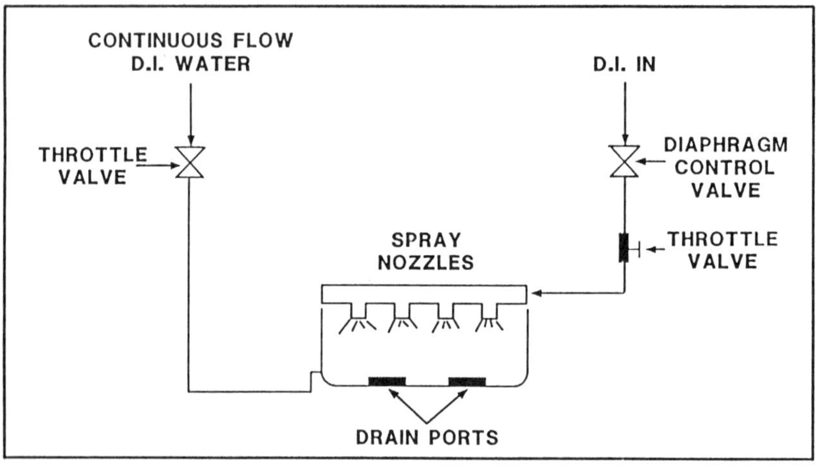

Figure 12-4. Schematic of a D.I. water rinse tank before modification.

overflow rinse tank configuration, water is fed at a moderately high flow rate into the bottom of the rinse tank. Water flows through the tank in an upward direction, then overflows at the top portion of the tank. In the overflow configuration, relatively high flow rates are required in order to provide enough turbulence to break the surface boundary layer and remove chemical from the wafer surface. When the tank is sitting idle, a low flow of ultrapure water moves through the tank to minimize bacteria growth. Some wafer fabrication areas use two or more overflow tanks in a cascade arrangement as shown in Figure 12-5. In this configuration, wafers are rinsed in sequential steps and moved to a bath of increasing purity as the acid is removed from the wafer surface.

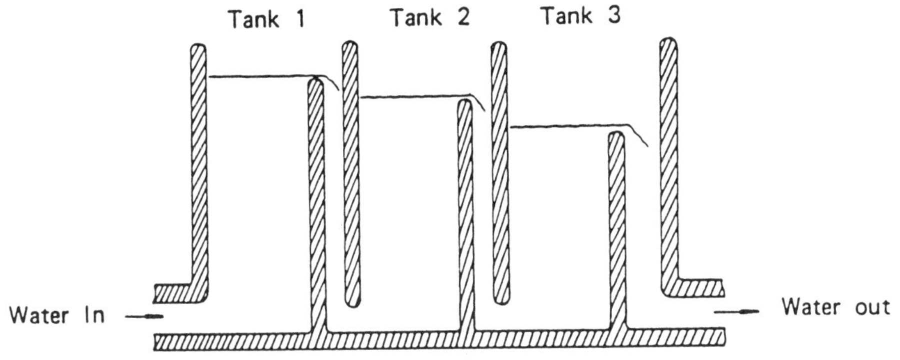

Figure 12-5. Schematic of cascade rinse tank.

There are four common contamination problems associated with an acid rinse tank (either configuration). These are:

- Excessive bacteria growth
- Tank and/or tubing material degradation
- Particle generation in valves, regulators, etc.
- Inadequate water flow rate to remove the acid.

Bacterial contamination is typically the most predominant problem within a rinse tank. In general, bacteria are a very pervasive contaminant within ultrapure water and can reproduce and colonize efficiently with only trace quantities of organic nutrients. Bacterial contamination has deleterious effects on semiconductor devices because of particle and trace metallic mechanisms (Gough et al., 1986). Their growth can be suppressed by optimizing flow velocities past surfaces (Siegman, 1987). Bacteria attach to pipe walls and other surfaces within the boundary layer of the water adjacent to the surface, where the flow velocity is very low (Harned, 1986). Higher flow velocities reduce the thickness of the boundary layer and thus the relatively stagnant area where bacteria are known to colonize. Figure 12-6 shows the number of particles added to bare silicon test wafers in a spray/dump rinse tank with high and low bacteria levels. Growth of bacteria within the tubes, nozzles, and vessels of a rinse tank results in a large number of particles being deposited onto a wafer surface during a rinse step. Bacteria growth in a rinse tank can be controlled by (1) optimizing hydraulic flow, and (2) using sanitizing agents.

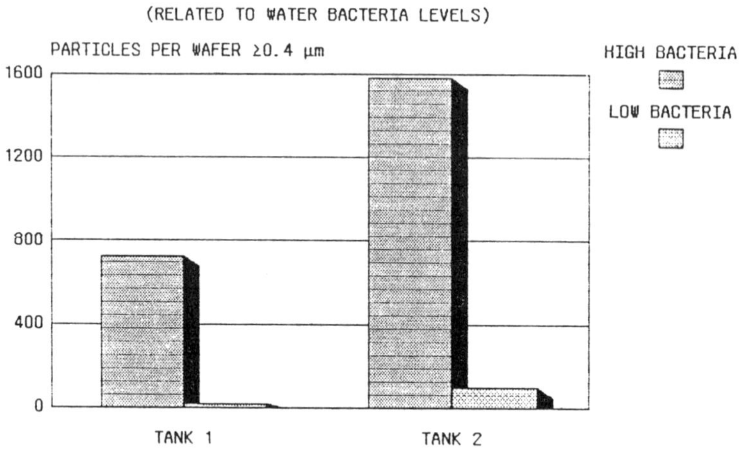

Figure 12-6. Particles added during a rinse operation.

THE SPIN DRYER

The last step of a chemical immersion process is the wafer-drying operation. A centrifugal spin dryer system is most commonly used. The wafer cassette is placed in a rotating chamber and heated nitrogen is allowed to flow through the chamber. One common type of spin dryer is shown schematically in Figure 12-7. Within such a unit, there are several

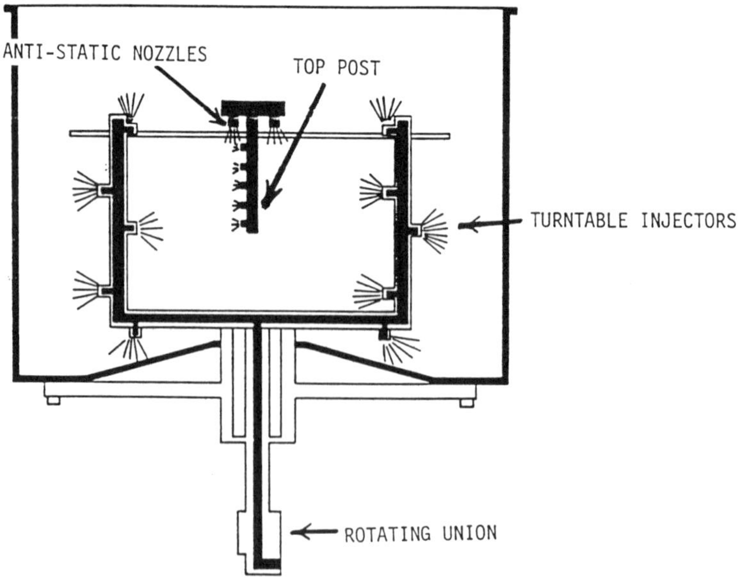

Figure 12-7. Nitrogen entry locations in a multiple-carrier spin dryer.

potential particle generation sources whose influence can vary widely as a function of time. Spin dryers are commonly recognized as an unpredictable source of particulate contamination. Because a spin dryer is the last environment that the wafer is exposed to within a wet etching process module, it often eradicates any positive effects of the chemical clean and/or water rinse steps.

There are many sources of particles within a spin dryer. Within the unit, a spray rinse step is sometimes used just before the dry step; the bacterial problems associated with the water spray can be quite disastrous (Dillenbeck, 1988). To avoid this, many wafer fabrication areas use the spin dryer in a dry-only mode. As a result, most particles that are generated are transported to the wafer by the nitrogen gas.

Common sources of particles in the nitrogen stream are:

- Gas solenoid or diaphragm valves
- Rotating drive mechanisms
- Gas entry manifolds and injectors
- Sputtering from air ionization emitters (Donovan, Lawless, and Smith, 1986)
- Imbalanced exhaust ports
- Leaking door seals.

Figure 12-8 presents an example of the components that can contribute large numbers of particles to the nitrogen stream within one particular type of spin dryer. Most of these particle sources are mechanical and result from abrasion of one solid surface against another. Figure 12-9 shows the nitrogen particle levels sampled with an aerosol particle counter at points downstream of the various gas handling components. Particle generation within a spin dryer can be minimized by careful selection of fluid handling components and filters, and by the implementation of a conscientious preventive maintenance program.

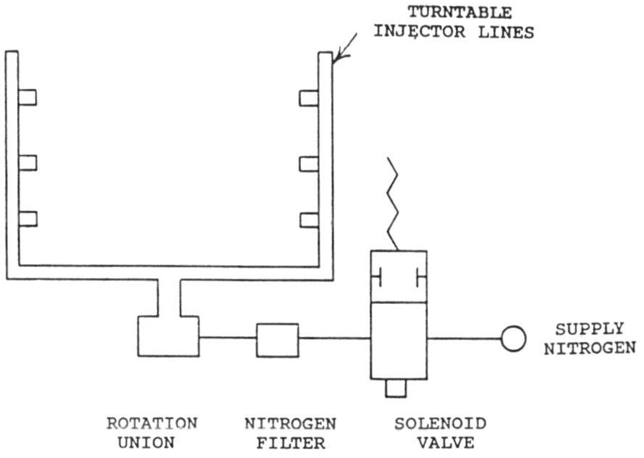

Figure 12-8. Potential particle sources in spin-dryer nitrogen stream.

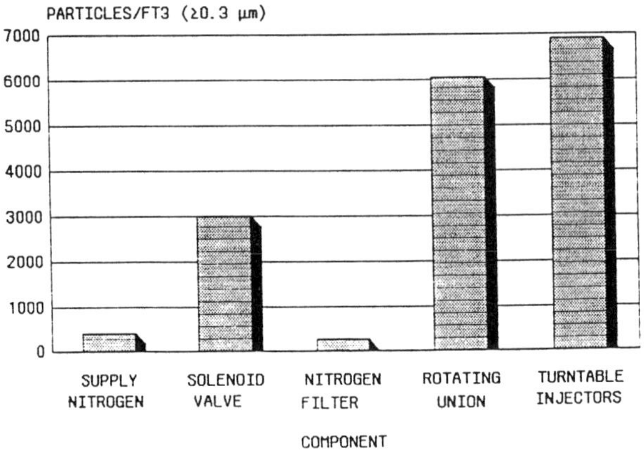

Figure 12-9. Particles generated by nitrogen stream components.

REFERENCES

Dillenbeck, Keith. 1987. Contamination Engineering Principles for Semiconductor Wafer Fabrication. Tutorial presented at the Microcontamination Conference and Exposition, October, 1987, Santa Clara, CA.

Dillenbeck, Keith. 1988. Application of Contamination Engineering Techniques for Solving Particle Problems in Process Equipment. *Proceedings of the 34th Annual Technical Meeting of the Institute of Environmental Sciences*, May, pp. 532-539. IES, Mount Prospect, Illinois.

Donovan, R. P., P. A. Lawless, and D. D. Smith. 1986. Polarity Dependence of Electrode Erosion Under dc Corona Discharge. *Microcontamination* 4(5):38-49.

Gough, A., R. W. Atwell, D.F.D. Hardy, and R. Caldwell. 1986. Microbial Contamination in Ultrapure Water. *Solid State Technology* 29(2):139-142.

Harned, Wayne. 1986. Bacteria as a Particle Source in Wafer Processing Equipment. *The Journal of Environmental Sciences* (May/June):32-34.

Jones, A. H., J. G. Hoffman, A. W. Jones, and W. Yuan. 1987. Improved Wafer Cleaning with Ultrapure Piranha. *Proceedings of the Microcontamination Conference*, October, pp. 69-80. Canon Communications, Santa Monica, California.

Siegman, R. P. 1987. Point-of-Use Contamination Control of Ultrapure Deionized Water for Semiconductor Processing. *Microcontamination* 5(3):22-26.

13. MEASUREMENT OF PARTICLES FROM EQUIPMENT AND PROCESSES: THE PARTICLES-PER-WAFER-PER PASS (PWP) METHOD

B. TULLIS Hewlett-Packard Company, Palo Alto, California

INTRODUCTION

World competitiveness in the semiconductor business, coupled with the rapid advance of semiconductor technologies into submicrometer geometries, has introduced a pressing need for ultraclean process equipment and process steps. To achieve ultraclean process equipment and process steps requires means of quantifying the amount of contamination that individual pieces or groups of process equipment—and the processes they run—contribute to product wafers (Osburn et al., 1988). Furthermore, users and vendors of process equipment require particle specifications that can be tested in a standard and reliable (i.e., repeatable) manner.

This chapter will discuss the particles-per-wafer-per-pass (PWP) methodology of measuring particles added to wafers when these wafers are passed through one or more pieces of process equipment. PWP is the number of particles added to wafers in one pass through one or more pieces of process equipment. Figure 13-1 depicts the concept of the PWP methodology. An initial particle count is first obtained on relatively clean wafers. These precounted

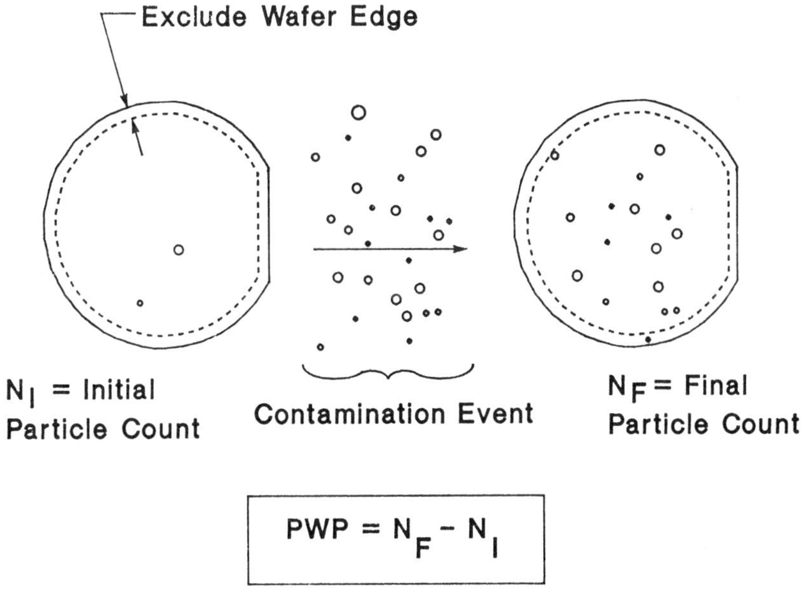

Figure 13-1. Particles per wafer per pass (PWP).

wafers are then passed through the equipment, process step, or process segment in question. Finally, these wafers are inspected again for the resulting particle counts. The differences between the initial and final particle counts yield counts of how many particles were added to each wafer. These differences are the PWP counts. Alternatives in the logistics and calculations of the PWP methodology include passing the wafers through the equipment or process multiple times and dividing the total accumulation count by the number of passes to obtain an average PWP. Or PWP can be measured many times over and then averaged. Many other alternatives and details are also presented in this chapter.

Depending upon the contamination problem to be investigated, the PWP method may be applied using any of a variety of tools and techniques for counting particles on wafer surfaces. The counts can be obtained, for example, with microscopes, laser scanners, vidicon imagers, or holographic counters (Billat, 1987). To supplement particle counts, often surface analyses are carried out to determine chemical signatures and visual categorizations are done to determine morphological signatures. Both procedures help to better determine the specific sources of the particles (see Chapter 20).

The principal topics of importance in PWP methodology are calibration and PWP characterization of the counting tool(s), measurement plan structure, analysis of variances, and the calculation of statistical confidences. Counter calibration is discussed in Chapter 5. The rest of these topics will be addressed in this chapter.

APPLICABILITY TO EQUIPMENT AND PROCESS SEGMENTS

The most traditional methods of quantifying contamination in clean rooms have been to use aerosol particle counters and witness wafers (see Chapter 20). These, however, have not been adapted well for precise measurement of contamination contributed to product wafers by specific equipment. They tend to provide only general indications of where and when equipment or process-step contamination may be of concern. The stream-tube method of measuring equipment-sourced contamination, for example, does not indicate directly what amount of contamination actually reaches and deposits onto wafers, and it is not well suited for measuring contamination levels that require long sampling times to obtain statistically significant counts (see Chapter 14). The stream-tube method is, on the other hand, well suited for quick and easy measurements when contamination levels are high, and its feedback can help to reduce counts by leading to efforts to eliminate or repair the particle sources (Cheung and Roberge, 1987). The reservoir method is, of course, uniquely suited for measuring the total contribution of particles from isolatable equipment components—and even from entire equipment—to the environment. But the reservoir method can not quantify the contributions of contamination directly to wafers (see Chapter 14). Instruments that measure particle flux directly have some applications in isolating specific particle sources in vacuum equipment, especially in determining the exact moment(s) during equipment or process cycles when particle sourcing occurs (Borden, Baron, and McGinley, 1987). Electrical test masks measure the final defect densities of entire processes and even of some small process segments, but generally cannot isolate particle sources to specific pieces of process equipment or process steps (Weber, 1988; see also Chapter 26). The uniqueness of the PWP method is its perspective on contamination directly to wafers and its specificity to whole pieces of processing equipment or specifically selected process steps or process segments.

Thus, for characterizing the contamination levels of specific processes and equipment, the PWP method has important advantages over measures of contamination of a clean room in general, and over other particle counting methods (Hemmes, 1987a; Hemmes, 1987b; and Baker, 1985). The most obvious of these advantages is that it directly measures particles that actually are deposited onto wafers and thereby have real potential to cause yield loss. The particle counts of other methods include particles that never deposit onto wafers and which may therefore be of no concern whatsoever. The PWP method can measure particle contamination in places in which probes (other than wafers) cannot go. It permits the evaluation of contamination from equipment in situ without requiring that equipment to be moved and without requiring any special equipment alterations or enclosure. And it can quantify the contamination contribution of specific process steps or specific wafer handling steps and storage steps. Even the impact of equipment maintenance on wafers in process can be measured, as can the impact on wafers when storage, transport, or shipping containers undergo pressure changes.

On the other hand, limitations of the PWP method include the difficulties of counter calibrations when the wafer surfaces are complex structures composed of films, either unpatterned or patterned. Although it is possible to calibrate counts on such complex surfaces, at least crudely, the PWP method sees most of its application today on unpatterned test wafers with only an insignificantly thin native-oxide film. In these tests with native-oxide test wafers, the equipment under test is made to run only a "dry" process. By "dry" we mean a process that (1) includes the handling and passage of wafers through the equipment(s) so that as many of the normal equipment functions as possible are exercised according to the process recipe in question, but (2) deletes chemical steps that would optically alter the surface of the wafer and make calibration for particle measurements difficult. (Refer to Chapter 5 for a discussion of how wafer surface characteristics alter the sensitivity of a particle counter.) By "wet" process we mean the full process by which the equipment normally processes product wafers, whether wet chemicals are involved or not.

A shortcoming of the PWP method as well as other particle-counting techniques is the inability to differentiate among particles that come from equipment, chemicals, process gases, the clean-room environment, and the process reactions, except by deduction from circumstantial evidence. In addition, the PWP method, like other methods, counts both those particles that may ultimately cause electrical defects and those that may be washed off in a subsequent cleaning or stripping step. Only electrical tests can measure actual defect densities; but these may include defects caused not only by some of the particle contaminants measured with the PWP method but also by process errors, insufficient process margins, and defects classified as non-visible (such as those caused by mobile ions from ionic contaminants or by adsorbed molecules). However, it is possible under certain circumstances to correlate PWP counts at specific process steps with a substantial proportion of electrical defect counts. One methodology for enhancing these correlations and developing sensitivity ratios between PWP counts and yields is to purposefully deposit (seed) controlled contamination densities onto wafers at specific steps in the process and to correlate these increased densities with the subsequent reductions in yield. Seeding techniques are especially attractive when used in combination with statistically designed experiments for response surface modeling and analysis, topics discussed briefly in a later section.

REVIEW OF STATISTICAL INFERENCES AND CONFIDENCE LIMITS

The principal reason to make PWP measurements is that they permit quantitative comparisons. Comparisons are usually made to answer the questions of whether a piece of equipment or a process step or process segment meets a specification, is better or worse or more tightly controlled than another, is stable, or has improved or deteriorated. To answer such questions meaningfully when particle count fluctuations are significant, it becomes necessary to use certain basic statistical concepts and techniques (Gibra, 1973; Box, Hunter, and Hunter, 1978; Alvarez, Welter, and Johnson, 1983; and Natrella, 1986). Understanding these statistical concepts and techniques is a prerequisite for use of PWP methods and results. These concepts and techniques are reviewed in this section.

The first understanding needed is of the terms "frequency distribution" and "histogram." Both a frequency distribution and a histogram, as depicted in Figure 13-2, express the relative frequencies with which the values of some variable occurs. The appropriate range of possible values is presented as an abscissa axis, called the variate axis. The ordinate axis is one of frequency or, in a relative sense, how frequently the values of the variate occur. An example would be the relative frequencies with which some particle counting results occur as a function of the count magnitudes.

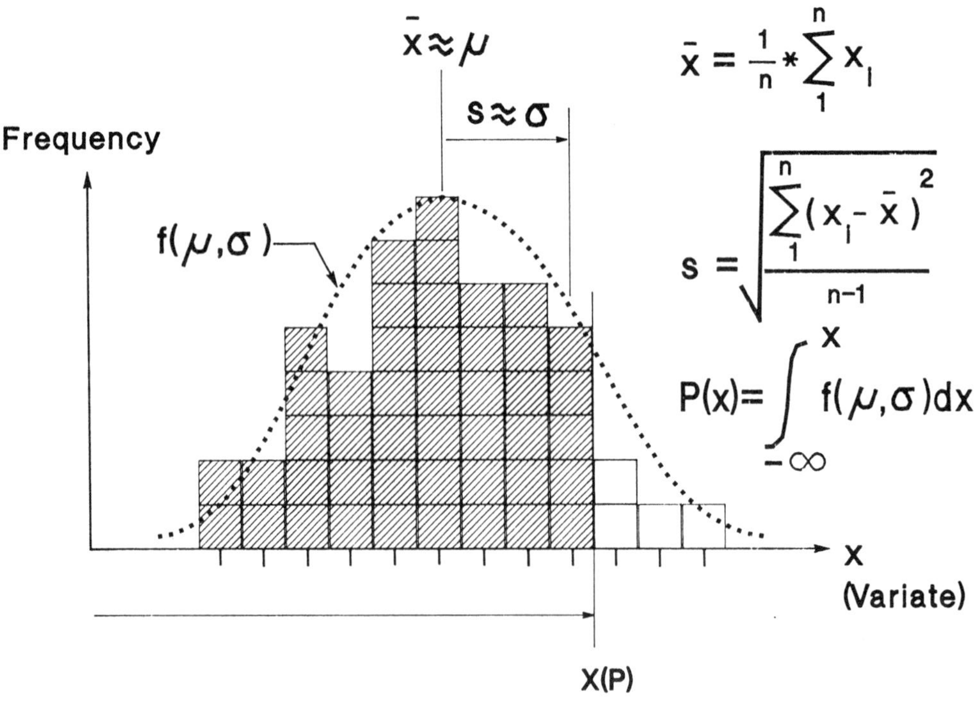

Figure 13-2. Frequency distribution and probability integral.

In Figure 13-2, a frequency distribution is shown as the dotted line labeled "f" and a histogram is shown as a collection of boxes or squares. The possible values of either a continuous or a discrete variable are shown as the abscissa and are represented here by the variable name "x." The boxes are centered over increments of equal-width along the x-axis. In a histogram, each measurement taken to obtain a value of the variate is represented by a box centered on the x-axis increment that includes the value measured. If boxes representing several measurements all belong in the same x-axis increment, they are drawn stacked along the ordinate direction. The ordinate of a histogram is then seen to represent the number of measurements that resulted in a given x increment. The sizes of the boxes can be chosen to make the total area of all the boxes (squares) unity. If a large enough collection of measurements is made of a variable having a suppressed random noise (as caused by instrument noise or process noise), then the number of measurements will be distributed over some range of variate values. The values along the x-axis can be either continuous or discrete. If the variate is discrete and the increment size chosen for the histogram is the step size of this discreteness, then a frequency distribution, which will also have to be discrete, will be identical to the histogram. If the variate is continuous, the increment size for a histogram can be made arbitrarily small, making the histogram asymptotically approach identity with the frequency distribution. (Note that it is customary to omit the display of horizontal line segments between adjacent boxes of any one vertical stack, but they are shown here for clarity.)

The average measurement result for the population of repeated measurements represented by the histogram in the example in Figure 13-2 is a good approximation of the average of the frequency distribution "f," where "f" represents the distribution of results from an infinite number of repeated measurements or the entire population of measurement results. The average for the variate of the histogram is designated by the symbol "$\bar{x}$." This is the average obtained from a finite sample. The true population mean, i.e., the average value of the variate of "f," is designated by μ.

The characteristic half-width of the histogram is designated by a small "s" and is called the "standard deviation of the sample," where the sample is depicted as the collection of boxes in the histogram. The half-width of the distribution "f" (i.e., the width of the total population, or the population of which the histogram is only a sample) is depicted by σ and is called "standard deviation of the population." The formula for calculating the half-width of the histogram (the width of a finite sample) is given in the figure. The indicator of half-width of a distribution is the square root of the average of the squares of deviations from the average, integrated over the entire variate axis. It should be noted for the histogram, in taking the average under the root sign, that the average is taken by dividing by n-1 (or one less than the number of measurements in the sample). To take an average, one would expect to divide by n in this case, but statistics requires that n-1 be used in order to inflate the resulting value of "s" to improve its value as an estimator of σ. The logic of applying this correction can be appreciated by considering that as larger and larger samples are taken, there is more and more likelihood that the rare outlying values of the total population will become possible values. Because small samples are less likely to properly estimate the width resulting from these rare outliers, the width of a smaller sample, calculated to be an estimator of the total population, will be undervalued if the formula divides by the usual n instead of n-1. The square of a standard deviation is called a "variance."

One last concept depicted in Figure 13-2 is that of a probability integral. Areas under the histogram and the frequency distribution "f" can be taken to represent probabilities. Probabilities can be represented as decimal numbers between 0 and 1 or as percentages from 0 to 100 percent. Note that if the total area under either the histogram or the continuous frequency distribution is defined to be unity, then the area over any range of variate values can represent the probability of a measurement value occurring with values within that range. The probability of some value, along the entire x-axis under the given histogram or frequency distribution, is certainty. Certainty is represented by unity, 100 percent, or the total area. The probability P(x) of a value occurring that is less than some particular value "x(P)" is numerically equal to that part of the area under the histogram or frequency distribution to the left of x(P). The area of the shaded boxes in Figure 13-2 is the probability of a measurement value resulting that is less than the value at the point marked as "x(P)" in the figure. Because x(P) represents a property of the histogram or distribution, it is justifiably called a "statistic" of the distribution, just as the mean or average and the standard deviation are statistics.

Figure 13-3 presents the frequency distributions most useful in analysis of particle counts. In Figure 13-3a, the origin of the abscissa of a frequency distribution has been shifted to correspond to the average frequency, and the scale has been normalized by the standard deviation of the frequency distribution. Also, as before, the frequency axis (ordinate) has been scaled to normalize the sum or integral of all frequencies to unity. Also shown in Figure 13-3a are two vertical lines delimiting upper and lower limits of a central area under the plot and at the same time delimiting upper and lower tail areas labeled "α." The particular plot shown in this example is roughly bell-shaped and is symmetrical about the origin. It is a particular form called a Gaussian distribution and describes the frequency distribution of many naturally random variables, such as the results of physical measurements that are not perfectly precise. The rescaled and normalized variable along the abscissa of a Gaussian distribution is called the "z-variate" or "z-statistic," and values are called "z-scores." Given such a distribution (whether Gaussian or not), and limits like those shown, the probability can be stated of a random measurement occurring with a value lying within these limits. Similarly, given a collection of random measurements, the percentage of these measurements that would be expected to lie within these limits could be stated. In both cases, whether for the probability of one random measurement or for the percentage of a collection of random measurements, the stated probability or percentage is given numerically as the area under the distribution between the two limits shown.

For the Gaussian distribution, the positions of the limits that bracket the central 95 percent of the area under the distribution are at the approximate z-score values +2 and -2, i.e., on the shifted and scaled abscissa. These limits correspond to positions at approximately $+2\sigma$ and -2σ above and below the average value of x. (Actually, the z-score or multiple of σ is closer to 1.96 to bracket 95 percent of the area.) Therefore, with a Gaussian distribution, it can be stated that there is a 95 percent chance that a random value will lie between the 2σ limits. Or, 95 percent of random measurements will lie between the 2σ limits. Or, extending these concepts to the probability integral involving minus infinity as the lower limit, there

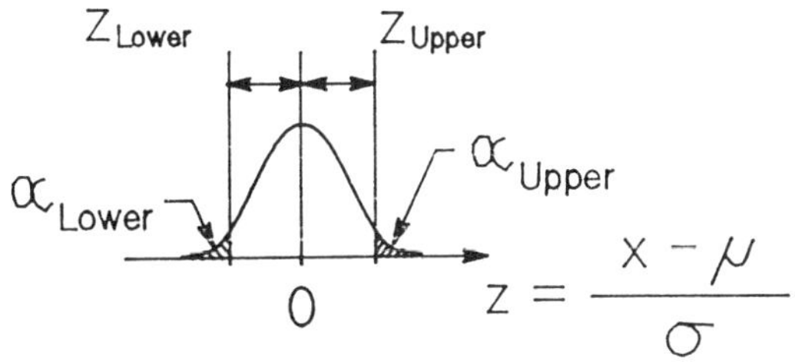

(a) Gaussian distribution

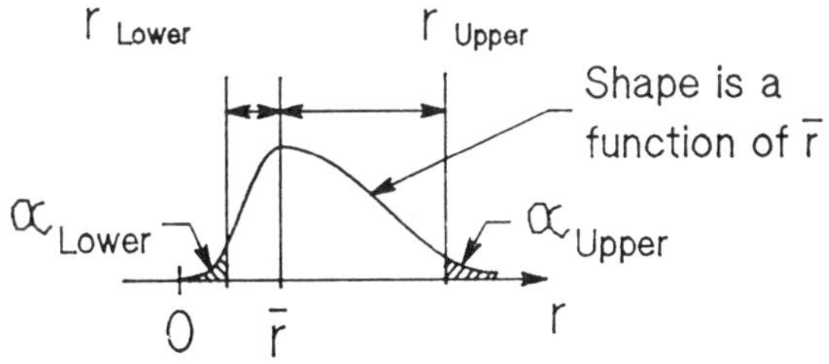

(b) Poisson distribution

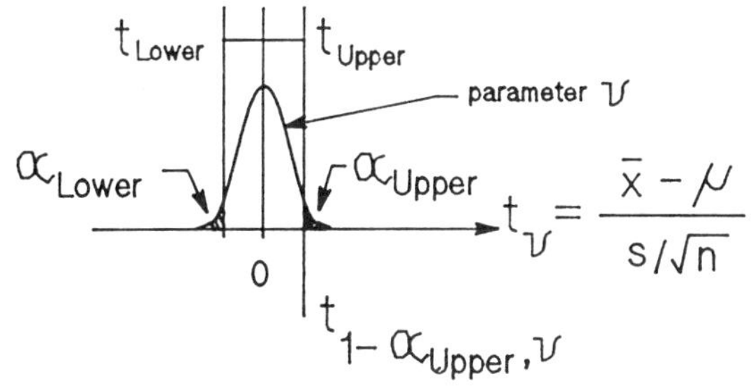

$$\bar{x} = \mu \pm t_{1-\alpha/2,\nu} * s/\sqrt{n} \quad \text{,for Confidence of } 1-\alpha$$

where ν = degrees of freedom = n−1

(c) Student distribution

Figure 13-3. Frequency distributions.

would be a 97.5 percent probability that a random measurement would result in a value lying below the $+2\sigma$ limit (or below $z = +2$). Mathematically, a complementary statement would be that there is 2.5 percent chance that a random measurement would result in a value above the upper 2σ limit, i.e., would result in the upper tail, or tail above $z = +2$.

A convention in statistics that it will be important to recognize is that whether both upper and lower limits are used together or are taken separately, the area or probability represented by the combined tails or one tail alone, respectively, is represented by α. In most situations involving upper and lower limits together, whether the distribution is symmetrical or not, the limits are placed such that the areas in the two tails are equal and each can be referred to as having an area $\alpha/2$. In the case when only one finite limit is involved, however, there is only one tail and that tail is referred to as α, not $\alpha/2$; this is easy to forget but important to remember.

The distribution in Figure 13-3b is not symmetrical. It describes the Poisson distribution, and is skewed to the right (positive skewness). The standard deviation of a Poisson distribution is a function of the average abscissa value: it is equal to the square root of the average value. Limits and probability statements can be made as before, but note that the limit defining a 2.5 percent tail of highest values (abscissa values) will be farther from the average value than will the limit defining a 2.5 percent tail of lowest values. Because the shape of the Poisson distribution is a function of the average value, the limits, expressed as multiples of the standard deviation, will also be a function of the average value, unlike the case with the Gaussian distribution. The Poisson distribution is well known in particle counting because it generally describes the distribution of small proportions of particles uniformly, randomly, and independently distributed in space. These proportions are expressed as particles per sample size, e.g., particles per volume or per area.

Often the distribution of particle counts for a piece of equipment or process step under test cannot be presumed to be known beforehand, as from historical data or an accurate assumption that the distribution will derive from particles uniformly, randomly, and independently distributed in space. In general it is hard to justify these assumptions, so a known Poisson distribution can rarely be safely assumed for PWP counts.

Fortunately, there is a well known distribution, called the student distribution or t-distribution, which often accurately describes distributions of sample averages. Although the student distribution was originally derived for distributions of sample averages taken from a parent population that is Gaussian, it is fairly robust to the actual parent distribution being sampled. This robustness is explained by the central limit therorem, which says that sample means are Gaussian for any parent population distribution. Therefore, although the Poisson distribution naturally fits many particle count phenomena and obeys a reproductive property, the distribution of averages of Poisson-distributed data should be student distributed. Even when counts of particles contaminating wafers are a function of non-Poisson-distributed variables, averages of random and independent samples of PWP counts can often be appropriately modeled by the t-distribution.

An example of the student distribution, for a selected but arbitrary sample size "n," is shown in Figure 13-3c. Its shape is a function of the sample size but remains conveniently symmetrical. In comparison to a Gaussian distribution, a student distribution with any sample size is more peaked, while its tails remain higher for larger absolute magnitudes of the abscissa. And it asymptotically approaches the Gaussian distribution when sample size is increased. The abscissa of a student distribution is defined similarly to that of the shifted and scaled Gaussian plot described above, except that the abscissa statistic is called "t" or "t-score" instead of "z" or "z-score," and the scaling factor is the standard deviation of values in a sample, divided by the square root of the sample size, instead of by σ alone. This is an important point, that the student distribution does not require a priori knowledge of a population's standard deviation (σ) as does the Gaussian distribution, but requires only the estimate "s" of standard deviation. Increasing sample size has the advantage of narrowing the student distribution, at a very significant rate when sample sizes are small, and more slowly when sample sizes are large. This ability to decrease the width of the student distribution by increasing sample size is very important for controlling the precision with which the average of a population of measurements can be reliably estimated.

Note that the scaling factor of the t-statistic is called "standard error." (The standard deviation of the sampling distribution of any statistic is called standard error rather than standard deviation. In this case the statistic is the average of a sample.) And note that the Student distribution is a function of a parameter represented by ν. This parameter is that of the "degrees of freedom" in the sample data. In most cases the degree of freedom equals the number of values comprising the sample, minus one. That is, one degree of freedom is removed in the calculation of the mean about which the distribution is centered. For a more complete discussion of how to calculate degrees of freedom in more complicated cases, refer to textbooks on statistics. Formulae for calculating degrees where needed in the rest of this chapter will be presented as appropriate.

Figure 13-3c includes a formula for the estimate ($\bar{x}$) of the true population mean (μ) and both positive (+) and negative (-) limits of a confidence interval. These limits represent a tolerance measure of statistical precision on the estimate $\bar{x}$. The true population mean, μ, would be expected to lie within these limits about the estimate $\bar{x}$. If repeated estimates were made, it would be expected ($1-\alpha$) percent of the time that the errors of the estimates would be equal to or less than these limits.

If distributions of measurement results are known, statements can be made about the probability of inference decisions being correct or incorrect. Simple inferences, often applicable to particle studies, are those that concern (1) whether a given measured estimate of an average value supports a decision that the true mean falls above, below, or to either side of the true mean of a reference distribution, (2) how precisely the true mean of a population is estimated by the mean of a sample, (3) whether a measured estimate of a difference of means supports a decision that the true difference of means falls above, below, or to either side of a prescribed difference, and (4) how precisely the difference between the respective means of two populations is estimated by the difference in their sample means. Each of these four types of inferences will require different treatments, which are discussed in the paragraphs that follow.

Figure 13-4a shows graphically how an inference is made of whether a given sample average, estimated by measurement, falls above or below a prescribed reference value. The distribution in the center of the figure represents a reference distribution about the reference value, is labeled H_0, and is called the "null hypothesis." The null hypothesis describes the distribution expected of sample means from the population represented by the reference distribution. (In the examples that follow, the distributions used are those of the student distribution for sample means, but the same logic can be applied to other distributions of a random and independent variable.) An alternative hypothesis might be that a given sample belongs not to the reference distribution but to another distribution that itself lies

(a) to the right of the reference distribution (e.g., hypothesis H_1),

(b) to the left of the reference distribution (e.g., hypothesis H_2), or

(c) to either side of the reference distribution (e.g., hypotheses H_1 or H_2).

The standard deviation or standard error of the reference distribution may be known or it may be estimated from that of the sample.

To reject a null hypothesis, a decision criterion is required. The two limit lines, drawn in Figure 13-4a, form a basis for such a criterion, either independently or collectively. Formally, one never accepts an hypothesis, but tests to see whether it can be rejected or not. If the sampled value in question lies outside of the limit line(s), then the null hypothesis will be rejected; if the sampled value lies inside of the limit lines(s), then the null hypothesis will not be rejected. Unbiased decisions usually require that decision criteria be made before samples are taken. The positions of these limits are defined by the tail areas they delimit in the reference distribution, and these tail areas, either collectively or singly as appropriate, are called the "α" criterion. Criteria corresponding to "a" or "b" (see previous paragraph) comprise what are called a "single-sided" test or "single-tailed" test. The criterion for "c" comprises what is called a "two-sided" or "two-tailed" test. The positions of the limits would be labeled $+t(\alpha)$ and $-t(\alpha)$ respectively for the two different single-sided tests, but would be labeled $+t(\alpha/2)$ and $-t(\alpha/2)$ for the two-sided test. The α criterion is the probability that a sample mean may have a value outside the limit(s) even though the population sampled is the same population represented by the reference distribution (i.e., the null hypothesis). If a desired limit for a single-sided test lies to the right of the center of the reference distribution, α corresponds to its tail area lying to the right of that limit. If the desired limit for a single-sided test lies to the left of the reference distribution, α corresponds to its tail area that lies to the left of that limit. If the desired limits form a two-sided test, then α corresponds to the sum of these two tails.

The probability of making what is called a "Type-I error" is represented by α. A Type-I error is the error of rejecting the null hypothesis when it is really true. In other words, if a sample mean were to lie outside a limit criterion, one would conclude that the sample did not belong to the reference distribution. Because there is a finite probability (equal to or less than α) that a sample belonging to the reference distribution might have such an extreme value, the possibility of an error (a Type-I error) always exists when the prescribed limit(s) is (are) used as a decision criterion. The probability of a Type-I error can be made arbitrarily small by

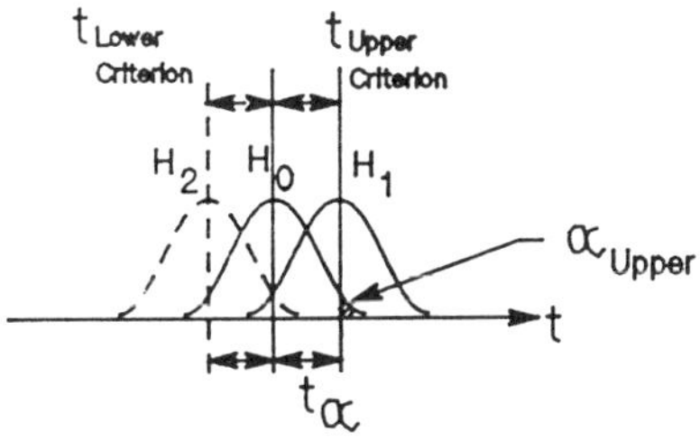

(a) Minimizing error in rejecting H_0

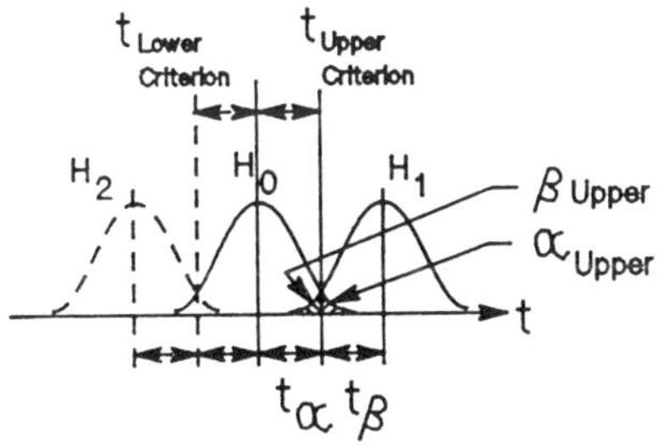

(b) Minimizing error in accepting H_1

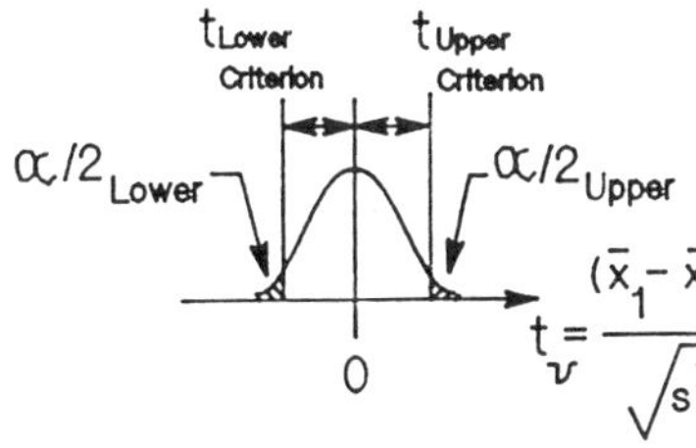

$$t_{\nu} = \frac{(\bar{x}_1 - \bar{x}_2) - (\mu_1 - \mu_2)}{\sqrt{s_1^2/n_1 - s_2^2/n_2}}$$

$$\text{If } \sigma_1 = \sigma_2: \begin{cases} s^2_{pooled} = \dfrac{(n_1 - 1) * s_1^2 + (n_2 - 1) * s_2^2}{n_1 + n_2 - 2} \\ \nu = n_1 + n_2 - 2 \end{cases}$$

$$\text{If } \sigma_1 \neq \sigma_2: \ \nu = (s_1^2/n_1 + s_2^2/n_2)^2 / \left[\frac{(s_1^2/n_1)^2}{n_1 - 1} + \frac{(s_2^2/n_2)^2}{n_2 - 1} \right]$$

(c) Student distribution of sample differences

Figure 13-4. Statistical inferences and confidence limits.

choosing a sufficiently small value of α as the decision criterion or by increasing sample size. A popular value to use in statistics for α is 5 percent; this value makes it unlikely that a difference will be perceived between the population sampled and the reference population when there really is no difference.

An α criterion, whether for a single-sided or two-sided test, allows a decision of whether or not to reject the null hypothesis, and controls the probability of a Type-I error. But when the null hypothesis is not rejected, an α criterion alone does nothing to control the consequent error in not rejecting an alternative hypothesis. This second type of error is called a Type-II error. Consider the alternative hypotheses H_1 and H_2 which are centered on the α limit lines in Figure 13-4a. If a measurement does not exceed the α limit line(s), we would not reject the null hypothesis, even though an alternative hypothesis may be correct. The probability of a Type-II error is called "β" and is that area under the distribution for an alternative hypothesis that falls inside of the limit line(s) defined by α. In Figure 13-4a, β for H_1 and H_2 are both 50 percent. A β of 50 percent means that if the alternative hypothesis is true, the α and β criteria (taken together) permit a 50 percent chance of making wrong decisions when an alternative hypothesis is rejected. This is synonymous with not being able to recognize real differences more than half the time if they really exist.

Figure 13-3b shows a situation with a value of β smaller than 50 percent, i.e., when the alternative hypothesis is farther removed from the null hypothesis. Suppose α is 5 percent and β is 10 percent. Then if the t-score of a sample mean does not exceed the α limit, the null hypothesis will not be rejected, but the alternative hypothesis would be rejected. Also the probability of making a Type-II error will be β, or 10 percent, but only if that alternative hypothesis is really true. If the sample falls outside the α criterion limit, and the null hypothesis is rejected accordingly and not the alternative hypothesis, the chance of a Type-I error is α, or 5 percent. But this latter event should not be interpreted as justification to accept the stated alternative hypothesis or any other. Rather, if the null hypothesis is rejected, the chance of Type-II errors can be evaluated as a function of different alternative hypotheses. Note that because the width of a student distribution for the mean of a sample can be made arbitrarily small by increasing sample size (recall the inverse relation between standard error and sample size n), a sufficiently large sample size can be chosen for any required α and β to achieve a selected tradeoff between Type-I and Type-II errors.

Upper and lower confidence limits, i.e., statistical confidence limits, can be used to express the probable precision of a sample mean in estimating a true mean. No α criterion is involved because no a priori test or reference distribution is involved, but the concept of β can be used. In Figure 13-4a, there is a lower tail to the hypothesis H_1 that lies below the center or mean of the central distribution. And there is an upper tail to the hypothesis H_2 that lies above this mean. Each of these tails can define a probability of $\beta/2$. Thus β becomes a measure of the probability that the true distribution could be as far away from the center mean (which can represent the value of the measured sample mean) as the alternative hypotheses and still result in the value measured for the sample mean. The positions of the limit lines drawn through the center lines of these two alternative hypotheses then become our tolerance limits for saying that the measured mean estimates the position of the true mean to within these limits to a confidence equal to $1-\beta$, i.e. one minus the combined area of the two tails described. The formula that expresses this statement was given in Figure 13-3c. Note that if the distribution shape is symmetrical, unlike that of the Poisson distribution, the two tails described coincidentally will be of the same area as the tails shown in Figure 13-3c. However, to use the tails shown in Figure 13-3c to explain the definition of confidence limits would not have been strictly correct.

Confidence limits are very good for quantifying the accuracy of measured PWP results. The absolute value of the limits that apply to the measurement are obtained by (1) calculating the standard deviation of the measurements in a sample, (2) then dividing this standard deviation by the degrees of freedom to obtain the standard error in the assumed t-distribution of sample means, and (3) finally multiplying the result by the value of the t-score corresponding to $1-\beta/2$ (t-scores for both α and β probabilities are commonly referred to in tables as α.)

To infer that the mean of a first sample of measurements would fall either above or below a second sample of measurements, a null hypothesis would state that their difference is zero. To infer that their difference would be a certain finite amount, the null hypothesis would state this amount instead of zero. In either case, if the differences are random and independent, the distribution shown in Figure 13-4c will be used. This distribution is also the student distribution. As in the previous discussion, inferences can be made directly with this distribution. Similarly, one can calculate confidence limits as the precision with which a sample of the difference between two means can be used to estimate the real difference. For these applications, the scale factor for the t-statistic is the square root of the sum of the squares of the standard errors of each sample. If it can be assumed that the standard deviations of the two populations from which the two samples were drawn were the same (whether or not the sample sizes were the same), then a pooled value of standard deviation can be used in place of the two sample standard deviations; however, since this is not a valid assumption with Poisson distributions that have different means, this pooling is not recommended for inferences of particle counts.

Whereas the Gaussian distribution requires only a value of probability for looking up a corresponding value of the z-statistic in a table (or vice versa), the student distribution requires the parameter of degrees of freedom. For testing a null hypothesis for a mean, the degrees-of-freedom parameter is simply equal to one less than the sample size. For testing a null hypothesis that the difference between two sample means is different from some stated value (which may or may not be zero), two appropriate formulae are given in Figure 13-4c. One of these formulae is used when the experimenter is confident, from prior knowledge, that the true standard deviations of the two populations are equal, although their common value may not be known. In this case the degree of freedom is two less than the sum of the two sample sizes. On the other hand, if it is not known that their standard deviations are the same, the more likely case with particle counts, the degree of freedom to use is given in the second formula. This second formula calculates degrees of freedom as the weighted average of the degrees of freedom in each sample and as a function of the two standard deviations of the samples. This second formula for degrees of freedom is called the Aspin-Welch formula (Natrella, 1986, pp. 3-27 to 3-30 and Duncan, 1986, p. 616). Another equation, much like this one but differing by replacement of the two minus signs with plus signs and by subtraction of 2 from the whole right-hand side of the equation, works almost as well and is called the Welch formula (Gibra, 1973, pp. 260 and 325; Natrella, 1986, pp. 3-27 to 3-30; and Beyer, 1984, p. 313).

One more statistical tool of value to review is analysis of variances, called "ANOVA" (Gibra, 1973; Box, Hunter, and Hunter, 1978; Natrella, 1986; Duncan, 1986; Beyer, 1984; and Rossi, 1984). Here the appropriate distribution to use is the F-distribution. The variate statistic to use is formed by taking the ratio of the respective quantities [(sample $s)^2$/(population $\sigma)^2$] for each of two variances in question. If it is hypothesized that the two population standard deviations are one and the same (null hypothesis), but the value of this ratio is sufficiently different from unity, then the variances must be from different populations and we would reject the null hypothesis. Therefore, just as inferences on means of populations are formed, so are inferences about their variances. Although the F-statistic will not be used in this chapter, the reader will do well to be aware of its utility, because variance analysis can play an important role in analyzing variance components that make up the total variance underlying the confidence limits stated with confidences in mean particle counts. That is, with variance analysis, one can compare the relative contributions to PWP uncertainty caused by the PWP of the counter itself, the analytical error of the counter, the variance in PWP of the equipment or process step or process segment being tested, and the variance of other factors that affect the measurements and the counts being measured.

It will also be helpful in some of the discussions below to recall that variances are additive when their respective populations vary randomly and independently of one another. Although purposefully not done in this chapter, this fact can be used to mathematically refine confidence measures on PWP measurements by removing the variance effects of the contamination by counters and by removing other random errors associated with the precision of counters, if estimates of these variances can be substantiated as accurate.

METHOD STRUCTURES, TRADEOFFS, AND CALCULATION OF STATISTICAL CONFIDENCE

Before we discuss the benefits of conducting particle counting experiments in different ways, we must first emphasize that the accuracy, precision, and meaningfulness of particle counts can be no better than that of the counting instruments and techniques used.

Fundamental to meaningful PWP measurements is accurate calibration of the particle counter instrument and counting technique to be used. Refer to Chapter 5 for discussion of the requirements of good analytical calibration. Especially note the dependence of calibration on the absorptive and reflective nature (refractive indices and thicknesses) of any films comprising the surfaces on which particles are to be counted. Also note how calibrations may be affected by statistics of overcounting and/or undercounting caused by the particular design of the counter's detection optics and processing electronics and software. And if counting efficiency or particle sensitivity varies with wafer rotation, either maintain wafers in a standard orientation when counting them, or assure that sufficient numbers of wafers are used in random orientations to avoid biasing the means of the counts.

Also important is agreement on the criteria for the setting of particle size thresholds. For example, if a threshold is to be set at a level corresponding to the signals produced by latex particles from a monodisperse distribution whose count median is at a diameter of 0.212 μm, then the percentage of these particles that will be detected when the threshold is set to this "set-point size" must be prescribed. Two of the most common choices are to set the threshold such that the counter counts either 50 percent or 95 percent of the monodisperse population. Because the 50-percent point most closely represents the signal amplitudes produced by diameters equal to the set-point size, this author argues in favor of the 50-percent criterion (See Chapter 5). Note that setting the threshold to permit the counter to count 95 percent of the population in the mondisperse distribution actually means setting the threshold to detect large fractions of population densities at sizes lower than the stated set-point size.

Given clear agreement on what counts mean, there are numerous scenarios for obtaining surface particle counts to characterize the cleanliness of processing equipment and/or process steps. To avoid confusion, it is helpful to categorize the six most obvious and useful of these scenarios and to give each a discriminating name:

[From this point on in this chapter, for convenience and brevity, the word "equipment" will be used to mean "process equipment, process steps, and/or process segments."]

1. "Monitoring Counts over Time at One Location": This method measures particle counts on wafers that have traveled through the normal process flow up to and including a piece of processing equipment in question. These wafers may even be patterned. This method has the advantage of detecting real changes when they occur, but it cannot determine whether it was the last piece of equipment that caused the change. Also, size calibrations are difficult and may remain in question when counting particles on patterned wafers or on wafers with films more than approximately 100 Å in net thickness.

2. "Comparing Monitored Counts at Different Locations": This method of measurement uses the first method (number 1 above) but adds the additional complexity of determining how much change counts undergo as they move from one equipment to the next. The measuring techniques and limitations are the same as before, only now the causes of change might be isolatable to a specific piece of equipment. The difficulty of calibrating sizes carefully for the wafer surface structures at different points in a process makes this method risky for determining how much particle densities differ from one point in a process to another. Also, if variabilities are large at respective locations, it is hard to recognize variability between locations.

3. "Looking for Accumulations (PWP) on Initially Clean Test Wafers": This method requires that the wafers used be initially clean test wafers, which is to say that it is known or assumed that the number of particles on the test wafers before they are put through the equipment will be insignificant compared to the final counts obtained when the wafers exit the equipment.

4. "Looking for Accumulations (PWP) on Test Wafers, Blocking by Wafer": This method requires counting the particles on test wafers both before and after they are passed through the equipment under test. For each wafer, the initial count is subtracted from the final count to obtain the accumulated count on that wafer. This "blocking" of initial and final counts by wafer permits what is called a paired test, because each final measurement is paired with the initial count on the respective wafer. This way, variabilities on initial counts do not obscure the data on what is added to the wafers by the equipment.

5. "Looking for Accumulations (PWP) on Test Wafers, Blocking Only by Lot or Batch": In this method, both initial and final counts are obtained for each test wafer, but then the initial and final counts are averaged separately before their averages are subtracted to obtain an average PWP. Recording data and making calculations are considerably easier than when each wafer difference is tracked independently as in method 4. However, the correlation between original wafer cleanliness and increased final counts is ignored, resulting in larger variance and therefore looser confidence limits in the average PWP.

6. "Looking for Accumulation Rates (PWP) by Regression": This method requires that the same wafers be passed through a piece of equipment sequentially many times, and that the measuring of particle counts on these wafers be synchronized to occur between certain of these multiple passes. A linear regression through the succession of generally increasing counts then produces an estimate of the slope of this increase, which can permit calculations to provide a good estimate of the average PWP. Either a plot of the increasing counts with the regression line included, or a calculation of variance about the regression line, reveals the pass-to-pass variability of PWP. If the equipment under test can be made available long enough to perform the necessary measurements, and if the net count does not accumulate enough to place the counter accuracy in question, then this method is the best one for obtaining one-time estimates of PWP and its variability.

In all of the six scenarios listed above, it must be remembered that the particle counts obtained include not only particles added to the wafer by the process equipment, but also particles added to the wafers while they were being inspected. And some of the scenarios give counts which, whether they include the particles initially on the wafers or not, result in confidence tolerances that are inflated by the variability of the initial counts. That this is true can be seen with the help of Figure 13-5, in which errors are shown as they accumulate in the process of starting with fresh wafers, counting them initially, putting them through a piece of equipment to be tested, and counting them afterward. The letters P, L, Q, M, and R all refer to how many times things are done and are defined in paragraphs below that discuss Figure 13-6. For present purposes the values of these letters are all set to unity.

Particle measurements on wafers (whether requiring both initial counts and final counts, a series of repeated counts, or just final counts) begin with wafers that have an undetermined number of particles (whether zero or not). The particles on these wafers originally may have an unknown number distribution (not to mention an unknown size distribution) whose population is certainly different from zero, as indicated by the deviation labeled ϵ_1 in Figure 13-5. Means may vary from one boat (or cassette or batch) to another according to a second distribution. Any one boat may have a mean that is different from the mean over all boats.

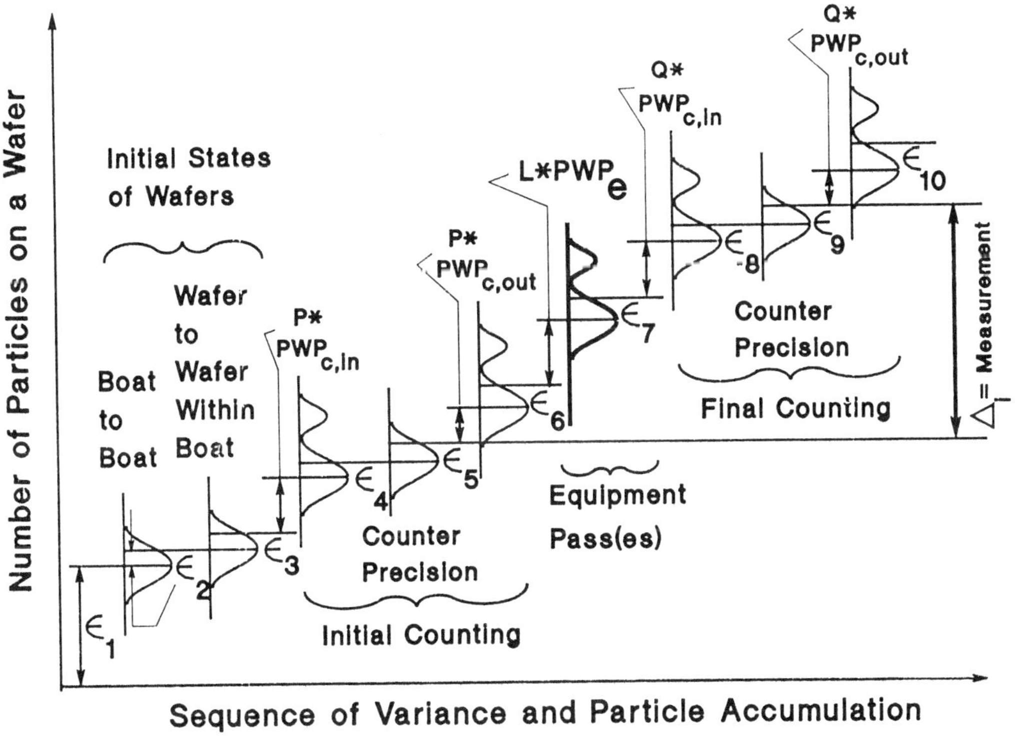

Figure 13-5. General analysis of variance (Ex: P=L=Q=M=R=1).

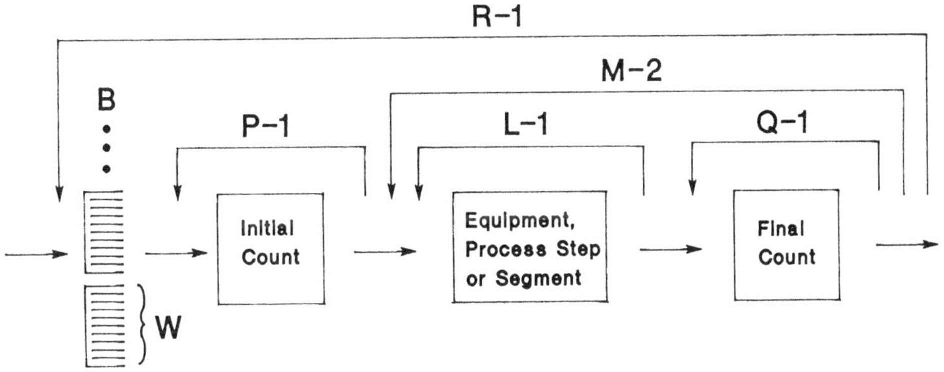

Figure 13-6. General structure for particle counting.

This second deviation is ϵ_2. The number of these original particles on each wafer within a boat may vary according to a third distribution, where the number on any one wafer may vary from the mean of that distribution by the deviation ϵ_3.

If the original particle numbers are checked with a particle counter—for example, one that is automatic and has both an input and an output cassette handler—then while the wafers are sitting on the input cassette, an unknown additional number of particles may deposit on these wafers. This leads to a new distribution with a new count mean of that boat. This shift in mean is labeled $PWP_{c,in}$ to indicate the mean number of particles that have been added to the wafers while they were being managed at the input cassette. The resulting distribution shown in the figure is depicted as bimodal to remind us that with cassette handlers, some wafers end up being exposed to the ambient environment longer than others. In counting the number of particles on one of these wafers, if the true count for one wafer is different from the cassette mean by the deviation ϵ_4, the counter itself may make an error in counting, indicated by the deviation ϵ_5. While the wafers are being handled by the output cassette handler of the counter, another shift in particle count is incurred due to new particles depositing from the environment, and this is labeled $PWP_{c,out}$.

The number of particles determined to exist on one wafer just before it is subjected to the contamination of a piece of equipment under test is different from the mean number for its cassette by the deviation ϵ_6. The contribution of the equipment under test is labeled PWP_e, but any one wafer will have an additional deviation, labeled ϵ_7.

Deviations ϵ_8, ϵ_9, and ϵ_{10} and contributions of particles going in and coming out of the counter in obtaining the final particle counts are defined as before for the initial counting.

These deviations may not all be additive in any one case as they are depicted in the figure; rather, most tend to vary independently. But the conclusion is that many variations are involved, and if they are not given proper attention, they can lead to incorrect estimates of particle contributions by the equipment under test. The tradeoff between (a) including initial counts or (b) skipping them is the tradeoff between (a) the total PWP_c of the counter (and its variability) along with the analytical error of the counter in the initial measurement and (b) the magnitude of the count of original particles. (The total PWP_c of the counter is the sum of both $PWP_{c,in}$ and $PWP_{c,out}$.) The ultimate choice therefore will depend upon whether the relative magnitude of the anticipated PWP_e of the equipment under test is significantly larger than either the PWP_c and analytic error of the counter and/or the probable number of particles on the original wafers. Naturally, the best percent accuracy and precision of counts will be obtained if the PWP_c of the counter is low, the analytic error of the counter is low, and the number of particles accumulated by the equipment is large.

In the scenarios numbered 3 through 6 above, all of which are implementations of the general PWP method, the number of particles accumulated by the equipment can be made arbitrarily large by subjecting the wafers under test to many passes (a factor denoted by L) through the equipment before the final particle counts are taken. Dividing the total accumulation by the number of passes measures the average PWP more precisely on that equipment. A disadvantage is that if most of the counted particles are added in only one or a few of the multiple passes, this pass-to-pass variability would not be detected, because this approach, only produces an average PWP. In a similar manner, multiple passes may subject the counting experiment to greater risk of wafer handling errors ("particle accidents"), and these could ruin the integrity of the final PWP_e estimate. This approach for making counter and original count errors insignificant will not work in those cases in which the particle contributions of the equipment are not additive (as in an effective cleaning step) or in which the equipment alters the surface characteristics of the wafers sufficiently that multiple passes would invalidate size sensitivity calibrations during final counting.

Because the standard deviation of a Poisson distribution is equal to the square root of its mean, and because particle counts are sometimes Poisson distributed, the original wafers should be as clean as possible to reduce uncertainty in the measured estimates of the original number, even if initial counts are to be taken.

Figure 13-6 describes a general structure for executing PWP particle measurements. Special structures can be generated by specifying the numerical values of the letters P, L, Q, M, and R, as well as of the letters B and W. P is the number of sequential times (0, 1, 2, 3, etc.) that wafers would be measured initially before being put through the equipment under test. L is the number of sequential times the wafers would be put through the equipment before being recounted. Q is the number of sequential times the wafers would be counted after leaving the equipment. M-1 is the number of times these same wafers would experience the sequence of exposure to the equipment followed by counting. If M-1 is of value 1 (i.e. M-2 = 0), then no iterations as described will be done, and the first instance of Q will become the final measurements. Note that M is the number of times a set of wafers newly arrives at the counter for counting, including the first arrival for P sequential, initial counts. R is the number of times the above activities would be performed with a new group of wafers, including the first time.

The details of wafer sequencing within the counter and within the equipment are not defined here, as they will depend largely on the characteristics of the counter and the equipment to be tested. One example of this detail is whether one wafer of a group is counted P times before the next wafer is counted or whether all wafers of a group are counted once before going on to a second counting, etc., until all wafers have been counted P times. Other examples are: in what sequence are wafers to be handled from a boat or cassette, how many wafers should be used in a boat or cassette, how many cassettes of wafers should be used, should the cassettes or boats themselves be reused for multiple passes in equipment, how should wafers be grouped and handled during testing of equipment with batch sizes different from numbers conveniently accommodated by the capacities of the cassettes or boats, should the wafer carriers used between the equipment and the counters be the same cassettes or boats used in the equipment under test, etc. The determination of these details is left to the experimenter but should be prescribed and documented in each case.

Figures 13-7, 13-8, and 13-9 show mathematics and tables useful for calculating estimates of means and 95 percent confidence limits for PWP_e of equipment, using favorable

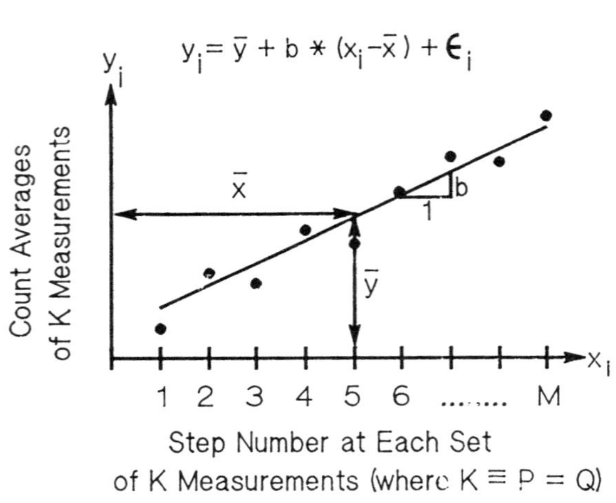

$$b = \frac{\sum_1^M (x_i - \bar{x}) * (y_i - \bar{y})}{\sum_1^M (x_i - \bar{x})^2}$$

$$s = \sqrt{\frac{\sum_1^M (y_i - \bar{y})^2 - b^2 * \sum_1^M (x_i - \bar{x})^2}{M-2}}$$

$$SE_b = s \Big/ \sqrt{\sum_1^M (x_i - \bar{x})^2}$$

$$K * \overline{PWP_c} + L * \overline{PWP_e} = b \pm t_{1-\alpha/2,\, M-2} * SE_b$$

1) First evaluate counter alone (Requires L=0): $\overline{PWP_c} = b/K \pm t_{1-\alpha/2,\, M-2} * SE_b / \sqrt{K}$

2) For evaluation of equipment in general: $\overline{PWP_e} = b/L - K * \overline{PWP_c}/L \pm t_{1-\alpha/2,\, M-2} * SE_b / \sqrt{L}$

M−2	1	2	3	4	5	6	7	8	9	10	20
t(97.5%,M−2)	12.7	4.30	3.18	2.78	2.57	2.45	2.36	2.31	2.26	2.23	2.09
$\sum_1^M (x_i - \bar{x})^2$	2	5	10	17.5	28	42	60	82.5	110	143	885.5

Figure 13-7. PWP by linear regression.

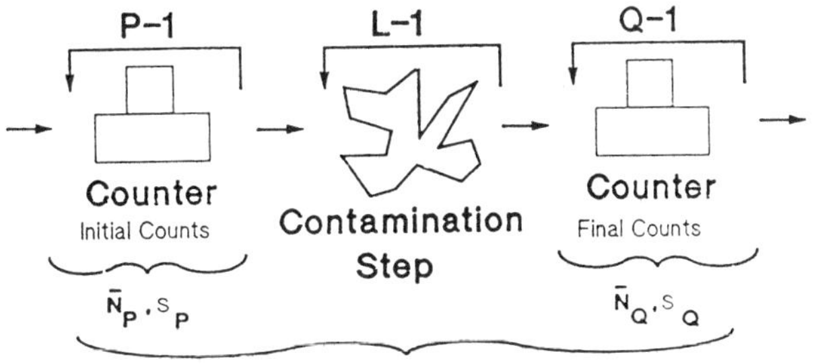

$$\Delta = (\bar{N}_P - \bar{N}_Q)$$

$$\Delta = L*\overline{PWP_e} + \left(\frac{P+Q}{2}\right)*\overline{PWP_c} \pm t_{97.5\%,\nu} * \sqrt{\frac{s_P^2}{P} + \frac{s_Q^2}{Q}}$$

$$\overline{PWP_e} = \Delta/L - \left(\frac{P+Q}{2*L}\right)*\overline{PWP_c} \pm t_{97.5\%,\nu} * \sqrt{\left(\frac{s_P^2}{P} + \frac{s_Q^2}{Q}\right)}/L$$

$$\text{where } \nu = \left\{\left[\frac{s_P^2}{P} + \frac{s_Q^2}{Q}\right]^2 \Bigg/ \left[\frac{(s_P^2/P)^2}{P-1} + \frac{(s_Q^2/Q)^2}{Q-1}\right]\right\}_{\text{nearest Integer}}$$

but when P = 0, then $\nu = Q-1$

ν	1	2	3	4	5	6	7	10	20	100	1000
t(97.5%,ν)	12.7	4.30	3.18	2.78	2.57	2.45	2.36	2.22	2.09	1.98	1.96

Figure 13-8. PWP from multiple counts, initially and finally, unpaired.

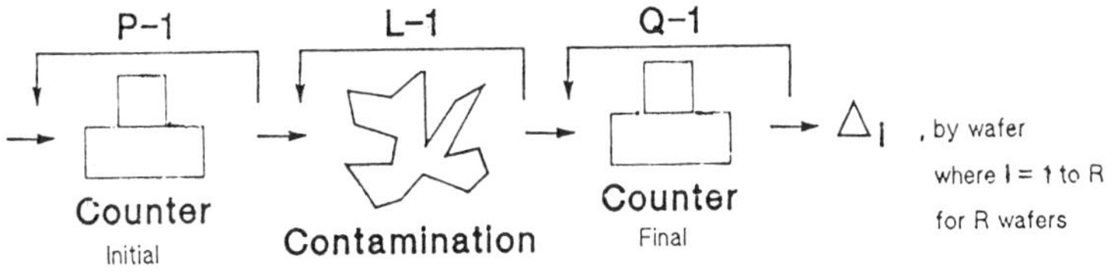

$$\Delta_I = L*\overline{PWP_e} + \left(\frac{P+Q}{2}\right)*\overline{PWP_c} + \epsilon_I(PWP_e, PWP_c, \text{Counter})$$

$$\bar{\Delta} = \frac{1}{R}\sum_{1}^{R}\Delta_I = L*\overline{PWP_e} + \left(\frac{P+Q}{2}\right)*\overline{PWP_c} \pm t_{97.5\%,R-1} * \frac{S}{\sqrt{R}}$$

$$\overline{PWP_e} = \bar{\Delta}/L - \left(\frac{P+Q}{2*L}\right)* \overline{PWP_c} \pm t_{97.5\%,R-1} * \frac{S}{\sqrt{R*L}}, \text{ where } S = \sqrt{\sum_{1}^{R}\frac{(\Delta_I - \bar{\Delta})^2}{R-1}}$$

R-1	1	2	3	4	5	6	7	10	20	100	1000
t(97.5%,R-1)	12.7	4.30	3.18	2.78	2.57	2.45	2.36	2.22	2.09	1.98	1.96
$t/\sqrt{R*L}$	8.99	2.49	1.59	1.24	1.05	.926	.834	.669	.456	.197	.062

Figure 13-9. PWP by pairing initial and final counts by wafer.

subsets of the general structure. Although only the formulae for confidence limits are given, the reader should be able to use the terms in these expressions for t-statistics, standard error, and degrees of freedom for forming other inferences as reviewed in the previous section.

Figure 13-7 applies to the situation when $R = 0$ and $(M\text{-}2) > 0$ and is the case when a given set of wafers is brought back to the equipment under test multiple times to be reprocessed and recounted. The measurements resulting from such a sequence of exposure and measurements permits a linear regression or fitting of a straight line through the successive data points. Each data point is the mean count obtained each time the wafer(s) was(were) counted P or Q times. Note that in the solution of the regression equations for the PWP_e component of the slope "b" (see equation 2 in Figure 13-7), the estimate of confidence limits is conservative because it is purposefully left unaltered and retains the effects of variances in the counter PWP_c and in the counter analytic error. Note that the principal advantage of using an L factor greater than 1 is that the confidence limits are tightened by the factor $1\sqrt{L}$). This approach to PWP evaluation as presented in Figure 13-7 is very eloquent and has the advantage of encouraging a graphical plot of the data to substantiate the goodness of fit. The disadvantage, however, is that it requires that the contamination incurred to the wafers with each subsequent repetition of exposure and recounting be sufficiently small that after many such repetitions, the net number of particles on the wafer is still within the range in which the counter will be accurate.

It is highly recommended that both the process depicted in Figure 13-7 and formula 1 be used to characterize and evaluate the PWP_c of the counter before it is used in to evaluate PWP_e of other equipment. Setting the factor L to zero to achieve this simply means that for this purpose, the wafers are not to be subjected to any equipment other than the counter itself. It is advisable that the factor K—which is the number of successive counts of a wafer that will be averaged to create one data point for the regression—be 4 or 5. This is in order to avoid skewing the regression toward higher slopes ("b") that might otherwise be caused by any Poisson-like distribution in the particle-counting error or particle accumulation rate incurred during the counting steps. Plotting these raw data points, used for the regression of the particle data, from exposure to the counter alone makes it easy to ascertain whether the PWP_c of the counter can be relied upon to be repeatable. If the counter can be demonstrated to be repeatable, then in using the counter to evaluate the PWP_e of other equipment, the known PWP_c of the counter can be subtracted from the raw PWP data of the equipment under test. This is done in equation 2 (Figure 13-7) in the formula for PWP_e.

Figure 13-8 describes a useful structure to use in obtaining a one-time estimate of PWP_e. This approach is best to use (a) when PWP_e is too large to permit the regression approach of Figure 13-7 along with its equation 2 (i.e., when the wafers become too dirty for accurate counting) and/or (b) when the equipment to be tested is available only long enough to carry out a single pass (or L passes) during a single test. It can be used even when the variability and count of the original wafers are significant compared to the numbers added by exposure to the equipment. The structure is particularly appealing because it uses the average of the initial counts and the average of the final counts as well as their respective sample standard deviations. It therefore does not require careful (sometimes tedious) management of individual wafers so as to pair each initial count with the appropriate final count by wafer. That is, it does not require blocking by wafer. It is important to repeat this whole process enough times to experience the type of equipment variance from batch to batch, day to day, and maintenance to maintenance that one hopes to include in any characterization of PWP_e. Confidence limits on the average PWP_e obtained by performing this test sequence once on one wafer or on a batch of wafers are given in Figure 13-8. Confidence limits on the average PWP_e obtained by performing this test more than once can be calculated using the formulae presented in Figure 13-9, which will be discussed in the next paragraph. Note again that the confidence limits calculated are conservative ones in that they include the variances in PWP_c or in the analytical errors of the counter; because of the structure of this test, they will also include the effects of any variability in the initial counts.

Figure 13-8 can be made applicable to the situation when P is equal to zero, as when wafers are presumed originally clean relative to the contamination expected of the equipment under test and therefore the initial counting is omitted. Only in this case, the degree of freedom is simply Q-1.

Figure 13-9 is best applied to the conditions just stated for Figure 13-8, except that in this case initial and final counts are matched by wafer to obtain a collection of differences, labeled Δ_i. The advantage of blocking the initial and final counts by wafer in this manner is that the confidence tolerances calculated for the mean over many such differences do not include the variability of the initial counts. As before, the confidence limits given here are made conservative by not removing the effects of variances in PWP_c or in the analytic error of the counter. Figure 13-10 depicts some of the complications that can make wafer management and recording of count data tedious and error-prone when initial and final counts are blocked by wafer.

Another approach combines elements of Figures 13-7 and 13-8. A test is conducted exactly as described in Figure 13-8, but regressions of the P initial count data and the Q final count data are performed to estimate the PWP_c of the counter. This requires more computational work than performing separate analyses of PWP_c and PWP_e but may be justified when the contribution of the counter is both large and variable. This combined approach was discussed in Tullis (1988) and will not be covered further here.

CONTROL CHARTS, STATISTICALLY DESIGNED EXPERIMENTS (SDEs), AND PRACTICAL OBSERVATIONS

Although a detailed discussion would be beyond the scope of this chapter, an overview of Control Charts and Statistically Designed Experiments is in order. The theory and background of these statistical tools and disciplines can be explored in the references (Gibra, 1973; Duncan, 1986; and Western Electric Co., 1956).

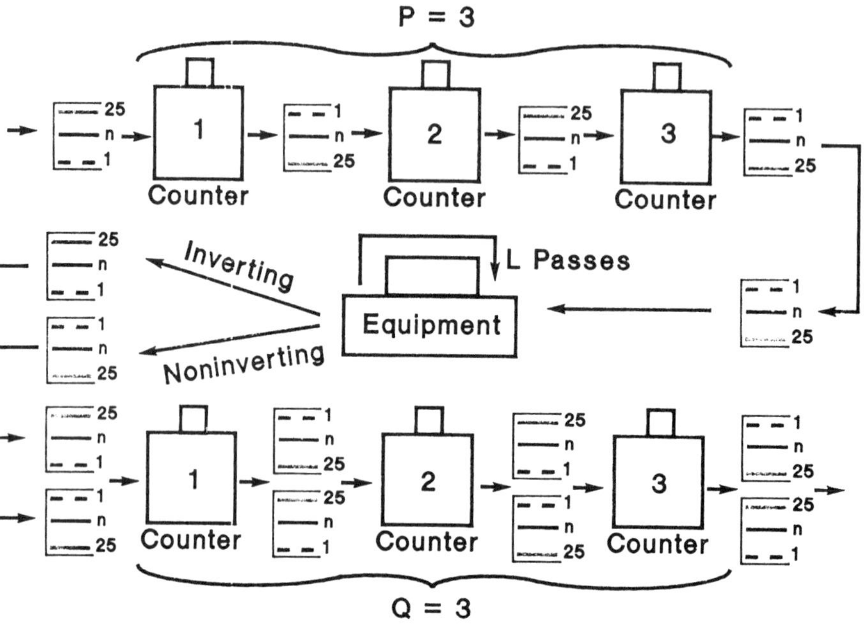

Figure 13-10. Tracking of wafer position.

The formulae presented to calculate confidence limits and to estimate mean values of PWP_c and PWP_e perhaps are handy but also can easily be misused and misinterpreted. The reason is simple: the calculation of confidence interval limits as above is based on assumptions of randomness, independence, and uniformity that do not always apply to the sampling techniques used or to the contamination being estimated. Even analyses of variance and inferences based upon plotted rather than assumed distributions are not immune to erroneous conclusions, for the same reasons. Control charts offer another means of analysis that can be used, in addition, to visualize the frequency distribution of data and its distribution with time. Also, they provide an objective set of rules with which to diagnose whether the charted data are distributed as it should be or is varying in more than a naturally random manner.

Several alternative and competing types of control charts exist that are applicable to particle counting. If PWP measures can be made for a piece of equipment on a routine basis (but with samples taken randomly and independently), then plotting the resulting data in control charts helps determine how stable and controlled the counts are. PWP data should be plotted as a C chart if the PWP data points correspond to single wafers for which the area of the wafer inspected remains constant. However, if the PWP points plotted are calculated as the total count on sets of wafers which vary in number, or calculated as counts on only fractions of a standard wafer area, then a U chart should be used. Even better, if sufficient data can be obtained, X-bar & R charts should be used. The advantages of control charts are that they allow one to visualize the stability of the PWP results over time, and to judge this stability objectively relative to special control limits calculated either from historical data or from the data on the chart itself. These judgments are made using established rules for detecting possible instabilities. Examples of C charts, U charts, and X-bar & R charts are given in Figures 13-11, 13-12, and 13-13. The rules for evaluating possible signs of unusual variance in the data plotted on these charts are given in Figure 13-14. These rules should be followed in marking areas of particular interest ("X") and interpreting the data on these charts. These charts not only can be used for initial characterization and acceptance testing, but also serve very well for routine inspection, give constant feedback relating to ongoing efforts at improvement, and provide rapid notification when jumps in particle sourcing occur. The control limits can be determined in a capability study and then frozen to create shop control charts. A shop control chart allows plots of new points to be compared to the variance and control limits determined earlier in the capability study. Control limits on a capability study control chart are calculated from the distribution of the data plotted on the same plot.

The reasoning behind C charts and U charts rests on an assumption that the particles accumulated by wafers under steady-state conditions will obey a Poisson frequency distribution. A Poisson distribution is what would be expected if the contamination particles were caused by a uniform, random, and independent dispersement of particles over the surface area of the wafer. As discussed before, these assumptions are not always valid.

It is safer to take not one but several measurements (four or five are strongly recommended) of PWP, randomly and independently of one another, and to identify these as a single sample. A succession of averages of such samples of PWP measurements can then be

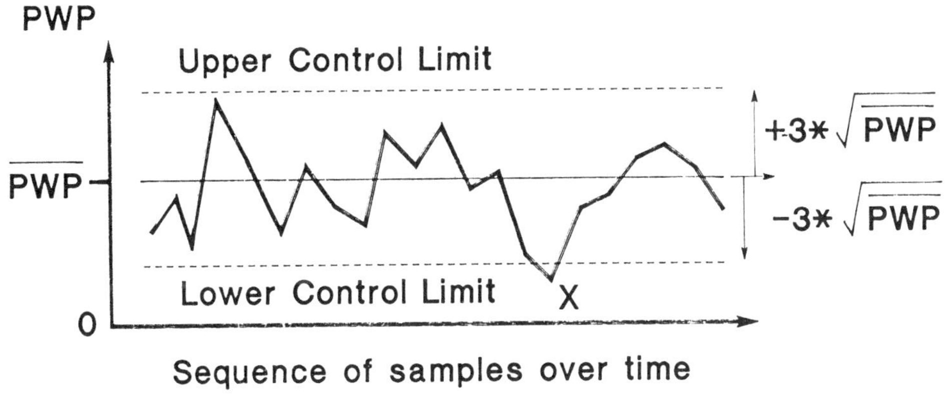

(One initially clean wafer per point)

Figure 13-11. Control chart: C chart.

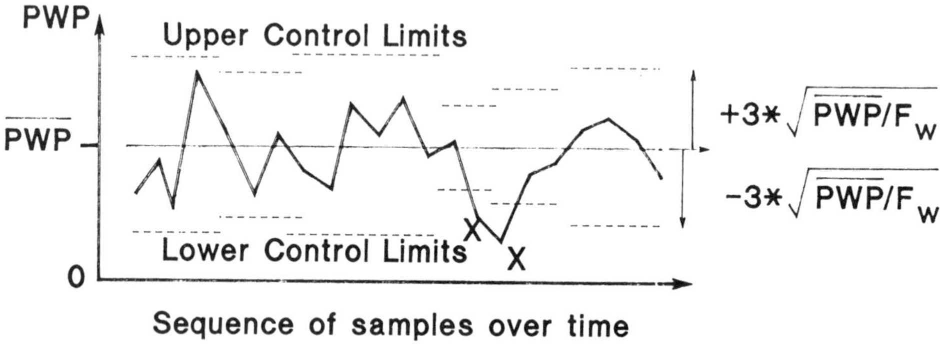

(When inspecting fractions F_W of initially clean wafers)

Figure 13-12. Control chart: U chart.

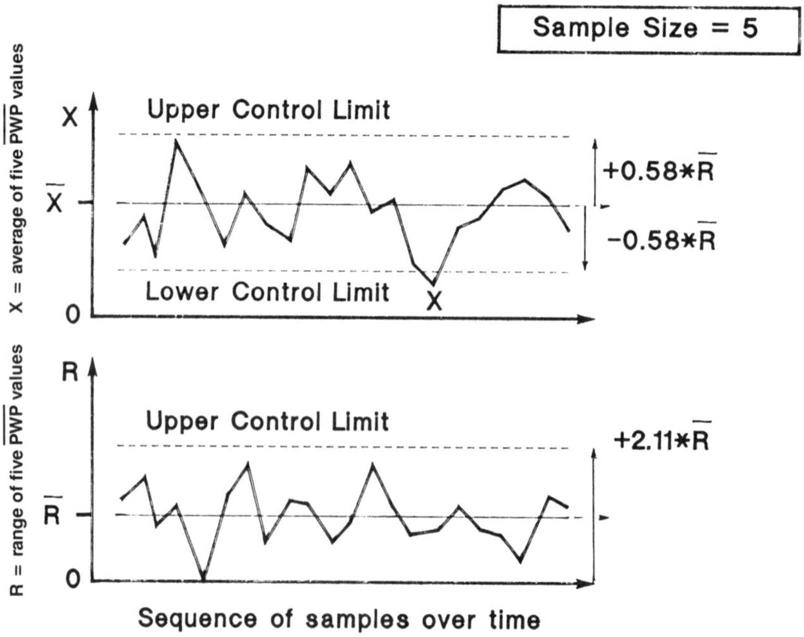

Figure 13-13. Control chart: X-bar and R chart.

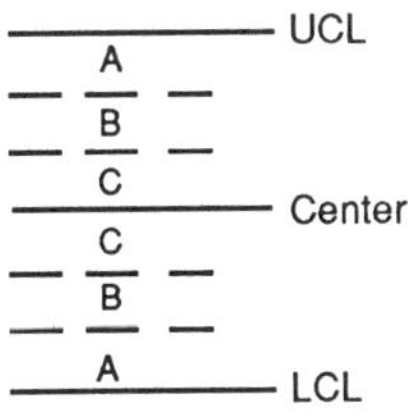

Three equal width zones between centerline and each control limit.

Label A, B, C from control limits.

Remember, control limits are not specification limits.

- Natural Patterns (in statistical control):
  Most points are near the center, with a few near control limits, but rarely outside control limits.
- Unnatural Patterns (Changing system):
  1. Single point outside control limit.
  2. 2 out of 3 successive points in zone A or beyond.
  3. 4 out of 5 successive points in zone B or beyond.
  4. 8 successive points on one side of center.
- Other Unnatural Patterns:
  1. Stratification — 15 consecutive points in zones C.
  2. Mixture — 8 consecutive points outside of zones C.
  3. Systematic Variable — a repeating, predictable pattern.
  4. Chart Following — two charts moving in unison.
  5. Trends — consecutive or alternate points trending in one direction.
  6. Level Shift — abrupt change in level.

Figure 13-14. Making and interpretation of control charts.

plotted onto the X-bar portion of an X-bar & R chart. Plotted on the R portion is the range within each sample. An X-bar & R chart, unlike the C chart and the U chart, is somewhat robust to the actual parent distribution of the individual data that make up the samples; this is what gives it greater utility. Estimates of control limits are drawn on both the X-bar and the R charts for evaluating the distribution of sequentially plotted data. The exact locations of these control limits are a function of the sample size and are multiples of the average of the range data, where the multiples to use are conveniently available in the references (Duncan, 1986; Western Electric Company, 1956). On the X-bar chart, the control limits are estimates of what will bracket 99.7 percent of the data (i.e., corresponding to $\pm 3\sigma$ limits of a Gaussion distribution). On R charts, the control limits are also estimates of what will bracket 99.7 percent of the data, but this does not correspond to $\pm 3\sigma$ of the R-data since R-data are not normally distributed. Figure 13-13 is an example for a sample size of five PWP values per plotted pair of X-bar and R points. The stability evaluation rules apply (see Figure 13-14), but should be applied first to the R chart, and only if that shows some measure of controlled stability should they then be applied to the X-bar chart. The "control limits" for diagnosing stability on control charts are not related to specification limits that would be determined on the basis of what control is required by manufacturing. Also, calculation of control limits on X-bar & R charts does not depend on the variability between the sample means plotted in the

X-bar chart, but is based only on the variability found within samples, that is, on the average of the R data. In the X-bar chart, the variability between samples is therefore compared with the "within-sample" variability represented in the X-bar control limits as defined.

Still another form of analysis of possible use to readers working with PWP measures of equipment and process performance is that of statistically designed experiments (Box, Hunter, and Hunter, 1978; Wolf and Tauber, 1986; Hays, Bishop, and White, 1987; and Bryce and Collette, 1984). Although often not practical because many contamination problems are sporadic, intermittent, nonlinear, and not easily modeled analytically, some more consistent and stable contamination processes can be experimentally analyzed to determine their dependencies on factors that cause and control them. Statistically designed experimental plans can be used to maximize the modeling information obtainable from a minimum number of PWP measurements. Full treatment of this topic is beyond the scope of this volume and is only offered here by way of a brief introduction. More information on this subject can be found in the references, under the headings "statistically designed experimentation," "multiple factor regression," and "response surface analysis."

The prescription and recording of complete PWP measurements should include not only their overall structure and analysis, but also careful description of the following:

1. Whether the process to be run is to be "dry" or "wet" (according to the definitions given earlier).

2. How many times each wafer and/or batch of wafers is to pass through the equipment under test.

3. How many wafers are in a batch, their positions, the order in which they are handled, and the use of any baffle or "bookend" wafers.

4. How long wafers reside in each step of the equipment sequences under test, including how long they sit in either input or output cassette elevators and how long they are on transport mechanisms such as belts, etc.

5. The number and nature of handling steps, such as cassette transfers, load-lock transfers, vacuum wand picks and places, etc.

6. The number of wafer-order reversals in cassettes.

7. Whether PWP calculations are to be conducted separately or together for top, bottom, middle wafers, etc.

8. Locations for control witness wafers in the environment near (or even inside of) equipment under test.

9. Class of the environment.

10. Clean-room gowning practices, especially what kind of gloves operators wear.

11. Airflow velocities and directions around and inside of equipment.

12. Temperature and humidity.

13. Interactions between operators and equipment as to how their presence and activities may impact the PWP results.

14. Instruments used, states of calibration and settings. Locations of probes and volume sampling rates if aerosol counters are used simultaneously.

15. Timing of activities and events; snafus, especially in wafer handling.

16. Controls on potential sources of particles; for example, relocation of pneumatic switches, cylinders, motors, frictional parts.

17. Directions and flow rates (volumes rates and velocities) of fans and exhausts within the equipment under test.

Probably the one error that vendors and users alike commit most often when they test PWP performance of equipment is not taking sufficient data to allow them to truly understand the variability of PWP with time, usage, and other factors that may significantly change PWP values. Testing should not stop after initial measurements, but should continue as an ongoing concern. The frequency of testing should change according to the criticality of the process step involved. The most blatant error is that of taking PWP data once and quoting confidence levels based upon the variability of the counter involved, rather than taking repeated PWP measurements and quoting confidence based on variability of the PWP values over a significant period of use.

A special practical matter of concern is how to handle the PWP values obtained from cassettes of wafers. First, Figures 13-15 and 13-16 remind us that particle depositions most often vary in time and that the amount of contamination collected by wafers while in cassettes can depend strongly on how long each wafer remains "unprotected" without another wafer positioned immediately above it in the neighboring slot(s). It is best to summarize the PWP results obtained on the end-wafers separately from those obtained from wafers that are protected by these "bookends." Second, the number of particles that accumulates on wafers is often significantly affected by the length of time wafers sit in the cassettes on the load and unload stations of a cassette-in/cassette-out piece of equipment. It should be obvious that in such a case, the number of wafers used in a test will directly affect the outcome. For example, if only three wafers are used, these wafers will only sit on the load and unload stations a short time relative to how long they will sit if a full cassette of wafers is used. And if less than a full cassette is to be used for counting, then one should clearly specify whether the remaining slots of the cassette are to be filled with dummy wafers, and exactly which slots are to be used for the wafers that will be measured.

And finally, good experimentation usually requires the use of what are called "controls." Controls are measures made in parallel with the measurements required by the test. Their purpose is to assure that additional factors did not affect the results of the test. Controls are usually designed to be influenced by any additional factors that would also influence the test. For example, a PWP test of a piece of equipment may require that five wafers be placed in the bottom slots of a cassette and be run through the process step performed by that equipment. A good control for this test would consist of a like cassette of five wafers that are counted and

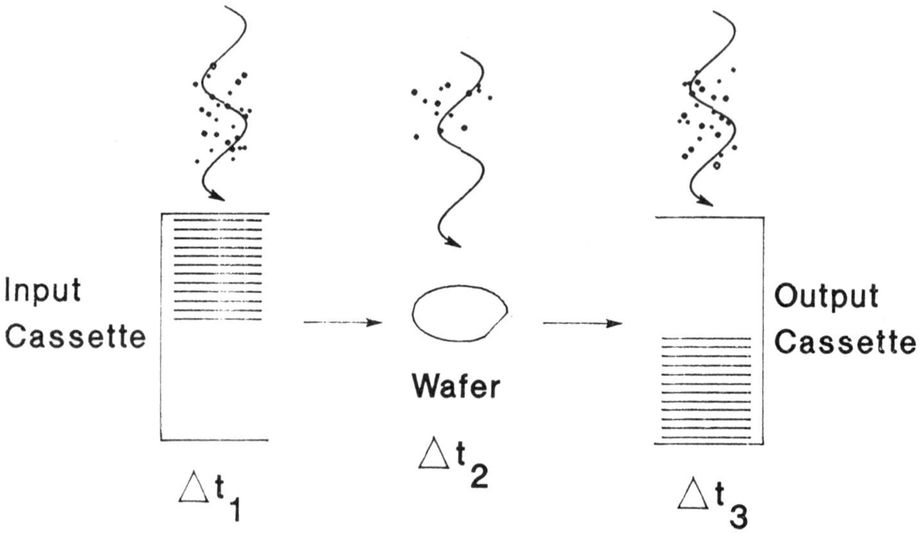

Figure 13-15. Particle fallout may vary with location during a PWP measurement.

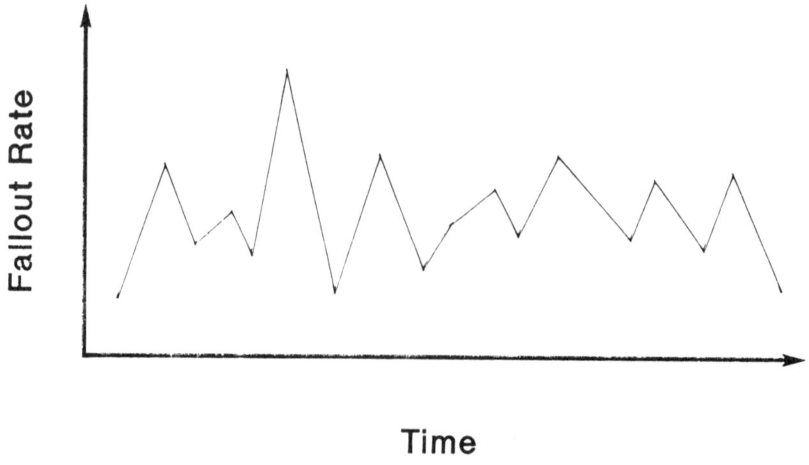

Figure 13-16. Particle fallout rate is a function of time.

carried along with the test cassette and positioned near the test cassette when the test wafers are processed. As a result, the control wafers undergo all the same handling and environmental experiences as the test wafers. The only difference between the two cassettes of wafers is that the test wafers go through the equipment during the equipment cycle, whereas the control wafers do not. Comparing the PWP results of the two sets of wafers, where the equipment exposure or process step experienced by the control wafers is really just a "null equipment" or "null process," enables a better interpretation of how much contamination to wafers the equipment is causing. The null equipment or null process used here might, for example, consist of leaving the control wafers exposed to the neighboring environment of the equipment under test, or placing them in a protective and clean box while the primary test wafers are inside the equipment under test.

The measurement of equipment and/or process contamination requires knowledge, skill, conscientiousness, perseverance, and more. It is hoped that this chapter has supplied some of the knowledge required and will encourage the reader to garner the rest.

REFERENCES

Alvarez, A., D. Welter, M. Johnson. 1983. "Problem Solving in the IC Industry Through Applied Statistics: Comparing Two Processes." *Solid State Technology*, July, pp. 127-133.

Baker, E. 1985. "Characterizing Particle Generation by Front-End Semiconductor Processing Equipment." *Microelectronic Manufacturing and Testing*, July, pp. 20-22.

Beyer, W., ed. 1984. *Standard Mathematical Tables.* CRC Press Inc., Boca Raton, FL, 27th edition.

Billat, S. 1987. "Automated Defect Detection on Patterned Wafers." *Semiconductor International*, May, pp. 116-119.

Borden, P., Y. Baron, and B. McGinley. 1987. "Monitoring Particles in Vacuum-Process Equipment." *Microcontamination*, October, pp. 30-34.

Box, G., W. Hunter, and J. Hunter. 1978. *Statistics for Experimenters.* John Wiley & Sons, New York, NY.

Bryce, F., and D. Collette. 1984. "Optimizing Wafer Processing via Statistical Design." *Semiconductor International*, March, pp. 71-77.

Cheung, S., and R. Roberge. 1987. "An Inside Look at Measuring Particles in Process Equipment." *Microcontamination*, May, pp. 45-50,94,96.

Duncan, A. 1986. *Quality Control and Industrial Statistics.* Richard D. Irwin Press, Homeward, IL.

Gibra, I. 1973. *Probability and Statistical Inference for Scientists and Engineers.* Prentice-Hall, Englewood Cliffs, NJ.

Hays, K., K. Bishop, and T. White. 1987. "Particle Reduction Investigation of a LPCVD Tungsten Reactor." *Microcontamination Conference Proceedings*. October, pp. 1-7.

Hemmes, D. 1987a. "Microcontamination Reduction in PECVD Systems - Part I." *Semiconductor International*, May, pp. 90-97.

Hemmes, D. 1987b. "Microcontamination Reduction in PECVD Systems - Part II." *Semiconductor International*, June, pp. 98-102.

Natrella, M. 1986. *Experimental Statistics*. National Bureau of Standards Handbook 91 (stock number 003-003-00135-0), October.

Osburn, C., H. Berger, R. Donovan, and G. Jones. 1988. "The Effects of Contamination on Semiconductor Manufacturing Yield." *The Journal of Environmental Sciences*, March/April, pp. 45-57.

Rossi, R. 1984. "LPCVD Process Equipment Evaluation Using Statistical Methods." *Solid State Technology*, November, pp. 227-232.

Tullis, B. 1988. "Particle Contamination by Process Equipment." Chapter 13 of *Handbook of Contamination Control in Microelectronics,* edited by D. Tolliver. Noyes Publications, Park Ridge, NJ, pp. 410-444.

Weber, C. 1988. "Standard Defect Monitor." *Proceedings of IEEE Conference on Microelectronics Test Structures, February 20-22*, vol. 1(1), pp. 114-119.

Western Electric Company. 1956. *Statistical Quality Control Handbook*. Select Code 700-444, AT&T Technologies (1-800-432-6600).

Wolf, S., and R. Tauber. 1986. *Silicon Processing for the VLSI Era*. Vol. 1, *Process Technology*. Lattice Press, Sunset Beach, CA.

14. MEASUREMENT OF PARTICLE EMISSION RATES FROM EQUIPMENT

R. P. DONOVAN Research Triangle Institute, Research Triangle Park, North Carolina

INTRODUCTION

The particle per wafer per pass (PWP) method discussed in Chapter 13 lends itself admirably to measuring the particulate contamination of most interest to the semiconductor technologist—particles on the wafer surface, a primary concern in wafer processing. The PWP method has some drawbacks, however: (1) It requires that a blank monitor wafer or wafers be added to a process batch, which adds to the processing costs, especially if every run is monitored by this method; (2) it is only a *near* real-time measurement method, capable of detecting unsatisfactory conditions only after a run is completed and then only as an integrated exposure (it cannot narrow the particle source to a specific action or time within the process); and (3) it is useful primarily in measuring particulate contamination from those tools and apparatus that process wafers directly. It is less suitable for measuring particle emissions from ancillary tools and equipment that are necessary for operating the fab line but do not handle wafers directly (i.e., air ionizing systems, garments, meters, scopes, particle measuring instruments, computers, motors, mechanical relays, electrical test equipment, etc.).

Various contemporary research and development activities respond to all these drawbacks. First, instruments for measuring particles on patterned product wafers continue to improve. Second, in situ methods for measuring particle concentrations in the processing atmosphere are being developed. This chapter addresses the third drawback cited above, by reviewing methods of measuring absolute particle emission rates from clean-room apparatus that do not directly process wafers. The methods described here would allow particle emission rate to be included in the performance specifications by which such apparatus are bought and sold, although particle-emitting specifications are not yet universally used.

CONTINUOUS FLOWING DUCT METHOD

The most direct and obvious method for describing emissions from a particle source is to measure the increase in particle concentration surrounding the source when it is operated. This approach is implicit in the characterization sometimes used by vendors—in terms such as "Class 100 compatible" or "Class 10 rated"—to describe the particulate burden added by apparatus. Presumably, in some undefined test configuration, the particle concentration detected in the vicinity of the operating apparatus did not exceed the limits required by FED-STD-209D (see Chapter 25) for the cited class of air quality. This type of description is unsatisfactory because the particle concentration measured depends on (1) sampling probe position with respect to the emission points of the source, and (2) the airflow direction and magnitude between the source and the probe. Without these additional conditions specified, reproducibility between measurements from one site to another is left to chance. What must be specified is the particle emission rate of the apparatus. This value is a true apparatus property independent of test configuration. It can be calculated from concentration measurements made in a specific source-detector configuration and compared with values calculated from concentration measurements made in a different configuration. If both measurements of concentration are correctly converted to emission rates, the emission rates will agree, even though the two concentrations from which each was computed are quite different.

In the continuous flowing duct method, the apparatus to be evaluated is placed in a duct through which a known airflow passes. For measuring particle emission rates from candidate clean-room apparatus, the air flowing in the duct would be HEPA-filtered (Chapter 7) or essentially particle-free. In any event, measuring the particle concentration in the airflow prior to activating the particle source is the first step. With HEPA-filtered air, this quantity will be near zero, or in the noise level of the particle counter.

The source is then activated and a second measurement of particle concentration is made in the flowing air, one duct width or more downstream of the particle source. The increase in steady-state particle concentration—the particle concentration does not change with time—can be related to the particle emission rate as follows:

$$S = (\Delta C)\, Q \quad , \tag{14-1}$$

where

S = the particle generation (emission) rate of the source (particles/time)

ΔC = the increase in steady-state particle concentration upon activation of the source (particles/volume)

Q = the airflow rate through the duct (volume/time).

Equation (14-1) makes the relationship between (ΔC) and Q obvious. For apparatus of constant S, ΔC will be high at low Q and vice versa. Thus, any attempt to describe particle emissions in terms of ΔC must also specify the Q at which the measurement was made. Using S to describe the particle emission rate avoids this complication.

A complication not obvious from Equation (14-1), however, is the position dependence of ΔC. The implicit assumption is that particle concentration is uniform across the duct both upstream and downstream of the particle source—that concentrations measured anywhere within the duct (or at least 10 percent of the duct dimension away from the wall) will be the same. For turbulent flow at high Reynolds number (Re $>$2,000-4,000), this assumption is probably valid. For airflow of 90 ft/min (46 cm/s) through a 4 ft^2 duct (61 cm on a side), the Reynolds number in the duct is above 18,000 (using Equation [2-6] with the duct dimension replacing the particle diameter). This type of flow is well into turbulent flow. In addition, the source itself usually is of a sufficient size to create additional turbulent mixing, thus further promoting uniformity. Inserting baffles of almost any size or description into the duct also promotes mixing.

As a final check, measurements can be made at various positions along duct traverses and averaged to obtain a mean value, as well as to check uniformity. Wall losses (particle capture by the duct walls) depend on particle concentration, which is usually low in a flowing duct supplied with HEPA-filtered air. Thus, wall losses can be ignored in the flowing duct method, whereas they must be considered in the closed-box method described later in this chapter.

The advantage of a continuous flowing duct measurement over the open area clean-room measurement is that the airflow is known and all the apparatus emissions can be assumed to be uniformly distributed in the downstream duct, or at least confined to well-defined streams in which particle concentration can be reproducibly measured.

Example 1. High-Voltage Air Ionizing Bars

(See Chapter 19 for a fuller discussion of air ionization.)

Air ionizing bars are lengths of conducting rods or bars to which sharp pointed needles or barbs are periodically attached. The entire assembly is electrically biased to voltage high enough (> 1,000 V) to create an electrical discharge between the sharp points deliberately incorporated into the bars and nearby conductive regions of different electrical potential—sometimes another rod or bar held at a specified potential, but often random ground points or regions in the vicinity. The purpose of the electrical discharge is to make ions out of air molecules. When the ionization is controlled so as to produce both positive and negative ions simultaneously, or alternatingly at high repetition rate, then electrically isolated surfaces downwind of the ion generation (including aerosol particles and silicon wafers that otherwise might accumulate electrical charge) are neutralized by the ions, regardless of the sign of the surface charge. Plots showing neutralization times of electrically isolated surfaces typically biased to 5,000 V graphically demonstrate the shortened surface neutralization times brought about by operating an air ionizing system. (The capacitance of the biased surface must be specified in these measurements for neutralization time data collected in differing configurations to be compared.) They also are a favorite promotional plot of virtually all vendors of air ionizing equipment (Figure 14-1).

Not so readily available are plots of particle generation by this equipment. The same action that produces large concentrations of air ions also can produce large concentrations of ultrafine aerosol particles (Liu et al., 1987; Donovan, Lawless, and Smith, 1986). Because these ionizing bars are typically installed immediately downstream of the HEPA filter bank in a clean room, their particle emission rate is an important performance parameter.

The continuous flowing duct method lends itself nicely to measurement of emission rate from this source. Using a commercial clean bench (an EACI 100) placed within a chemical hood (to control ozone emissions under accelerated test conditions) and a custom-built duct in front of the 2 x 2 ft (0.6 x 0.6 m) HEPA filter face creates a continuous flowing duct in which a section of an ionizer bar can be mounted (Figure 14-2). The clean bench, which consists of a blower and its housing, drives air through a 2 x 2 ft (0.6 x 0.6 m) HEPA filter at an adjustable rate, set for these measurements at 100 fpm (50 cm/s). The ionizer bars to be evaluated are mounted about 6 in. in front of the HEPA filter exhaust plane. Particle measurement is by a condensation nucleus counter (CNC) located at the midpoint of the duct and about 2 ft (0.6 m) downstream of the electrodes.

Under these conditions, particle concentration measured without the ionized bars operating was about 0.01/cm^3 (280/ft^3). With the bars operating, the average concentration rose to 1/cm^3 (28,000/ft^3). Thus, the particle emission rate is, from Equation (14-1):

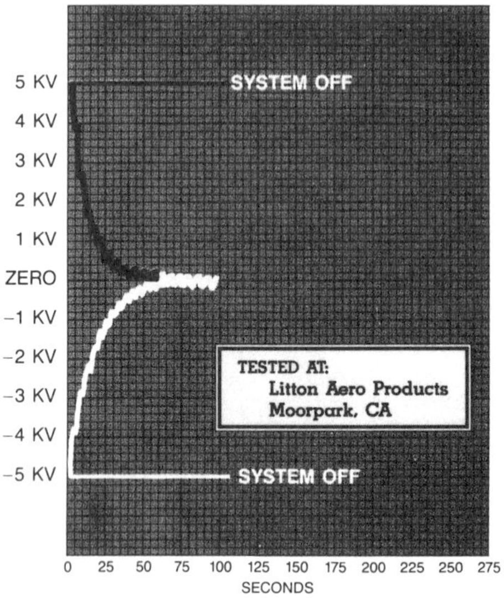

Figure 14-1. Improvement in surface charge neutralization time brought about by an area air ionization system (Static Control Services).

Figure 14-2. Continuous flowing duct configuration for measuring particle emission rate from air ionizing bars.

$$S = (\Delta C)\ Q = (0.99/\mathrm{cm}^3)(400\ \mathrm{ft}^3/\mathrm{min})\ (2.8 \times 10^4\ \mathrm{cm}^3/\mathrm{ft}^3)$$

$$= 1.1 \times 10^7\ \mathrm{CNC\ particles/min}\quad .$$

This emission rate applies to this short section of ionizer bar operating at a current level approximately 250 times that of a typical installation.

Example 2. Particle Emissions From Paper Under Flex Testing (Michaels, Fickel, and Donovan, 1986)

Paper used in the clean room can be a source of particle emissions. Specially prepared papers designed for clean-room use emit fewer particles than, say, conventional bond paper. One method used to document this improvement is the Texwipe modified flex tester, which is based on a continuous flow duct method of measurement (Havel and Sheridan, 1985). This test method consists of mechanically stressing a sheet of paper in a box through which the only airflow is that drawn by a particle counter. Make-up air entering the test box comes from a HEPA filter so that the background count prior to and during stress flexing is low.

Figures 14-3 and 14-4 illustrate the setup. The paper sample, an 8-1/2 x 11 in. (22 x 28 cm) sheet, is wrapped around two spindles and held in place with Velcro™ straps. The box is closed and a sampling tube inserted. When the particle counter is turned on, air drawn from the box is replaced by HEPA-filtered make-up air entering through holes cut in the side of the box. Typically, counts are low when the paper sample is stationary. To flex-test the paper, an accordionlike crushing action is introduced by moving one spindle toward the other (Figure 14-5). The frequency of the oscillating motion is typically 1 Hz.

Figure 14-6 shows representative data collected by flex testing six different papers in this apparatus. Four of these papers are "clean room papers"; the other two are bond paper. These data were collected using a 1-cfm (500-cm^3/s) optical particle counter (the PMS 525). The cumulative particle emission rates plotted in Figure 14-6 are based on particle concentrations measured in five size bins: 0.2, 0.3, 0.5, 1.0 and >5 μm. Total particle counts in 1 min yielded a concentration, ΔC, in ft^{-3}, which was numerically equal to the emission rate S, in particles/min because the sampling rate, Q, was 1 cfm (500 cm^3/s).

These data assume that all particles emitted by the flexed-paper specimen are swept into the sampling tube and that fallout or wall losses are negligible. These assumptions are more valid at higher flow rate sampling than at low sampling flow (1 cfm [500 cm^3/s] is better than 0.1 cfm [50 cm^3/s] or lower flow rate). A duct geometry that minimizes dead space, such as a cylindrical tube with tapered entry and exit—no corners—would be preferable. Large particles, such as paper fibers or lint, will probably not be sampled and counted by this measurement configuration. A fallout monitor wafer or a filter catch sample would be necessary to measure these large particles. Nevertheless, for the submicrometer component of particle emissions, this nonstandard version of a continuous flowing duct provides reproducible data.

Figure 14-3. Continuous flowing duct used in the Texwipe modified flex tester.

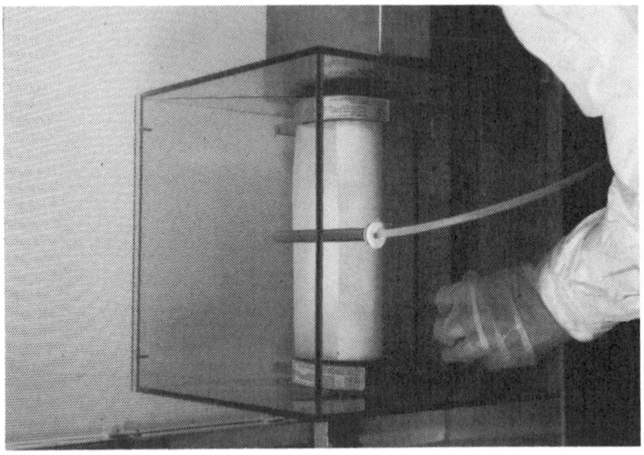

Figure 14-4. Paper mounted in flex tester; sampling probe in place.

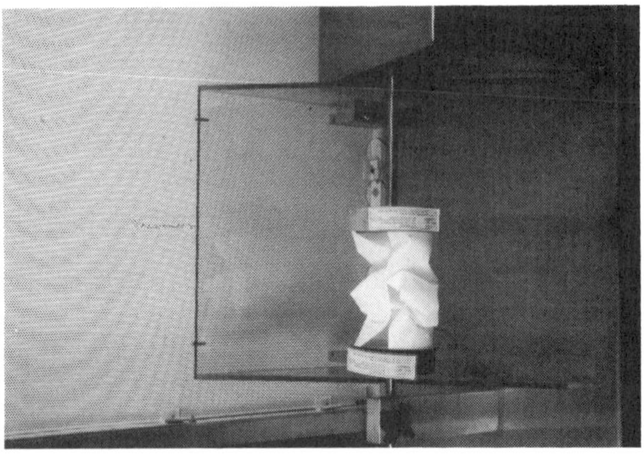

Figure 14-5. Accordion-stress applied to paper.

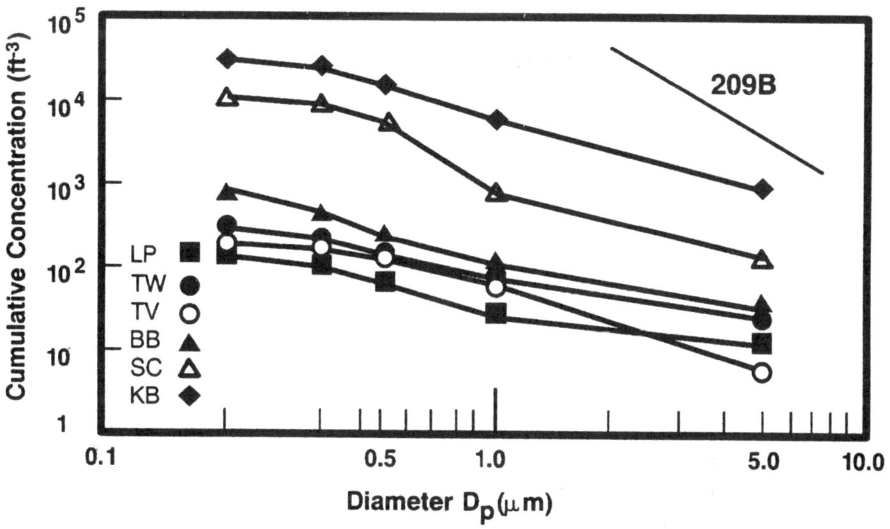

Figure 14-6. Size distribution of particle emission from various papers.

CLOSED BOX METHOD (Donovan, Locke, and Ensor, 1987; Rogers and Bailey, 1987)

One potential drawback of the continuous flowing duct method is the particle dilution introduced by the airflow. That is, particles emitted into a high airflow may not significantly increase downstream particle concentration. In the first example of the previous section (air ionizing bar emissions), if the emission rate was actually 100 times less, or 10^5/min, as in a nonaccelerated test, the increase in particle concentration would have been 0.01/cm^3 (280/ft^3), the same concentration as that of the background, thus complicating the data collection and interpretation but not necessarily preventing the method from making the measurement. An emission rate lower still by a factor of 10, however, probably would. Changes would have to be made. This section describes one such change.

In the closed-box method, there is no dilution airflow. Rather, the particle-emitting source fills a box or chamber. The initial filling rate is one measure of particle emission rate, and the steady-state concentration eventually reached after long-term filling is a second, independent measure of particle emission rate. This determination of emission rate, based on the measurement of a steady-state concentration, also requires a measurement of wall loss. Therefore, it takes somewhat longer than either the box method, based on initial concentration buildup, or the continuous flowing duct method described in the previous section. However, wall loss can easily be measured as part of the measurement procedure (described below).

The initial measurement method assumes that all particles emitted by the source and enclosed in the box contribute to an increase in the particle concentration within the box. Thus a measured increase in particle concentration of ΔC in a source operating time, Δt, is assumed to be related to particle generation rate, S, as follows:

$$S = \frac{\Delta C\ V}{\Delta t} \quad , \tag{14-2}$$

where

V = the box volume

ΔC = the increase in particle concentration over time, Δt, that the source is operating

S = particles emitted/unit time.

Equation (14-2) assumes that the particles emitted by the source are uniformily distributed throughout the box and that only an insignificant number of them are captured by surfaces within the box, such as box walls.

Applying Equation (14-2) to the example of the 10^5/min source that was deemed just marginally measurable by the continuous flowing duct method shows that 10 min operation in a box of 1 m^3 will increase particle concentration by 1/cm^3 (28,000/ft^3), a concentration well above the particle instrument background. This increased sensitivity to low particle emission rates is an important advantage of the closed-box method.

Determining the relationship between particle emission rate and steady-state particle concentration within the closed box (the second closed-box measurement method) requires solving the rate equation:

$$V\frac{dC}{dt} = S - K_r C - Q_x C \quad , \tag{14-3}$$

where

K_r = the wall loss rate (volume/time)

Q_x = the sampling rate of the particle-measuring instrument (volume/time) and other symbols are as before.

Equation (14-3) simply says that the change in box particle concentration is the difference between particle generation and particle loss. The solution of this differential equation is:

$$C(t) = \frac{S}{Q_x + K_r} - \left[S - \frac{C_o\,(Q_x + K_r)}{Q_x + K_r}\right] \exp\left(-\frac{(Q_x + K_r)}{V}\right) \quad . \tag{14-4}$$

The concentration within the box at any time depends on the initial concentration, C_o, the rate at which particles are emitted, S, and the rate at which they are removed, $Q_x C + K_r C$. In steady state, C(t) approaches C_s, the steady-state value of C, and the exponential term on the right-hand side of Equation (14-4) approaches zero. Thus, for steady-state conditions,

$$C_s = \frac{S}{Q_x + K_r}$$

or (14-5)

$$S = C_s \left(Q_x + K_r\right) \quad .$$

The particle generation rate equals the steady-state concentration times wall loss plus sampling rate. Sampling rate is known but wall loss is not. It must be measured.

To measure wall loss, the source is simply turned off after reaching steady-state, C_s, and particle concentration within the box is again measured as a function of time. Under these conditions, S = 0 and the rate equation (Equation [14-3]) becomes:

$$V\frac{dC}{dt} = -C\left(K_r + Q_x\right) \tag{14-6}$$

or

$$\frac{d\ln C}{dt} = -\frac{K_r + Q_r}{V} \quad . \tag{14-7}$$

Thus, the slope of a plot of ln C vs. t (after source deactivation) equals the quantity on the right-hand side of Equation (14-7) from which K_r can be determined. This value is then substituted in Equation (14-5) to calculate the particle emission rate from measured values of K_r and C_s and the known Q_x value of the particle sampling rate.

Measurement of particle emissions by the closed-box method is aided by the capability to reduce the starting particle concentration in the box to near zero. This starting point is easily achieved by incorporating a HEPA filter and blower inside the closed box. After the source is loaded (but not activated), the box is closed and the blower is turned on to recirculate the box air through the HEPA filter. This step reduces the starting particle concentration, C, to near zero. Thus, when the source is activated and the HEPA fan turned off, the C term in Equation (14-3) is near zero and Equation (14-3) reduces to Equation (14-2).

With time, C increases, eventually reaching a steady-state, C_s, for which Equation (14-5) provides a second independent measure of S based primarily on wall loss K_r. As just described, K_r is measured by deactivating the source and measuring the slope of ln C vs. time (Equation [14-7]).

Once a HEPA filter is incorporated into the closed box, it is a simple matter to also incorporate a well-defined duct in which the source is placed. When the blower is recirculating duct air through the box and back into the duct and the source is activated, particle removal from the box is dominated by the filter. Equation (14-3) becomes

$$V\frac{dC}{dt} = S - \eta\, Q\, C \quad , \tag{14-8}$$

where

η = the filter efficiency, assumed to be 1

Q = the recirculating flow rate through the filter.

In steady state ($C = C_s$; $dC/dt = 0$) Equation (14-8) becomes:

$$S = C_s Q \ ,$$

which is similar to Equation (14-1), where wall losses and instrument losses are neglected with respect to filter losses.

In summary, both the continuous flowing duct and the closed-box methods of measuring particle emission rate can be incorporated into a single test configuration as shown in Figures 14-7 and 14-8. The sequence and procedures for carrying out these measurements are detailed in Donovan, Locke, and Ensor (1987). Box volume and duct dimensions can be varied to accommodate source size and the magnitude of the particle emission rate.

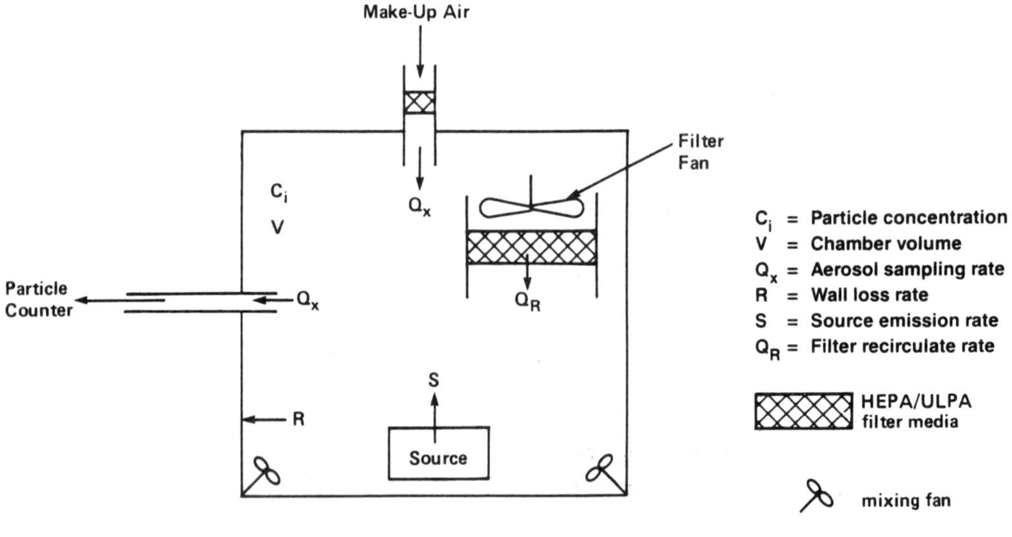

Figure 14-7. Schematic of test chamber for measuring the particle emission rate from a source S.

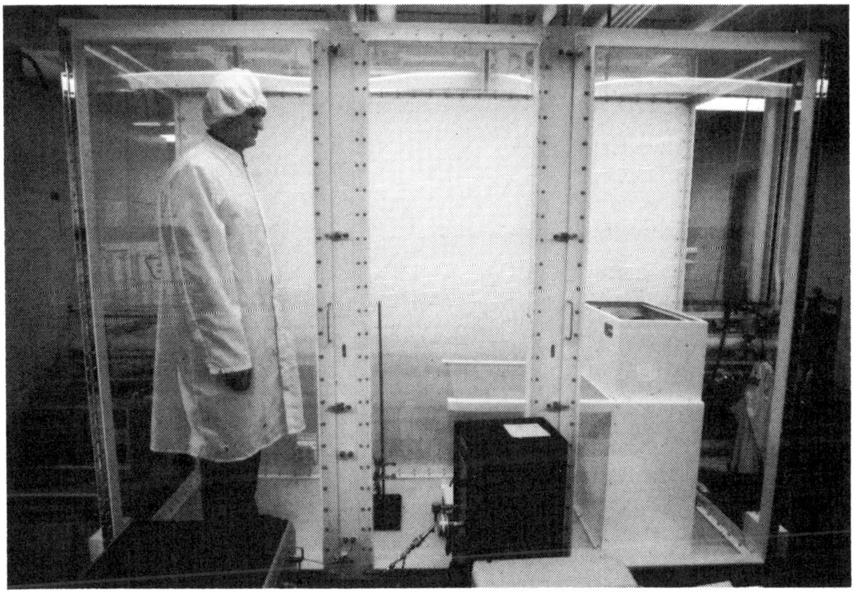

Figure 14-8. RTI version of Figure 14-7 test chamber.

A word of caution: These methods are not yet accepted as test methods and hence should be spelled out in any procurement based on them.

REFERENCES

Donovan, R. P., B. R. Locke, and D. S. Ensor. 1987. Measuring Particle Emissions from Cleanroom Equipment. *Microcontamination*, 5(10):36-39, 60-63.

Donovan, R. P., P. A. Lawless, and D. D. Smith. 1986. Polarity Dependence of Electrode Erosion Under dc Corona Discharge. *Microcontamination*, 4(5):38-49.

Havel, W. J., and C. Sheridan. 1985. Texwipe Modified Flex Test for Particulate Analysis of Dry Wipes. Proceedings of the 31st Annual Technical Meeting of the IES, Las Vegas, Nev., May, pp. 80-83.

Liu, B.Y.H., D.Y.H. Pui, W. O. Kinstley and W. G. Fisher. 1987. Aerosol Charging and Neutralization and Electrostatic Discharge in Clean Rooms. *J. Environ. Sci.*, 30(2):42-46.

Michaels, L., L. M. Fickel, and R. P. Donovan. 1986. An Experimental Study of the Particle Emissions from Cleanroom Papers. *J. Environmental Sci.*, 29(6):38-40.

Rogers, G. G., and L. G. Bailey. 1987. A Closed Chamber Method of Measuring Particle Emissions from Process Equipment. *Microcontamination*, 5(2):42-47, 66-67.

Static Control Services. n.d. Total Atmosphere Room Ionization. SCS, 870 Research Drive, Bldg. #9, Palm Springs, CA 92262.

15. SURVEY OF PROCESS-INDUCED EFFECTS

KEITH DILLENBECK Ashland Chemical Company, Columbus, Ohio

INTRODUCTION

The predominant sources of defects introduced during the manufacturing of integrated circuits have changed significantly during the past decade (Martin, 1987). In the 1970s and early 1980s, many critical circuit defects could be attributed to the clean room facility and to people within the clean room. Significant progress has been made in recent years in controlling contamination caused by people and the facility. Today, the predominant sources of defects lie within the wafer processing equipment. Defects introduced during individual process steps originate from particles generated by the process tool itself and/or by the process performed within the tool. The degree to which the hardware and the process can be fully integrated largely determines the defects that a specific process will contribute.

In surveying levels of defects from the various unit processes throughout the integrated manufacturing process, one quickly realizes that each process can potentially contribute an alarming number of defects to device wafers. Some generic processes typically generate more defects than others; for example, chemical vapor deposition (CVD) processes are known to be significant particle generators because of flaking of large particles from thermally stressed films. Particle problems from CVD processes are quite pervasive; however, there are effective techniques for minimizing defect generation of this type. On the other hand, processes generally felt to be "clean" can produce disastrous levels of defects unless a great deal of care is exercised.

PROCESS TOOL FACTORS

The number of defects generated by a specific unit process can vary greatly. As a result, one cannot generalize as to the "typical" defect levels from various processes. Considerable attention to processing details is required on the part of semiconductor process engineers in order to arrive at and maintain low defectivity levels within a unit process. Some of the important elements that determine the defect contribution from a process or process tool are:

- **Design of equipment**—The vintage of the process tool plays a key role in determining the particle generation levels. Equipment designs during recent years have improved to the point that the particle generation rates are orders of magnitude lower than they were in previous generations. Advancements have come through improvements in mechanical design, component selection (e.g., valves and gauges), and selection of materials.

- **Design of chemical processes**—The specific processes used within a tool can have a significant impact on the level of defects for a unit process. Common examples are plasma etch processes, in which the plasma chemistry can generate a polymer byproduct that manifests itself as particulate defects. This problem can be quite severe, but can be minimized through careful process design and chamber cleaning procedures (Shaw, 1985).

- **Purity of process fluids**—The gases, chemicals, and ultrapure water used in various unit processes can be a very significant source of defects within those processes. The initial purity of the fluid and the integrity of the distribution system are both important factors. Depending on the type and vintage of the process tools, it is not uncommon to find that the purity of the process fluid has been significantly degraded by the pipes, valves, and tubing *within* the process tools (Chapters 10 and 11). Proper tool design and material selection can minimize these types of problems.

- **Degree of maintenance**—The frequency and quality of preventive maintenance procedures have a tremendously influencing effect on process tool defectivity levels. Preventive maintenance procedures include chamber cleaning, filter replacement, mechanical adjustment, component replacement, and quartzware cleaning or replacment. A properly selected and highly regimented preventive maintenance program can have a very positive influence on the severity and consistency of defect problems throughout the integrated process.

- **Process tool usage rates**—The number of wafers processed through a process tool in a given period of time can influence the number of particles deposited onto wafers during the process. Process tools that are heavily used usually have higher defect generation levels than tools used relatively infrequently. This is because of the increased number of actuations of components and the thickness of films deposited on reactor surfaces, among other factors. Heavily utilized process tools usually require stepped-up preventive maintenance schedules for their reliability and defectivity levels not to be adversely affected. As the trend within manufacturing operations moves toward increased usage of process equipment, the "care and feeding" of the process tools **must** be adjusted accordingly.

- **Characterization and modification**—It is a fact that many process tools used in current manufacturing operations were not designed or manufactured with regard for particle generation characteristics. It is a common practice to take process tools out of production for a period of time to modify specific components and subsystems and improve the overall particle performance (Dillenbeck, 1988). Equipment modification will not be effective unless the tool is characterized thoroughly—that is, unless specific particle generation sources within the tool are identified and ranked. A methodical approach to characterizing particle-generating mechanisms, combined with the effective use of particle-detection instruments, is extremely important (Dillenbeck, 1984). In some manufacturing operations, the requirements for increased manufacturing consistency have resulted in the appointment of dedicated teams of specialists to characterize and modify process equipment, one tool at a time.

PARTICLE LEVEL SURVEY

In the following sections, an overview is presented, showing the range of "typical" particle levels from the various unit processes. Tables 15-1 through 15-5 present the particle survey for the five major categories of unit processes. These processes are wet cleaning, photolithography, plasma etch, CVD/physical vapor deposition (PVD), ion implant, and thermal oxidation. The particle levels listed represent the average number of particles added to a 5-in. wafer, $\geq 0.5\ \mu m$, in two categories. The first category, referred to as "as high as," represents processes and process tools for which the level of defects has not been minimized. In this category, the process tools may be old and not modified for particle reduction or may be poorly maintained. The second category, referred to as "as low as," represents processes that have been adjusted to minimize the number of defects. In this category, the process tools may be newer, modified for low-particle operation, and/or cleaned very frequently. Process tools in this category certainly must be well maintained, including preventive maintenance. The particle levels listed for each of these two categories are only order-of-magnitude numbers,

Table 15-1. Particle Survey for Wet Cleaning and Etching Processes (particles per 5-in. wafer, >0.5 μm)

Process	Typical particle level: As high as	Typical particle level: As low as	Methods for particle control
Nitride etch	>2000	<50	- Pure acid - Pure tank material - Proper tank heater - Hot acid recirculation/ filtration
Piranha, photo-resist strip	>1000	<10	
Pre-sputter clean	>3000[a]	<50	- Acid recirculation - Ultrapure water - Clean spin dryer - Preventive maintenance of spray processor - Acid filtration - Pure acid
HF oxide etch	>3000[a]	<25	
Prediffusion clean (RCA chemistry)	>2000	<25	
Brush scrubber	Adds >1000	50% removal efficiency	- Pure fluid (water or isopropyl alcohol) - Proper brush material - Brush cleaning practice - Mechanical adjustment
Spin dryer	>1000	<100	- Proper mechanical components - Pure nitrogen - No water - Mechanical adjustment
Solvent vapor dryer		<5	- Proper materials - Pure solvent

[a]Hydrophobic wafer surface.

Table 15-2. Particle Survey for Photolithography Process (particles per 5-in. wafer, >0.5 μm)

Process	Typical partical level: As high as	Typical partical level: As low as	Methods for particle control
Photoresist spin	>500	<20	- Clean resist supply - Resist filtration - Proper track exhaust - Edge bead removal - Proper mechanical design
Pattern exposure	>2000 (Projection aligner)	<5 (Stepper)	- Proper mechanical components - Proper preventive maintenance schedule - Clean reticles/pellicles - Clean air/nitrogen
Pattern develop	200	<10	- Thorough rinse - Pure developer - Developer filtration - Proper mechanical components

Table 15-3. Particle Survey for Plasma Etch Processes (particles per 5-in. wafer, >0.5 μm)

Process	Typical particle level: As high as	Typical particle level: As low as	Methods for particle control
Oxide etch	>2000	<100	- Low polymer formation - In situ cleaning
Poly etch	>1000	<50	- Pure gases - Proper mechanical design - Clean wafer transport
Aluminum etch	>3000	<100	- Slow pump rate - Slow vent rate - Leak-free gas components
Nitride etch	>1000	<10	- Vertical wafer orientation
Photoresist strip	>5000	<50	- Slow pump - Slow vent - Proper materials - Proper mechanical design

Table 15-4. Particle Survey for Chemical and Physical Vapor Deposition Processes (particles per 5-in. wafer, >0.5 μm)

Process	Typical particle level: As high as	Typical particle level: As low as	Methods for particle control
LTO deposition	>3000	<200	- Adequate preventive maintenance schedule - Quartzware cleaning/ replacement schedule - Tube cleaning and replacement schedule - Tube vacuuming - Slow pump/vent - High vacuum integrity - Cantilever/soft land - Coating of boats (PECVD)
Poly deposition	>1500	<100	
Silicon nitride deposition	>2000	<100	
PECVD oxide	>5000	<300	
Aluminum sputtering	>1500	<20	- No ion milling process - Slow pump/vent - Pure argon/nitrogen - Proper wafer handling - In situ chamber clean - High vacuum integrity

LTO = Low-temperature oxide
PECVD = Plasma-enhanced chemical vapor deposition

Table 15-5. Particle Survey for Ion Implant and Thermal Processes (particles per 5-in. wafer, >0.5 μm)

Process	Typical particle level: As high as	Typical particle level: As low as	Methods for particle control
Ion implant	>2000	<150	- Separate pump/vent ports - Slow pump/vent - Gentle wafer handling - Edge bead removal - Proper mechanical design - Cleaning of load-lock
Atmospheric oxidation	>300	<10	- Cantilever/soft land - Pure gases - No atmospheric backstreaming - Proper gas filtration - Quartz cleaning and replacement

but serve to illustrate the wide variation between a unit process that is optimized for low defectivity levels and one that is not. The last column in Tables 15-1 through 15-5 presents the common Methods for Particle Control for each unit process. These generic attributes certainly clarify the difference between a unit process with low defect levels and one with high defect levels. (Some of the details of these low-particle qualities for the unit processes may be found in Tolliver, 1984; Hemmes, 1987; Baker and Fishkin, 1986; Bedini, 1986; Bowling and Larabee, 1986; Cheung and Roberge, 1987; Hancock and Yarling, 1984; Harned, 1986; and Pollock et al., 1984.)

REFERENCES

Baker, E. J., and B. Fishkin. 1986. Particle Control in Process Equipment: A Case Study. *Microcontamination* 4(9):52-58.

Bedini, T. 1986. Practical Techniques for Installing Semiconductor Processing Equipment and Troubleshooting Particulate Problems. Proceedings of the Institute of Environmental Sciences 32nd Annual Technical Meeting, Dallas, Texas, May 6-8, pp. 415-423.

Bowling, R. Allen, and Graydon B. Larabee. 1986. Particle Control for Semiconductor Processing in Vacuum Systems. *Proceedings of the Microcontamination Conference*, Santa Clara, California, November, pp. 161-168. Canon Communications, Santa Monica, California.

Cheung, S. D., and R. P. Roberge. 1987. An Inside Look at Measuring Particles in Process Equipment. *Microcontamination* 5(5):45-50, 90.

Dillenbeck, Keith. 1987. Contamination Engineering Principles for Semiconductor Wafer Fabrication. Presented at the Microcontamination Conference and Exposition, Santa Clara, California, October.

Dillenbeck, Keith. 1988. Application of Contamination Engineering Techniques for Solving Particle Problems in Process Equipment. *Proceedings of the 34th Annual Technical Meeting of the Institute of Environmental Sciences*, May, pp. 532-539. IES, Mount Prospect, Illinois.

Hancock, S., and C. Yarling. 1984. Particle Measurements in Medium-Current Ion Implanters. *Semiconductor International* (May):144.

Harned, Wayne. 1986. Bacteria as a Particle Source in Wafer Processing Equipment. *The Journal of Environmental Sciences* (May/June):32-34.

Hemmes, D. G. 1987. Microcontamination Reduction in PECVD Systems. *Semiconductor International* 10(6):90-97.

Martin. 1987. Management of Contamination Control as Related to Product Defects. Presented at the IEEE 21st Videoconference on Microcontamination Control, December.

Pollock, Turner, Milgate, Resnek, and Hertel. 1984. Particle Performance of a Wafer Handler for Serial Process Ion Implantation. Presented at the 5th International Conference on Ion Implantation Equipment and Techniques, Vermont, July, 1984.

Shaw. 1985. Particulate Control in Dry Oxide Etching. Presented at the SEMICON Southwest Technical Conference, Dallas, Texas, October.

Tolliver, D. L. 1984. Contamination Control: New Dimensions in VLSI Manufacturing. *Solid State Technology* 27(3):129-137.

16. PARTICLE CONTROL METHODS: II

WAYNE G. FISHER Texas Instruments, Inc., Sherman, Texas

PART 1
CLEAN-ROOM GARMENTS

INTRODUCTION

The most important purpose of clean-room garments is that they protect wafers and processing equipment from particles emitted by clean-room personnel. These particles may be particles from the workers' bodies, mostly hair and skin flakes, or from their clothes and shoes. A second most important purpose of clean-room garments is that they establish and reinforce the attitude among workers that special behavior is required in the clean room to limit particle contamination by their activity.

In the first section of this chapter, the requirements for clean-room garments related to particle performance will be discussed. Then the types of garments currently used in clean rooms will be described, followed by a discussion of ways to measure their particle performance. Finally, the present state-of-the-art and future requirements for clean-room garments will be outlined.

PERFORMANCE REQUIREMENTS FOR CLEAN-ROOM GARMENTS

First, clean-room garments should keep particles from peoples' bodies inside the garments so that they are not emitted to areas where the particles might contaminate product wafers or the equipment they are processed in. The garments should act as a filter to keep particle-laden air from passing through the fabric. They should also prevent bursts of air containing particles from passing through openings in cuffs, collars, closures, and around the face. In some cases, the garments may act to direct emitted particles away from areas where sensitive material might be. For example, the main flow of air from smocks is from the bottom of the smock, toward the floor, and generally "downwind" of the normal working height in clean rooms.

A second requirement of clean-room garments is that they should not produce particles themselves. The fabric they are made from should be inherently noncontaminating. They should be constructed without pockets, pleats, or other features that may trap particles that are reemitted later. The cut edges of woven materials should be sewn under with continuous filament thread.

Other requirements are less closely related to particle protection or production. Major factors are the comfort and convenience of the garments. Included in this are the temperature of the wearer, an important factor considering that clean-room garments are usually worn over street clothes. This is closely related to the vapor transmission properties of the material because cooling is largely dominated by the evaporation of perspiration. The fit of the garments is important to comfort, and garments should be easy to put on and take off. Comfort and convenience are related to particle protection because garments are likely to be worn incorrectly if they are too much of an inconvenience.

Another important requirement of clean-room garments is that they must be safe. They must protect the workers from chemical hazards in the clean room, and they should not introduce new hazards themselves. The chemicals encountered in clean rooms include a wide range of inorganic acids, bases, and oxidizers, solvents, and photoresist. Ideally, the garment material should shed liquid chemicals rather than absorb them, and they should be chemically inert. Other desirable safety features include nonflammability and nonslip soles on shoes. A final requirement of clean-room garments is control of the static charge on workers' bodies.

CLEAN-ROOM GARMENT MATERIALS

The three basic types of clean-room garment materials are woven and nonwoven filament polymers and expanded PTFE (see Figure 16-1) laminate. The nonwoven polymer materials are spun-bonded polyolefin. These materials are lightweight and have small pore sizes, so they provide good barrier protection. They generally have poorer vapor transmission than woven garments and expanded PTFE garments, so they may be perceived as less comfortable than the others in spite of their light weight. They were first used in the health industries, primarily hospitals, because they can be sterilized and are disposable. When developed for these industries, the particle sensitivity of electronics processing was not a concern. In fact, the garments sold for use in hospitals may shed enormous numbers of small particles, generally not as visible lint, but in the micrometer or submicrometer size range that semiconductors are sensitive to. When these materials are used in semiconductor clean rooms, the user must be very careful to ensure that garments from the particular source selected are suitable for these areas. These garments are generally considered to be of limited use or disposable.

Woven garments suitable for clean-room use are made and sewn with continuous filament synthetic threads. The polymer is usually polyester, and the fabric comes in various weaves. The tighter, thicker weaves have smaller openings between the threads, so they are better filters for particles in the air passing through the fabric. Woven polyester garments are very durable in use and will stand up to repeated washings without loss of their nonparticle-generating nature. Most of the materials used for clean-room garments have high surface resistivities and so are subject to the buildup of static electricity, but these materials may have conductive threads woven into the fabric to prevent static charge. An advantage of fabric with these threads is that the antistatic properties are not removed by washing.

The third type of clean-room garment material is expanded PTFE. For strength, this material is laminated to one or two layers of polyester. The advantages of this material are that it releases very few particles and has good vapor transmission. The pore sizes are very small, so it has excellent filtration properties. A disadvantage is that garments made from this material are relatively expensive. These garments are washable.

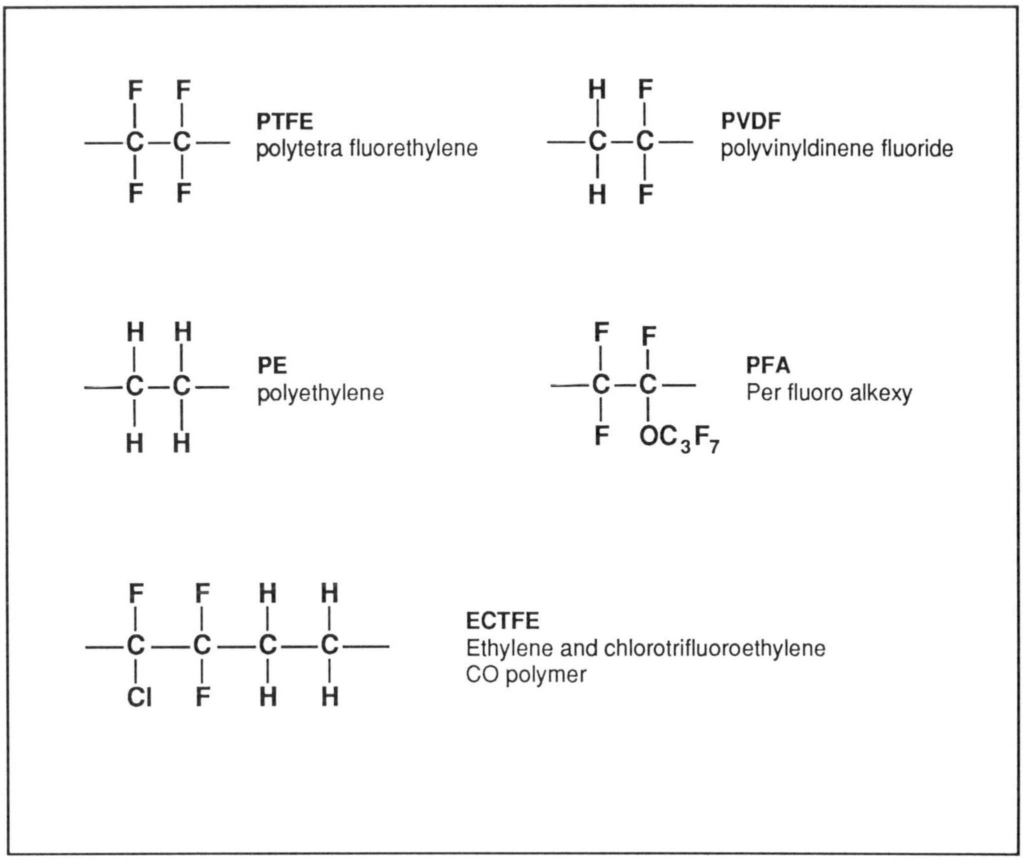

Figure 16-1. Common polymers used in microelectronics.

MEASUREMENT OF CLEAN-ROOM GARMENT PARTICLE PERFORMANCE

A number of methods, some very elaborate, have been reported in the literature for measuring particle contamination from clean-room garments (Brinton and Swick, 1984; Hayakawa, Fujii, and Yoshida, 1985; and Hirasawa et al., 1984). The elaboration in these experiments is an attempt to quantify the results of the tests, generally a comparison between the styles of garments, the construction or materials, or their use.

These experiments measure three basic properties of garments. First is the particle generation rate of the fabric. This is basically a test for detachable particles, particles either held in the fabric from manufacturing or left in the fabric after washing, or particles coming from the material itself. Such tests will tell if the fabric is suitable for use in clean rooms. If the garments have been washed, it is a test of the number of particles left on the garment by the clean-room laundry.

One testing method for releasable particles from garments is ASTM F51-68 "Sizing and Counting Particulate Contaminant in and on Clean Room Garments." In this testing method, air is drawn through a section of the garment and particles liberated from the material are collected on a filter. Disadvantages of the method are that it is labor intensive and limited to detection of larger particles. Also, it is limited to the testing of materials that air can be drawn through.

A less rigorous test is the simple inspection of the material under an optical microscope. Loose fibers are easiest to detect by this method on woven materials because the underlying fabric has a very regular pattern of continuous filaments. Foreign material is generally easy to detect on all types of clean-room garment fabrics if the foreign material is not white or transparent.

Most other tests for releasable material are an attempt to circumvent the use of an optical microscope by measuring the emission of particles with an optical light scattering particle counter. Some of these particles may be difficult to detect with an optical microscope inspection because of their small size. In one such type of test, the garment is tumbled in a rotating drum. Air is sampled from the drum with an optical particle counter. A simple modification of this test is to shake the garment over the inlet of the particle counter while counting particles in the air around the garment for a set period of time.

These tests must be done in a clean-room area, an area with an airborne particle concentration at least lower than that measured while testing the garments. Also, while these tests provide a quantitative measure of the numbers of particles released by the garments, these numbers are only relative. They depend on the specific experimental conditions, and, in the case of testing by shaking, the absolute numbers of particles obtained may be highly dependent on the person conducting the test.

A second method of testing the particle performance of clean-room garment material is measurement of its filtration properties. In this type of test, the fabric is tested in the same way that filter media are tested. The fabric is challenged with an aerosol with a measured concentration, then the concentration of the filtered aerosol is measured. The ratio of the two concentrations is the filter efficiency of the fabric (Brinton and Swick, 1984).

The final method of testing the particle performance of clean-room garments is to measure particle generation and emission under actual or simulated clean-room use conditions. In this type of test, people perform actual or simulated manufacturing tasks while the particle generation is measured in one or more ways. One way is to count the numbers of particles added to witness wafers in the area of the tests. Another method is to count particles liberated into the air with an airborne particle counter. An advantage of witness wafers is that large particles that might be lost to an airborne particle counter can be detected on wafer surfaces. However, most current-generation airborne particle counters will detect particles smaller than those detectable with wafer scanners. A variation of this method is the closed chamber method described in Chapter 14.

An advantage of use-simulation tests over other types of garment tests is that the entire garment system is measured simultaneously. The particle generating nature of the fabric and its filtration properties both affect the outcome of the test. Also tested are the garment's ability to prevent particles from escaping around collars, cuffs, the face, and closures. A disadvantage of this type of test is that the outcome may depend on the people subjected to the tests. There are differences in the way that garments fit individuals and the way that people move, even when performing the same task. At the very least, the number of particles escaping the garment depends on the challenge from the inside, a factor that may vary a great deal from person to person.

In spite of the test-to-test variability and the lack of absolute values for results of most garment tests, the relative performance of the three types of clean-room garments and garment fabrics follows the general pattern in the table below.

	Releasable particles	Filtration (0.5 μm)	Simulated use tests
Nonwoven filament	3	2	1
Woven filament	1	3	3
Expanded PTFE	1	1	1

In this table, the best performing material is rated 1 and the worst 3. Although it appears that the only choice is the expanded PTFE garment, other factors not related to garment performance must be taken into account. A major factor is cost per use over the life of the garment. It is generally recognized that the expanded PTFE is much more expensive than the other types of materials, so its use may be limited to situations demanding the performance it achieves.

HEAD COVERS AND FACE MASKS

Up to this point, the requirements for head covers and face masks have only been referred to indirectly. Covering the head is complicated because people in clean-room garb need to see, talk, breathe, and hear without too much difficulty to perform their jobs correctly. Testing the air with an airborne particle counter around people without head coverings shows that the particle emission from hair varies a great deal from person to person, and that particle emission largely occurs when hair moves. To contain particles from hair, it should be completely covered, it should be prevented from moving around inside the head covering, and the head covering should fit closely around the face.

Sullivan and Trimble (1986) have published a study of the effectiveness of various face coverings. They tested both a live subject breathing normally and simulating a cough or sneeze and a mannequin breathing titanium tetrachloride smoke. They found that all face coverings eliminated the disruption of the downward flow of air from ceiling-mounted HEPA filters that occurred when an unmasked person simulated a cough or sneeze. They probably also eliminate most of the large particles or droplets expelled during a cough or sneeze. However, most of them did not filter the small particles from titanium tetrachloride very well and some emitted particles themselves. Full face coverings of solid clear plastic eliminated particles from the face completely. One full face shield is the Dryden Particle Control Shield (Dryden Engineering, Santa Clara, CA). In this system, a clear plastic bubble covers the face. Air is drawn in around the face through a tube by a belt-mounted blower. The air is exhausted through a filter at waist level. A similar competitive system is made by W. L. Gore and Associates, Elkton, MD.

CLEAN-ROOM GLOVES

Gloves are a special type of clean-room garment because a person's hands touch objects that the rest of the body normally does not. Hands cause contamination on products because they shed skin cells and bacteria, because they carry contamination from previous contacts, and because they leave oil and salt residues on surfaces they contact. A second reason for the importance of gloves in the clean-room environment is safety. Hands reach into areas where chemicals are used, and they may touch or pick up objects that are hot or cold enough to injure the skin.

Gloves fall into three categories: barriers that protect the hands from objects and objects from the hands, knit and woven gloves, and gloves that insulate the hands from hot and cold objects. Except for some special applications, gloves that are barriers are the choice for use in clean rooms.

Protection of the hands is so important that safety issues may determine whether a particular glove may even be considered for use in a clean room. The factors to be considered when determining the safety of clean-room gloves include:

- Chemical resistance—the ability of the gloves to resist degradation when exposed to inorganic chemicals, acids, bases, and oxidants, and to solvents
- Barrier integrity—the lack of holes or tears in the gloves
- Permeability—the ability of the gloves to resist chemicals soaking through the glove material
- Insulation—the ability of the gloves to protect the hands from hot and cold objects

- Durability—resistance to cutting, tearing, or wearing holes in the gloves
- Tactility—the ability to retain a sensitive feeling of touch while wearing the gloves.

Here we are primarily concerned with the control of particles. However, the particle performance of clean-room gloves is so closely tied to their performance with other contaminants that particles cannot be considered separately. The three sources of contaminants from gloves are particles and other contaminants from the gloves themselves, contaminants from the hands that come through the gloves, and contaminants carried by the gloves from previous contacts. The relative importance of the last of these is determined by how the gloves are used. The importance of the first two categories is determined by the properties of the gloves themselves.

Particles from clean-room gloves may come from powder on the gloves, debris from manufacturing or packaging, or flaking or other degradation after contact with chemicals. The Institute of Environmental Sciences RP5 Committee on Clean Room Gloves and Finger Cots recommends the following test for particles from clean-room gloves. First, a glove is placed in a container of clean water, then the container is sealed and shaken. Particles in the liquid are counted by filtering the liquid, then counting the particles on the filter with the aid of a microscope. The particles in the liquid may also be counted with a liquid particle counter. Other tests are not recommended by the IES RP5 committee but are useful for quick testing. These are rubbing the gloved hands over the inlet of an airborne particle counter while counting the particles in the air or over a clean witness wafer, then counting the particles with a surface scanner. Another is to perform the same type of tests, but while shaking the tested glove over the particle counter or witness wafer.

Extractable material is tested by soaking the gloves in solvent, then analyzing the solvent for residue by evaporation and weighing or chemical analytical techniques such as infrared absorption. Solvents used for the test include all those used where the gloves will be used. Organic solvents may extract plasticizers and oils from gloves while aqueous solvents are more likely to extract inorganics such as chloride and alkali ions.

Permeability and integrity determine the resistance to the passage of material from the hands to product or equipment surfaces. Gloves may be tested for pinholes and other leaks by filling them with water and observing the outside for leakage.

Chemical resistance is determined primarily by the material the glove is made of. For most materials, the chemical compatibility is known, and users can consult with manufacturers or their literature to determine acceptable usage. When necessary, users may test a material for chemical compatibility by exposing the glove material to the subject chemical, then examining the material for signs of degradation.

Other factors to consider when choosing a type and source of clean-room gloves include:

- Strength and durability—the resistance to tear and wear
- Resistance to the buildup of static charge
- Fit and comfort—the availability of different sizes and cuff lengths
- Ease of putting on and taking off
- Heat resistance
- Manufacturer's or distributor's quality control
- Packaging
- Cost.

In summary, clean-room gloves reduce transfer of contaminants from peoples' hands to products. In doing so, they should not introduce new contaminants themselves. One way to reduce the amount of contamination is to use cleaner gloves and to use gloves with greater integrity. A better way is to reduce the chance for contamination by reducing human involvement with the product or surfaces that come into contact with the product.

PRESENT USAGE AND FUTURE REQUIREMENTS

Goodwin (1987) and Salvo (1985) and others have published recommendations of the extent of body coverage required as a function of the air cleanliness class of the clean room. Generally, smocks are sufficient for clean rooms with classes greater than 1000, with jumpsuits recommended for cleaner areas. A covering for the hair is recommended for all cleanliness levels transitioning to hoods and face masks as the cleanliness approaches class 10. Foot coverings should always be used, and gloves appropriate for the work being performed.

It should be remembered that it is the product that is being protected, not the clean room. The need for clean rooms with better air quality is driven by the need for a cleaner environment to manufacture the product. These better clean rooms have more favorable air-flow patterns, quicker flushing of particles from the room, and better isolation of the product from the workers.

In the future, the most advanced products will need cleaner processing equipment and handling procedures that isolate products from workers. The integration of sequential processing steps, as in the lithography cell proposed by Oberai (1987) or in vacuum processing as proposed by O'Hanlon (1987), reduces the need for operator handling. The SMIF concept, discussed in Chapter 17, is one way to isolate product from the environment. A study at National Semiconductor (1988) is investigating whether a clean room is really necessary if wafers can be isolated inside clean equipment and transported between processes in SMIF containers.

The realization of clean isolated process equipment and wafer transport between processes will result in less and less exposure of product to people and to the environment that people contaminate by their presence. The cleanliness of the clean room and the need for clean-room garments will decline in importance because the wafers will be less exposed to the environment and people. The end result may be that clean-room garments will be required to protect processing equipment from contamination more than to protect product wafers directly.

REFERENCES

ASTM F51-68. "Sizing and Counting Particulate Contaminant in and on Clean Room Garments." American Society for Testing and Materials, Philadelphia, PA.

Brinton, S. J., and R. W. Swick. 1984. "Evaluation of the Elements of Clean Room Garments for Particle Protection and Comfort." In: *Proceedings of the 30th Annual Technical Meeting, Institute of Environmental Sciences, Orlando, FL, May 1-3.* pp. 163-165.

Goodwin, B. 1987. "Packaging People to Protect Products—Particulate Contamination Control Apparel." In: *Proceedings of the 33rd Annual Technical Meeting Institute of Environmental Sciences, San Jose, CA, May 5-7.* pp. 483-485.

Hayakawa, I., S. Fujii, T. Yoshida. 1985. "Studies on Sizing and Counting Particulate Contaminant in and on Clean Room Garments." In: *Proceedings of the 31st Annual Technical Meeting, Institute of Environmental Sciences, Las Vegas, NV, April 30 - May 2.* pp. 84-88.

Hirasawa, K., E. Sudo, Y. Isobe, M. Kuroyanagi. 1984. "A New Method for Contamination on Clean Room Garments and Comparison with Clothes Made in Japan." In: *Proceedings of the 30th Annual Technical Meeting, Institute of Environmental Sciences, Orlando, FL, May 1-3.* pp. 176-181.

National Semiconductor. 1988. "Study Investigates Wafer Processing Without a Clean Room." *Semiconductor International, 11*(4):16.

Oberai, A. S. 1987. "Lithography – Challenges of the Future." *Solid State Technology, 30*(9):123-128.

O'Hanlon, J.F. 1987. "Advances in Vacuum Contamination Control for Electronic Materials Processing." *Journal of Vacuum Science Technology, A5*(4):2067-2072.

Salvo, N. M. 1985. "Practical Considerations in Contamination Source Control: How to Make the Best of a Retrofit Clean Room." *Microcontamination, 3*(8):43-45.

Sullivan, G., and J. Trimble. 1986. "Evaluation of Face Coverings." *Microcontamination, 4*(5):64-70.

PART 2
POINT-OF-USE FILTRATION

INTRODUCTION

Part 2 of this chapter discusses point of use filtration of gases and liquids. Gases are discussed first, then deionized (DI) water as a special case of liquid filtration. In the last section, differences between the filtration of DI water and the filtration of other liquid chemicals are outlined.

Two general comments about point-of-use filtration are appropriate here. First, the liquid stream flowing to the point of use should be fairly clean. The point-of-use filter should not be expected to compensate for errors of commission or omission in the upstream plumbing. Second, there should be as little hardware as possible between the point-of-use filter and the point of use. Ideally, the filter should be the last component of the distribution system to contact the fluid before it flows to the reaction chamber or to the wafer.

POINT-OF-USE GAS FILTRATION

Point-of-use gas filtration is easy compared to point-of-use liquid filtration. One reason is that certain particle capture mechanisms are more effective during the filtration of gases than they are during the filtration of liquids (Chapter 7). In particular, small particles in gases are efficiently captured by diffusion, so particles much smaller than the pore size of the filter medium are removed from the gas and retained on the filter. Commercially available point-of-use gas filters have measured efficiencies greater than 99.9999997 percent, with most penetrating particle sizes less than 0.1 μm (Accomazzo, Rubow, and Liu, 1984). Using these filters, gas delivery systems with 10^{-5} particles cm^{-3} (less than 0.1 particle ft^{-3}) have been constructed (Kasper and Wen, 1986).

The same physics works just as well in gas prefilters before the point of use, so the particle challenge to point-of-use gas filters is small compared to that of liquid filters. The low challenge to point-of-use gas filters ensures that they will not become loaded quickly, so the pressure drop across the filters does not increase rapidly. Point-of-use gas filters last a long time.

Another reason gases are easy to filter is that gases have low viscosities compared to liquids. High flow rates can be obtained at low pressure drops. Only a small filter package is required for high flow rates.

A final reason that gases are easy to filter is that the testing of gas filters is easy relative to the testing of liquid filters. This is primarily because of the availability of gas-phase particle counters that will count smaller particles than will liquid particle counters. Optical particle counters will count particles as small as 0.1 μm with low background counts, and condensation nucleus counters will count particles as small as 0.01 μm. These counters make it relatively easy to verify the efficiency of new filter designs. And filters may be tested in the field, so particle problems are relatively simple to troubleshoot.

But this does not mean that the filtration of gases is simple. Process gases commonly used for semiconductor manufacturing may be flammable, toxic, pyrophoric, or corrosive. The filters for these gases must be installed correctly not only to prevent contamination of the gas, but also to avoid hazardous leaks or mixing with air. Even when highly efficient filters are correctly installed, particles may be entrained from downstream components or from the surfaces of tubing, especially when there are pressure surges in the gas or vibration of the gas-handling hardware (Grant, 1987).

Point-of-use gas filters are often the first component blamed when a particle problem occurs in semiconductor processing equipment. It is tempting to immediately change all the point-of-use filters when such a problem is found, but when this is done it cannot be determined whether the problem has been fixed. A solution is to test suspect filters for leaks and for contamination on surfaces on the downstream side of the filter media. Elaborate systems may be constructed for these measurements (Rubow and Liu, 1984; Rubow, Liu and Grant, 1987; and Remiarz and Nelson, 1984). However, a quick test may also be done with less sophisticated equipment to detect gross leaks and contamination. Room air is a suitable filter challenge for such tests because it contains particles with a broad spectrum of sizes. The concentration of particles in indoor air with sizes between 0.01 and 3 μm may be 10^3 to 10^4 per cm^3. Drawing this air through the suspect filter with a pump and measuring the downstream particle concentration with an optical particle counter or condensation nucleus counter will easily reveal leaks. Downstream surface contamination may be detected by tapping the filter housing. Particles dislodged from downstream surfaces will be detected by the particle counter. Loading of the used filter may be detected by measuring the pressure drop across the filter during these tests and comparing it with a new filter or with the manufacturer's specifications.

Representative types of point-of-use filters are the Pall Gaskleen, Pall PFA Gaskleen, Millipore Wafergard, and Millipore Wafergard F (see Figure 16-1). The Pall Gaskleen has PTFE filter media and polypropylene hardware. The Pall PFA Gaskleen has an all-Teflon™ filter element. The Millipore Wafergard uses a PVDF filter medium on polysulfone. The Millipore Wafergard F has all-fluorocarbon construction with a PTFE filter medium on an ECTFE support. Both companies make smaller filter elements in smaller housings for use where space is restricted or lower flow rates are required. These small filters, the Millipore Wafergard F Mini In-Line and Pall Mini-Gaskleen, both have all-Teflon filter media and hardware. The housings of all of these filters are welded 316 stainless steel. For minimum leakage and dead space at the inlet and outlet connections, metal-gasketed VCR fittings are recommended.

In summary, point-of-use gas filters are available with efficiencies so high that they can be considered absolute filters for particles of all sizes. They generally have high flow rates and low pressure drops compared to liquid filters of comparable size. Generally, the particle challenge at the point of use is low, so the filters have a long service life. Quick integrity testing can be done in the field with a vacuum source and a particle counter. Filters are available with several types of filter media, media support hardware, and housing configurations. The manufacturer should be consulted to determine the proper filter medium and hardware for each gas to be filtered.

POINT-OF-USE DEIONIZED WATER FILTRATION

Point-of-use filtration of liquids is more complicated than point-of-use filtration of gases because the mechanisms of filtration are more complex, DI water typically has a higher particle concentration than occurs in gases, and there are more choices of filter media and filter systems for water than for gases.

The most important particle capture mechanisms operative during the filtration of gases are interception and diffusion. In contrast, the mechanisms operative during the filtration of liquids are determined by the physical chemistry of the particle-fluid-filter media system. When conditions are unfavorable, the only mechanism for the capture of particles is sieving; i.e., particles are retained that are too large to pass through the pores of the filter media. When conditions are favorable, the mechanisms of diffusion, interception, gravitational settling, and electrostatic attraction may also be operative. Grant et al. (1988) have compared filtration mechanisms in gases and liquids.

DI water typically has a higher particle concentration than gases because liquids are more likely than gases to entrain particles from surfaces. This is because liquids reduce the adhesive forces between particles and surfaces, they exert higher drag forces on particles, and they have thinner boundary layers. Prefilters are less likely to be efficient in liquids, so fewer particles are removed before the liquid reaches the point of use. Also, high-purity water is an aggressive solvent, and will attack the surfaces of the water delivery system. Finally, DI water supports the growth of bacteria, which reproduce to make more particles if active measures are not taken to prevent their growth. DI water filters in particular are excellent breeding grounds for bacteria. The filters hold particles that the live bacteria can live on, including fragments of dead bacteria. Bacteria control techniques include not allowing water to stagnate and sterilizing periodically with ozone, hydrogen peroxide, or ultraviolet light. The service life of DI water filters will be drastically reduced unless the particle challenge to them is reduced by prefilters, selection of clean system components, and the control of bacteria.

A final reason that point-of-use DI water filtration is complex is that there are many choices for filter media. Included are symmetric and asymmetric membrane media and membrane media with cylindrical pores. Filter media may be made from nylon, polysulfone, polyester, polycarbonate, polypropylene, or various fluorocarbons. The media may be charge-modified to produce large concentrations of positively or negatively charged groups at the surface to enhance particle capture by electrostatic attraction. The media may be naturally hydrophilic or hydrophobic, or hydrophobic media may be chemically modified to make them hydrophilic. Finally, beyond all of these well-known choices for filter media, ultrafiltration and reverse osmosis are beginning to be used for DI water filtration at the point of use.

Microporous membranes typically have a tortuous network of passages through which the fluid must flow. Symmetric membranes have the same density and average pore size through the depth of the media. Asymmetric membranes have larger average pore sizes on the upstream side of the filter. The average pore size decreases through the depth of the filter. Cylindrical pore membranes are made of either polycarbonate or polyester. They are made by exposing the polymer to high-energy nuclear particles, then etching the area of the polymer damaged by the nuclear particle track to produce a cylindrical pore. The pore size may be closely controlled by controlling the etch time, and measured with optical or scanning electron microscope techniques.

Naturally hydrophilic materials include nylon and polycarbonate. These materials are spontaneously wet by water, so do not require prewetting with a low-surface-tension liquid before use. Naturally hydrophobic materials include PTFE and PVDF. Polysulfone and PVDF may be surface-modified to make them hydrophilic. Although hydrophobic materials make perfectly good filter media for DI water and other high-surface-tension liquids, if air bubbles form on them or if air gets into the system, the dry area will not rewet. This area will be lost for filtration.

The pore size rating of filter media is not strictly related to the size of passages in the media. Historically, it is based on a filter's ability to quantitatively retain bacteria of a given size. The pore size is now often measured using a bubble point test, performed in the following manner. First, the filter is prewet and placed in the filter housing, then air pressure is gradually applied to the upstream side of the filter. The air will push the liquid through the filter until only a film of liquid remains on the media. As the air pressure is increased, air will slowly diffuse through the liquid film so that a slow flow of air will be visible on the downstream side of the filter. If the air pressure is raised further, the largest pores will blow out first, allowing a steady and continuous flow of air. The pressure at which the first pore blowout occurs is the bubble point pressure. The pore size is inversely proportional to the bubble point pressure.

Filters are commonly rated by manufacturers by their bubble point pressure. A low bubble point pressure indicates a large pore size and a high bubble point pressure indicates a small pore size. Bubble point pressures significantly lower than specified by the manufacturer for that filter indicate a damaged membrane or seals, or a leak in the system.

Charge-modified filters will retain particles smaller than their pore size by way of electrostatic attraction of charged particles. For example, the media may be chemically treated to produce a high concentration of positively-charged amine groups at its surface. Such a filter is said to have a positive zeta potential. Because many of the particles, bacteria, and colloidal material in water have a negative charge, these materials are removed by media with a positive zeta potential. An advantage of charge-modified filters is their ability to remove particles much smaller than their pore size from water, even particles with molecular dimensions (Krygier, 1986; James et al., 1988). For example, total organic carbon (TOC) measurements upstream and downstream of charge-modified filters have shown a reduction in the TOC level in DI water. A disadvantage of these filters is that as the active sites are neutralized, the capacity for electrostatic adsorption decreases. This capacity may be exhausted well before the filter shows signs of plugging. One suggestion is to use charge-modified filters just upstream of a finely pored final filter at the point of use.

Filter performance is measured by manufacturers using the following tests. These tests may also be performed by users to help choose a type of source of filter for point-of-use water filtration, or to test the integrity of the source:

- Bacterial challenge
- Bubble point, or diffusional flow (the rate of flow of air through the media below the bubble point)
- Downstream TOC measurement
- Ionic extractables
- Time or volume to recover resistivity
- Pressure drop as a function of flow rate
- Particle cleanup
- Particle efficiency.

In use, the failure (or imminent failure) of a point-of-use DI water filter may be detected by an increased drop in pressure across the filter or a slowing of the flow through the filter, an increase in particle counts in the filtered DI water or on wafers rinsed in the water, or an increase in the bacteria concentration in the filtered water. Suspected filter failures may be verified by bubble point testing or by measuring the diffusional flow of air through the filter media at a pressure just less than the bubble point pressure.

Selecting a point-of-use filter system for DI water is more complicated than can be described in a simple recommendation. In addition to all the filter performance tests described or listed above, several other points must be determined: the sensitivity of the product to particles and other contaminants, the cleanliness of the water flowing to the point of use, the frequency of sterilization, and the acceptable interval between filter changes. Then manufacturers' literature should be studied, manufacturers' engineers should be consulted, and experimentation should be done to determine the appropriate filter for each application.

POINT-OF-USE CHEMICAL FILTRATION

Point-of-use filtration of chemicals is very similar to that of DI water. In general, the same filtration mechanisms apply. Chemical filters are tested by manufacturers and users in the same ways that DI water filters are tested. DI water is often used for integrity tests and for the determination of pore size by bubble point pressure.

A major difference between chemical filters and DI water filters is that the filter must be carefully selected to be compatible with the chemical being filtered. This concern must also extend to the membrane supports, the o-rings used to seal the element to the housing, and the filter housing. Filter manufacturers have determined the chemical compatibility for most media with a wide variety of chemicals, including acids, bases, oxidants, solvents, adhesion promoters, and resists. So it is seldom necessary for users to perform their own compatibility studies. When in doubt, PTFE has been shown to withstand almost any chemical at temperatures to 125 °C.

Another difference between chemical filters and DI water filters is that bacteria will seldom be a problem in chemicals. However, chemicals are still likely to carry large concentrations of particles to point-of-use filters for the same reasons that DI water will. Bottled chemicals repeatedly have been shown to have high particle concentrations (Dillenbeck and DeVore, 1985), and bulk liquid chemicals should be expected to have equivalent concentrations unless proven otherwise. The high particle concentrations in chemicals at the point-of-use filter may result in a short service life.

The high viscosity of some chemicals compounds the problem of clogging of filters because the pressure drop across the filter may be great even when the filter is clean. A solution is to increase the effective surface area of the filter by using parallel arrays of filters.

Ionic extractables from point-of-use chemical filters may be measured in the same way as those from DI water filters, but they should be tested with the chemical that they will filter. This is most important for mineral acids and bases.

An advantage when filtering solvents is that no prewetting is necessary. Most solvents will spontaneously wet hydrophobic media. They will also rewet portions of the media dried by air bubbles, so this area is not lost to filtering when the air is removed.

A disadvantage of filtering chemicals is that charge-modified media will be ineffective in strong acids or bases, or in solvents in which there are no ions. The mechanism of electrostatic attraction cannot be used to filter these liquids.

Failures of chemical filters are detected by a decrease in flow or an increase in pressure drop, or by the appearance of particles downstream of the filter. Integrity failures may be detected by bubble point pressure tests.

SUMMARY

It is easy to obtain low particle concentrations in gases at the point of use because gases are relatively easy to filter. Particle capture mechanisms are effective for removing particles of all sizes, and particle concentrations are low in gases as compared to liquids. Gases have low viscosities, and gas filters are easy to test, although care must be taken when filtering hazardous gases to avoid leaks or mixing with air.

Filtering liquids is more complicated than filtering gases. Particle-capture mechanisms in liquid filters are more complex than those in gas filters, and liquids typically have higher particle concentrations. DI water, in particular, supports the growth of bacteria. There are more different types of media available for filtering liquids than gases, including hydrophobic, hydrophilic, and charge-modified media. Some liquid chemicals are aggressive toward filter media, supports, and housings, so filters for these chemicals should be carefully chosen in consultation with the manufacturer.

REFERENCES

Accomazzo, M. A., K. L. Rubow, and B. Y. H. Liu. 1984. "Ultrahigh Efficiency Membrane Filters for Semiconductor Process Gases." *Solid State Technology, 27*(3):141-146.

Dillenbeck, K., and S. DeVore. 1985. "Characterization of Particle Levels in Incoming Chemicals." In: *Proceedings of the 31st Annual Technical Meeting, Institute of Environmental Sciences, Las Vegas, NV, April 30 - May 2*, pp. 127-134.

Grant, D. C. 1987. "Particle Shedding from Semiconductor Process Gas Filters." In: *Proceedings of the 33rd Annual Technical Meeting, Institute of Environmental Sciences, San Jose, CA, May 5-7*, pp. 370-374.

Grant, D. C., B. Y. H. Liu, W. G. Fisher, and R. A. Bowling. 1988. "Particle Capture Mechanisms in Gases and Liquids: An Analysis of Operative Mechanisms." In: *Proceedings of the 34th Annual Technical Meeting, Institute of Environmental Sciences, King of Prussia, PA, May 2-6*, pp. 464-473.

James, J. B., G. Howard, P. D. Blanden, and V. Krygier. 1988. "Bacterial and Endotoxin Retention by Charge-Modified Filters." *Semiconductor International, 11*(4):80-82.

Kasper, G., and H. Y. Wen. 1986. "A Gas Filtration System for Concentrations of 10E-5 Particles/cm^3." *Aerosol Sci. Technol., 5*(2):167-185.

Krygier, V. 1986. "Rating of Fine Membrane Filters Used in the Semiconductor Industry." *Microcontamination, 4*(12):20,22-26.

Remiarz, R. J., and P. A. Nelson. 1984. "A New, Automated Method for Testing 'Point of Use' Gas Filters." In: *Proceedings of the 30th Annual Technical Meeting, Institute of Environmental Sciences, Orlando, FL, May 1-3* , pp. 69-73.

Rubow, K. L., and B. Y. H. Liu. 1984. "Evaluation of Ultra-High Efficiency Membrane Filters." In: *Proceedings of the 30th Annual Technical Meeting, Institute of Environmental Sciences, Orlando, FL, May 1-3*, pp. 64-68.

Rubow, K. L., B. Y. H. Liu, and D. C. Grant. 1987. "Characteristics of Ultra-High Efficiency Membrane Filters." In: *Proceedings of the 33rd Annual Technical Meeting, Institute of Environmental Sciences, San Jose, CA, May 5-7*, pp. 383-387.

17. CLEAN EQUIPMENT DESIGN RULES AND THE SMIF ISOLATION CONCEPT

B. TULLIS Micronaut, Hewlett-Packard Company, Palo Alto, California

INTRODUCTION TO MICRONAUTICS

Micronautics is evolving both as a science and as an engineering discipline. Astronautical, aeronautical, and nautical sciences and engineering have to do with safe navigation in space, through the atmosphere, and on water. In an analogous fashion, micronautical sciences and engineering (micronautics) have to do with navigation of products, sensitive to microscopic contamination or materials defects, safely through manufacturing environments.

In micronautics, as applied to the world of integrated circuit manufacturing, the voyagers are wafers of integrated circuits. The solar winds, auroras, and lightning storms of the larger world are the high-temperature diffusion gases, the electrically active plasma atmospheres, and high-energy implant beams of the wafer world. The re-entry heat shields to protect space vehicles are the masks of resist and nitride to protect wafers. The thick fogs are the vapors of vapor deposition equipment. The oceans are the process chemicals of hydrofluoric, phosphoric, and sulfuric acid, ammonium hydroxide, developer, and deionized water. The reefs and rocks are cassettes and wafer handlers. The perils of intergalactic particles, of meteorites, flocks of birds, wash of high seas, or whales are the dust particles, spittle, resist flakes, nozzle sprays, and bacteria along the courses traveled by microcircuits.

Micronautics, as applied to microelectronics, is motivated by the following objectives: to gain a competitive edge in international markets, to win manufacturing leadership, to obtain high yields, to provide contamination-free product environments, to be a minimized factor in robustly designed process chemistries, to be a minimized factor in the manufacturing of automation hardware, and not to be the limiting factor in complex process integrations.

It is hoped that this introduction to the concept of micronautics will help readers to see that the scientific disciplines for its understanding and the engineering disciplines for its application are many and diverse. Micronautics is a special branch of manufacturing that, in its own arena, can be as new, as complex, and as adventure-filled and exciting for its development pioneers as are ventures into outer space.

Micronautics combines the knowledge and practices being developed in manufacturing to improve product yields by controlling and reducing microcontamination.

This chapter addresses that portion of micronautics that has to do with designing integrated-circuit process equipment to free wafers from defect-causing contamination throughout the semiconductor manufacturing process.

Specifically, the subjects of clean equipment design and product isolation and protection through the application of a Standard Mechanical Interface (SMIF) will be discussed.

New Technology Trends: Clean Equipment

Figure 17-1 depicts the delicate balance that has prevailed throughout the history of semiconductor manufacturing from the late fifties through the eighties. Wafer yield has teetered up and down depending on the net balance of effects between clean-room unidirectional flow and the infinite sources of particles and vapors from operators and today's process equipment. The concept behind clean rooms with unidirectional flow (often called "laminar flow") has been to bathe process equipment and operators alike in a "river" of filtered air, to sweep away contaminants, with the hope that these contaminants will not be swept over the product wafers. This yield fluctuation is no longer acceptable if manufacturers are to compete successfully in the high-stakes world markets of quality microelectronics. New approaches are required and are now beginning to be developed.

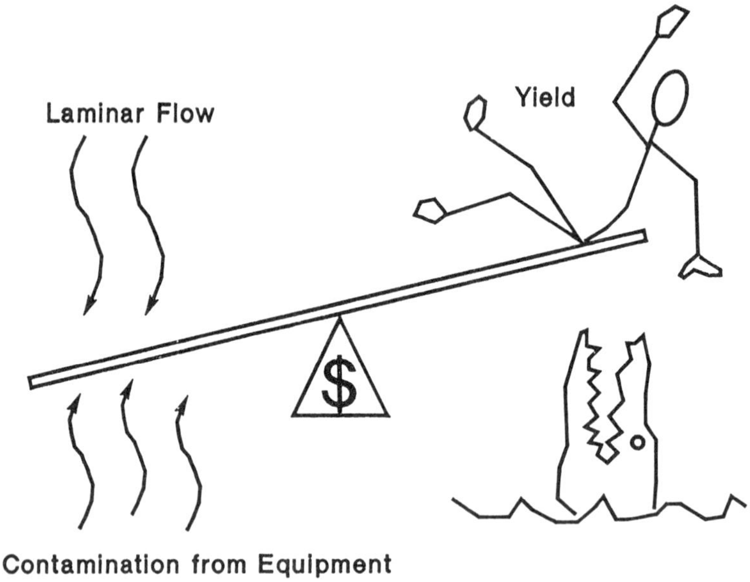

Figure 17-1. Delicate balance.

Figure 17-2 reveals the weaknesses of relying solely on unidirectional flow to prevent wafer contamination. Although the state of the art in filter design is admirably good, a quick calculation reveals that even in air filtered to Class 1 cleanliness levels (1 particle/ft^3 of air), the approximately 80 to 100 ft/min flow rate of the unidirectional flow can carry approximately 12,000 particles per day through a cross-sectional area equal to that of a 100 mm diameter wafer. This isn't to say this number will deposit onto such a wafer held or stored in this flow. It does mean that the best clean rooms of today cannot prevent "large" fluxes of particles into the environments of wafers. For instance, eddies and deviations from perfect unidirectional flow can cause particle transport between wafers and their environment in a clean room (Osburn et al., 1988). Figure 17-2 illustrates that operators, who easily walk at velocities exceeding that of the unidirectional flow, can create eddies with enough strength to carry particles upstream and significant distances across stream. Even operators speaking near wafers can easily contaminate them with spittle, in spite of the flow in a cleanroom.

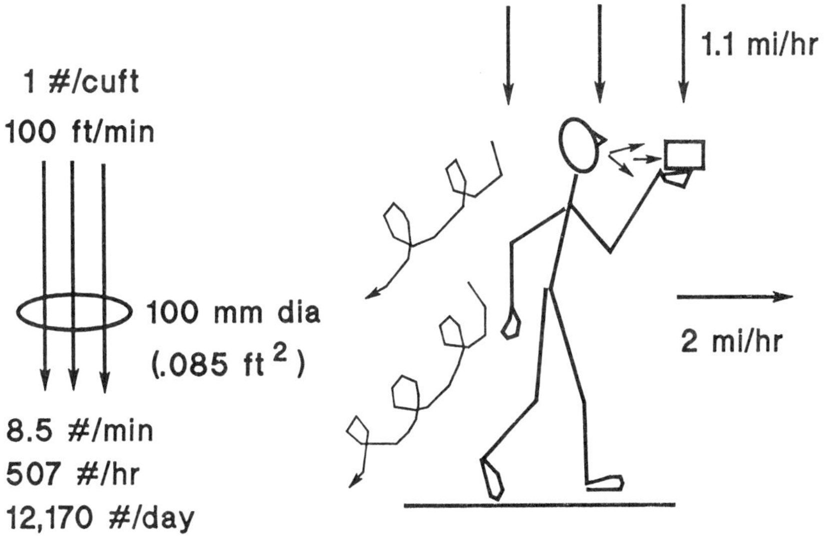

Figure 17-2. "Clean" "laminar" flow?

Some of the contamination problems associated with clean rooms have been controlled by requiring operators to wear restrictive garments and to follow very rigid protocol in handling and moving about the wafers. This tradeoff has been to the disadvantage of worker comfort and operational flexibility in general. Nevertheless, positive improvements in contamination levels have been earned. In spite of significant filtering and dilution of the contamination that operators and equipment emit into the clean room, however, significant levels persist within the equipment. It is this remaining contamination that currently limits yields. The sources of this remaining contamination lie inside the process equipment where the unidirectional flow of cleanrooms is unable to penetrate and to help significantly. Figure 17-3 illustrates these sources. They include chemicals and process gases, such as compressed air and nitrogen, used for pneumatic controls, air tracks, purging, and venting. Contaminating particles also are generated by process reactions, by wear and lubrication of mechanisms, by outgassing of hot parts like resistors and plastic-encapsulated transistors of control circuitry, and by mechanisms that touch and transport the wafers being processed. And of course they enter on the bodies of operators and maintenance personnel.

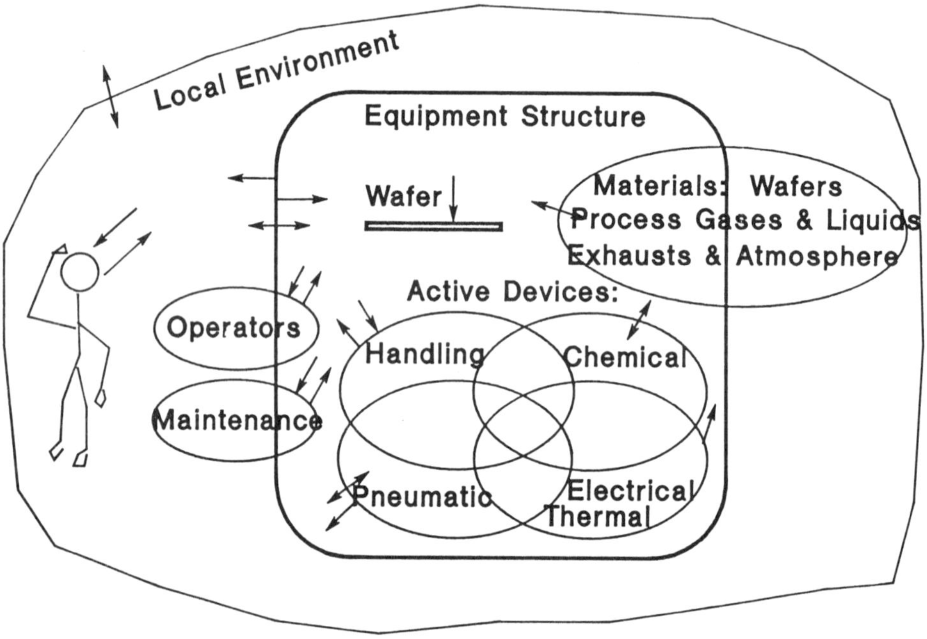

Figure 17-3. Generation and transfer of contamination.

The most recent technology of contamination control in semiconductor processing has moved toward particle prevention and away from strict reliance on the purging effect of clean rooms (Cheung and Roberge, 1987; Jensen, 1987; Schmidt, 1983; Hemmes, 1987; Hays, Bishop, and White, 1987; Hardegen and Lane, 1985; and Baker, 1985). This naturally turns the focus to process-related causes and to the inner design of the process equipment itself (Monkowski, 1983; Pearce, 1983). Because clean rooms cannot significantly reduce contamination occurring within process equipment, it is becoming obvious to most that in the future, manufacturers will no longer rely solely, and possibly not at all, on clean rooms to attain maximum yields. The underpinnings of such a radical change will be three achievements. These three are:

- Clean equipment design technology
- The shrinkage of contamination-controlled spaces from the global approach of today's cleanrooms inward to the significantly smaller process spaces of the environments immediately surrounding wafers

- The improvement of automation reliability so that wafers can be transported, transferred, positioned, and inspected without human personnel having to violate these wafer environments by physically assisting inadequate wafer-handling mechanisms.

Experiences in the early and mid-eighties have demonstrated significant improvements in the cleanliness of some equipment. Most of these improvements have been made by modifying existing equipment. Others have come about as a result of completely new designs. Still, it can be said that no piece of equipment has been designed from the ground up with cleanliness as one of its highest priorities, and by designers suitably competent in micronautics.

For example, Figure 17-4 illustrates some of the errors commonly included with today's equipment designs. Notice that the cassette in the illustration is close to an inside corner of the equipment's outer enclosure, putting the wafers in the path of a nonunidirectional flow. A control panel is located above the cassette, requiring an operator to reach over the cassette. A cassette elevator (not shown) would lower the cassette gradually down into the inner region of the equipment, where it would no longer be protected (even poorly) by the unidirectional flow of the clean room. A wafer-handling mechanism has a pneumatic air cylinder as part of its mechanism, and compressed air leaking through the rod seal can carry oil laden air across the wafers. Pneumatic control valves and exhausts are located below the wafer handler where the dirty air emanating from them can reach the wafers. A dc motor with carbon brushes is located in the equipment enclosure for opening and shutting a vacuum load-lock to an etch chamber (chamber not shown); lubrication and carbon wear particles from this motor can disperse to the immediate environment of the wafers. Electronics in an enclosure near the bottom of the equipment emit organic vapors from hot electrical components, again into the same space shared by the wafers. A cooling fan in this electronics unit aids in circulating all this dirty air throughout the equipment enclosure. This accumulation can stir up dirt from the floor and entrain it as well; note there is no bottom on the equipment enclosure. And an exhaust unit on the back wall, designed to collect potentially dangerous fumes and gases resulting from the process chemistry of the etch chamber (when the chamber is opened for unloading wafers) also pulls dirt up from the floor and across the wafers.

Another drawback in this design is that maintenance work done anywhere within the equipment could contaminate those regions occupied by wafers. A more subtle drawback is that even when the contamination of wafers is discovered, it is difficult to figure out what caused it; a user very likely would not be able to discriminate the various contaminating sources, because all share a common, interconnected environment. How could the user, by measurement, separate the contributions to contamination from the clean room itself, the operator, and the host of sources within the equipment enclosure? It would not be easy.

The above illustration demonstrates some designs with flaws obvious to most designers who have been asked to give a little thought to cleanliness needs. There are many other more subtle problems of design that are less obvious but must be solved. These issues include material selection and surface finishes; outgassing from plastics and adhesives; decomposition and erosion of surfaces; dispersion of particles of wear, paint chips, and specks; static cling; and all sorts of other phenomena that feed particles into wafer environments.

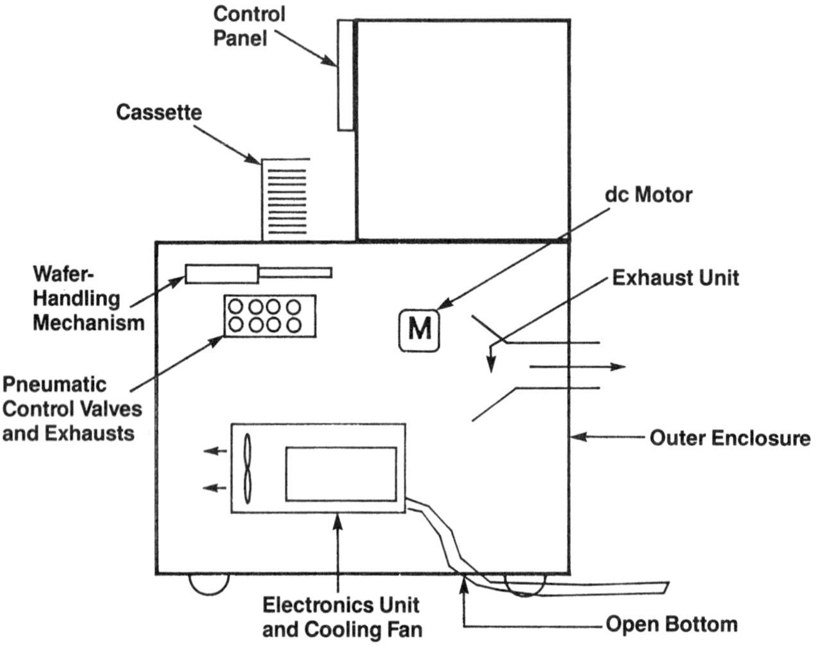

Figure 17-4. Contamination within process equipment.

Two whole areas of technology in which process-equipment designers in the integrated-circuit (IC) manufacturing business simply must become more skilled and knowledgeable are those of tribology and triboelectricity.

Tribology is the science and engineering of lubrication and wear. Both generate particles in the IC business. (Bierlein [1975] and Scott, Seifert, and Wescott [1974] are excellent and very readable introductions to the subject of tribology.) Translational mechanisms with four-bar designs are an example of how good design approaches can be used to reduce sources of wear particles, in this case by replacing sliding mechanisms with rotary ones. Particles generated by friction in rotary mechanisms are more easily sealed within the bearing than are particles generated by sliding linear bearings, for example.

Triboelectricity is the mutual charging that takes place between bodies in contact with one another. It is one of the major factors in the generation of electrostatics in the IC manufacturing environment and plays a principal role in particle transport and adhesion (Huffman, Nichols, and Bossard, 1987; Turner, 1983). Figure 17-5 is an abbreviated list of common materials, ordered according to their relative charging properties. Those materials at the top of the list become positively charged when brought into contact with materials below them on the list. Those materials lower on the list become negatively charged when in contact with materials above them on the list. Note that the increasing dielectric constant of these materials corresponds approximately with an increasing tendency to become positively charged. Also, either decreasing work function or decreasing length of polymer chains (in plastics) corresponds approximately with an increasing tendency to become positively charged. One outstanding relationship to note is the one between Teflon™ at the bottom of the list and air and hands at the top of the list. The implications should be remarkably clear to most IC makers: almost anything, and hands or flowing air in particular, will charge up most plastic cassettes. This charging not only can cause electrostatic damage, but also can attract particles to the wafers in the cassette.

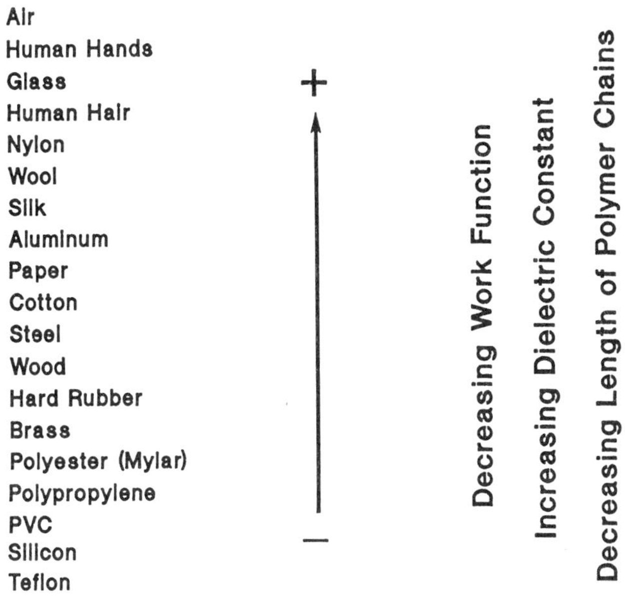

Figure 17-5. Triboelectricity in material selections.

Triboelectricity can be explained by the work-function potentials at the surfaces of materials; it exists whenever two bodies are in contact and have different work-function potentials. The net potential difference at their mutual interface drives electrons across that interface until a charge balance develops to oppose this driving force. When the two bodies separate, each is left with the charge it accumulated while the two were touching. The body having the higher dielectric constant will be left positively charged, while the other body will be negatively charged (Gallo and Lama, 1976b).

An effect which is not as well known is that this mechanism may also cause particles of either different materials or different sizes to become charged (see Figure 17-6). For example, droplets formed in a liquid spray may become charged as a result of size differences. One explanation of why particles of the same material but different sizes may become charged is that the work-function potential at the surface of the particle depends on the size of the particle. An explanation is attempted in Gallo and Lama (1976a), where a supporting model for insulating materials is developed using three classical electrostatic forces: (1) the mutual attraction between residual charges, which results from the work-function mismatch, (2) the mutual repulsion between a residual charge in a first particle and a polarization dipole induced within the second particle by its residual charge, and (3) attraction between the residual charge in the first particle and an image dipole within a second particle.

Note that larger particles, or particles with higher dielectric constant, tend to become positively charged; whereas smaller particles, or particles with lower dielectric constant, tend to become negatively charged.

Figure 17-7 portrays a not-too-unfamiliar situation in photolithography resist and developer spin cups. A silicon wafer is shown here in contact with a vacuum spin-chuck, made of stainless steel, which may or may not be electrically grounded. A spray or drip nozzle of stainless steel overhangs the wafer and also may or may not be electrically grounded. An inverted plastic dome is positioned to contain the wayward spray mist that scatters when the dispensed fluid from the nozzle strikes the spinning wafer. Not shown is a circular trough or cup that runs around the perimeter on the underside of the wafer to catch excess fluid as it flies off the spinning wafer. Also not shown are the complicated airflow patterns that develop because of the spinning wafer, and because of a combination of (1) an inflow of purge nitrogen introduced above the wafer and (2) an exhaust flow removed from beneath the wafer. The electrical charges shown can suspend and later precipitate photoresist droplets. This generation and precipitation onto wafers—either before, during, or after the intended cycle—is not well understood. No design is known to exist, and therefore none has been proven, that would minimize this source of contamination on the processed wafer.

Considerable data and experience exist on these and other detailed subjects important to micronautics, but designers of process equipment have not yet been exposed to, or used, much of it. One such example is an interesting and extensive list of common design materials for which NASA has quantified outgassing properties used in the spacecraft industry (Campbell, Marriott, and Park, 1980).

Experience with a number of procedures—including (1) monitoring particles per wafer per pass (PWP) measures of wafer contamination on wafers passed through state-of-the-art process equipment (see Chapter 13), (2) retrofitting improvements to this equipment, and (3) remeasuring—has generated general design rules that can be applied to almost all of today's process equipment. These rules are in two sets. The first has more to do with strategy; the second, with actual design guidelines.

The Self-Charging of Objects through Mutual Contact and Dissimilar Work Functions:

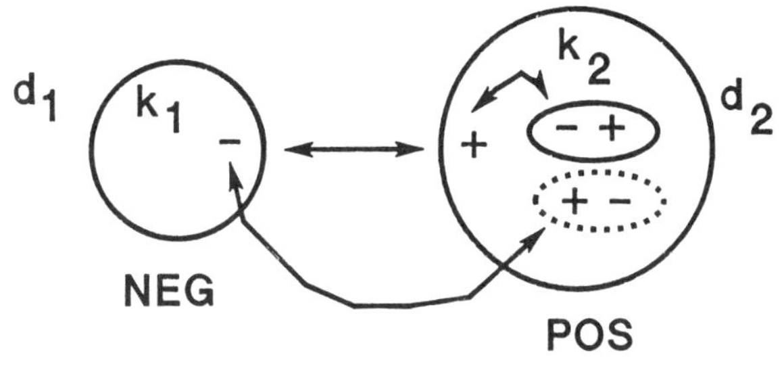

$k_1 < k_2$ or $d_1 < d_2$

k = dielectric constant
d = particle diameter

Figure 17-6. Triboelectric charging of particles.

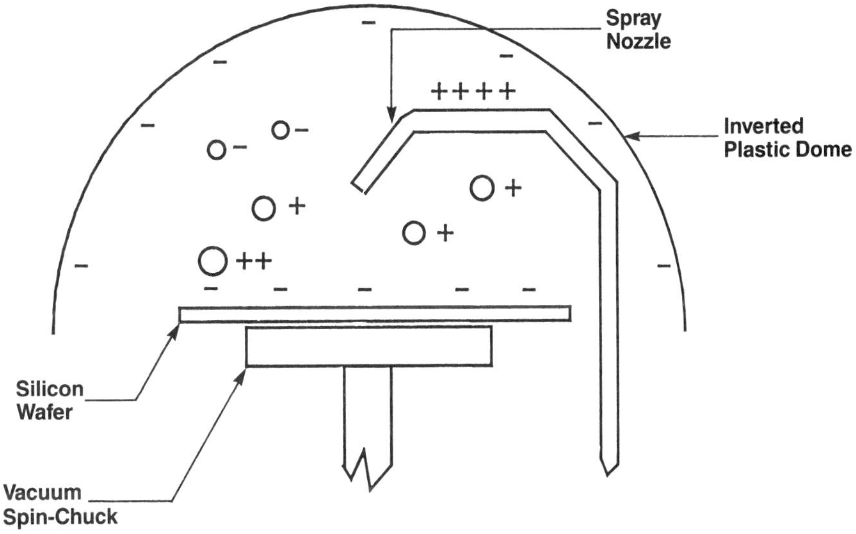

Figure 17-7. Triboelectricity: an example.

The strategies that have been found to produce clean processing equipment consistently are the following:

- Maximize the reliability of wafer-handling automation used within process equipment
- Design equipment and its controls to make contamination measurements easy, and to make particle-source isolation and diagnosis easy
- Design access for rapid maintenance and cleanup inside of equipment. Plan so that maintenance can be done cleanly. Make cleaning of all hardware easy, both inside and out
- Provide a means for the equipment to monitor its own contamination levels
- Avoid the temptation to apply solutions that only suppress a contamination problem when it could be eliminated entirely
- As much as possible, isolate wafers from contamination sources
- Eliminate, contain, or otherwise manage contamination sources, all as part of the equipment engineering assignment

Overall guidelines for designing clean processing equipment are the following:

- Choose designs that minimize wear. Use rotary bearings if any mechanical motions are required. Design minimum-wear fasteners and use minimum-wear tools
- Choose materials and finishes that will not spall or shed and that are nonporous, smooth and easy to clean
- Minimize outgassing of structural and sealing materials, components, adhesives, and lubricants
- Pay attention to triboelectricity effects, using them advantageously where possible, but otherwise eliminating them through choice of materials
- Prevent turbulences. Use smooth and gently contoured aerodynamic shapes to define the boundaries to any spaces the wafers must share. Use laminar internal air filtering and internal laminar flow where necessary; select optimum flow velocities for minimum contamination on wafers
- Minimize surface areas inside of equipment when those surfaces share an interface with wafer environments. Where equipment components must share environments with wafers internal to equipment, seal these components into enclosures having smooth and simple external surfaces. Seal complexes of components, or components with complicated external shapes. Jacket bundles of cables and tubing together. Avoid or seal recesses that would be hard to clean or that could prevent thorough cleaning

- Design critical equipment interior spaces and accesses to them to facilitate authorized "white glove" inspections and total cleanups

- Control vapor and fumes with an exhausted, positive-pressure inner space surrounded by an exhausted, negative-pressure outer shell.

No set of rules alone can guarantee clean operating process equipment. Often contamination can be quite process-specific. Mastery of the tools of contamination investigation is as important as the rules of clean design. These tools include not only instruments that can count particles and defects on either unpatterned or patterned wafers but also surface analysis tools (Chuan and Chiang, 1985; Wood, 1985; Barnard, 1987) and electrical defect monitors. The key advantage of surface analysis is that it can sometimes be so specific in describing the nature of the contamination material that the source becomes obvious. The value of process-step-specific electrical defect monitors (Weber 1988) is that they can determine the relative importance of other contamination measures, such as particle counts from various equipment used in the same process. The value of particle counters that can be used on patterned wafers (Billat, 1987) is that they can measure contamination occurring on wafers that have been subjected to the full rigors of the process steps, and any number of integrated steps; whereas, counting particles on bare wafers can amount to no more than counting particles that will be removed naturally in subsequent processing steps before those particles can cause any damage. Refer to Chapter 13 for a discussion of statistics necessary for dealing with contamination measurements.

This section has presented insights on improvements that need to be made to equipment designs for the semiconductor processing equipment market. Reducing the contaminating properties of the process chemistry is an integral part of the challenge to provide clean equipment. This requires that process developers and equipment designers work closely together. However, assuming that clean equipment can and will be designed, there would no longer remain any reason to operate such equipment in a clean room, if contamination from the operator and the air in the clean room is kept out of the equipment. Keeping out contamination that derives from sources external to the process equipment is the subject of the next section.

SMIF

Much has been written already about SMIF since its invention at Hewlett Packard Company in 1982 (Gunawardena et al., 1984; Gunawardena et al., 1985; Harada and Suzuki, 1986; Hughes, Moslehi, and Castel, 1988; and Sayre et al., 1989). SMIF is a Standard Mechanical InterFace between sealed, protected environments. A SMIF system is the collection of many such individually protected environments, each capable of being connected by way of the SMIF interface technique.

In a SMIF system, the protected space should be the minimum required for wafer processing, inspecting, and transportation, throughout the entire wafer fabrication process.

The future and success of SMIF would seem to depend entirely on the success of three prerequisites:

- Development of reliable wafer handling automation within equipment
- Development of clean equipment
- Development of convenient means to clean these inner spaces initially and after both scheduled or unscheduled maintenance, without requiring a complete and monolithic clean room to contain the entire process.

What is meant in the SMIF concept by "a protected environment" (see Figure 17-8) is an environment the inside of which is engineered for minimizing contamination to the wafers which may at times share that environment. A "SMIF'ed" process would be one in which individual process tools are contained within, or contain within them, their own respective protected environments. In the future, each process tool would conceivably come from a vendor with the protected and contamination-controlled environment actually built into the equipment as an integral element of that tool.

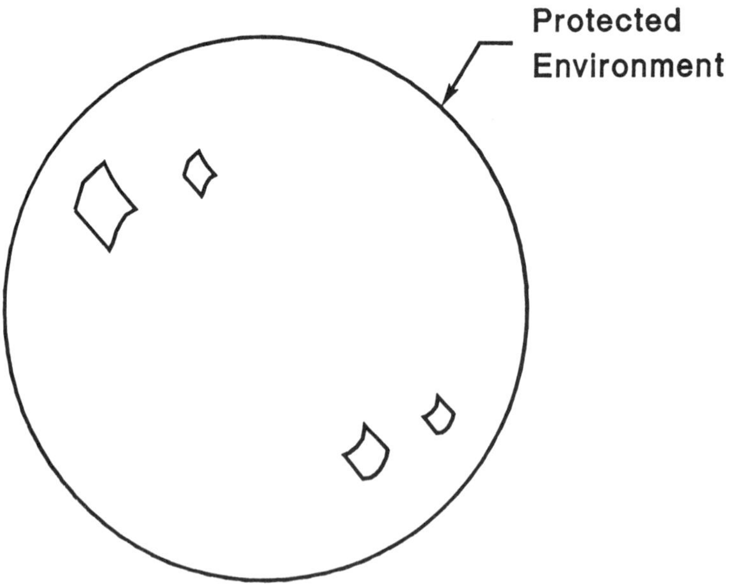

Figure 17-8. Analogy for a SMIF-protected environment: a soap bubble, which is clean and enclosed (i.e., protected).

Figure 17-9 shows how wafers would get from one tool to another, by temporarily connecting the two by way of a "standard mechanical interface." Cassettes of wafers, or even single wafers, would travel from tool to tool via a portable protected space, more commonly known as a SMIF cassette box. Figure 17-9 together with Figure 17-10 shows the SMIF concept of environment interfacing. Two protected spaces, one containing a process tool and one containing the cassette(s) of wafers arriving or leaving the tool, are brought into contact with one another at a specially designed interface where each has a door as a portion of its defining shell or enclosure. These two doors, once in contact with one another, are then moved into thc interior space of one of the two enclosures. This opens a port that permits the wafer(s) to be passed from one space into the other. So long as the doors are "opened" in this manner, they are kept together in mutual contact as one unit. Note that this is a clean interfacing technique. Whatever dirt may have been on the exterior of either door is therefore trapped between the two "opened" doors as long as they remain "opened."

SMIF has great promise for IC equipment users. Some of its advantages are:

- Equipment operators are freed from the restrictive garments that clean room protocol imposes upon them
- Operators and maintenance personnel will no longer affect the cleanliness of wafers merely by being close to and operating the equipment
- Operators, engineers, and managers are free to move about the equipment and to communicate with one another freely and easily
- What goes on in the rest of a facility is of little concern to any one piece of equipment, at least from a contamination point of view, because that equipment is isolated from any particles moving about the atmosphere outside of its protecting enclosure
- The cost of SMIFing the first or each additional piece of equipment is relatively inexpensive in risk, in commitment, and in actual cost of the SMIF enclosure
- SMIF can facilitate flexible siting of equipment. This flexibility stems from the fact that each equipment is its own clean space. Pieces of equipment do not have to be positioned underneath a laminar flow outlet in a cleanroom, or according to constraints regarding potential contamination effects from neighboring equipment or nearby traffic and activity of operators, or according to constraints on flow balance throughout a cleanroom
- SMIF implementations can be retrofitted onto existing equipment to get started in an existing facility.

SMIF directly suggests the possible elimination of clean rooms. Because major shifts in any technology take some time to be absorbed and to allow for careful management of change, SMIF was first used in an existing cleanroom where the cleanroom filters provided a "safety net" for the "early explorers." This approach is being used by most first-time users. Some bolder first-time users (Hughes, Moslehi, and Castel, 1988) have started without the safety net, but only in an R&D facility. Whichever approach is used for the introduction of SMIF, much progress has been made, and more will continue until the future sees a downturn in the building of clean-room spaces that cost many hundreds of dollars (or more) per square

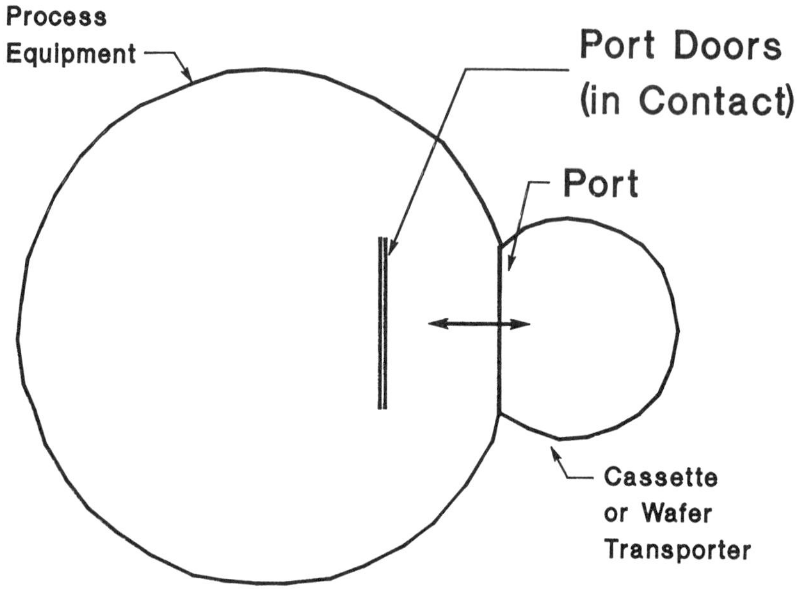

Figure 17-9. SMIF system.

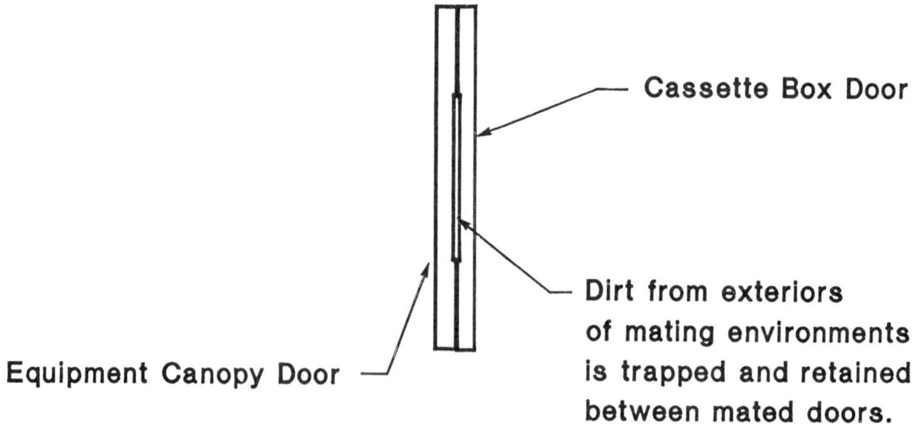

Figure 17-10. SMIF's clean interface with dockable doors.

foot. It makes good sense that this expensively clean air should be purchased and maintained for the small environments required by the wafers themselves but not for entire manufacturing facilities.

An important and valuable observation made by SMIF users is that once a piece of process equipment has been isolated in an enclosure (with a SMIF port), if that equipment has particle sources within, and no internal air filtering or flow control has been added, then the particle counts within the enclosure and on processed wafers can increase (see the closed box analysis in Chapter 14). This phenomenon has been very beneficial in speeding the discovery of particle sources that would otherwise have defied attempts at systematic detection, identification, and removal. If the equipment is enclosed within an isolated space, any particle count increases can only be associated with internal sources. This removal of the operator, experimenter, and external environment from the list of possible sources makes it significantly easier to focus on the ultimate causes. Unlike the traditional clean room, in which the detection and subsequent removal of particle problems is often difficult or nearly impossible, the SMIF approach permits rapid feedback on contamination problems, with more sensitivity for detecting particle sources, and with much more precision in isolating their cause(s). When as many significant contamination sources as possible have been identified and engineered out of the equipment by replacing parts or changing designs, then and only then should air flow control and filtering be installed within the enclosure, if needed.

Figure 17-11 testifies to the success of the above implementation strategy in the refinement of the hardware in one of the very first SMIF ports. The test setup consisted of two enclosures, each equipped with a SMIF port. One enclosure functioned as a cassette transporter, the other as a representative enclosure around a piece of imaginary process equipment. The equipment enclosure contained no process equipment. The only equipment it contained was the hardware latch and the elevator mechanisms, which were to unlatch the closed doors of both enclosures and lower both the cassette and the contacted doors down into the inner space of the "equipment" enclosure.

The challenge was to determine whether the port, port latches, or elevator mechanisms generated any particles when operated together as a system. Figure 7-11 shows a bar chart of the particle counts contributed by various components of the system both before and after all improvements had been made. The smaller counts are those finally achieved after the causes of particle generation were found and eliminated by design. The larger counts shown are those obtained after each particle source was returned to its original state. These components were changed to their original states one component at a time, to prevent any confusion in the measurements with interactions among the various sources. At first, before design changes were made, the particle counts were high enough to detect with an aerosol particle counter. This made it possible to find two of the particle sources by detecting repeatable increases in the particle concentrations whenever the aerosol probe was put in close proximity to either source. Toward the end of the project the residual particle counts could only be detected by surface counts on wafers that had undergone hundreds of repeated cycles of port, port latch, and elevator operations. In the latter case, identification of the source with a single mechanism was only achieved through design replacements and remeasurements. This level of

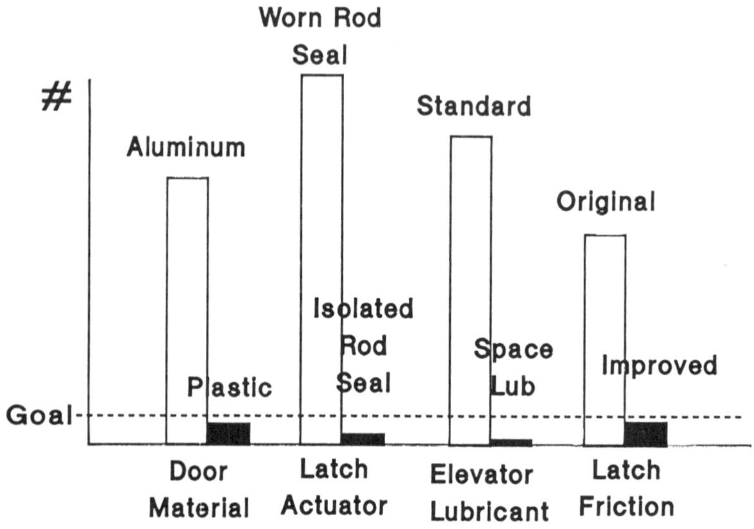

Figure 17-11. Particle reduction efforts (early '84).

precision in detective work and degree of improvement simply could not have been accomplished with the hardware open to the airflow in a cleanroom. The final outcome was an improved design for the hardware and mechanisms of both SMIF ports and cassette handlers. These new designs were then used successfully to implement cleanup and SMIFing of real process equipment.

Opportunities and Limitations

The needs for cleaner processing equipment and new approaches to controlling contamination in the immediate environments of wafers present rich and new opportunities. These will attract entrepreneurs and established suppliers alike.

Opportunities exist for equipment manufacturers to compete by selling self-contained and self-monitoring clean equipment that will not depend on the quality of the clean room into which users install them.

As mentioned at the beginning of the previous section, the promise for cleaner equipment and production-worthy SMIF systems, with or without clean rooms, will depend heavily on significant progress being made first (or in parallel) to improve the reliability of wafer-handling mechanisms used within protected wafer environments. No longer can the industry afford to put up with operators needing to manually assist stuck wafers. Such actions not only contaminate wafers but greatly reduce equipment utilization and operator productivity. In general, poor reliability impedes the development of cleaner process equipment.

Robotics and mechanisms designers can affect positively the yields and productivity of integrated circuit manufacturers if they will design wafer and cassette handlers to run for days rather than minutes without mishandlings. Their progress in improving reliability will contribute directly to progress in reducing sizes of protected wafer environments.

Opportunities exist for users to achieve higher yields, at both lower initial capital costs and lower operating costs. Workers may have the opportunity to return to normal work environments, where they can communicate easily, unhindered by restrictive garments and separated physically and emotionally from the "normal" community outside the cleanroom. Managers and engineers may have the opportunity to return to productive managing and engineering by moving closer to the equipment, the operators, and the process operations.

On the other hand, there are limitations to be overcome. Clean design is not an old and established discipline. Schools and universities do not teach it. There are no textbooks. Managers must be brave enough to risk the needed change and skilled enough to manage it well. A well-entrenched clean-room and garment industry must manage change from products for cleaning large spaces to products that are smaller in scale, but perhaps larger in sales volume, that will (1) keep very small spaces clean and (2) protect the wafer environments from the contamination effects of routine intrusions for maintenance. Clean-room and facilities architects may have to give up large air-handling systems and through-the-floor flow designs, and focus instead on providing greater flexibility for the users in equipment and process-flow site planning.

The Challenge

In summary, the greatest challenge in microcontamination control today is the awakening to a new vision. This new vision is a giant leap forward in advancing the technology of clean equipment design and a giant leap backward in the amount of air that must be moved and filtered during the manufacture of integrated circuits.

Lastly, just as this chapter began on the theme of micronautics, let us conclude with it. Micronautics must draw together the many and diverse branches of physics, chemistry, and engineering that will contribute to microcontamination control. Reducing process defects to allow processes to achieve their highest yields requires more contributions than can be meaningfully represented by the phrases "yield improvement," "microcontamination control," "clean equipment design," or "isolation technology" and "SMIF." What is needed is an industry focus, a rallying point, a single word that encourages interest and innovation. The word proposed here is micronautics.

REFERENCES

Baker, E. 1985. Characterizing Particle Generation by Front-End Semiconductor Processing Equipment. *Microelectronic Manufacturing and Testing* (July):20-22.

Barnard, L. 1987. Using Particle Analysis to Solve Microcontamination Problems. *Microcontamination* (August):34-47.

Bierlein, J. 1975. The Journal Bearing. *Scientific American* 233(1):50-52 and 57-64.

Billat, S. 1987. Automated Defect Detection on Patterned Wafers. *Semiconductor International* (May):116-119.

Campbell, W., Jr., R. Marriott, and J. Park. 1980. Outgassing Data for Spacecraft Materials. NASA Reference Publication 1061, August.

Cheung, S., and R. Roberge. 1987. An Inside Look at Measuring Particles in Process Equipment. *Microcontamination* (May):45-50,94,96.

Chuan, R., and W. Chiang. 1985. Cascade Impactor Collects Airborne Samples for Particle Identification. *Solid State Technology* (July):133-137.

Gallo, C., and W. Lama. 1976a. Classical Electrostatic Description of the Work Function and Ionization Energy of Insulators. *IEEE Transactions of Industry Applications* (January/February):7-11.

Gallo, C., and W. Lama. 1976b. Some Charge Exchange Phenomena Explained by a Classical Model of the Work Function. *Journal of Electrostatics* (February):145-150.

Gunawardena, S., R. Haven, U. Kaempf, M. Parikh, B. Tullis, and J. Vietor. 1984. The Challenge to Control Contamination: A Novel Technique for the IC Process. *The Journal of Environmental Sciences* 27(3):23-30.

Gunawardena, S., U. Kaempf, B. Tullis, and J. Vietor. 1985. SMIF and Its Impact on Cleanroom Automation. *Microcontamination* (September):55-62 and 108.

Harada, H. and Y. Suzuki. 1986. SMIF System Performance at 0.22 μm Particle Size. *Solid State Technology* (December):61-65.

Hardegen, B., and A. Lane. 1985. Testing Particle Generation by a Wafer Handling Robot. *Solid State Technology* (March):189-195.

Hays, K., K. Bishop, and T. White. 1987. Particle Reduction Investigation of a LPCVD Tungsten Reactor. *Microcontamination Conference Proceedings*, pp. 1-7, October.

Hemmes, D. 1987a. Microcontamination Reduction in PECVD Systems - Part I. *Semiconductor International* (May):90-97.

Hemmes, D. 1987b. Microcontamination Reduction in PECVD Systems - Part II. *Semiconductor International* (June):98-102.

Huffman, T., G. Nichols, P. Bossard. 1987. Reduction of Particle and ESD Damage by Room Ionization. *Solid State Technology* (November):127-130.

Hughes, R., G. Moslehi, and E. Castel. 1988. Eliminating the Cleanroom: Experiences with an Open-Area SMIF Isolation Site (OASIS). *Microcontamination* (April):31-37.

Jensen, D. 1987. Reducing Microcontamination Generation and Entrapment within High-Purity Gas Distribution Systems. *Microcontamination* (May):52-62,97,99.

Monkowski, J. 1983. Role of Particulate Contamination in Device Yield and Reliability. *Semiconductor Processing and Equipment Symposium Technical Proceedings*, Zurich, pp. 304-310, March.

Osburn, C., H. Berger, R. Donovan, and G. Jones. 1988. The Effects of Contamination on Semiconductor Manufacturing Yield. *The Journal of Environmental Sciences* (March/April):45-57.

Pearce, C. 1983. Defect and Contamination Control in VLSI Fabrication Processes. *Semiconductor Processing and Equipment Symposium Technical Proceedings*, Zurich, pp. 293-303, March.

Sayre, S., T. H. Lin, P. Sanchez, and S. Bhola. 1989. SMIF Reduces Defect Density in Class 100 Production Facility. *Semiconductor International* 12(10):104-107, September.

Schmidt, P. 1983. Furnace Contamination and Its Remedies. *Solid State Technology* (June):147-151.

Scott, D., W. Seifert, and V. Westcott. 1974. The Particles of Wear. *Scientific American* 230(5):88-97.

Turner, T. 1983. Static in a Wafer Fabrication Facility: Causes and Solutions. *Semiconductor International* (August):122-126.

Vander Wood, T. 1985. Identification of Particulate Contaminants in IC Manufacture. *Solid State Technology* (August):177-182.

Weber, C. 1988. Standard Defect Monitor. *IEEE Conference on Microelectronics Test Structures*, Vol. 1(1), pp. 114-119, February 20-22.

18. AIRFLOW MODELING AND PARTICLE CONTROL BY VERTICAL LAMINAR FLOW

TOSHIAKI YAMAMOTO Research Triangle Institute, Research Triangle Park, North Carolina

INTRODUCTION

The general clean-room ambient (the air entering the fab from the high-efficiency particulate air [HEPA] filter bank) is no longer a major contributor to chip particulate contamination because state-of-the-art clean rooms at rest typically have less than 1 particle/ft^3 (3.5 x 10^{-5} particle/cm^3) greater than 0.5 μm in diameter. Particles from clean-room processing equipment or workers now play the dominant role in wafer contamination. The particles generated by any source within the clean room are continuously swept away by the airflow, typically designed to be a vertical laminar flow. Therefore, it is essential to understand the interaction of the flow within the clean room and the particle deposition mechanisms in the work area. Airflow modeling is the first step toward gaining further understanding of particle contamination problems.

Many publications on clean-room airflow modeling have appeared recently (see bibliography). In this chapter, computational methods for airflow modeling in clean rooms and mechanisms of particle deposition on silicon wafers are reviewed. The flow fields are calculated by solving the two-dimensional Navier-Stokes equations, and clean-room case studies are presented to help understand the role of the flow field. Theoretical guidelines for describing clean-room airflows are also discussed. Finally, the deposition of particles on silicon wafers under idealized flow conditions and the effect of electrostatic forces and thermal gradients on particle deposition are discussed.

FLOW FIELD FORMULATION

The governing equations of two-dimensional, incompressible flow in rectangular geometry are the continuity equation and the momentum (Navier-Stokes) equations. The Navier-Stokes equations result from applying the role of viscosity, Newton's second law, and mass and momentum conservation to a volume element through which the fluid is flowing. The detailed derivation of these equations is described in many textbooks (e.g., Knudsen and Katz, 1958; Schlichting, 1968).

$$\frac{\partial u}{\partial t} + u\frac{\partial u}{\partial x} + v\frac{\partial u}{\partial y} = -\frac{1}{\rho}\frac{\partial P}{\partial x} + \nu\left(\frac{\partial^2 u}{\partial x^2} + \frac{\partial^2 u}{\partial y^2}\right) , \qquad (18\text{-}1)$$

$$\frac{\partial v}{\partial t} + u\frac{\partial v}{\partial x} + v\frac{\partial v}{\partial y} = -\frac{1}{\rho}\frac{\partial P}{\partial y} + \nu\left(\frac{\partial^2 v}{\partial x^2} + \frac{\partial^2 v}{\partial y^2}\right) , \qquad (18\text{-}2)$$

where u and v are the x and y components of the fluid velocity, P is the pressure, ν is the kinematic viscosity, and ρ is the gas density. The continuity equation in two dimensions for an incompressible fluid is:

$$\frac{\partial u}{\partial x} + \frac{\partial v}{\partial y} = 0 \quad . \tag{18-3}$$

The continuity equation follows from the conservation of mass, which states that the change of fluid mass in any fixed volume element equals the rate of mass flow in or out of that element. When the fluid density is constant in time and position (an incompressible fluid), the continuity equation takes the form shown in Equation (18-3).

The pressure in Equations (18-1) and (18-2) is eliminated by cross-differentiation with respect to y and x and substitution. Using the definition of vorticity, ς:

$$\varsigma = \frac{\partial u}{\partial y} - \frac{\partial v}{\partial x} \tag{18-4}$$

and the definition of the stream function, ψ, the vorticity transport equation, and the elliptic Poisson equation can be written as:

$$\frac{\partial \bar{\varsigma}}{\partial \bar{t}} = - \bar{u}\frac{\partial \bar{\varsigma}}{\partial \bar{x}} - \bar{v}\frac{\partial \bar{\varsigma}}{\partial \bar{y}} + \frac{1}{Re}\left(\frac{\partial^2 \bar{\varsigma}}{\partial \bar{x}^2} + \frac{\partial^2 \bar{\varsigma}}{\partial \bar{y}^2}\right) \tag{18-5}$$

$$\bar{\varsigma} = \frac{\partial^2 \bar{\psi}}{\partial \bar{x}^2} + \frac{\partial^2 \bar{\psi}}{\partial \bar{y}^2} \quad , \tag{18-6}$$

where

$$Re = \frac{U_o W}{\nu} \quad , \quad u = \frac{\partial \psi}{\partial y} \quad , \quad \text{and } v = -\frac{\partial \psi}{\partial x} \quad ,$$

and where all quantities are nondimensionalized with reference to the half-width of the clean room, W, and the average inflow velocity, U_o:

$$\bar{\varsigma} = \frac{W\varsigma}{U_o}, \ \bar{\psi} = \frac{\psi}{U_o W}, \ \bar{t} = \frac{U_o t}{W}, \ \bar{u} = \frac{u}{U_o}, \ \bar{v} = \frac{v}{U_o}, \ \bar{x} = \frac{x}{W}, \text{ and } \bar{y} = \frac{y}{W} \quad . \tag{18-7}$$

Many solution methods such as finite difference, finite element, direct solution methods, etc., are available for solving Equations (18-5) and (18-6). The solution presented in this chapter is based on a finite difference approximation using a standard explicit method such as the successive-over relaxation (SOR) method. The upwind difference scheme of the convective term in Equation (18-5) was used because it has been found to eliminate the numerical instability that occurred with central differences (Taylor, Ndefo, and Masson, 1972; Roache, 1972). In all cases, the flow calculations started with an appropriate initial condition, and time was advanced until a steady state was reached. For numerical stabilities, the time increment should be less than the minimum cell transit time over all cells. The time increment used here was:

$$t = 0.8 \left[\frac{\Delta x}{|u|}\right]_{min} \quad \text{or} \quad t = 0.8 \left[\frac{\Delta y}{|v|}\right]_{min} , \qquad (18\text{-}8)$$

whichever is less and where Δx and Δy are the grid size in the x and y direction, respectively.

Boundary Conditions

As with the solution of any time-dependent partial differential equation, it is necessary to specify the boundary conditions and the initial state of the system. Most of the boundary conditions given are either Dirichlet conditions (specified function value) or Neumann conditions (specified normal gradient). No mathematically rigorous relations to specify the boundary conditions are available for many of the fluid problems. Moreover, the large number of contingencies make analytical studies difficult and limited in applicability. Computational expectation and experimental data are often required to specify realistic boundary conditions.

A typical boundary condition used for this simulation is the following: (1) The nonslip boundary conditions are used for wall boundary conditions (fluid velocity = 0 at the wall); (2) as inflow boundary conditions, a uniform flow is assumed through the filter banks; (3) for outflow boundary conditions, a less restrictive condition, such as parallel flow, is appropriate and allows a considerably shorter domain for comparable accuracy; and (4) a first-order formulation consistent with the wall boundary conditions is used at the sharp corner. The detailed boundary conditions are described by Yamamoto et al. (1987a; 1988a).

NUMERICAL RESULTS AND DISCUSSION

[Case 1] Tabletop in Center of Room

A configuration of no obstacles in the clean room is assumed for the first modeling. Figures 18-1a and 18-1b show the fluid streamlines and corresponding velocity plots for half the clean room, respectively (AB is the centerline of the clean room). Note that the velocity is nondimensionalized with a uniform inflow velocity of 0.46 m/s (90 ft/min) throughout this discussion. As can be expected, the flow is very smooth and ideal; no dead zone or zone of recirculation is formed.

When a 1-m (3.3 ft) long tabletop is placed in the center of the room, the tabletop being 1-m (3.3 ft) above the raised floor, the flow streamlines and velocity plots are as shown in Figures 18-2a and 18-2b. In these figures, a pair of flow circulation zones under the table is observed.

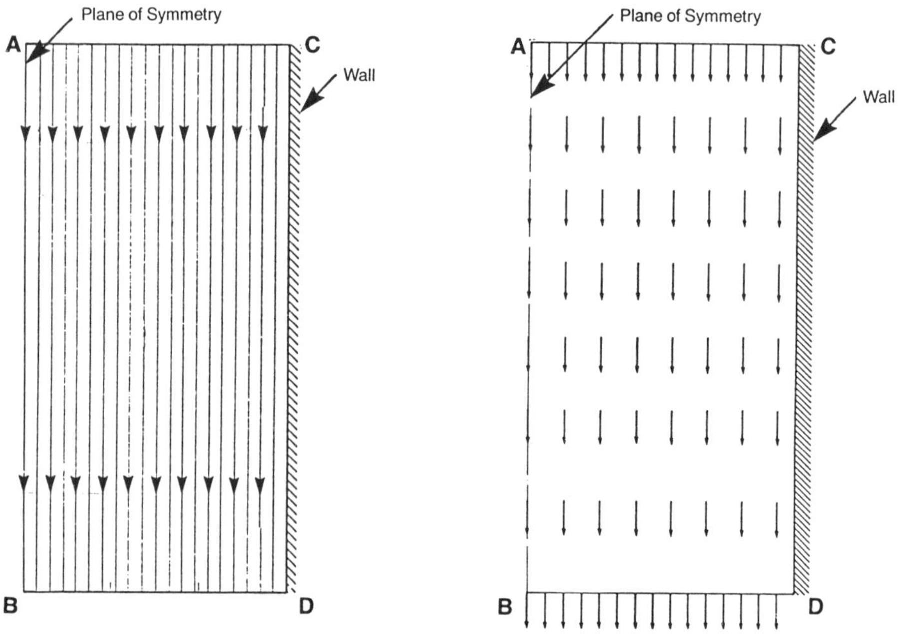

Figure 18-1. (a) Flow streamlines and (b) velocity plots for an empty clean room.

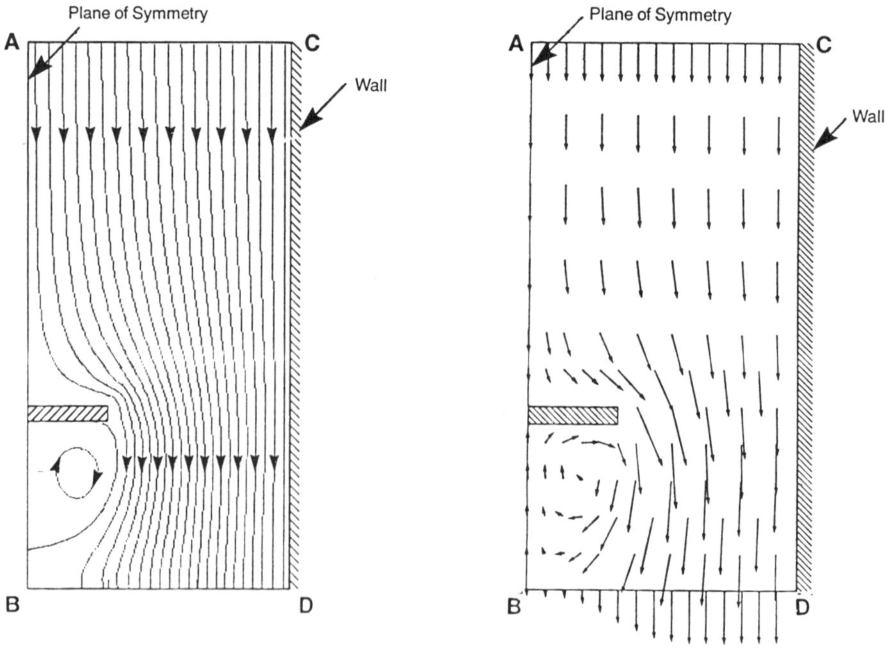

Figure 18-2. (a) Streamlines and (b) velocity plots for a table to the center of the room.

This flow recirculation is not a favorable condition because it may entrain and trap particles that originate near the work area. Once particles are trapped in such a recirculating zone, they cannot escape unless inertia effects, gravitational settling, turbulent diffusion, Brownian diffusion, or other external forces (i.e., electrostatic forces) are present. This plot recommends the use of a perforated table to reduce the flow recirculation under the tabletop. (A perforated table, however, is expected to increase particle deposition velocity on a horizontal wafer resting on the table (Bunkofske, 1988).)

[Case 2] Solid Workbench in Center of Room

Figures 18-3a (raised floor) and 18-3b (sidewall return) show the fluid streamlines for a centered solid workbench in a uniform vertical flow. As seen in Figure 18-3a, the flow field for the raised floor configuration is smooth and stable throughout the clean room. On the other hand, the flow field for the sidewall return configuration (Figure 18-3b) shows a large recirculation zone or flow instability near the side face of the workbench. Using a workbench with rounded corners (high radius of curvature) would minimize the flow recirculation. Obviously, the raised floor configuration is preferable for this geometrical configuration.

[Case 3] Solid Workbench Below Isolating Curtain

Figures 18-4a and 18-4b illustrate the streamlines gained from adding a vertical isolating curtain that hangs 0.7 m (2.3 ft) from the room's centerline (AE = 0.7 m [2.3 ft]) and extends down to the same level as the workbench (1.0 m [3.3 ft] above the floor). The inlet air velocity across the ceiling is uniform. For both return duct configurations, the flow is straightened up to near the bottom of the curtain and then reflected more sharply sideways. The air velocity between the bottom of the isolating curtain and the workbench becomes 3.5 times higher than the inlet air velocity. A large flow recirculation is induced near the side face of the workbench for the raised floor configuration, which is not seen without the curtain (Figure 18-3a). The higher air velocity near the bottom of the curtain causes a low local static pressure, resulting in flow instabilities and recirculation. With the sidewall return configuration (Figure 18-4b), the recirculation zone is significantly reduced so that the sidewall return configuration is actually superior to that of the raised floor, when used with this curtain configuration. For both return configurations, it would be expected that any perturbation or disturbance occurring on the right-hand side of the isolating curtain has less chance to affect the workbench area than when the curtain is absent; thus, the curtain isolation technique should be effective in shielding the work area from disturbances adjacent to the curtained bench.

Figures 18-5a and 18-5b show the streamlines when the isolating curtain is raised to 0.5 m (1.6 ft) above the solid table (1.5 m [4.9 ft] from the floor). The recirculation zone is significantly reduced for the raised floor configuration. However, as shown in Figure 18-5b for the sidewall return configuration, a recirculation zone reappears at the same location that it did without the curtain (Figure 18-3b). The height of the curtain above the workbench clearly affects the recirculation zone below the tabletop.

[Case 4] Isolating Curtain Defining Regions of Different Entry Velocity

Different entry air velocities to the two zones are sometimes employed as an energy conservation measure where one section uses well-filtered air at a high flow rate and the other uses filtered air at a lower flow rate. With the section of inflow, AE, maintained at 0.46 m/s (90 ft/min) and an air velocity of section EC one-half that of section AE (or 0.23 m/s [45 ft/min]), the results are as shown in Figures 18-6a (raised floor configuration) and 18-6b

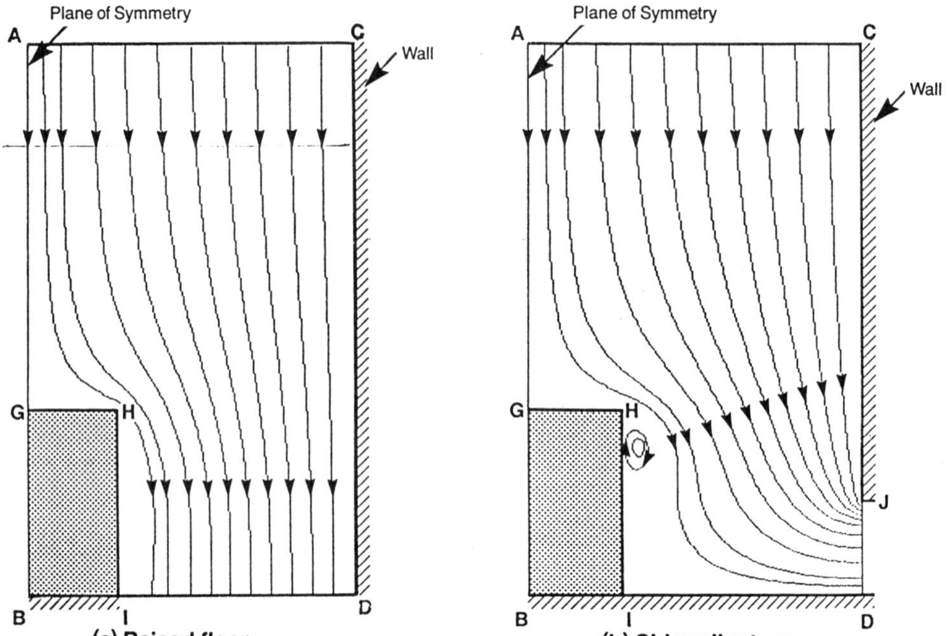

(a) Raised floor (b) Sidewall return

Figure 18-3. Flow streamlines for a centered solid workbench having a uniform inflow (0.46 m/s).

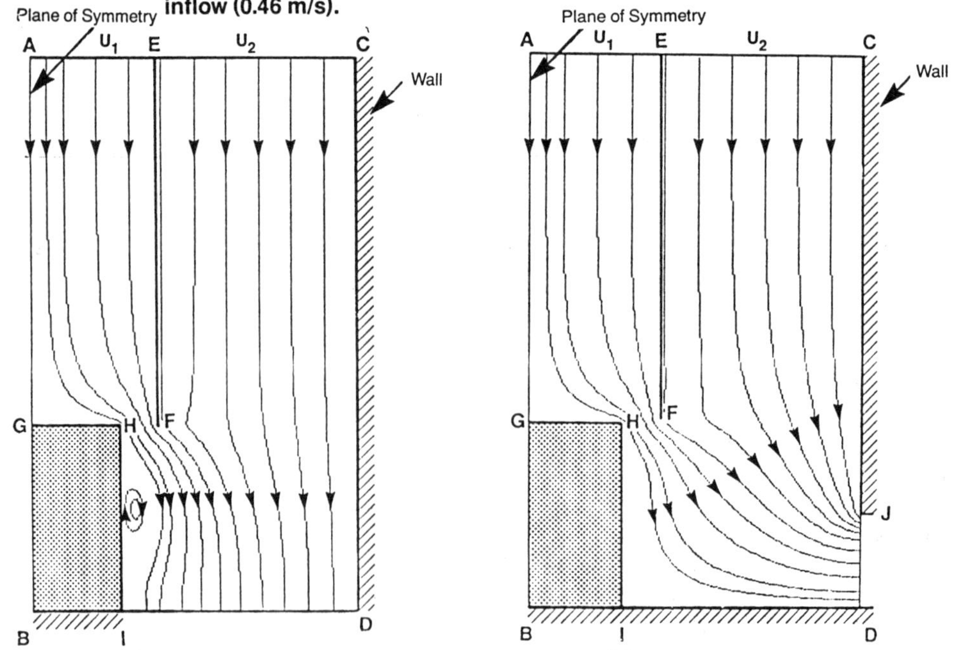

(a) Raised floor (b) Sidewall return

Figure 18-4. Flow streamlines for the addition of an isolating curtain having a uniform inflow (0.46 m/s).

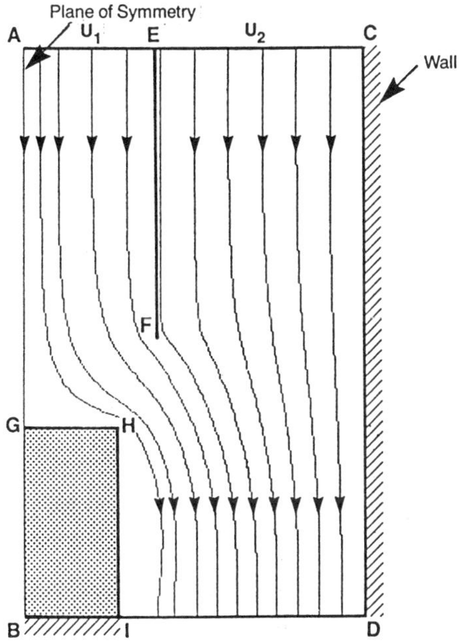

(a) Raised floor

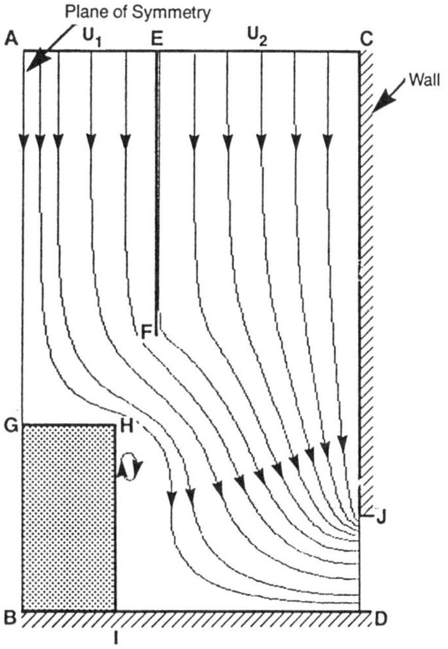

(b) Sidewall return

Figure 18-5. Flow streamlines for the shorter curtain having a uniform inflow (0.46 m/s).

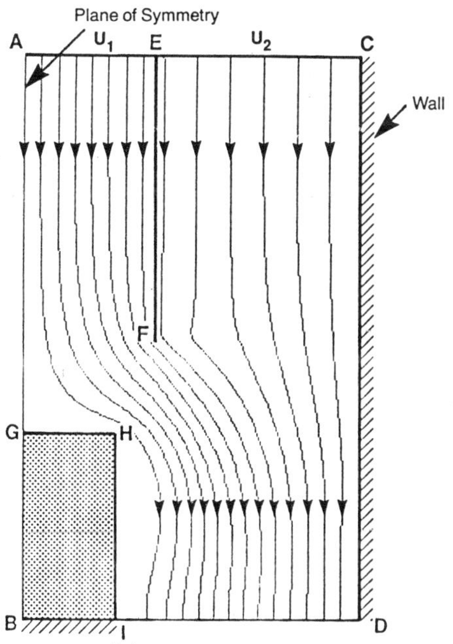

(a) Raised floor

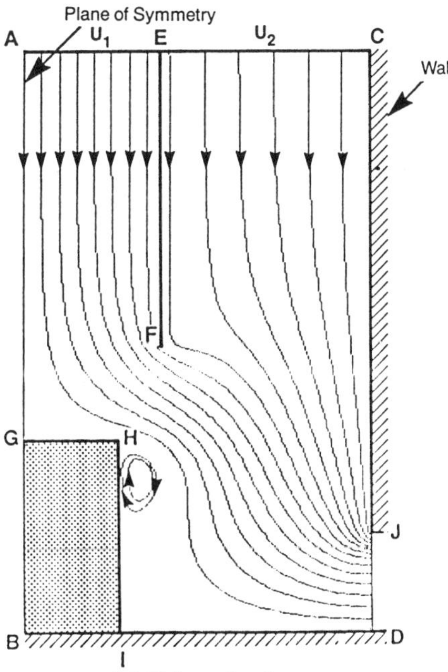

(b) Sidewall return

Figure 18-6. Flow streamlines for the shorter curtain having different entry air velocities (0.46 m/s for AE and 0.23 m/s for EC).

(sidewall return configuration) when the bottom of the curtain is 0.5 m (1.6 ft) above the workbench. A large amount of recirculation is observed for the sidewall return configuration. Thus, the raised floor configuration remains more desirable for the higher curtain height. If the curtain height is further increased (0.8 m [2.6 ft] above the workbench), the flow distributions remain similar to those in Figures 18-6a and 18-6b.

Figures 18-7a and 18-7b show the airflow streamlines and velocity plots where the section EC has an entry velocity that is 1/100 that of section AE (0.0046 m/s [0.90 ft/min]). There are two distinct recirculation zones: the side face of the workbench and the outside of the bottom of the curtain, which is not desirable at all. Thus, simply increasing the ratio of the flow rates in the two zones is of itself not always a desirable design approach.

[Case 5] Clean Room Exhaust Hood

Figure 18-8a shows the computed streamlines in a clean-room exhaust hood when there is a sliding door opening (hood opening) of 0.40 m (1.3 ft). The entry air velocities across the ceiling (U_1 for AE and U_2 for EC), and the exhaust velocity at the grating workbench (U_3 for GH) in Figure 18-8a are assumed uniform: U_1 = 0.46 m/s (90 ft/min), U_2 = 0.1 m/s (20 ft/min), and U_3 = 0.2 m/s (40 ft/min). No computations were performed downstream of the grating inside the workbench (GHIJ). As seen in Figure 18-8a, the flow field is smooth and stable throughout the clean room. The flow rate through the hood opening is an important parameter for protecting workers from hazardous fumes, and the flow rate is usually controlled by the damper opening, located downstream of the exhaust port, and the hood opening (FG). The maximum velocity becomes 4.1 m/s (810 ft/min) located near the return air exhaust (DJ).

Figure 18-8b illustrates the streamlines formed by increasing the exhaust fan capacity (U_3 = 0.6 m/s [120 ft/min]) without changing the room configurations. A larger percentage of the incoming room flow now exits through the workbench. A large flow recirculation, induced near the inside of the bottom sliding door, is now observed. If hazardous fumes are generated in the workbench area, there is now less of a chance for them to penetrate into the workers' area; and if the workers' area generates particles, fewer particles are returned to the return air. The flow streamlines inside the bottom of the sliding door are pushed closer to the sliding door because the local pressure at the high-velocity zone outside the bottom of the sliding door is lower than the local pressure at the low-velocity zone, which results in the local flow instabilities. This suggests that the fume hood effectiveness could be limited by the formation of flow recirculation. The maximum local velocity becomes 1.75 m/s (344 ft/min), which occurs between the sliding door and the workbench.

[Case 6] Clean Bench

When the entry velocities are reversed (U_1 = 0.1 m/s [20 ft/min] and U_2 = 0.46 m/s [90 ft/min]) and the exhaust velocity U_3 is chosen to be 0, the airflow streamlines are as shown in Figure 18-9. This configuration is now no longer that of an exhaust fume hood, but rather that of a clean bench such as is used primarily in industrial processing facilities. A large recirculation zone appears at the bottom of the sliding door. The flow is pushed farther

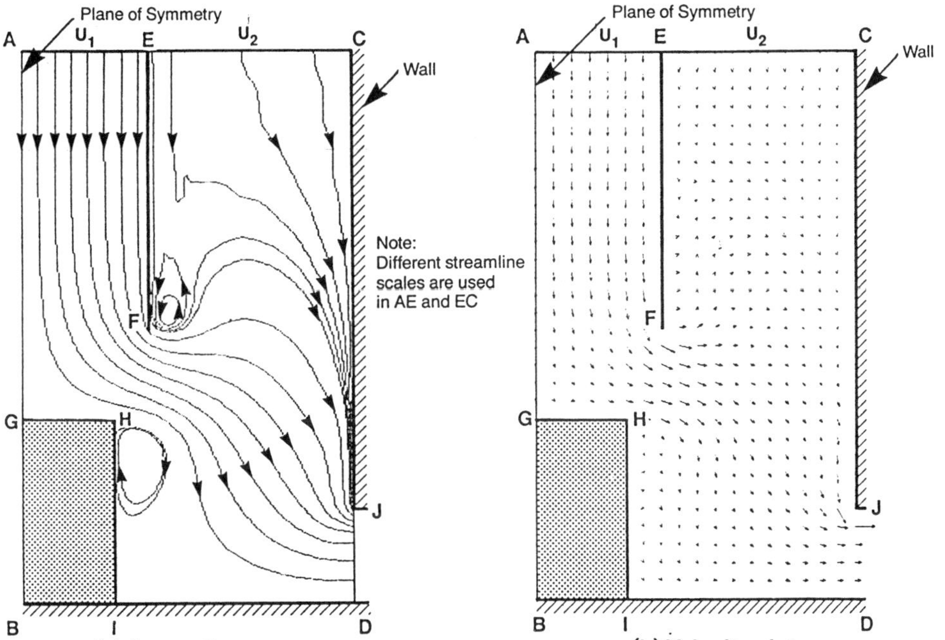

(a) Streamlines (b) Velocity plots

Figure 18-7. Flow streamlines for different entry velocities (0.46 m/s for AE and 0.0046 m/s for EC).

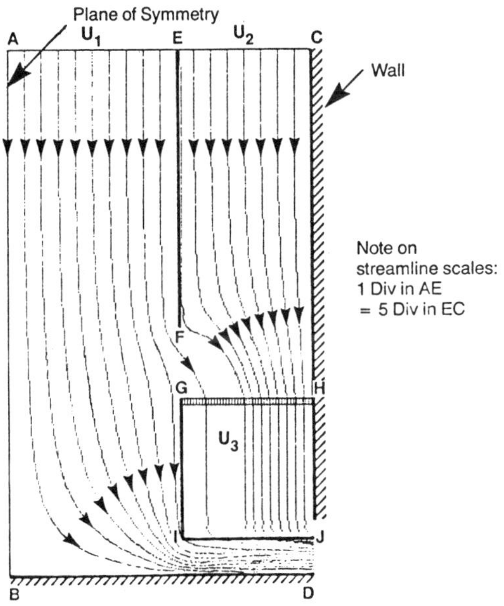

Figure 18-8 (a). Streamlines for U_1 = 0.46 m/s, U_2 = 0.1 m/s, U_3 = 0.2 m/s, and hood opening of 0.4 m.

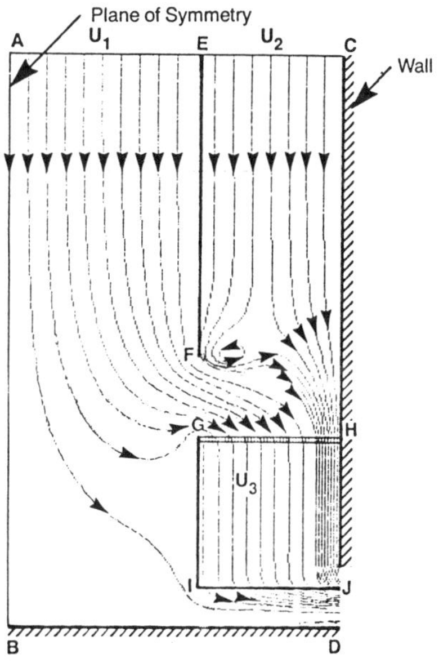

Figure 18-8 (b). Streamlines for U_1 = 0.46 m/s, U_2 = 0.1 m/s, U_3 = 0.6 m/s, and hood opening of 0.4 m.

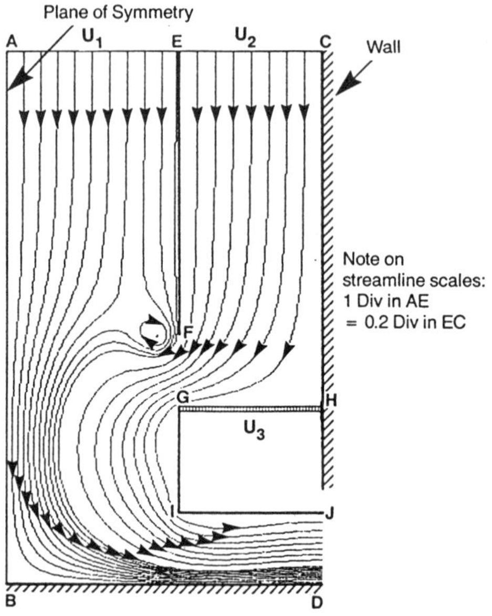

Figure 18-9. **Streamlines for U_1 = 0.1 m/s, U_2 = 0.46 m/s, U_3 = 0 m/s, and hood opening of 0.4 m.**

toward the centerline of the clean room, and a large amount of flow passes through the return air duct. The maximum velocity becomes 15.2 m/s (3,000 ft/min). Increasing U_1 (or decreasing U_2) will reduce the flow recirculation, and increasing the return air duct dimension (JD) also reduces the local high velocity, which is usually beneficial in the sense of stabilizing the flow. Installation of guide vanes at the return air duct should improve the outflow distribution and will also be helpful for flow stabilization.

PARTICLE DEPOSITION VELOCITY UNDER VERTICAL LAMINAR FLOW

In a typical semiconductor manufacturing environment, fine particles are deposited on the wafer surface by sedimentation (gravitational settling), diffusion, and/or electrostatic attraction. The mechanisms of particle deposition depend in a complicated manner on physical and electrical properties and the fluid flow in which the particles are suspended. Sedimentation alone is important for particles larger than 1 μm in diameter, while diffusion alone is important for particles smaller than about 0.1 μm in diameter. Both sedimentation and diffusion are important in the intermediate range. When the particles are electrostatically charged, particle deposition is generally enhanced. In all cases, the deposition rate is influenced by the airflow around the wafer.

Particle Deposition by Sedimentation

Particle deposition velocity to a wafer surface because of sedimentation (gravitational settling) is the same as the particle settling speed, V_s, which is given by the Stokes law with the slip correction (Hinds, 1982, Equation (2-10) in Chapter 2):

$$V_s = \rho\, gC_c d^2/18\, \eta \qquad (2\text{-}10)$$

where C_c is the Cunningham correction factor, d is a particle diameter, η is the gas viscosity, and g is the acceleration of gravity. C_c is always greater than 1.0 and reduces the Stokes drag force, thereby increasing the deposition velocity.

Particle Deposition Due to Diffusion

Liu and Ahn (1987) used the analogy between mass transfer and heat transfer to determine particle deposition rate by diffusion. Mass transfer from a disk surface to the ambient air over a wide range of air velocities can be correlated by the following relationship (Sparrow and Geiger, 1985):

$$\overline{Sh} = 1.47\ Re^{1/2} \qquad (18\text{-}9)$$

where $\overline{Sh} = \overline{K}D_d/D$ is the mean Sherwood number, $\overline{K}$ is the mean mass transfer coefficient, D_d is the diameter of the wafer, and D is the diffusion coefficient. $Re = U_o Dd/\nu$, where U_o is the airflow velocity and ν is the kinematic viscosity.

The above relationship can also be expressed in terms of the Schmidt number, Sc, which is defined as $Sc = \nu/D$. Further, $\overline{Sh}$ is proportional to $Sc^{1/3}$, resulting in:

$$\overline{Sh} = 1.08\ Sc^{1/3}Re^{1/2}\ . \qquad (18\text{-}10)$$

The diffusion coefficient of the particle is given by the Stokes-Einstein equation:

$$D = kTC_c/3\pi\eta\ d \qquad (18\text{-}11)$$

where k is the Boltzmann constant and T is the absolute temperature. The diffusion coefficient has a dimension of $(\text{length})^2/\text{time}$. It increases with temperature and is inversely proportional to particle size. For smaller particles, the Cunningham slip correction factor is also important. The mean deposition velocity, V_d, which is the same as the mean mass transfer coefficient, $\overline{K}$, can be calculated from Equation (18-10), using the appropriate definitions. V_d (or $\overline{K}$) is defined as the deposition flux, J (the number of particles diffusing and sticking to the wafer surface per unit surface area per unit time) divided by the concentration of particles in the free stream, n_o:

$$V_d = J/n_o\ . \qquad (18\text{-}12)$$

The total particle deposition velocity, V_{tot}, attributable to diffusion and sedimentation combined, is simply the sum of the deposition velocities for sedimentation and diffusion:

$$V_{tot} = V_s + V_d\ . \qquad (18\text{-}13)$$

Particle Deposition Due to Electrostatics

Electrical forces can play an important role in particle deposition. There are three major electrical forces between particles and a wafer: coulomb force, image force, and dielectrophoretic force.

The coulomb force is the electrical force between a charged wafer and a charged particle. It can be either repulsive or attractive, depending on the sign of the charges. Alternatively, the coulomb force can be expressed as the product of particle charge and the electric field created by the charged wafer. The charged particles move under the influence of the wafer's electrostatic field and can precipitate on the wafer's surface. The deposition velocity due to the coulomb force, V_{coul}, can be written as:

$$V_{coul} = n_e eE/3\pi\eta\, d \tag{18-14}$$

where n_e is the net number of elementary charges on the particle, e is the elementary charge of an electron, and E is the electric field. The electric field along the center above the wafer can be expressed as:

$$E = \frac{C\,V_o}{2\pi\epsilon_o r_w^2}\left[1 - \frac{z/r_w}{\sqrt{1 + \left(z/r_w\right)^2}}\right] \tag{18-15}$$

where C is the capacitance, V_o is the surface potential, r_w is the wafer radius, z is the distance normal to the wafer surface, and ϵ_o is the dielectric constant of vacuum.

The image force is the attractive force created between a charged particle and a charge polarization induced in the neutral wafer. The deposition velocity due to the image force can be calculated by treating the wafer polarization as a fictional charge located at an appropriate image point. The deposition velocity due to the image force, V_{im}, for a point particle near an infinite plane surface can be expressed as:

$$V_{im} = -\frac{n_e^2 e^2}{4z^2}\left(\frac{\epsilon_2 - \epsilon_1}{\epsilon_2 + \epsilon_1}\right) \tag{18-16}$$

where ϵ_1 and ϵ_2 are the dielectric constants of the surface and gas media, respectively.

The dielectrophoretic force also results from polarization. This force results when charge separation in an electrically neutral body takes place in a region of a nonuniform electric field. The charge displaced toward the region of a higher electric field experiences a larger coulomb force than does the equal charge displaced toward the region of a lower electric field. An electrically neutral particle will always experience an attractive dielectrophoretic force pulling it toward the charged wafer, regardless of charge sign. This force is usually very small compared to the coulomb force and thus can be neglected.

RESULTS FOR PARTICLE DEPOSITION VELOCITY

Sedimentation and Diffusion

Calculations of the deposition velocity (Liu and Ahn, 1987) as a function of particle size and flow velocity, based on Equations (2-10) and (18-9) through (18-13), are shown in Figure 18-10 for a horizontal wafer in a vertical laminar flow. The deposition velocity decreases with increasing particle size in the diffusion regime, reaches a minimum at the size of approximately 0.2 μm, and then increases with increasing particle size when sedimentation becomes important. The effect of air velocity on particle deposition velocity is very small for particle sizes of 0.5 μm or larger, but it is significant for particles 0.2 μm or smaller. The minimum deposition velocity varies from approximately 0.3×10^{-3} to 0.7×10^{-3} cm/s. These values of deposition velocity are about 10^{-4} times the air velocity approaching the wafer and reflect the fact that only a small fraction of the particles contained in the air column approaching a wafer are expected to actually end up on the wafer.

It is also interesting to note that the particle deposition velocity of the critical killer particles has been decreasing in the past--as chip dimensions dictate a killer particle threshold size lowering from 1 μm to 0.5 to 0.3 μm, deposition velocity has decreased, thus partially shielding chip yield from the full effect of the higher particle concentrations typically encountered at these smaller particle sizes. As the critical particle size decreases below 0.2 μm, however, the deposition velocity increases because of diffusion, and the fraction of particles in this size range that end up on the chip will increase. Nature will no longer be so benevolent—at least from a particle deposition viewpoint.

Similar particle deposition velocity calculations have also been carried out by Liu and Ahn (1987) for vertically oriented wafers.

Electrostatics

Electrical forces have been predicted to be important in the capture of submicrometer particles (Liu, Fardi, and Ahn, 1987), and some experimental evidence in support of such predictions now exists (Donovan et al., 1987a; 1987b). Liu, Fardi, and Ahn (1987) calculated the deposition velocities for electrically neutral particles; for particles charged to the Boltzmann root-mean square (rms) level; and for particles charged to 10 times the rms Boltzmann levels. All calculations assumed a surface electric field of 100 V/cm. These results, as shown in Figure 18-11, indicate that the deposition velocity is very sensitive to particle charge over the submicrometer size regime. The addition of the Boltzmann charge to neutral submicrometer particles is predicted to increase the deposition velocity by a factor of 5 to 10. For large particles, the influence of electrical charge is much smaller. It should be noted that the rms Boltzmann charge is about 3 elemental charges for a 1 μm diameter particle and 10 elemental charges for a 10 μm diameter particle. More detailed information regarding particle charge levels can be found in Liu and Pui (1974).

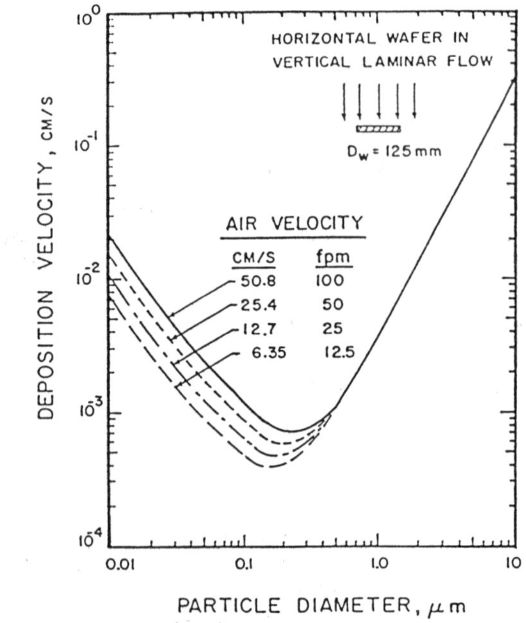

Source: Liu and Ahn (1987)

Figure 18-10. Mean deposition velocity for a 125 mm diameter, horizontal wafer in a VLF clean room.

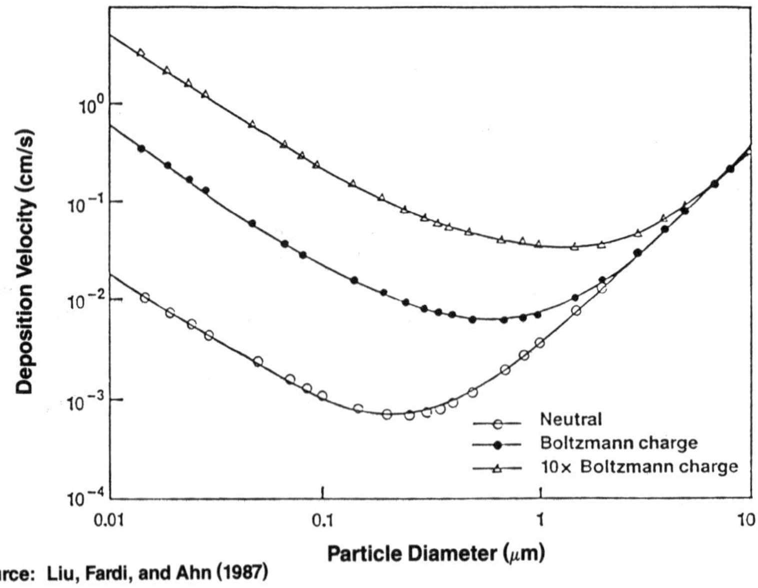

Source: Liu, Fardi, and Ahn (1987)

Figure 18-11. Calculated deposition velocity including effects of diffusion, settling, and electric field (E = 100 V/cm).

The assumption of 100 V/cm used in these calculations is quite modest. Blitshteyn and Martinez (1986) and Wilson (1987) measured voltages of 5 to 10 kV and 1 to 5 kV, respectively, at various stations in a semiconductor manufacturing area; those voltages almost assuredly create electric fields well in excess of 100 V/cm. A higher electric field, of course, increases the calculated deposition velocities of the charged particles to values greater than shown in Figure 18-11.

The experimental data reported by Donovan et al. (1987a and 1987b) showed the effect of wafer potential upon the deposition velocity measured in a deposition chamber using monodisperse polystyrene latex (PSL) spheres (Figure 18-12). The particle charge in most of the measurements was that associated with the Boltzmann distribution as achieved by passing PSL spheres through a Kr^{85} neutralizer. As can be seen from this figure, an electrical potential of either polarity increased the deposition velocity, although a positive potential was more effective than a negative potential. This experimental work was repeated recently, and the deposition velocities measured were about one-half of the values shown in Figure 18-12. In all cases, the negatively charged particles are greater in concentration or have a higher mobility than do the positively charged particles. With an alternating potential on the wafers, as plotted in Figure 18-12, no dependence on surface potential exists. This observation is consistent with the dominant electrostatic capture force being a coulomb attraction rather than a polarization interaction.

Figure 18-13 shows the increase in deposition velocity brought about by removing the neutralizer from the aerosol feed line. No measure of the actual particle charge was made, but it was obviously much higher as evidenced by the increased values of particle deposition velocity.

Fissan and Turner (1988) carried out an analysis of particle deposition velocity over a wide range of electrostatic parameters using plane stagnation flow analysis. Figure 18-14 shows their values of particle deposition velocity calculated for zero electric field and particle charge as indicated by the parameter $\psi\sqrt{f(\epsilon)}$ (an image force). Two fluid velocity gradients ($a = 1\ s^{-1}$ and $a = 100\ s^{-1}$) are plotted, assuming the fluid kinematic viscosity, ν, remains constant. The effect of the image force is strongest in the neighborhood of the minimum deposition velocity. As the fluid velocity gradient increases, the minimum in the deposition velocity shifts to larger particle sizes, and the contribution of the image force also increases. A comparison of Figures 18-11 and 18-14 demonstrates that even relatively weak coulombic forces will dominate over the image force. However, as the wafer surface potential is reduced, the relative role of the image force increases.

Thermophoresis

When a temperature gradient exists in a gas, the aerosol particles in that gas experience a force in the direction of decreasing temperature. The motion of the aerosol particles that results from this force is called "thermophoresis." Recently, this phenomenon has been proposed as a mechanism for reducing particle deposition velocity on silicon wafers (Donovan et al., 1987b; Ensor et al., 1987). Theoretical studies for calculations of the dust-free space attributable to thermophoresis have been made by Stratman et al. (1988).

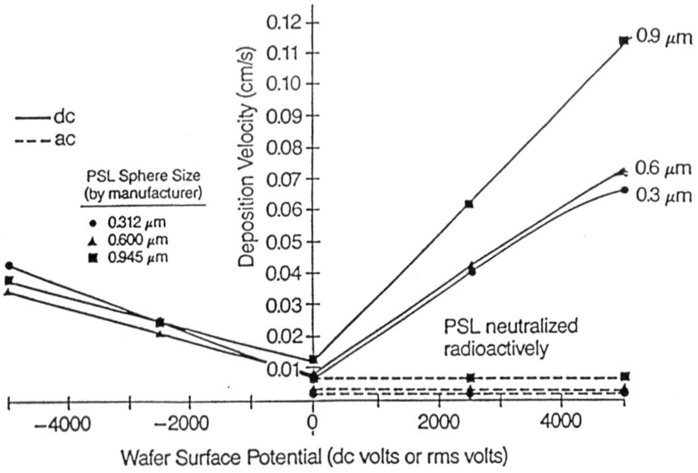

Figure 18-12. Mean values of deposition velocity as a function of wafer surface potential.

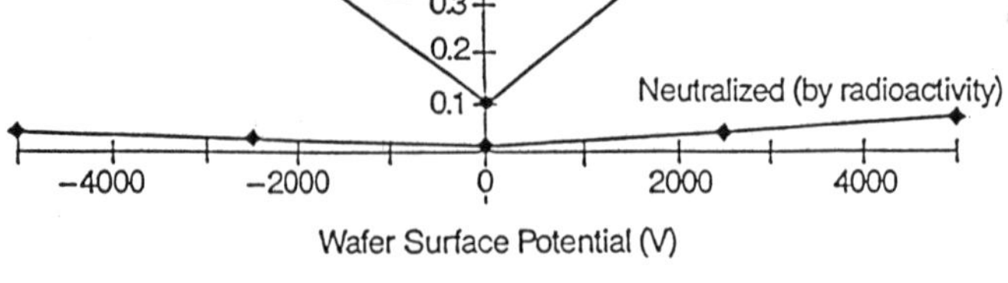

Figure 18-13. Increase in deposition velocity brought about by increased electrical charge on 0.6 μm polystyrene latex spheres.

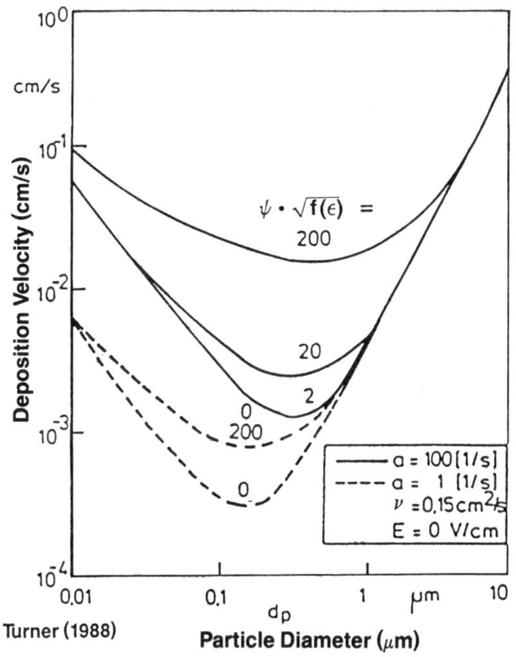

Source: Fissan and Turner (1988)

$\psi \cdot \sqrt{f(\epsilon)}$ = **(Particle charge expressed in multiples of the Boltzmann equilibrium charge)**

$\times\left(\frac{\epsilon_2 - \epsilon_1}{\epsilon_2 + \epsilon_1}\right)^{1/2}$, **where ϵ_1, ϵ_2 are the dielectric constants of the gas medium and the surface, respectively.**

Figure 18-14. Particle deposition onto a horizontal flat surface from plane stagnation flow—various image force strengths and fluid velocity gradients.

Thermophoretic velocity for particle diameter less than 0.1 μm is given by Hinds (1982):

$$V_{th} = -2.8 \times 10^{-4}\ \nabla T\ (cm/s) \qquad (18\text{-}17)$$

where ∇T is the temperature gradient.

For larger particles (greater than 0.1 μm), the velocity of thermophoresis is given as:

$$V_{th} = \frac{-3\eta\ C_c\ H\ \nabla T}{2\ \rho\ T} \qquad (18\text{-}18)$$

where H is the coefficient that includes the effect of the temperature gradient inside the particle (Hinds, 1982), and T is the absolute temperature of the particle. Unfortunately, ∇T proves elusive to quantify, and Equations (18-17) and (18-18) have not yet been verified in wafer deposition experiments.

Donovan et al. (1987b) measured particle deposition velocity onto silicon wafers heated up to 140 °C (284 °F). The normalized deposition velocities for horizontally (Figure 18-15) and vertically (Figure 18-16) oriented wafers are presented for 0.3, 0.6, and 0.9 μm PSL aerosols. An order of magnitude reduction of deposition velocity was measured with just a 30 °C (86 °F) temperature difference. Two mechanisms are hypothesized to contribute to the reduction of particle deposition velocity with wafer temperature: thermophoretic repulsion of particles from the vicinity of the heated wafer and thermal convective currents that sweep particles from the surface in a buoyant flow. The relative contribution of thermophoretic force and convection force remains unsettled at this stage.

Combined Effects of Electric Fields and Thermal Gradients

That an electrically charged wafer gets dirtier, and that a heated wafer remains cleaner. suggest that thermal forces can compensate for the presence of electrical charges and forces with respect to particle deposition. This suggestion was investigated by Donovan et al. (1988). The wafer temperature selected was 50 °C (122 °F), which is just 30 °C above nominal room temperature. This modest wafer temperature is easily achieved and yet has been shown to bring about one order of magnitude decrease in deposition velocity (see Figures 18-15 and 18-16).

Donovan et al. (1988) measured particle deposition velocity as a function of wafer potential and wafer temperature in the clean-room deposition chamber. Figure 18-17 shows that, below a wafer potential of ±1,000 V, the 50 °C (122 °F) wafer clearly acquires particles at lower rates than the 20 °C (68 °F) wafer—the 30 °C (54 °F) temperature difference more than offsets the electrical bias effects for radioactively neutralized 0.6 μm PSL spheres over this range of wafer surface potential. Between a 1,000 and 2,000 V surface potential, the curves merge; above 2,000 V, no significant difference in deposition velocity exists between a 20 °C (68 °F) and 50 °C (122 °F) wafer.

Similar results appear to hold for 0.9 μm PSL spheres although insufficient data were collected for drawing plots. Using nonneutralized 0.6 μm PSL spheres appears to eliminate the thermal protection altogether, but too few measurements were made to allow any statistical statement to be made.

CONCLUSIONS

Mechanisms of particle deposition on silicon wafers are briefly reviewed by reference to and citation of results from several recent theoretical and experimental papers.

Particulate contamination of silicon wafers depends on particle transport to the wafer surface and particle capture by the surface in addition to the presence of particle generators. Several actions that have been shown to minimize particle deposition are: (1) minimizing flow recirculation zones that often occur at high air velocity gradients, (2) eliminating particle and wafer electrical charge, and (3) elevating wafer temperature above the ambient temperature.

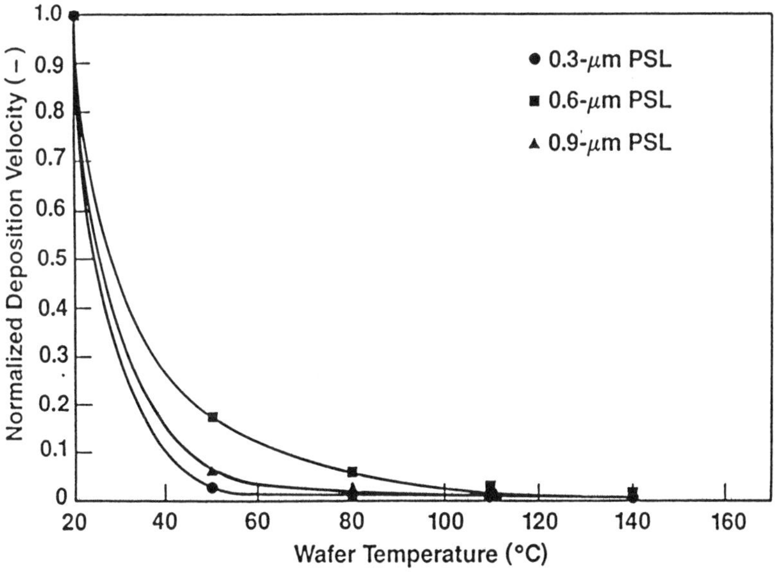

Figure 18-15. Normalized deposition velocity, horizontal wafers.

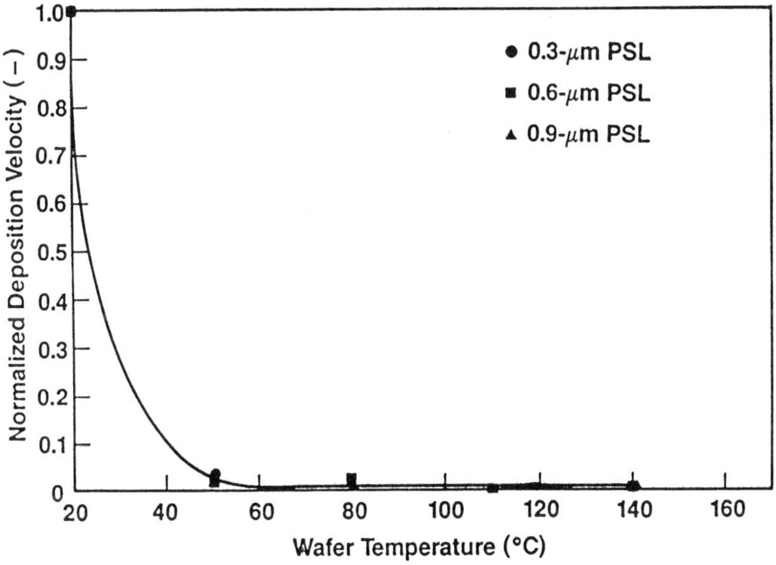

Figure 18-16. Normalized deposition velocity, vertical wafers.

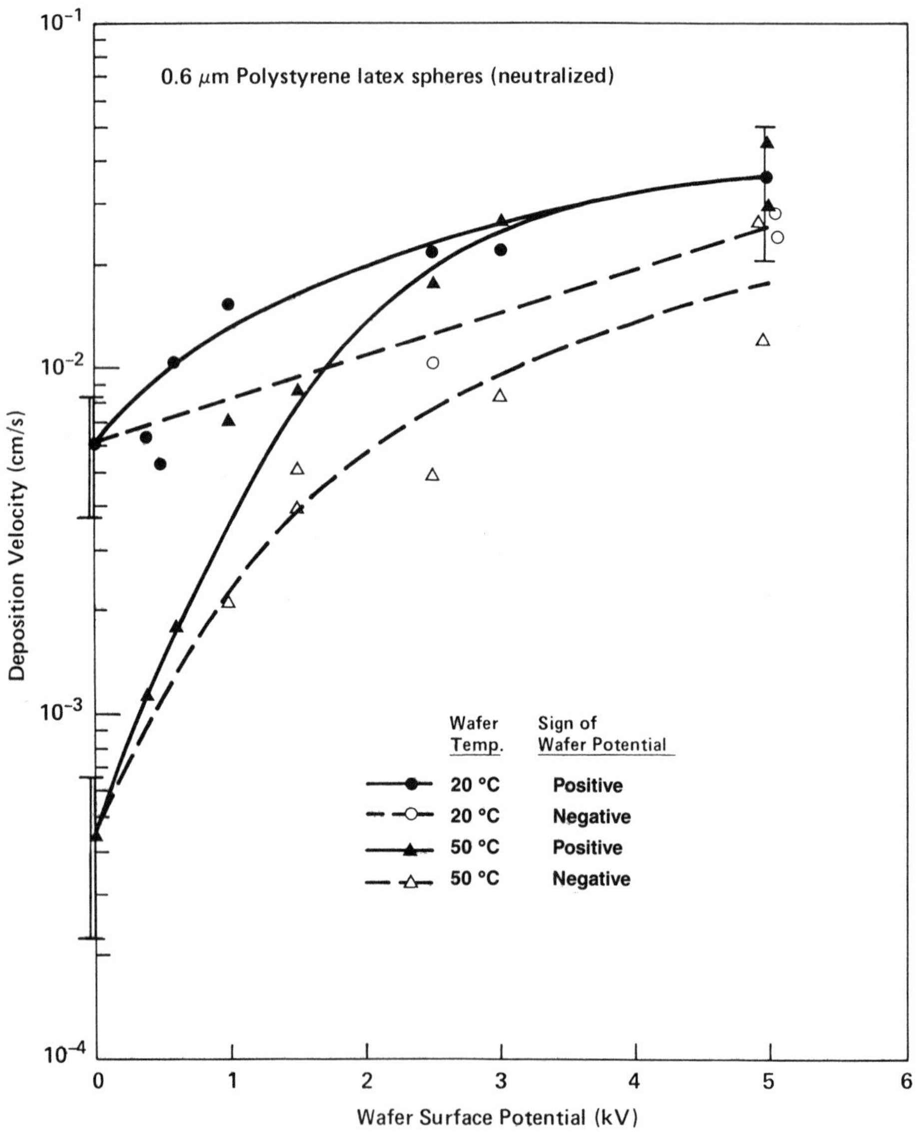

Figure 18-17. Deposition velocity as a function of wafer potential, wafer temperature as a parameter.

REFERENCES

Blitshteyn, M., and A. M. Martinez. 1986. Electrostatic Charge Generation on Wafer Surfaces and Its Effect on Particle Deposition. *Microcontamination*. 4(11):54-60, 132.

Bunkofske, R. J. 1988. Performated versus Solid Top Work Stations. In: Proceedings of IES Annual Technical Meeting, pp. 458-463.

Busnaina, A. A. 1987. Modeling of Clean Rooms on the IBM Personal Computer. In Proceedings of the Institute of the Environmental Sciences Annual Technical Meeting, pp. 292-297.

Busnaina, A. A., S. Abuzeid, and M. A. R. Sharif. 1988. Three-Dimensional Numerical Simulation of Fluid Flow and Particle Transport in a Clean Room. In: Proceedings of IES Annual Technical Meeting, pp. 326-330.

Cooper, D. W. 1986. Particulate Contamination and Microelectronics Manufacturing: An Introduction. *Aerosol Science and Technology*, 5:287-300.

Deaves, D. M., and D. Malam. 1985. Advanced Analysis Techniques for the Optimum Design of Clean Rooms. *Journal of Environmental Sciences*, 28(5):17-20.

Donovan, R. P., A. C. Clayton, and D. S. Ensor. 1987a. The Dependence of Particle Deposition Velocity on Surface Potential. In: Proceedings of the Institute of Environmental Sciences Annual Technical Meeting, pp. 473-478.

Donovan, R. P., T. Yamamoto, A. C. Clayton, and D. S. Ensor. 1987b. Dependence of Particle Deposition Velocity on Silicon Wafer Surface Temperature. In: Proceedings of the Fall Meeting of the Electrochemical Society, Honolulu, Hawaii, October.

Donovan, R. P., T. Yamamoto, A. C. Clayton, and D. S. Ensor. 1988. Experimental Study of Particle Deposition on Silicon Wafers Under the Combined Effects of Electric Fields and Thermal Gradients. Presented at the 9th International Symposium on Contamination Control, Los Angeles, CA, September 26-29.

Ensor, D. S., A. C. Clayton, T. Yamamoto, and R. P. Donovan. 1987. Particle Deposition Velocity Studies in Silicon Technology. Presented at the 18th Annual Meeting of the Fine Particle Society, Boston, MA, August.

Fissan, H. J., and J. R. Turner. 1988. Electrostatic Effects in Particle Deposition onto Product Surfaces. In: Proceedings of the Institute of Environmental Sciences Annual Technical Meeting, pp. 400-404.

Fujii, S., I. Hayakawa, and K. Iwase. 1984. Studies on Design Theory of Laminar Flow Type Clean Room. In: Proceedings of the Institute of Environmental Sciences Annual Technical Meeting, pp. 107-111.

Hinds, W. C. 1982. *Aerosol Technology*. John Wiley and Sons, New York, NY.

Hoenig, S. A., S. R. Martin, and I. J. M. Slinski. 1987. Electrostatic Applications in Contamination Control. In: Proceedings of the Institute of Environmental Sciences Annual Technical Meeting, pp. 479-482.

Jinno, N. 1986. Study of Airflow Pattern in the Clean Room for Semiconductors. In: Proceedings of the Institute of Environmental Sciences Annual Technical Meeting, pp. 556-564.

Kato, S., S. Murakami, and S. Chirifu. 1986. Study on Air Flow in Conventional Flow Type Clean Room by Means of Numerical Simulation and Model Test. In: Proceedings of the 8th International Symposium on Contamination Control, Milan, Italy, pp. 781-791.

Knudsen, J. G., and D. L. Katz. 1958. *Fluid Dynamics and Heat Transfer*. McGraw-Hill Book Co. Inc., New York, NY, pp. 23-28.

Kuehu, T. H., V. A. Marple, H. Han, D. Liu, I. Shanmugavelu, and S. W. Youssef. 1988. Comparison of Measured and Predicted Airflow Patterns in a Clean Room. In: Proceedings of IES Annual Technical Meeting, pp. 371-336.

Liu, B. Y. H., and K. Ahn. 1987. Particle Deposition on Semiconductor Wafers. *Aerosol Science and Technology*, 6(3).

Liu, B. Y. H., B. Fardi, and K. Ahn. 1987. Deposition of Charged Aerosol Particles on Semiconductor Wafers. In: Proceedings of the Institute of Environmental Sciences Annual Technical Meeting, pp. 461-465.

Liu, B. Y. H., and D. Y. H. Pui. 1974. Equilibrium Bipolar Charge Distribution of Aerosols. *Journal of Colloid Interface Science*, 49:305-312.

Marple, V., and D. J. Rader. 1987. Three-Dimensional Flow Field Within a Box. In: Proceedings of the Institute of Environmental Sciences Annual Technical Meeting, pp. 310-318.

Roache, P. J. 1972. *Computational Fluid Dynamics*. Hermosa Publications, Albuquerque, NM, pp. 64-112.

Schlichting, H. 1968. *Boundary-Layer Theory*. McGraw-Hill Book Co., New York, NY, pp. 44-64.

Shanmugavelu, I., T. H. Kuehn, and B. Y. H. Liu. 1987. Numerical Simulation of Flow Fields in Clean Rooms. In: Proceedings of the Institute of Environmental Sciences Annual Technical Meeting, pp. 298-303.

Sparrow, E. M., and G. T. Geiger. 1985. Local Average Heat Transfer Characteristics for a Disk Situated Perpendicular to a Uniform Flow. *Journal of Heat Transfer*, 127:321-326.

Stratmann, F., S. Friedlander, H. Fissan, and A. Papperger. 1988. Suppression of Particle Deposition to Surfaces by the Thermophoretic Force. *Aerosol Science and Technology*, in press.

Taylor, T. D., E. Ndefo, and B. S. Masson. 1972. A Study of Numerical Methods for Solving Viscous and Invicid Flow Problems. *Journal of Computational Physics*, 9:99-119.

Wilson, R. 1987. Nuclear Air Ionization for Contamination Control. In: Proceedings of the Institute of Environmental Sciences Annual Technical Meeting, pp. 466-472.

Yamamoto, T., D. S. Ensor, and R. P. Donovan. 1987a. Clean Room Air Flow Modeling. In: Proceedings of the Institute of Environmental Sciences Annual Technical Meeting, pp. 304-309.

Yamamoto. T., R. P. Donovan. and D. S. Ensor, 1987b. Numerical Simulation of Flow Fields in Clean Rooms. In: Proceedings of Pharm Tech Conference '87. Aster Publishing Corp., Eugene, OR, September, pp. 117-128.

Yamamoto, T., R. P. Donovan, and D. S. Ensor. 1988a. Numerical Simulation of Clean Room Air Flows. *International Journal of Engineering Fluid Mechanics*, 1 (Spring):134-150.

Yamamoto, T., R. P. Donovan, and D. S. Ensor. 1988b. Modeling of Flow in Clean Room Hoods. In: Proceedings of the Institute of Environmental Sciences Annual Technical Meeting, pp. 321-325.

19. PARTICLE DEPOSITION DATA: ROOM AIR IONIZATION AS A CONTROL METHOD

R. P. DONOVAN Research Triangle Institute, Research Triangle Park, North Carolina

INTRODUCTION

This chapter addresses the importance to wafer processing of the previous chapter's calculations of particle deposition velocity and the supporting measurements carried out in carefully controlled laboratory deposition apparatus. In the absence of electrical or thermal forces, the dominant mechanisms of submicrometer particle deposition were shown to be gravitational settling and diffusion. These mechanisms are always present; airflow contributes a superimposing convective or turbulent transport mechanism that also affects deposition velocity. The addition of electrical or thermal forces (or both) further modifies predictions and results.

However, it is in the real world, where conditions such as airflow and aerosol particle size distribution are neither well-controlled nor well-specified, that wafer processing takes place. The significance of the modeling work in this environment is discussed in the first part of this chapter. A specific control technology for minimizing electrical forces in clean rooms—room air ionization—is reviewed in the second part. And finally, beginning studies of particle deposition on wafers dipped or immersed in liquids are summarized in the third section.

REAL-WORLD PARTICLE DEPOSITION VELOCITY

The calculations of Liu and Ahn (1987) for wafers oriented either horizontally or vertically in a vertical laminar flow of 100 fpm (50 cm/s) predict deposition velocities (V_D) of 10^{-3} to 10^{-2} cm/s for submicrometer aerosol particles—velocities about 10^4 less than the nominal air velocity. For a V_D of 10^{-2} cm/s exposing a wafer for 1 h to Class 1 air ($\leq$1 particle/ft^3 > 0.5 μm [see Chapter 25]) is predicted to add about 0.001 particles/ cm^2. Monitor wafers in an operating fab verify that this increase in particle areal density is observed to be of the correct order of magnitude (Osburn et al., 1988). To further reduce this particle burden requires a reduction in the aerosol particle concentration of the ambient air, or a reduction in particle deposition velocity, or both. Chapter 18 identified wafer heating as one readily available method of reducing V_D; this concept has been verified in ambient laboratory air and also in a vacuum chamber operation, as will be discussed next.

Figure 19-1 presents a linear plot of normalized V_D for heated wafers in both a laboratory deposition chamber and an operating fab. These data are normalized because of the uncertainty and instability in the aerosol concentration within the fab area. Normalized behavior can be easily tracked, however, by simply comparing particle areal density on heated wafers with that on room-temperature wafers exposed simultaneously to the same laboratory aerosol. The Microelectronics Center of North Carolina's (MCNC's) laboratory datum point in Figure 19-1 was collected in that manner. This figure shows that, for exposure at a specific MCNC site, the wafers heated to 85 °C experienced a decrease in the deposition of laboratory aerosol, although not as great a decrease as was seen for aerosols in the deposition chamber. Nonetheless, it is clear that the thermophoretic and thermal convective forces described in Chapter 18 can be useful in reducing wafer-particulate contamination during storage or exposure to the fab ambient.

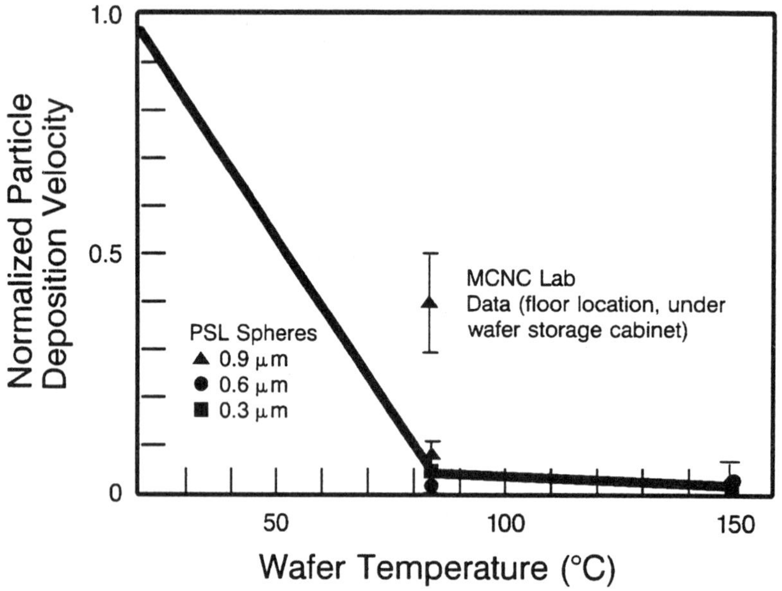

Figure 19-1. The effect of wafer temperature upon particle deposition velocity.

Perhaps more important is the use of thermal forces to shield wafers during processing. For example, vacuum processes are part of every fab line, and wafer particulate contamination often occurs during initial pump down and final venting back to atmospheric pressure. Table 19-1 shows the effect of wafer temperature upon particulate burden added to process wafers in a plasma deposition vacuum system. Two sets of wafers were sequentially processed through a pump down and bleed up in the same apparatus, the only difference being that one set of wafers was at room temperature and the other was 30 °C above room temperature (50 °C). These data show that the heated wafer in this apparatus acquired an order of magnitude fewer particles than wafers left at room temperature. (In the actual process wafers are much hotter during these steps.) Just a modest temperature increase can afford significant particle protection against deposition of these "real-world" particles.

Chapter 18 also presents calculations and experimental deposition chamber data that show that electrical forces, whether originating from particle charges, wafer charges, or both, can increase the particulate burden acquired by wafers. Placing charged (electrically biased) wafers in the fab environment shows that particle deposition velocity varies with wafer charge (Figure 19-2) but not as strongly as it does for charged wafers in the deposition chamber (Figure 19-3). Not only is the magnitude of the electrically enhanced deposition reduced, but the dependence on the sign of the wafer charge is reversed—the "real-world" fab apparently has more positively charged than negatively charged particles because the negatively charged

Table 19-1. Particle Deposition During Vacuum Pumpdown and Venting

Room Temp (20 °C)	Heated (50 °C)
5.66 ± 2.73	0.61 ± 0.42

wafers acquire more particles than do the positively charged wafers. Indeed, for this particular data set, there is no significant difference in deposition on positively biased wafers and zero bias wafers. Similar observations have been reported by Blitshteyn and Martinez (1986), who investigated just the effect of positively biased wafers on particle deposition because their survey of "naturally" existing biases in a fab revealed only positively biased surfaces. They concluded that electrically enhanced particle deposition is a minor effect in a contemporary wafer fab, a surprising conclusion in view of the large body of theoretical and experimental work, such as that touched upon in Chapter 18, suggesting otherwise. Note, however, that one of these authors (Blitshteyn) works for a manufacturer of air ionizing systems whose marketing opportunities would be best served by convincing semiconductor manufacturers of the importance of electrical neutralization throughout the fab. And indeed there are researchers who tout the need for such ionizing systems and present impressive numbers to support their claim (Mykkanen and Blinde, 1983; Huffman, Nichols, and Bossard, 1987). Thus, disagreement exists as to the need for and value of elaborate room air ionization systems in the fab. This topic is the subject of the next section.

ROOM AIR IONIZATION SYSTEMS

Room air ionization systems have been promoted as a means of electrically neutralizing particle and surface charges over a large area. This achievement not only reduces particle deposition velocity (Chapter 18) but also reduces device vulnerability to electrostatic discharge failures. This latter role will not be discussed in this section—just the particle role. Electrostatic discharges are less of a problem at the wafer level than at the chip level and are, in fact, a topic different from particle control.

Room ionization systems are systems designed to flood the ambient air of large rooms with a high concentration of bipolar ions—an ion cloud composed of both positive and negative ions. Table 19-2 compares ion concentrations measured in the ambient air of various locations. The major point is to illustrate the function of an air ionizing system: to create high concentrations of bipolar ions in the ambient air. Note also, however, that the clean-room data reported here show very low ion concentrations, suggesting that an efficient HEPA filter system removes not only particles but ions as well. Electrical neutralization of surfaces in a clean room relying only on ambient ions is likely to be a very slow process.

The most common method of generating these ions is by corona discharge between two high-voltage electrodes. Positive ions are air molecules that have been stripped of a valence electron either by the high-energy electric field itself or by collision with a charged species, almost always an electron, accelerated in that field. Negative ions are formed from air

Wafer No.	Z-15	Z-17	Z-18	Z-19	Z-20
Electrical Bias During Exposure to Clean Room Ambient	−5000 volts	−2600 volts	0 volts	+2600 volts	+5000 volts

Post Exposure Image on WIS 150

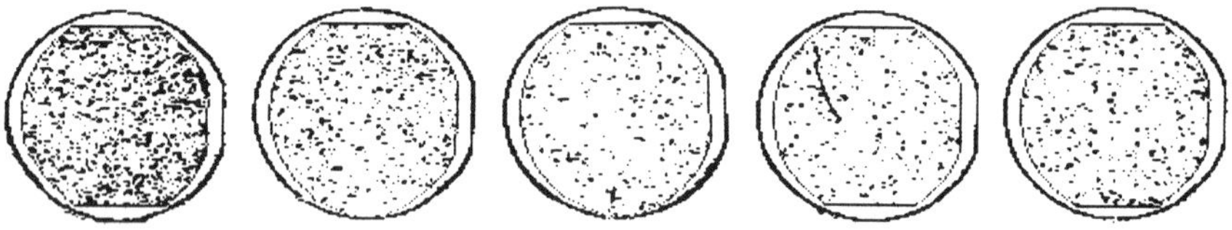

Total No. of Occupied Pixels After Exposure	1512	775	470	471	642

Figure 19-2. Laser scanner images of electricity biased wafers exposed to ambient air for 90 hr in a clean room at rest.

Wafer No.	T-15	T-14	T-13	T-12	T-11
Bias During Exposure	−4900 volts	−2500 volts	0 volts	+2700 volts	+5200 volts

WIS Wafer Image After Exposure

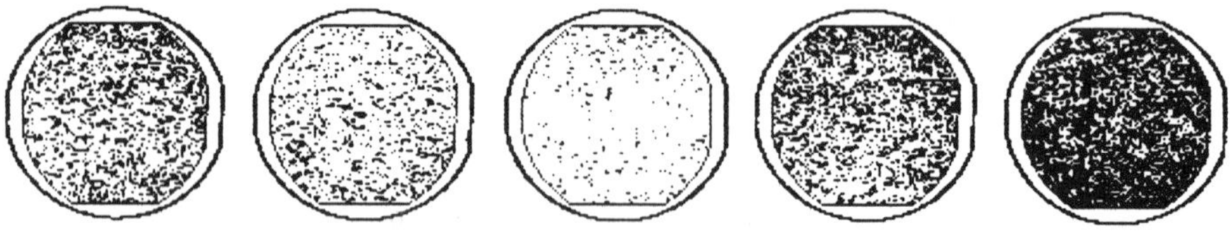

Total No. of Occupied Pixels After Exposure	1823	1094	357	2181	3534

Figure 19-3. Electrically biased wafers after an 11 min exposure to a PSL aerosol concentration of 2.2 particles/cm^3.

TABLE 19-2. Measured Ion Concentrations in the Ambient Air

	Positive Ions cm^{-3}	Negative Ions cm^{-3}
Hotel room	484	726
Conference room	218	303
Clean room	38	0
Under ionizer	1.5×10^5	1.5×10^5

Source: Liu et al., 1987.

that have captured a free electron or, more likely, another negatively charged ion. Free electrons do not last long in the air; they recombine with positive ions or are captured by neutral molecules in less than 1 μs after being freed. Negative ions have much longer lifetimes.

Various electrical bias schemes exist for creating bipolar ion clouds:

- Double direct current (dc) system
- Alternating current (ac) configuration
- Pulsed dc operation.

Double dc System

The double dc system refers to a configuration consisting of two high-voltage electrodes, one positively biased with respect to ground and one negatively biased with respect to ground. Both operate simultaneously and are approximately uniformly distributed across the ceiling of a clean room. The electrodes are positioned downstream of the HEPA filters. These electrodes are fabricated to have regions of small radius of curvature, typically needles or barbs, creating regions of high electric field that can be above air breakdown at voltages as low as 1,000 V.

In the vicinity of a biased needle point, both positive and negative ions are created under the action of the high-intensity electric field. At some threshold value of the electric field, the acceleration of free electrons between collisions becomes sufficient to ionize air molecules, creating additional free electrons that are accelerated and capable of ionizing other molecules (Meek and Craggs, 1978). (A small free electron population normally exists in the air because of cosmic radiation, radioactivity, other excitations.) This avalanche process creates high densities of both positive and negative ions in the vicinity of the high-voltage electrode, regardless of the sign of the high-voltage electrode. Those ions of sign opposite that of the high-voltage electrode are accelerated into the electrode and neutralized. However, those ions of the same sign as the high-voltage bias are repelled away from the electrode and into low field regions remote from the electrode (Figure 19-4). These repelled ions are those that are swept throughout the clean room to neutralize clean-room charges of opposite sign.

Thus, to establish a bipolar ion cloud, both positively and negatively biased electrodes can be operated separately and simultaneously in the same room—the double dc system. Recombination of positive and negative ions is maximized under this configuration, meaning that

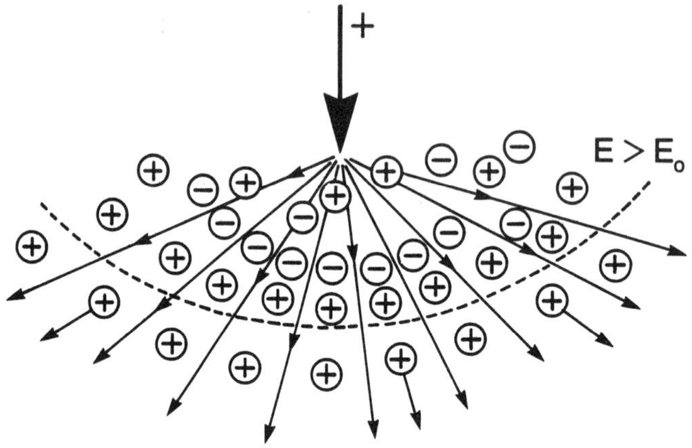

Source: Jonasson, 1985

Figure 19-4. Ion generation in the vicinity of a high voltage point.

a higher input power is required for a given downstream density of ions. Independent control of the power supply to each electrode allows good downstream ion balance (equal number of positive and negative ions). Electrode density and distribution are also important in achieving ion balance at the work level. Imbalance can lead to surface charging, exactly the condition the air ionizing system is supposed to eliminate.

ac Bias Configuration

In the ac configuration, the same electrode produces both the negative and positive ions, but it does so in alternating periods of applied bias. Pulses of negative and positive ions are emitted corresponding to the applied high-voltage ac. Ion formation occurs only during that peak fraction of the cycle in which the applied voltage exceeds the threshold voltage. Both positive and negative ions are generated in approximately equal quantities.

Again, ion recombination is high, and convective transport of the bipolar ion cloud to the work station is important for effective neutralization.

Pulsed dc Operation

Pulsed dc operation is similar to ac operation except that the pulse shape and repetition rate can be controlled to optimize ion density at a downstream work station. Pulse frequency is typically in the 0.5 to 3 Hz range rather than the 60 Hz typical of ac systems. In some designs the same electrode alternately pulses positive and negative ions; in others, separate electrodes are used for positive and negative ion emission. All pulsed dc designs create separate pulses that are longer in time than the 60 Hz ac pulses, reducing ion recombination in transit between the electrodes and the work station but also producing small charging pulses at the work station, corresponding to periods of alternate positive and negative charging.

As discussed earlier, the effectiveness of room ionizing systems for a fab area remains unsettled. Although both theoretical and experimental work, as cited in Chapter 18, show that electrical forces can be important in determining particle deposition velocity, measurements carried out in actual clean rooms are far less dramatic. Under reasonably clean fab operating conditions, witness wafers show small or insignificant reductions in concentration of deposited particles (Dillenbeck, 1985; Blitshteyn and Martinez, 1986). On the other hand, these same investigations report significantly reduced particle concentration on the surface of wafers exposed to relatively dirty ambient conditions, say Class 10,000 or dirtier, when a room air ionizer is activated. Similar improved wafer cleanliness when the wafer is exposed to clean-room ambient protected by a room air ionization system has also been reported by Huffman, Nichols, and Bossard (1987) and by Mykkamen and Blinde (1983).

The question appears to reduce to one of cost-effectiveness and control strategy. Although a "dirty" clean room will deposit fewer particles on test wafers when an air ionization system is operating, the optimum solution may be to clean up the area (install more filters and/or identify and eliminate major particle sources) rather than to install an air ionization system. If a deposition problem remains once these other obvious control steps have been taken, an air ionizing system may be a worthwhile option.

One other factor to be considered is the particle contribution of the air ionizing system itself. Chapter 14 reports significant emissions from an air ionizing bar operated at high currents in an accelerated test configuration. Such emissions were reported by Liu et al. (1987), who showed that most of these particles were in the size range below the size detection limit of the optical particle counter but were well within the range of the condensation nucleus counter. Figure 19-5 is a normalized differential size distribution plot of such ionizer emissions collected with a differential mobility particle sizer (DMPS) (see Chapter 6). The Figure 19-5 size distribution is characteristic of the 10^7/min emission rate measured for the ionizer bar example of Chapter 14. (This DMPS measurement took about 35 min to step through all the mobility channels. The emission sample was taken from a closed chamber that had been operated to a steady-state concentration; the ionizer was then turned off and the DMPS measurement begun. A high concentration of particles ($\sim 10^5/cm^3$) remained in the closed box at the end of the DMPS sampling period some 35 to 40 min after the ionizer had been shut down.)

Thus, it is clear that room air ionizer systems based on corona discharge are sources of ultrafine particles. The magnitude of these emissions depends on electrode composition (Donovan, Lawless, and Smith, 1986). Some manufacturers claim to have discovered a material that reduces these emissions to immeasurable concentrations (Sebald, Stiehl, and Sigusch, 1986) and data now exist which show only small or immeasurable increases in the ultrafine particle concentration of clean areas protected by air ionizing systems operating at normal current levels (Yost, Steinman, and Lieberman, 1989; Steinman, 1989; Kabaker and Daroff, 1989).

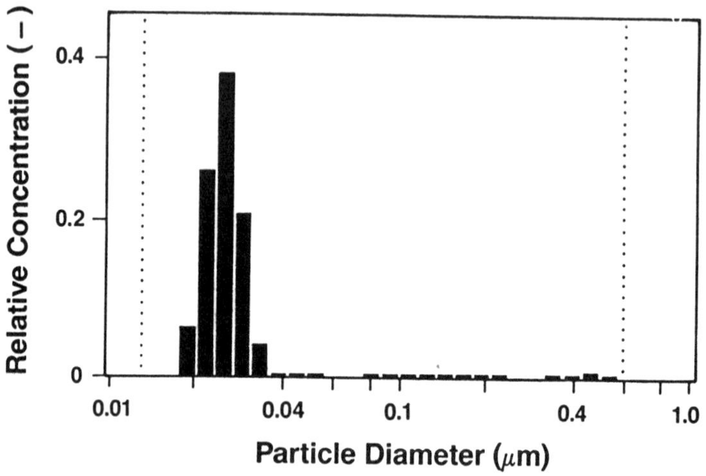

Figure 19-5. Differential size distribution of corona air ionizer particle emissions.

A second particle-related property of air ionizing systems is the small *decrease* in the downstream aerosol concentration of *OPC-sized* particles that has been observed (Huffman, Nichols, and Bossard, 1987; Hanley and Donovan, 1986). Huffman, Nichols, and Bossard described a decrease in downstream aerosol particle concentration as measured by an optical particle counter; they reported this decrease to a factor of 1.6. Hanley and Donovan, who used a challenge aerosol upstream of the HEPA filter, reported decreases in downstream concentration as high as a factor of 2 or 3 when the air ionizing system, a double dc design, was activated.

The reason for this decrease in OPC-sized particles is hypothesized to be electrostatic precipitation in the vicinity of the charging electrodes, although definitive proof does not exist.

The particle-generating properties of ionizers can be summarized as follows:

- Particles in the 0.3 μm and greater size ranges are generated at low rates. Aerosol particle concentration downstream of the ionizers may in fact decrease because of electrostatic precipitation.

- Particle generation in the 0.01 to 0.1 μm size range is high, and large increases in downstream concentration are observed by condensation nucleus counters when corona-based air ionizers are activated. This emission rate may be reduced by suitable choice of electrode composition, as some manufacturers have claimed.

- Particle deposition on monitor wafers exposed to dirty air is decreased by operation of an air ionization system; in state-of-the-art clean rooms, the effect of an air ionization system upon particle deposition on monitor wafers is difficult to measure.

Other methods of air ionization employ radioactive sources. These sources avoid the ultrafine particle generation drawbacks of corona systems but have a short range, which make them less appropriate for large-area clean rooms (Wilson, 1987). For local-area ionization these may be highly suitable, although the use of radioactive isotopes requires additional safety measures in personnel training and protection.

PARTICLE DEPOSITION FROM LIQUIDS

Liquid processing of silicon wafers, especially in cleaning procedures, remains a universal step in silicon chip manufacturing. Wafer contact with these processing liquids constitutes an additional threat of particulate contamination that could be especially grave in view of the relatively large particle concentrations found in processing liquids. As noted in Chapters 11 and 12, particle concentrations in liquids are expressed in numbers/mL for liquids instead of numbers/ft^3 as for air samples. On a number/volume basis, particle concentrations in processing liquids often exceed those in the clean-room ambient air by a factor of 10^3 to 10^4. It is these particle concentrations that wafers undergoing processing in liquids are not only exposed to but deliberately immersed in.

Fortunately, particle deposition velocity in liquids is lower than in air. For submicrometer polystyrene latex (PSL) spheres suspended in deionized water, deposition velocity on immersed wafers is in the 10^{-4} (0.3-μm PSL spheres) to 10^{-6} (1-μm PSL spheres) cm/s range rather than the 10^{-2} to 10^{-3} cm/s range characteristic of neutral particles in air. One immediately obvious air-liquid difference is that particle deposition velocity onto horizontal wafers decreases with increasing particle sizes in liquids, whereas it increases with increasing particle size in air. This dependence shows that gravitational settling is less important as a deposition mechanism in liquid systems than in air systems.

Figure 19-6 shows data collected in an aerosol deposition chamber. For the horizontal wafers, gravitational settling dominates and, qualitatively, the size dependence predicted by Equation (2-11) and plotted in Figure 2-4 is observed—the larger particles settle more rapidly than the smaller particles; the deposition velocity of the 0.9-μm spheres exceeds that of the 0.6 or 0.3-μm spheres.

For vertically oriented wafers, gravitational settling should not deposit any particles. The particles move parallel to the wafer surface, and only by diffusion, which is independent of wafer orientation, do they cross flow streamlines and the thin boundary layer adjacent to the wafer surface. Hence, vertically oriented wafers exhibit the size dependence characteristic of diffusion (Equations [2-15] and [2-16] and Figure 2-6). Large particles diffuse to a wafer surface more slowly than small particles because of the inverse dependence of diffusion coefficient on particle size (Equation 2-15). Below 0.3 μm, the two curves in Figure 19-6 should coincide because deposition on both the horizontal and the vertical wafer will be dominated by diffusion.

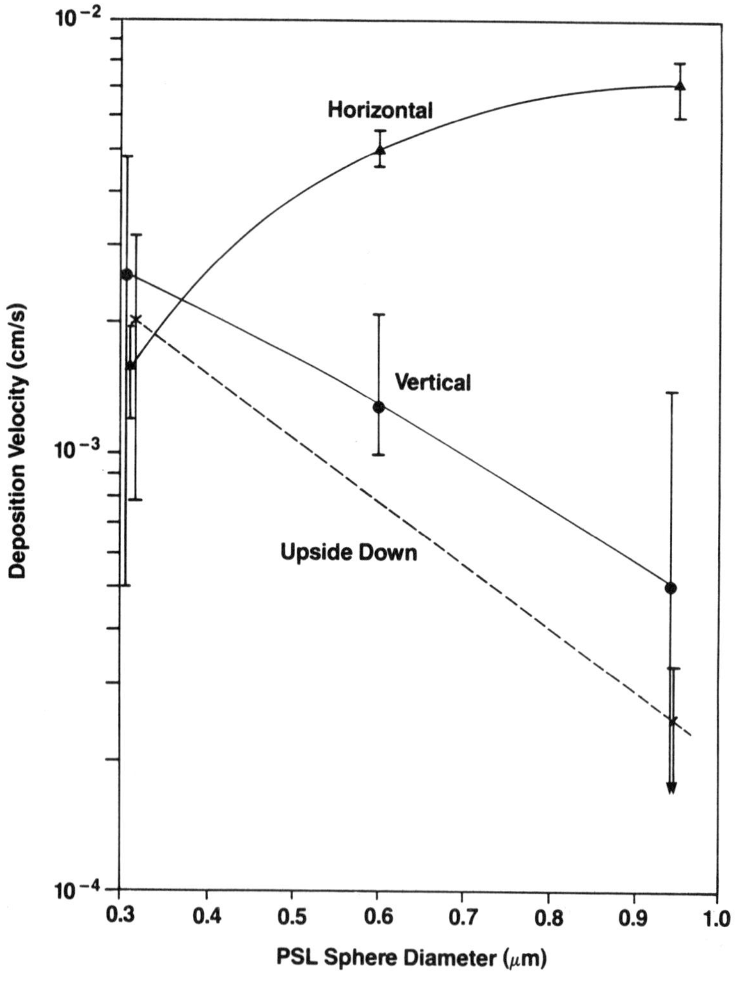

Figure 19-6. Size dependence of aerosol particle deposition velocity; wafer orientation as a parameter.

That horizontal wafers in liquids collect 1-μm PSL spheres at slower rates than 0.3-μm spheres, for example, and that wafer orientation (horizontal vs. vertical) makes no difference in particle deposition velocity in liquids support the conclusion that gravitational settling is not an important particle deposition mechanism in liquids over this PSL sphere size range. This same conclusion could be predicted from the settling velocity equation (Equation [2-11]) presented in Chapter 2. The density term in that equation is actually a density difference term in which ρ_p should be written ρ_p-ρ_{medium}. When the medium is air, ρ_{medium} is on the order of 10^{-3} g/cm^3, a value that is generally negligible with respect to typical aerosol particle densities (1 to 2 g/m^3). However, when the medium is water (ρ = 1.0 g/cm^3) and the particle PSL (ρ = 1.05 g/cm^3), the media density cannot be ignored. Using these values in Equation (2-11) makes clear the low settling velocities expected of such particles and particles in general—in water and in other liquids.

Observations from everyday experience also reinforce this conclusion. Objects that have substantial terminal velocities in air settle very slowly, if at all (some float), in water. Thus, particles in liquids settle onto wafer surfaces at lower rates than they do in air.

In addition, because of reentrainment, particle deposition on wafers immersed in a liquid bath is more complex than particle deposition on wafers exposed to air or other gases. A postulated equation describing the time change of particle areal density, N_s, on the surface of a wafer immersed in a liquid bath of particle concentration, N_l is:

$$\frac{dN_s}{dt} = -k_s N_s^a + k_1\ N_1^b \qquad (19\text{-}1)$$

where

N_s = areal density of particle on the wafer surface (l^{-2})

N_l = particle concentration in the liquid bath (l^{-3})

k_s = a rate constant for particle removed (t^{-1})

k_l = a particle deposition velocity (l/t)

a,b = exponent > 0.

This equation reduces to the aerosol particle deposition velocity equation (Equation 18-12) when the first term is negligible with respect to the second and b = 1 (to see this identity set $\frac{dN_s}{dt}$ = J and k_l = V_d). In liquid systems, however, the first term is not always negligible, especially at high values of N_s, making the first term the dominant term. When the first term dominates, the particle flux becomes negative and the wafer-liquid bath interaction is then a cleaning operation rather than a deposition step. Indeed, dirty wafers will oftentimes emerge cleaner after immersion in the same liquid bath that deposits particles on a relatively clean wafer.

If Equation (19-1) were in fact an adequate model of particle transport to wafer surfaces exposed to liquid baths, the problem would simply be one of calculating or measuring the value of k_s and k_l. However, Equation (19-1) is not an adequate model. For example, an implicit assumption of Equation (19-1) is that particle flux does not change with time. While this assumption may be valid over the time period that wafers are immersed in the bath, experiments show that a major contribution to the particle load on a wafer is made as the wafer is withdrawn through the liquid-air interface.

Consider the experiment outlined in Figure 19-7 designed to isolate the particle deposition at three distinct stages of wafer exposure to a liquid bath: insertion (C_I); immersion (C_T); and withdrawal (C_0) [Menon et al., 1988]. The experiment begins with two clean wafers in a "particle-free" water bath (whether the wafers are inserted into the bath or whether the water is poured on top of the wafers is immaterial—the water is assumed to be adequately particle-free so that no significant particle deposition is attributed to it). At some later time polystyrene latex (PSL) spheres are added to the bath. Wafer 1 is then immediately withdrawn, the assumption here being that all deposition on this wafer is attributable to withdrawal through the water-air interface. A third wafer is then lowered into the bath and immediately withdrawn. This wafer contains particles deposited during both entry and withdrawal through the water-air interface. Finally, after a suitable immersion time—3 min in this experiment—wafer 2 is withdrawn from the bath. Its particle load is attributable to immersion time and withdrawal.

These three wafers, each containing a different combination of contributions to total particle load, allow the isolation of each of these contributions. The results of this contribution breakdown are shown in Figure 19-8 for deposition on bare wafers having just a native oxide layer. Different wafer surfaces, for example, a freshly oxide-stripped surface, will behave differently. Clearly, for the wafers in Figure 19-8 the withdrawal step makes an important contribution to total particle load, especially from the more heavily contaminated baths. Leaving the wafer within the solution for a length of time approximately 10 times longer than the lowering and raising time increases the particle deposition somewhat but not proportional to the total time in solution. The more important exposure period for these deposition experiments is the exit from the bath.

A physical model that could explain this observation is that the particle concentration in the bath is greatest at the water-air interface. Particles introduced into the bath diffuse to the surface where they are captured by surface tension and are removed from the bulk liquid. A wafer completely submerged in a bath containing only floating particles would acquire no particles while immersed because the particle concentration adjacent to the wafer is zero.

To investigate this hypothesis a direct measurement of particle concentration in a surface region was made and compared with the concentration in the bulk (Carbonell and Riley, 1989). Figure 19-9 shows a cross section of the bath. Because of the negative meniscus,

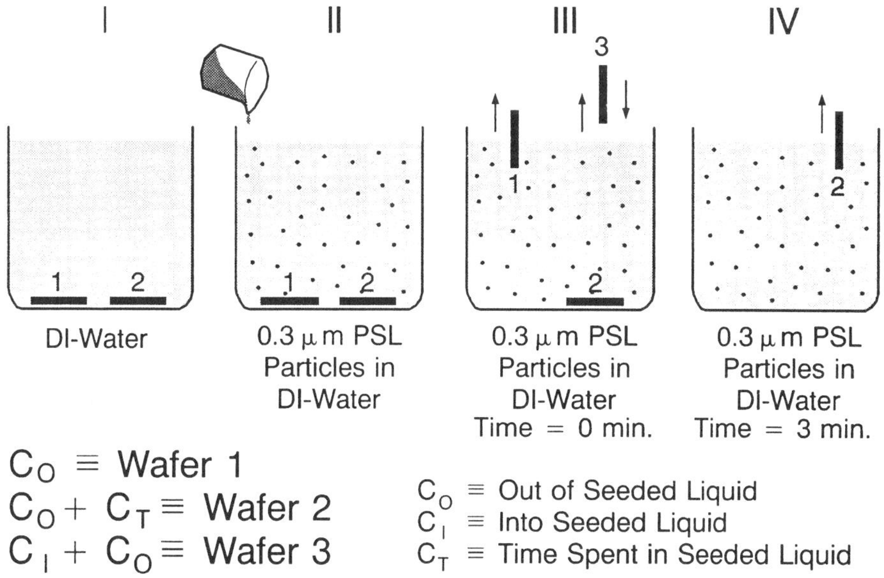

Figure 19-7. Particle deposition at the air/liquid interface.

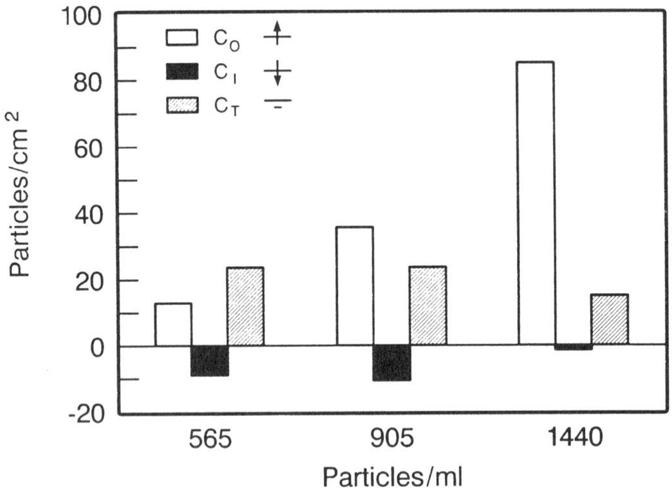

Figure 19-8. Contribution of air-liquid interface to particle deposition.

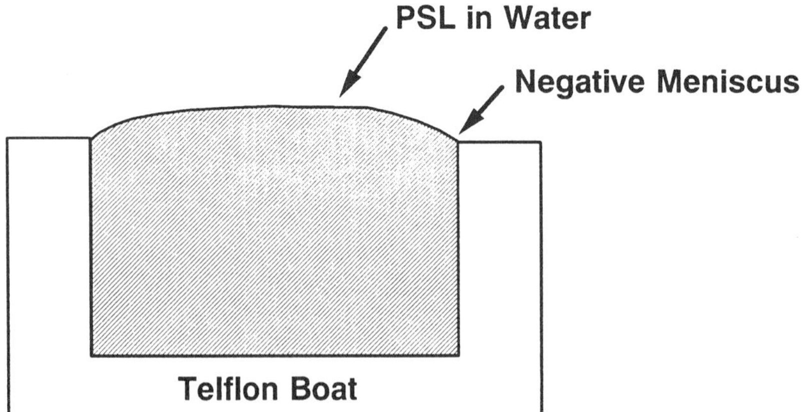

Figure 19-9. "Scrape" experiment to measure particle build-up at the air-liquid interface.

a surface portion of the bath protrudes above the Teflon container walls. This protruding surface portion of the solution was scraped off with a Teflon doctor blade and the particle concentration in that surface volume was measured. Within experimental error no difference in particle concentration was measured between the surface volume and the bulk volume, refuting enhanced surface concentration as the explanation of particle deposition during wafer withdrawal from the bath (Carbonell and Riley, 1989). Indeed, if enhanced surface concentration were the whole explanation, inserting wafers into the bath should be an equally contaminating action. The data of Figure 19-8 showed that wafer insertion was, in actuality, a cleaning step—particle concentration on the wafer was reduced by inserting it into the bath through the water-air interface.

Interactions other than enhanced surface concentration are needed to explain deposition during wafer withdrawal. Possibly water film adhesion to the withdrawing wafer and subsequent drying are involved in the interfacial contribution, but this interaction is not understood. Additional study is needed.

The importance of the interfacial contribution implies that the concept of a deposition velocity—particles added/unit area and time divided by bulk particle concentration—does not adequately describe particle deposition from a liquid. Therefore, while a deposition velocity can be calculated for a given time of immersion, as was done at the beginning of this section, the time dependence of this quantity makes it a poor variable for basing comparisons. Particles added per unit area for a given step without specifying when or over what time period during the exposure these particles were collected is the more appropriate variable for expressing particle deposition or removal.

REFERENCES

Blitshteyn, M., and A. M. Martinez. 1986. Electrostatic-Charge Generation on Wafer Surfaces and Its Effect on Particulate Deposition. *Microcontamination*, 4(11):54-60, 132.

Bowling, R. A., and G. B. Larrabee. 1987. Behavior and Detection of Particles in Vacuum Processes. In: *ULSI Science and Technology*, S. Broydo and C. M. Osburn, eds., ECS, Pennington, N.J., pp. 786-797.

Carbonell, R., and D. Riley. 1989. Direct Measurement of Particle Concentration at the Air-Liquid Interface, *Proceedings of the 1989 Annual Technical Meeting of the Institute of Environmental Sciences*, pp. 426-427, IES, 940 East Northwest Highway, Mount Prospect, IL 60056.

Dillenbeck, K. 1985. Characterization of Air Ionization in the Cleanroom. *Microcontamination*, 3(6):41-49.

Donovan, R. P., P. A. Lawless, and D. D. Smith. 1986. Polarity Dependence of Electrode Erosion Under dc Corona Discharge. *Microcontamination*, 4:(5)38-49.

Hanley, J. T., and R. P. Donovan. 1986. Documentation of Test Conditions and Particle Concentrations During Clean Room Tests of 28 February, 1986. Unpublished note, Research Triangle Institute, NC.

Huffman, T. R., G. H. Nichols, and P. R. Bossard. 1987. Reduction of Particle and ESD Damage by Room Ionization. *Solid State Technology*, November, pp. 127-130.

Jonasson, N. 1985. The Physics of Air Ionization. Technical University of Denmark, 2800 Lyngby, Denmark.

Kabaker, B., and I. Daroff. 1989. The Effects of Ionization on Airborne Particle Levels in a Class 100 Clean Room. Presentation to the 20th Annual Meeting of the Fine Particle Society, Boston, MA, August 21-25.

Liu, B.Y.H., and Kang-ho Ahn. 1987. Particle Deposition on Semiconductor Wafers. *Aerosol Science and Technology*, 6(3):215-224.

Liu, B.Y.H., D.Y.H. Pui, W. O. Kinstley, and W. G. Fisher. 1987. Aerosol Charging and Neutralization and Electrostatic Discharge in Clean Rooms. *J. Environmental Sci.*, 30(2):42-46.

Meek, J. M., and J. D. Craggs, eds. 1978. *Electrical Breakdown of Gases*. John Wiley and Sons, New York.

Menon, V. B., L. D. Michaels, L. A. Hollar, and R. P. Donovan. 1988. Interfacial Effects Associated with the Separation of Particles from Semiconductor Surfaces. Presentation to the AIChE Annual Meeting, Washington, DC, November.

Mykkanen, C. F., and D. R. Blinde. 1983. Room Air Ionization System. Final Test Report, May 31, Honeywell Underseas System Division, 600 2nd Street NE, Hopkins, MN 55343.

Osburn, C., H. Berger, R. Donovan, and G. Jones. 1988. The Effect of Contamination on Semiconductor Manufacturing Yield. *J. Environmental Sci.*, 31(2):45-57.

Sebald, T., H. H. Stiehl, and R. Sigusch. 1986. How to Avoid Particle Generation from Needle Electrodes of Ionizer Systems for Clean Rooms. Sorbios GmbH, Ackerstrasse 71-76, 1000 Berlin 65.

Steinman, A. 1989. Characterization of Particle Emissions from Air Ionizer Electrode Materials. Presentation to the 20th Annual Meeting of the Fine Particle Society, Boston, MA, August 21-25.

Wilson, R. 1987. Nuclear Air Ionization for Contamination Control. Proceedings of the Annual Technical Meeting of the IES, pp. 466-472.

Yost, M., A. Steinman, and A. Lieberman. 1989. Apparatus for Measuring Ultrafine Particles Emissions from Air Ionization Equipment. Presentation to the 20th Annual Meeting of the Fine Particle Society, Boston, MA, August 21-25.

20. PARTICLE IDENTIFICATION

WAYNE G. FISHER Texas Instruments, Inc., Sherman, Texas
JEFFREY M. DAVIDSON The BOC Group, Inc., Murray Hill, New Jersey

INTRODUCTION

The efforts directed toward particle minimization in the semiconductor fabrication environment are generally complemented with rigorous particle monitoring in all areas of concern. For example, particle levels in clean-room air, and process gases and chemicals, are usually measured with dynamic particle counters on a continuous basis. More recently, particle counters have been used to measure particle generation from semiconductor processing equipment. This particle monitoring is performed primarily for quality assurance/control. Some insight into possible particle origins may also be obtained through the location of particle detection, e.g. beneath a particular clean-room ceiling grid or in a particular batch of chemicals. Additional clues regarding particle origins may be deduced from the time-dependency of particle counts, e.g., the movement of a robot arm or the switching of a fluid valve. However, a much clearer indication of particle origins may be obtained through the compositional and physical analyses of particles (Duffalo and Monkowski, 1984; Davidson, 1987; Davidson and Ruane, 1987; VanderWood, 1985a and 1985b; Heinrich, 1980). Through this particle identification, the compositional and physical characteristics can be compared to those of potential particle sources in the fabrication environment. The offending material or component can then be removed or upgraded.

Much of the early work on microanalytical techniques focused on environmental pollutants; this work has more recently been extended to the area of microcontamination control in the semiconductor industry. These analytical techniques are often the same as traditionally used for bulk samples. In some cases, special modifications of the instrumentation or data analysis have been required to adapt these techniques to microsamples. This chapter focuses on those techniques specifically applicable to the semiconductor fabrication environment. These techniques will be briefly reviewed in the next section, and examples of their use to solve specific microcontamination problems will be provided in a subsequent section.

PARTICLE ANALYSIS

Two general types of analyses can be applied to particles: (1) single particle analysis, and (2) bulk analysis. In single particle analysis, the compositional and physical characteristics of each individual particle are determined. Bulk analysis averages out the compositions of all particles obtained in a sample and is generally incapable of determining physical characteristics of particles.

Single Particle Analysis

A prerequisite for single particle analysis is that the particles of interest rest on a substrate that does not interfere with the particle image or chemical signal. Obviously, the optimum substrate cannot be chosen in cases where particles present on processed wafers are being analyzed. However, suitable collection substrates can be selected if particles are being sampled from gases, liquid chemicals, or clean-room air. Frequently, this particle collection is achieved through filtration. Depth filters and most porous membrane filters are unsuitable collection substrates because particles can become buried within the filter pores. In contrast, flat membranes with discrete holes, such as those manufactured by Nuclepore Corporation,* provide smooth collection surfaces on which particles rest, with minimal interference. Some potential drawbacks to these membranes are that they can be burned by high-energy beams, they require conductive coatings for some analytical techniques, and their composition may obscure the analysis of organic particles with similar compositions. Silver membranes, such as those manufactured by Selas Corporation,** can overcome these problems, but their surfaces are not as flat and smooth as Nuclepore membranes. Either type of membrane may contain residual particles from the manufacturing and packaging procedures, and these particles need to be fully characterized as membrane background levels.

Cascade impactors are also finding increasing applications for particle sampling in gases and clean-room air (Chuan and Chiang, 1985; Kasper and Wen, 1985; Kasper, Wen, and Chesters, 1986). A cascade impactor consists of a series of orifices and impaction plates on which particles are collected. With cascade impactors, particle samples can be concentrated into smaller areas than with filters, and scanning electron microscope (SEM) specimen holder stubs can be used as impaction plates for direct particle analysis. The impactor materials can be selected to be compatible with a given gas or to minimize interference with chemical analyses. Drawbacks to cascade impactors include particle loss due to bounce and adherence to walls and the special cleanliness requirements.

In either case, a number of analytical techniques can be used. These are briefly discussed in the following items and summarized in Table 20-1:

- **Optical Microscopy** (VanderWood, 1985a and 1985b; Heinrich, 1980; Bayard, 1985; McCrone, 1986; McCrone and Delly, 1973)

  The oldest and simplest single particle analysis technique is optical microscopy. The optical microscope does not provide direct elemental analysis, but it can be used to characterize other physical properties through which particles can be identified. The most determinant of these properties are morphology, color, and opacity. Additional information regarding crystallinity and refractive indices can be obtained with polarized light microscopy (McCrone, 1986; McCrone and Delly, 1973). Several wet chemical analysis techniques have been adapted to microanalyses observed with optical microscopy. The major limitation of optical microscopy is the lower detection limit of around 1 μm.

- **Scanning Electron Microscopy (SEM)/Electron Beam Microprobe** (VanderWood, 1985a and 1985b; Heinrich, 1980; McCrone and Delly, 1973; Davidson, Gerristead, and Ruane, 1986; Gerristead, Sherman, and Davidson, 1987; Johnson et al., 1984; Hanna, Karcich, and Johnson, 1980; Kelly, Lee, and Lentz, 1980; Barnard, 1987; Kasher, 1987; Phillips, 1983)

  Higher magnification imaging and direct elemental analysis can be achieved with the SEM or electron beam microprobe, two instruments that were historically quite different, but now are quite similar in overall capabilities. In this technique,

*Nuclepore Corporation, Pleasanton, CA 94566-3294.

**Selas Corporation of American, Huntington Valley, PA 19006.

Table 20-1. Characteristics of Single Particle Analytical Techniques

Method	Approximate minimum particle diameter (μm)[a]	Probe particle/ radiation	Detected particle/ radiation	Information gained
Optical microscopy	1.0	Visible light	Visible light	Size, shape, morphology, color, opacity, crystal-linity, index of refraction
SEM/Microprobe with EDS/WDS	0.05-0.1	Electrons	Electrons x-rays	Size, shape, morphology, elemental composition
TEM with EDS/WDS	0.001	Electrons	Electrons	Size, shape, elemental composition, crystal structure
FTIR	10-20	Infrared	Infrared	Organic structural groups
Raman	0.5	Monochromic light (laser)	Light	Organic and inorganic structural groups
AES	0.05	Electrons	Electrons	Surface and bulk (depth profiling) elemental compositions
SIMS	1.0	Ions	Ions	Surface and bulk (depth profiling) elemental compositions, isotope ratios, organic fragmentation patterns
LAMMA	0.5	Laser	Ions	Elemental compositions, isotope ratios, organic fragmentation patterns

[a]Minimum particle diameter for compositional analysis.

Source: Vanderwood 1985a and Heinrich 1980.

the sample is bombarded with electrons, and the image is obtained through scattered electrons. Elemental analysis, from qualitative to quantitative, is obtained through either energy-dispersive (EDS) or wavelength-dispersive (WDS) spectroscopic analysis of X-rays generated from the sample. EDS is a quicker analysis, but WDS provides greater sensitivity, spectral resolution, and detection of all elements heavier than Be (conventional EDS cannot detect elements lighter than F, but newer thin-window and windowless detectors can analyze light elements). With SEM/microprobes, combined morphological and compositional data can be obtained on particles as small as 0.01 to 0.05 μm.

One drawback to analysis of particles with SEM/microprobe is the effort and time required to study a sufficiently representative particle population. One can appreciate this by noting that a single 47 mm sampling filter contains approximately 5 million fields-of-view at 5,000X magnification, at which 0.1 μm particles are detectable. SEM/microprobe analysis of a small fraction of this filter requires one or more days, and the manually accumulated data are not in a form suitable for statistical analysis. These difficulties have recently been overcome with the development of computer-interfaced SEM/microprobes for automated particle analysis (Davidson, Gerristead, and Ruane, 1986; Gerristead, Sherman, and Davidson, 1987; Johnson et al., 1984; Hanna, Karcich, and Johnson, 1980; Kelly, Lee, and Lentz, 1980). In a system designed specifically for aerosol analysis (Davidson, Gerristead, and Ruane, 1986; Gerristead, Sherman, and Davidson, 1987), dedicated computers control the specimen stage automation, so that any number of preselected areas on a collection filter can be scanned without user intervention. The computer also provides automated particle image recognition and sizing, as well as deciphering of spectrometer outputs for quantitative compositional information. Data for each recognized particle are displayed on a video monitor and simultaneously transferred to a mainframe computer. This more powerful system summarizes data for all particles analyzed on a filter and can be queried to evaluate more subtle trends, such as correlations between particle sizes and compositions. With this system, data for hundreds of particles can be obtained and filed into a data base within hours.

- **Transmission Electron Microscopy (TEM)** (VanderWood, 1985a and 1985b; Heinrich, 1980; Steen et al., 1983; Beaman, 1978)

In TEM, electrons are transmitted through the sample, as opposed to scattered from the sample as in SEM. TEM provides very high magnification and can easily detect particles as small as 0.001 μm. Modern TEMs are generally equipped with EDS capabilities for microanalysis and selected area diffraction (SAD) capabilities for crystal structure determination. The major disadvantage of this technique is that particles must be transferred from a collection substrate to a metal grid used in the TEM sample stage.

- **Micro-Fourier-Transform Infrared Spectroscopy (Micro-FT-IR)** (VanderWood, 1985a and 1985b; Heinrich, 1980; Humecki, 1985; Graeder, Flagan, and Seinfeld, 1987; Carpenter and Kisken, 1984)

In Micro-FT-IR, modulated infrared light passed through a sample is absorbed at specific frequencies related to the molecular vibrational modes of the material. The resulting interferogram is translated into an absorption spectrum by a Fourier transform.

Micro-FT-IRs are ideally suited to identifying organic materials, for which extensive libraries of reference spectra exist. One advantage of Micro-FT-IRs is that they are adaptable to conventional optical microscopes. However, current Micro-FT-IRs are limited to particles larger than 10 to 20 μm.

- **Raman Microprobe** (VanderWood, 1985a and 1985b; Heinrich, 1980; Klainer and Milanovich, 1985; Muggli and Anderson, 1985)

  Although similar to Micro-FT-IR, the Raman microprobe offers significant advantages in that it can analyze both organic and inorganic materials, as well as particles of these materials as small as 0.5 μm. In Raman spectroscopy, the sample is irradiated with monoenergetic, visible light, and the frequency shifts of the inelastically scattered light are measured. These frequency shifts yield information on molecular vibration modes, molecular symmetry, number and strength of chemical bonds, and atomic mass (Heinrich, 1980). The use of high-powered lasers and improved optical systems enable the Raman microprobe to be used with discrete microsamples.

- **Auger Electron Spectroscopy (AES)** (VanderWood, 1985a and 1985b; Heinrich, 1980; Bakale and Bryson, 1983; Cavallini, 1981)

  With AES, a sample is bombarded with electrons, and emitted auger electrons are energy-analyzed for elemental information. AES can be applied to particles as small as 0.05 μm, and it can yield compositional information on the particle surface and the interior through depth profiling.

- **Secondary Ion Mass Spectroscopy (SIMS)** (VanderWood, 1985a and 1985b; Cavallini, 1981; Gavrilovic, 1985)

  In SIMS, the sample is bombarded with energetic ions, and the ejected ions from the sample are mass-analyzed. SIMS has greater elemental sensitivity than AES and can detect all elements in the periodic table. One drawback to SIMS is that it cannot be used for particles smaller than 1 μm.

- **Laser Microprobe Mass Analysis (LAMMA)** (VanderWood, 1985a and 1985b; Heinrich, 1980; Kaufmann, Wieser, and Wurster, 1980; Schmidt, Beaudet, and Madore, 1987)

  The relatively new technique of LAMMA provides mass spectrometry of material that is vaporized and ionized by a high-power laser. Although this technique is similar in many ways to SIMS, LAMMA offers the advantages of speed and a smaller particle-size detectability of 0.5 μm.

Bulk Analysis

Perhaps the most commonly used bulk analytical technique for studying environmental aerosols is that of capturing particles on filters, dissolving the particles in acid, and then performing chemical analysis on the acid (Butler, 1979; Katz, 1977; Jacobs, 1960; Fogg and Seeley, 1984; Mitchell, 1973). This technique has been extended to process gas studies (Hara, Luzrezi, and Martinez de Pinillos, 1985) and could be applied to particles collected from air or liquid.

After particle collection, the filter is ashed and/or dissolved in acid, or the particles alone are dissolved in acid and the insoluble filter removed from the solution. In either case, complete dissolution of the particles is necessary. Chemical analysis can then be performed on the acid analyte with a variety of conventional techniques such as atomic absorption spectroscopy (AAS), inductively coupled plasma atomic emission spectroscopy (ICP), ICP with mass spectrometry (ICP-MS), microwave-induced plasma (MIP), and metastable transfer emission spectroscopy (MTES). Procedures have also been specifically tailored for the chemical analysis of single elements or compounds in the sample.

In all of these techniques, extreme care must be taken to avoid contamination from particles in the air, impurities in the reagents, and residual contaminants on the test apparatus (Mitchell, 1973). Blanks for baseline comparison must be analyzed, and these must be reproducibly controlled at chemical levels much lower than the elements being studied. This, of course, becomes a primary challenge when sampling clean air, gases, or liquids, and it may require prohibitively long sampling times. Another drawback of this technique is sample loss during the ashing and acid dissolution procedure.

In the cases of air and gases, particles can also be dissolved in acids by directly impinging the aerosol into an acid solution, which is subsequently analyzed by one of the aforementioned analytical techniques. This procedure precludes the problems associated with collection filters. However, particle impaction and scrubbing efficiencies tend to decrease for particles smaller than 0.5 μm. Dissolution efficiencies can be increased by using a series of gas washing bottles (Hara, Luzrezi, and Martinez de Pinillos, 1985).

Some techniques less commonly used in environmental studies allow direct determination of the average composition of all particles collected on a filter, or resting on a substrate, without particle dissolution. Techniques that could be extended to analysis of particles in the semiconductor processing environment include proton-induced X-ray emission spectroscopy (PIXE) (Wutjin, 1986), neutron activation analysis (NAA) (Katz, 1977; Fogg and Seeley, 1984), X-ray photoelectron spectroscopy (XPS) (Heinrich, 1980; Bakale and Bryson, 1983), X-ray fluorescence (XRF) (Johnson et al., 1984; Butler, 1979; Fogg and Seeley, 1984; Cullity, 1956), and X-ray diffraction (XRD) (Cullity, 1956).

EXAMPLES

Analysis by Shape

The identity and origin of a particle may sometimes be easily determined from its shape. Optical and scanning electron microscopy are both appropriate tools to see the shape of particles. Figure 20-1 is a scanning electron micrograph of particles collected from a high-purity deionized (DI) water system. It is easy to see that these are bacteria, a common cause of particle problems in DI water systems. Once identified, bacteria populations in DI water systems are controlled by periodic sterilization with hydrogen peroxide, ozone, or hot water. The dead bacteria and their fragments must be removed from the water by filtration. Then the regrowth of bacteria colonies is monitored by taking a periodic culture of water samples from the system.

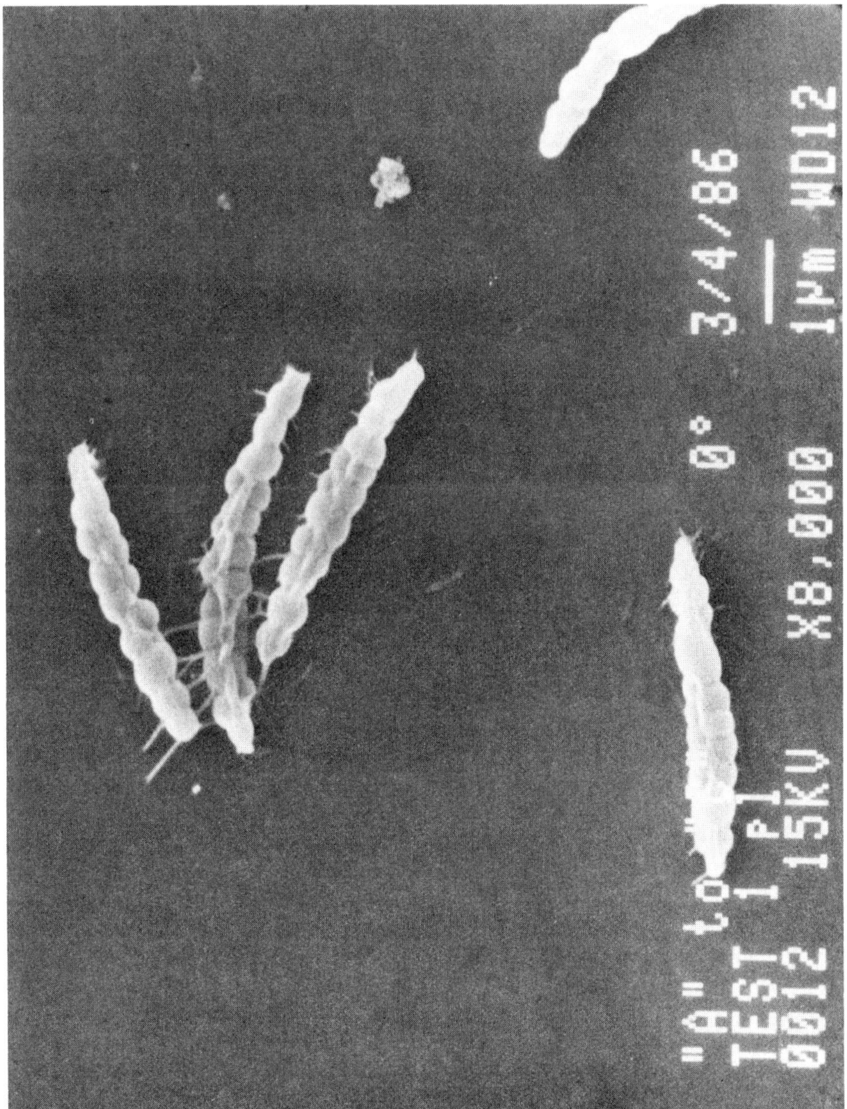

Figure 20-1. Bacteria collected by filtering from a high-purity deionized water system.

Scanning Electron Microscopy (SEM)/Electron Beam Microprobe

Figure 20-2 illustrates the power of the SEM/microprobe equipped with EDS and WDS capabilities. This study was performed on a non-VLSI nitrogen distribution system consisting of a 9-percent Ni-steel liquid storage tank, an aluminum vaporizer, and a predominantly copper piping system. No filtration was present throughout the distribution system. Membrane samples obtained at the point-of-use indicated particle levels of between 1,000 and 10,000 particles larger than 0.1 $\mu m/ft^3$. The greatest percentage of these particles were rich in iron (i.e., Fe > 50 wt.%) with 5 to 10 percent nickel, suggesting their origin was back at the liquid storage tank. Smaller quantities of aluminum-rich and copper-rich particles were also found, indicating that the aluminum vaporizer and copper plumbing were also contaminating the gas.

WDS analysis revealed these particles to be mostly oxides, which would suggest their formation through corrosion reactions at the hardware surfaces. Based on results such as these, stainless steels are increasingly being specified for gas delivery systems, as discussed in Chapter 10. Virtually no corrosion-related particles are found in these systems.

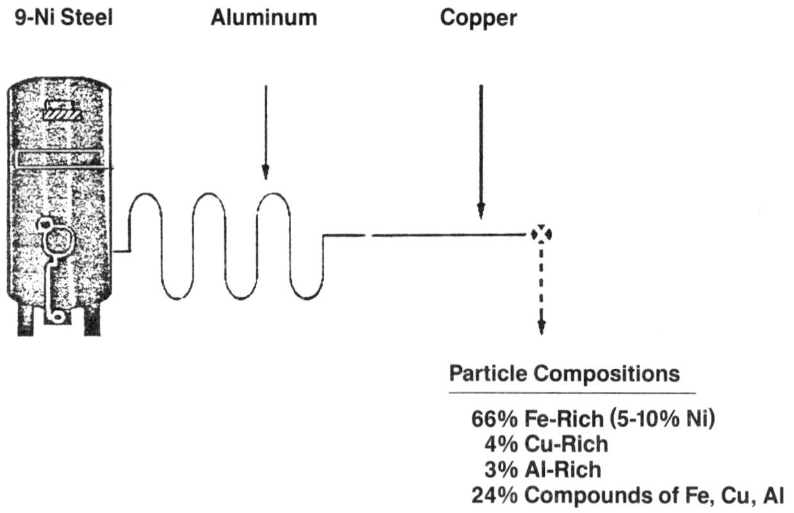

Figure 20-2. Compositions of particles collected at end of non-VLSI nitrogen storage and distribution system.

Transmission Electron Microscopy (TEM)/Selected Area Diffraction (SAD)/ Energy-Dispersive X-Ray Analysis (EDX)

Figure 20-3 illustrates the solution of a particle problem by TEM-based analysis techniques. The problem was first observed as particles on silicon wafers processed in a furnace used to deposit silicon-nitride films. Although these particles were very small, their location on the wafers could be determined with a wafer surface scanner. After determining the location of the particles, the wafer was coated with a thin layer of collodion to trap the particles on the wafer and to protect the surface for transport to the TEM lab. In a clean room adjacent to the TEM lab, the collodion film was peeled from the wafer surface and placed on a TEM sample grid. The sample was viewed in transmission in the TEM through the thin collodion film. A particle with dimensions of about 1,000 by 2,000 Å is illustrated in the figure.

The selected area diffraction pattern of the particle showed that it was crystalline. Subsequent elemental analysis by EDX showed that the particle was pure silver. Processing engineers immediately traced the problem to a silver-plated stainless steel screen supporting a filter in a gas line to the furnace. The coating was flaking from the screen, and the resulting silver particles were being carried into the furnace by flowing gas. Replacement of the filter support screen with an unplated stainless steel screen eliminated this particle problem.

Micro-Fourier Transform Infrared Spectroscopy (Micro-FT-IR)

The origin of particles detected on product wafers after etching in hydrofluoric acid was traced to the acid by before-and-after surface scanning of test wafers. A sample of the particles filtered from the acid was analyzed by micro-FT-IR spectroscopy. The spectrum obtained from the particles is shown as the lower trace in Figure 20-4.

Various parts of the acid delivery system were analyzed to determine which material had a similar IR spectrum. A sample of the material used to make the liner of the acid storage tanks, obtained from the tank manufacturer, gave the similar spectrum shown in Figure 20-5. When the material used to make the liner was treated with hydrofluoric acid, the spectrum of the resulting material (Figure 20-4, upper trace) was nearly identical to the spectrum from the particles in the acid. The particle problem was eliminated by changing the composition of the acid tank liner.

Raman Microprobe

Raman spectroscopy was used to solve the particle problem illustrated with the spectra in Figure 20-6. The problem was observed on the surface of silicon wafers as groups of particles with diameters of 1 to 5 μm. These particles could be seen on the surface of the wafer on the microscope stage of a micro-Raman spectrometer.

The spectrum from one particle is shown in the upper half of Figure 20-6. It is a close match to the spectrum of Teflon™ shown at the bottom of the figure. The origin of the particles was traced to Teflon™ wafer boats with loose particles on them. Replacement of the defective boats with new ones eliminated the particle problem.

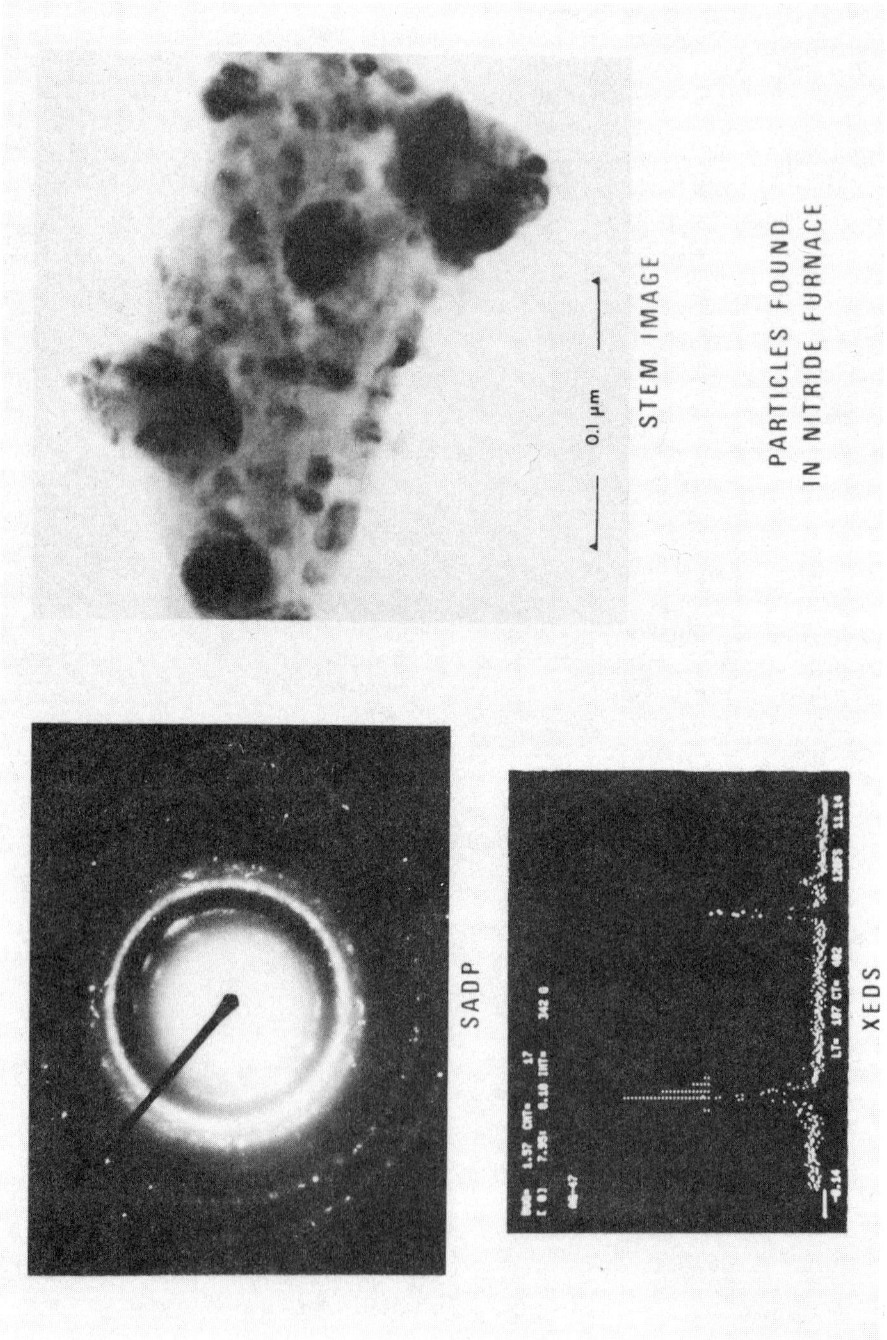

Figure 20-3. Scanning transmission electron micrograph, selected area diffraction pattern, and energy dispersive x-ray spectrum of a particle from a furnace.

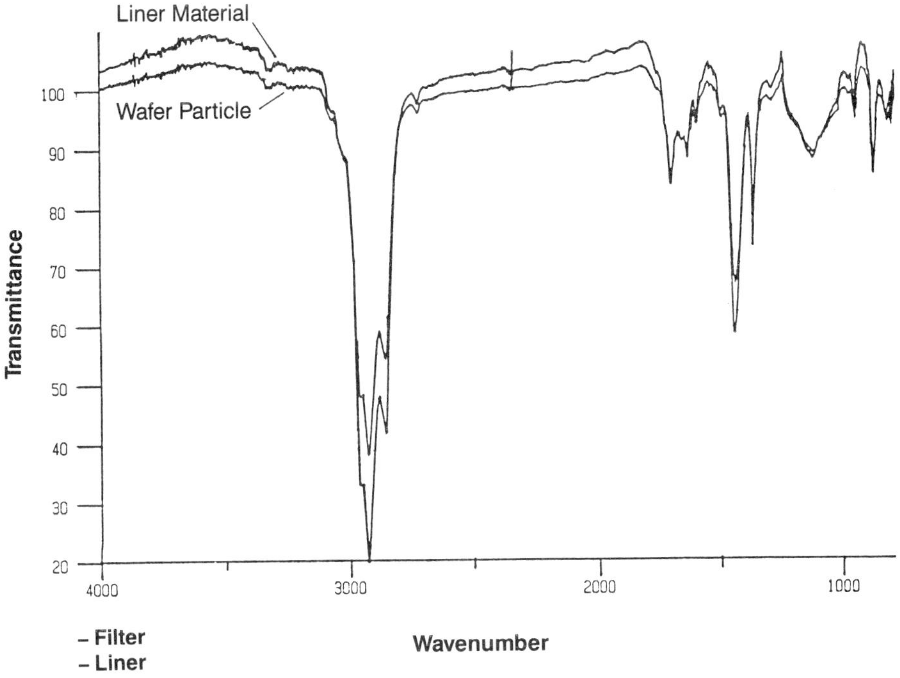

Figure 20-4. Micro-FTIR spectrum of particles filtered from hydrofluoric acid (lower trace), and of hydrofluoric acid tank liner after exposure to acid (upper trace).

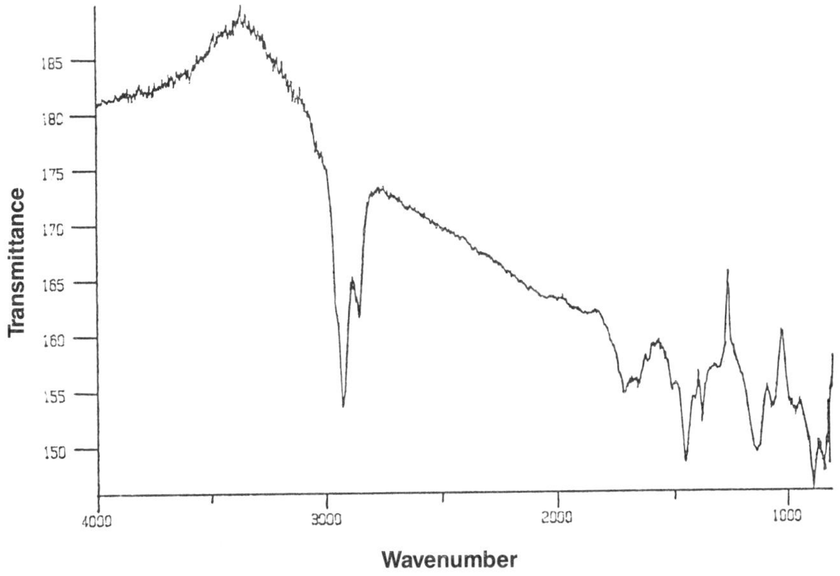

Figure 20-5. Micro-FTIR spectrum of hydrofluoric acid tank liner before exposure to acid.

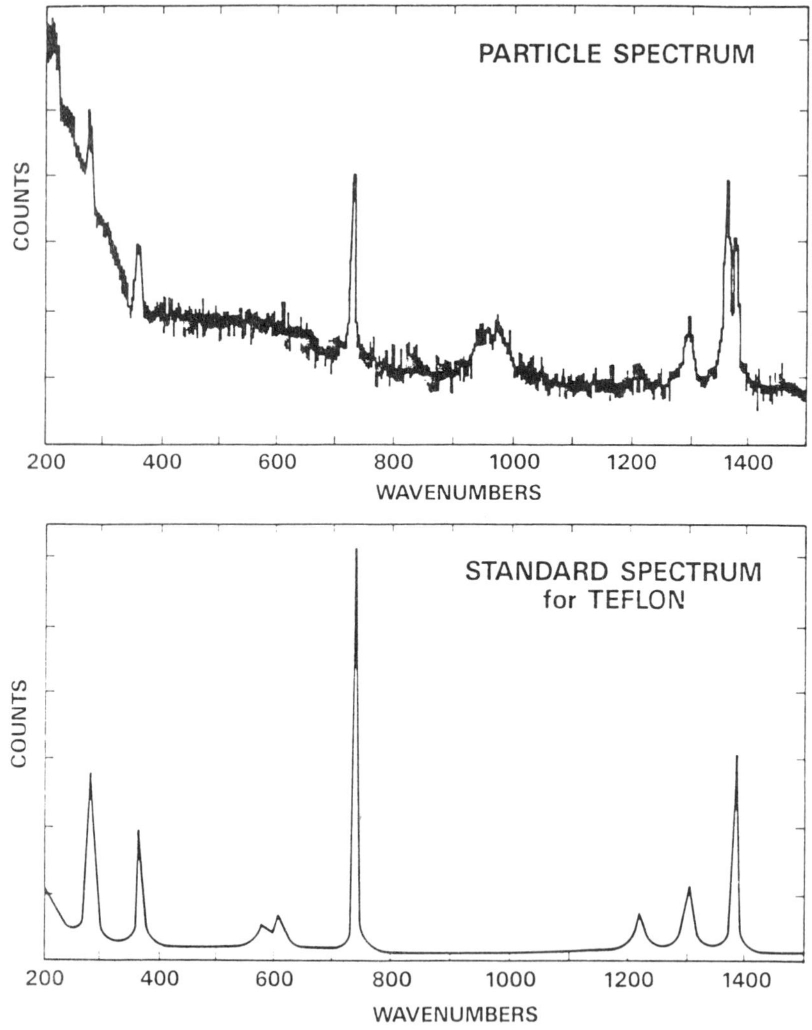

Figure 20-6. Micro-Raman spectrum of a particle on the surface of a silicon wafer (upper spectrum). Reference spectrum for Teflon^{TM} (lower spectrum).

X-Ray Diffraction

X-ray diffraction is not usually considered a single particle analysis technique unless the particles are relatively large. Typically the sample size must be at least 30 μm for a reasonable diffraction pattern to be obtained. Occasionally, though, a particle problem occurs where the sample is large enough for diffraction analysis.

In the last example of particle analysis in this chapter, a white powder was found in a reactor where silicon-nitride films were deposited by chemical vapor deposition. In this particular process, the silicon-nitride was deposited from silane and ammonia gas by reaction at elevated temperature and in the presence of a plasma. The X-ray powder diffraction pattern of the white material is shown in Figure 20-7. An automated search of powder diffraction patterns produced a perfect match with ammonium fluorosilicate, $(NH_4)_2SiF_6$.

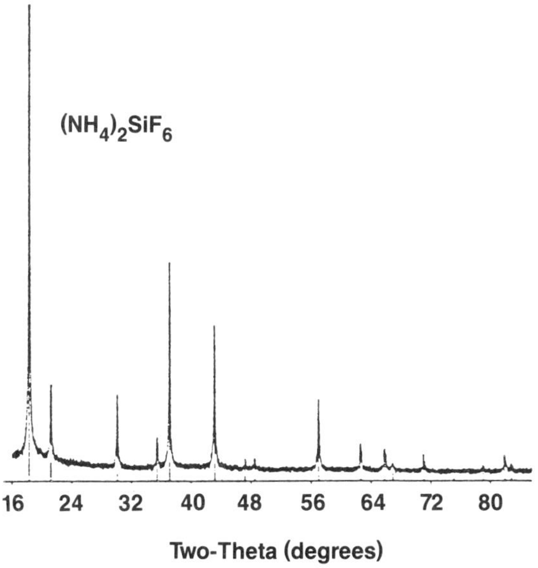

Xray diffraction identified $(NH_4)_2SiF_6$
Porous graphite parts retain Freon from etchback
Nitride deposition (SiH_4 + NH_3) + Freon → $(NH_4)_2SiF_6$
Problem solved by replacing parts with pyrolytic graphite

Note: Vertical lines under each peak illustrate the positions and intensities of peaks expected from ammonium fluorosilicate.

Figure 20-7. X-ray powder diffraction pattern of white powder from a chemical vapor deposition reactor.

An analysis of the procedure for cleaning the reactor revealed that the fluorine came from a fluorocarbon used periodically with a plasma to clean the internal parts of the equipment. Traces of the fluorocarbon had been absorbed by a porous graphite susceptor in the reactor. When the ammonia and silane were introduced into the reactor, the fluorocarbon reacted to form the white powder.

FUTURE REQUIREMENTS

A major problem in the analysis of particles found during semiconductor processing is that it is hard to find the particles when there are only a very few to analyze. For example, on a silicon wafer there may be fewer than 10 submicrometer particles resulting from a particular manufacturing process, but each of these particles may cause the loss of one integrated circuit.

The lack of clean particle analysis tools and techniques is another barrier to analysis. Few of the available analysis tools have been designed with the need to locate scarce particles in mind. In some instruments, the particle contamination is masked by the analytical technique because the equipment introduces additional contamination onto the sample during the analysis. The sample preparation technique itself may produce particles, such as the cutting of silicon wafers into pieces that will fit into an instrument vacuum chamber. Samples may also be contaminated by the environment because instruments that will analyze clean samples are often not located in clean-room areas.

Analytical techniques for small organic particles are lacking. In the example of the analysis of silver particles described above, there were no other analytical techniques to turn to if the particle had not been inorganic. In a processing environment where much of the equipment and facilities are constructed with organic polymers, the inability to analyze small organic particles is a serious handicap.

Future integrated circuits will require the control of smaller and scarcer particles in the processing environment. Elimination of these particles will require innovative application of particle analysis technology along with intimate knowledge of the physics and chemistry of semiconductor processing.

REFERENCES

Bakale, D. K., and C. E. Bryson. 1983. Surface Analysis: Detection and Measurement of Microcontaminants. *Microcontamination*, 1(5):33-35, 63.

Barnard, L. 1987. Using Particle Analysis to Solve Microcontamination Problems. *Microcontamination*, 5(8):34-37.

Bayard, M. 1985. Applications of Optical Microscopy to Fine Particle Analysis. In: *Proceedings - Inst. Env. Sci. - 1985*, pp. 96-98.

Beaman, D. R. 1978. Analytical Transmission Electron Microscopy and Its Application to Environmental Science. In: *Environmental Pollutants*, T. Y. Toribara, ed., pp. 255-294. Plenum, New York, NY.

Butler, J. D. 1979. *Air Pollution Chemistry*. Academic Press, New York, NY.

Carpenter, T. K., and L. A. Kisken. 1984. Particle Identification for Contamination Control by Infrared Microspectrophotometry. In: *Proceedings - Inst. Env. Sci. - 1984*, pp. 44-45.

Cavallini, M. 1981. Molecular Beam, SIMS and Auger Spectrometry for Analysis of Particulates. *IAEA-CN*, 40/103:179-195.

Chuan, R. L., and W. Chiang. 1985. Cascade Impactor Collects Airborne Samples for Particle Identification. *Solid State Tech.*, 28(7).

Cullity, B. D. 1956. *Elements of X-Ray Diffraction*. Addison-Wesley Publishing Company, Reading, MA.

Davidson, J. M., W. R. Gerristead, Jr., and T. P. Ruane. 1986. Automated Microprobe Characterization of Particles from High Purity Gas Streams. In: *Proceedings - Inst. Env. Sci. - 1986*, pp. 468-474.

Davidson, J. M. 1987. Determination of Particle Origins in High Purity Gases. *Solid State Tech.*, 30(7):63-67.

Davidson, J. M., and T. P. Ruane. 1987. Particle Control in VLSI Gases. In: *Emerging Semiconductor Technology*, ASTM-STP 960, D. C. Gupta and P. H. Langer, eds. American Society for Testing and Materials, Philadelphia, PA.

Duffalo, J. M., and J. R. Monkowski. 1984. Particle Contamination and Device Performance. *Solid State Tech.*, 27(3):109-114.

Fogg, T. R., and R. C. Seeley. 1984. ICP-OES Analysis of Atmospheric Aerosol Particles. *Amer. Lab.* (Dec.):36-39.

Gavrilovic, J. 1985. Secondary Ion Mass Spectroscopy for Bulk and Surface Analysis of Particulate Contaminants. *Solid State Tech.*, 28(4):299-302.

Gerristead, W. R., Jr., R. Sherman, and J. M. Davidson. 1987. Automated SEM/Microprobe Finds Gas Contaminants, Determines Sources. *Research & Development*, 29(6):93-100.

Graeder, G. S., R.C. Flagan, and J. H. Seinfeld. 1987. Fourier Transform Infrared Spectrometer for a Single Aerosol Particle. *Rev. Sci. Instrum.*, 58(4):584-587.

Hanna, R. B., K. J. Karcich, and D. L. Johnson. 1980. Determination of Particle Identities Via a Computer Assisted SEM/EDXA System. In: *Scan. Elec. Microscopy 1980*, p. 323.

Hara, L. Y., F. A. Luzrezi, and J. V. Martinez de Pinillos. 1985. Metallic Contaminants in Gases. In: *Microcircuit Pure Materials Conf. 1985*, pp. 32-39.

Heinrich, K. F. J., ed. 1980. *Characterization of Particles*. NBS Special Publication 533. NTIS, Springfield, VA.

Humecki, H. J. 1985. Infrared Spectroscopy and Its Application to Contamination Analysis. *Solid State Tech*, 28(4):309-313.

Jacobs, M. B. 1960. *The Chemical Analysis of Air Pollutants*. Interscience Publishers, New York, NY.

Johnson, D. L., B. McIntyre, R. Fortman, R. K. Stevens, and R. B. Hanna. 1984. Particle Analysis/Bulk Analysis. *Chem. Tech.* (Nov.):678-683.

Kasher, L. 1987. Techniques for Identifying Trace Contaminants. *Microcontamination*, 5(8):38-41.

Kasper, G., and H. Y. Wen. 1985. An Inertial Sampling Technique for Identification of Submicron Particle Content in High-Purity Gas Streams and Clean Rooms. In: *Proceedings - Inst. Env. Sci. - 1985*, pp. 99-101.

Kasper, G., H. Y. Wen, and S. Chesters. 1986. Identification of Particle Sources in Clean Gases. In: *Proceedings - Inst. Env. Sci. - 1986*, pp. 475-477.

Katz, M., ed. 1977. *Methods of Air Sampling and Analysis*. American Public Health Association, Washington, DC.

Kaufmann, R., P. Wieser, and R. Wurster. 1980. Application of the Laser Microprobe Mass Analyzer (LAMMA) in Aerosol Research. In: *Scan. Elec. Microscopy 1980*, pp. 607-622.

Kelly, J. F., R. J. Lee, and S. Lentz. 1980. Automated Characterization of Fine Particulates. In: *Scan. Elec. Microscopy 1980*, pp. 311-322.

Klainer, S. M., and F. P. Milanovich. 1985. Micro-Raman Spectroscopy, The Analyses of Micrometer and Sub-Micrometer Atmospheric Aerosols. NASA Contractor Report 177964.

McCrone, W. C. 1986. Microscopy and Ultramicroanalysis. *Amer. Lab.* (April):22-33.

McCrone, W. C., and J. G. Delly. 1973. *The Particle Atlas*. Ann Arbor Science Publishers, Ann Arbor, MI.

Mitchell, J. W. 1973. Ultrapurity in Trace Analysis. *Anal. Chem.*, 45(6):492A-511A.

Muggli, R. Z., and M. E. Anderson. 1985. Raman Micro-Analysis of Integrated Circuit Contamination. *Solid State Tech.*, 28(4):287-291.

Phillips, Q. T., W. D. Auser, J. M. Baldwin, and G. J. Washington. 1983. Cosmetics in Clean Rooms. *J. Inst. Env. Sci.* (Sept/Oct.):27-31.

Schmidt, J., S. Beaudet, and G. Madore. 1987. Laser Microprobe Mass Analysis of Microcontaminants. *Microcontamination*, 5(8):42-45.

Steen, D., M. P. Guillemin, P. Buffat, and G. Litzistorf. 1983. Determination of Asbestos Fibres in Air; Transmission Electron Microscopy as a Reference Method. *Atm. Env.*, 17(11):2285-2297.

VanderWood, T. B. 1985a. Identification of Particulate Contaminants in IC Manufacture. *Solid State Tech.*, 28(8):177-182.

VanderWood, T. B. 1985b. Methods of Microparticle Identification. In: *Proceedings - Microcontamination Conf. 1985*, pp. 26-26.

Wutjin, U., E. Bombelka, F. W. Richter, and H. Ries. 1986. PIXE Analysis of High Volume Aerosol Samples and Intercomparison of Results with Different Analytical Methods. In: *Tenth Ann. Conf. of Assoc. for Aer. Res.*, pp. 305-308.

21. PARTICLE ADHESION TO SURFACES: THEORY OF CLEANING

VENUGOPAL B. MENON Sematech, Austin, Texas

INTRODUCTION

A necessary prerequisite to obtaining high yields in very large-scale integrated (VLSI) fabrication is the ability to keep wafers scrupulously clean. Particles and films represent the two major categories of contaminants encountered in the microelectronics industry. Prevention of particle deposition onto surfaces is the best approach to maintaining a clean surface; however, so long as particle deposition cannot be completely eliminated, methods to clean surfaces will have to be used to improve product yield. Deposition of particles on silicon wafers occurs (in spite of most precautions) during various stages of the fabrication process, and the removal of these micrometer- and submicrometer-sized particles as well as other contaminants represents one of the most frequently repeated steps in the manufacturing sequence.

The nature of the particles commonly encountered in the microelectronics processing environment generally depends on the type of operation involved. For example, when silane is used for silicon epitaxy or chemical vapor deposition, and traces of oxygen are in the system, the silane will react to form particles of SiO_2 (Brown and Kamins, 1979). Other examples include rust particles formed in stainless steel plumbing used for HCl transport, metallic particles formed during epitaxial growth, etc. (see Chapter 20). In general, particulate contaminants can be classified as being inorganic, organic, or organometallic in origin. Organic contaminants include residues of photoresist or developer, skin flakes, and hair; inorganic contaminants range from silicon and quartz dust to silicon dioxide, silicon carbide, and silicon nitride. It is important to note that the forces of adhesion that hold particles on surfaces are strong functions of the particles' chemical composition, size, shape, and environment. Hence, the selection of an appropriate cleaning technology depends on a correct understanding of these factors.

The history of cleaning (which runs parallel to the history of dirt) can be traced back many centuries. Detergents have been used for surface cleaning for many years. A detergent (technically called a surfactant) molecule consists of a hydrophilic (water-loving) portion and a hydrophobic (water-hating) portion. The hydrophilic part of most effective detergents is usually an ionic group. Ions have a strong affinity for water owing to their electrostatic attraction to the water dipoles. Figure 21-1 depicts the mechanism of detergency (Kushner and Hoffman, 1951). A detergent removes dirt from a surface by altering the surface chemical properties of the dirt particles and the substrate; i.e., detergent molecules adsorb on the material to be cleaned, change the surface tension and wettability of the system, and cause dirt particles to be repelled from the surface. This method of cleaning is suitable for the removal of all soluble impurities and insoluble particles larger than a few micrometers. For silicon wafers, the use of detergents has two major disadvantages: they are not very effective for removing micrometer- and submicrometer-sized particles, and they tend to leave an undesirable residue (adsorbed surfactant molecules) on the wafer surface which is extremely difficult to remove. However, small amounts of surfactants are still used in some commercial scrubbers, mostly for improving the wettability of the wafer to the cleaning medium.

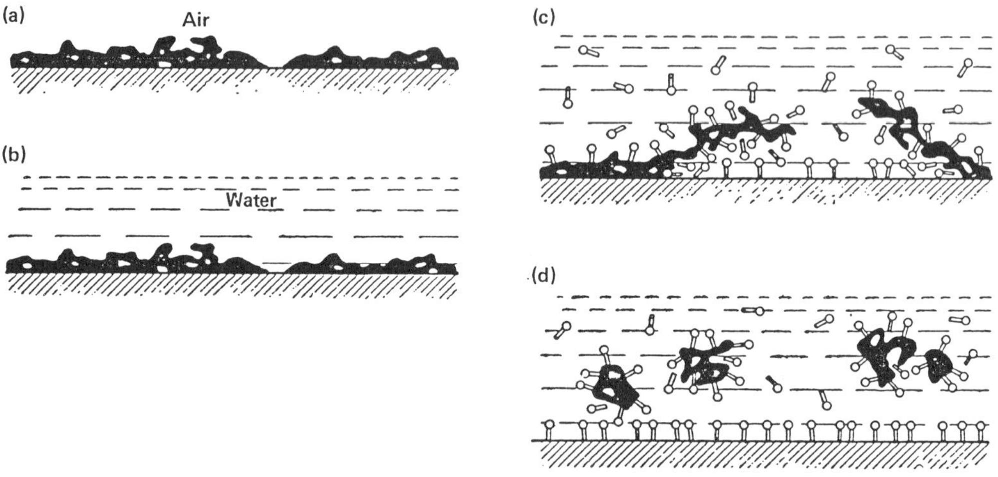

Source: Kushner and Hoffman, 1951

Figure 21-1. Removal of dirt from a solid surface by detergent and mechanical action. (a) Surface covered with greasy dirt. (b) Water by itself fails to dislodge the dirt mainly because of its high surface tension and inefficient wetting action. (c) Detergent is added to the water. The hydrophobic parts of the detergent molecules line up both on the dirt and on the solid surface, thus reducing the adhesion of the dirt to the solid. The dirt may now be dislodged by mechanical action. (d) Dirt is held suspended in the solution because the detergent molecules form an adsorbed layer on the cleaned surface and around the dirt particles.

The most prevalent approach to wafer cleaning is to either immerse or spray the wafer with liquid chemicals, the objective being the removal of organics, alkali ions, and metals. These chemicals react selectively with surface contaminants to cause either dissolution in the solvent or dispersion in the bulk liquid (Ruzyllo, 1987). An example of such chemical cleaning is the hydrogen peroxide-ammonium hydroxide solution developed at the RCA laboratories in the 1960s (Kern and Puotinen, 1970). The RCA clean strips mobile ions and metals trapped on the wafer surface and includes a step for regrowth of the native oxide layer on the silicon surface, which passivates the surface. Such techniques generally do not work for insoluble particulate contaminants because the force required to detach particles from the surface is not supplied by the immersion process. In fact, some of these chemical cleaning steps have been known to add particles to the wafer surface (Cooper, 1986; Bansal, 1984). It is only in recent years that systems designed specifically for particulate removal from silicon wafers have become commercially available.

Table 21-1 depicts the influence of particle size on the force of adhesion (Monkowski, 1987). Particles larger than a few micrometers are quite readily removed by the various wet rinsing operations. The removal of smaller particles, however, requires the expenditure of a considerable amount of energy. Table 21-2 shows the size of particles that will lead to killer defects within the next few years. Comparison of the data in Tables 21-1 and 21-2 reveals the magnitudes of force that will have to be supplied for cleaning these submicrometer particles from wafers.

In this chapter, a discussion of the forces of adhesion of particles on surfaces is provided, followed by a description of some of the commercially available cleaning techniques.

THEORY OF PARTICLE ADHESION

The principal interactions associated with particle adhesion include molecular (van der Waals) interactions, electrostatic interactions, hydrophobic/hydrophilic interactions, steric interactions, and chemical bonding. A large portion of the discussion on theories of adhesion has been adapted from recent review articles by Ranade (1987), and Bowling (1985).

Table 21-1. Relative Adhesional Force (Force/Particle Weight) for Quartz Bead on Glass Slide in Air at T = 25 °C, R.H. = 95%

Bead size (μm)	Relative adhesional force (gravity units)
100	510
50	2,159
10	57,716
1	674,600
0.1	749,552,300

Source: Monkowski, 1987.

Table 21-2. Defect Size vs. Design Rule Size (μm)

Size	Year				
	1985	1987	1988	1989	1990
Design rule	2.50	1.25	1.00	0.80	0.60
Defect	1.00	0.25	0.14	0.10	0.05

Source: Irwin, 1987.

van der Waals Molecular Interactions

Atoms in interacting bodies are instantaneous dipoles, and the summation over all atoms of the dispersive interactions between these dipoles and those induced in neighboring atoms is referred to as the van der Waals dispersive interaction. For any material, the van der Waals interaction is characterized by a descriptor known as the Hamaker constant. Thus, every interacting system can be described in terms of a Hamaker constant that is a function of the Hamaker constants of the interacting bodies and the intervening medium. The force of adhesion is a function of the Hamaker constant, the geometries of the interacting bodies, the distance of separation between them, and the properties of the intervening medium (Menon et al., 1987, 1988; Ranade et al., 1986).

Two major theories—the London and the Lifshitz—have been developed for expressing the Hamaker constant in terms of the molecular properties of a medium. The London theory (also referred to as the microscopic theory) is based on the pairwise addition of interactions between atoms of neighboring bodies, and the Lifshitz approach (also known as the macroscopic approach) has its basis in quantum electrodynamics (Zimon, 1982). For most practical situations, the two theories predict the same values of the Hamaker constant.

The van der Waals force of adhesion between particles of various geometries and a planar surface is depicted in Figure 21-2 (Mullins et al., 1988; Ranade, 1987). In Figure 21-2, d_p and d_c represent the major dimensions of the particle, Z_o is the smallest separation distance between the particle and the surface, and A_{132} is the Hamaker constant for the composite system. The relationship between the Hamaker constants of two dissimilar materials is usually expressed as the geometric mean of the individual Hamaker constants. Thus,

$$A_{12} = (A_{11} \bullet A_{22})^{1/2} , \qquad (21\text{-}1)$$

where A_{11} and A_{22} are the Hamaker constants for substances "1" and "2." When the two interacting bodies are separated by a medium "3," the Hamaker constant is given by:

$$A_{132} = A_{12} + A_{33} - A_{13} - A_{23} . \qquad (21\text{-}2)$$

Table 21-3 presents values of Hamaker constant for a variety of solids and liquids that are commonly encountered in the microelectronics industry. The constant is a function of particle and surface compositions, the hardness of the materials involved, and the shape and surface topography of the interacting bodies. Metals have high constants, while polymers have low constants, and semiconductors lie between these values. It should be noted that there could be large variations in the measured values of A_{11} depending on the whether the value of Z_o was directly measured or assumed, and depending on the chemical purity of the material. Glass is a typical example of such a material. A large number of glasses are available that exhibit minor differences from each other in their chemical content. These differences have resulted in a wide range of Hamaker constants, as seen in Table 21-3. The effect of medium on the Hamaker constant and the force of adhesion for a sphere-planar surface configuration is shown in Table 21-4 for 1-μm silica particles on silicon (corresponding to a bare wafer) and silica (corresponding to a silicon wafer with a thick oxide layer). The force of adhesion

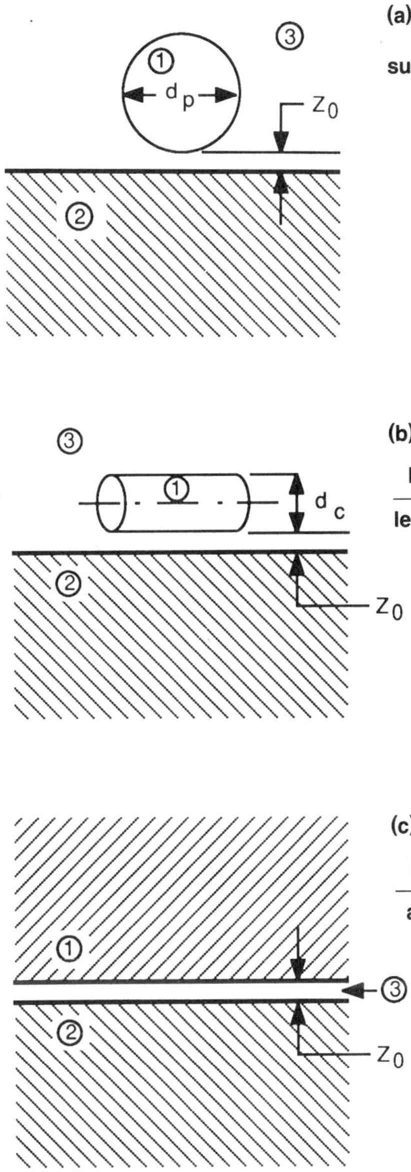

(a) sphere–planar surface, $F_{Ad} = \dfrac{A_{132} d_p}{12 Z_0^2}$

(b) cylinder–planar surface, $\dfrac{F_{Ad}}{\text{length}} = \dfrac{A_{132} d_c^{1/2}}{16 Z_0^{5/2}}$

(c) two planar surfaces, $\dfrac{F_{Ad}}{\text{area}} = \dfrac{A_{132}}{6\pi Z_0^3}$

Source: (b) Langbein, 1972
(c) Ranade, 1987

Figure 21-2. van der Waals molecular interactions.

Table 21-3. Hamaker Constants for System Components

Material	$A_{11} \times 10^{13}$ ergs	References
Silicon	24.8	Van den Tempel (1972) Boehme, Krupp, and Schnabel (1969)
Silica (SiO_2)	8.5	Visser (1976), Buettner and Gerlack (1970)
Glasses	2.0 3.4 6.3	Zimon (1982) Van den Tempel (1972) Israelachvili (1985)
Polystyrene latex	6.5	Visser (1976) Van den Tempel (1972)
Water	4.4	Visser (1976) Van den Tempel (1972)
Ethanol-acetone	3.3	Estimated from Fowkes' (1964) equation
Freon-TMS	1.8	Estimated from Fowkes' (1964) equation
Freon-TF	1.7	Estimated from Fowkes' (1964) equation

Table 21-4. Force of Adhesion for Test Systems

	Hamaker constant $(A_{132})^a$ (ergs x 10^{13})		F_{AD}[b] (mdyn)	
Fluid	Bare Si surface	Oxidized surface (0.5 μm)	Bare Si surface	Oxidized surface (0.5 μm)
	1 μm glass microspheres[c]			
Air	14.51	8.5		
Water	2.42	0.63	0.4	0.1
Ethanol-acetone	3.58	1.21	0.6	0.2
Freon-TMS	5.89	2.49	1.0	0.4
Freon-TF	5.99	2.50	1.0	0.4

[a]Hamaker constant for the system calculated using Equation 21-2.
[b]F_{AD} calculated using Equation (a) in Figure 21-2 and Z_o = 7.0 Å
[c]Density of glass, 2.65 g/cm^3.

decreases by about one order of magnitude when air is replaced by water as the intervening medium. This decrease in force of adhesion on immersion in a liquid is one of the reasons for the choice of wet cleaning over dry cleaning. The magnitudes of van der Waals forces in air or vacuum for a 1-μm particle can range from 1 mdyn to 1,800 mdyn, which corresponds to a force per unit area from about 0.2 x 10^9 to 3.0 x 10^9 dyn/cm^2 (Bowling, 1985). These are tremendous pressures, and most particles and/or surfaces can be deformed by such forces. The amount of deformation depends on the hardness of the particle and surface, and this deformation can increase with the length of time that the force acts on the particle and surface. Deformation affects the adhesion process by changing the contact area.

Electrostatic Interactions

The electrostatic force of adhesion is made up of coulombic attractions and electrostatic contact potentials. Coulombic interactions are possible only for particles and surfaces that are charged. The electrical interaction between a charged particle and an electrically neutral surface on which it rests is equivalent to a coulomb interaction between the particle and its "image," as shown in Figure 21-3.

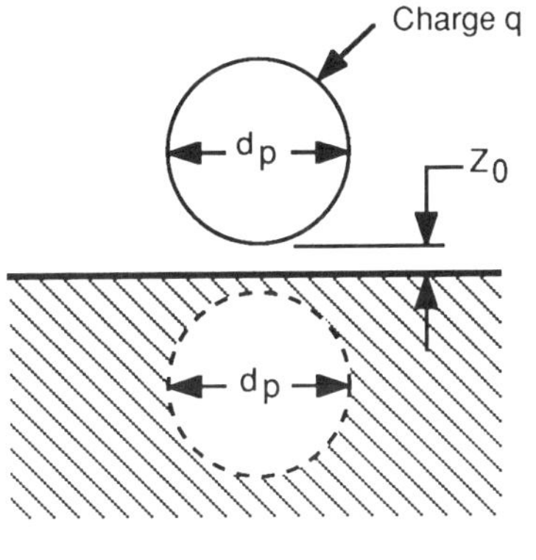

Source: Ranade, 1987

Figure 21-3. Electrostatic interaction between a charged particle and its "image."

The electrostatic adhesive force attributable to particle charge alone is given by the expression (Bowling, 1985):

$$F_{Ad} = \frac{q^2}{4\pi\epsilon \bullet \epsilon_o (d_p + 2Z_o)^2} , \qquad (21\text{-}3)$$

where ϵ is the dielectric constant of the medium between the particle and the surface, ϵ_o is the permittivity of free space, q is the particle charge, and $(d_p + 2Z_o)$ is the distance between the particle center and its image charge. The inverse dependence on the dielectric constant of the medium implies that the force will decrease as the system is moved from a gas to a liquid. In the case of air with a dielectric constant of about 1 and water with a dielectric constant of about 80, the effect is a factor of 80 reduction in image force in water compared to air (Grant et al., 1988).

The more important electrostatic force for very small particles is the electronic double-layer force induced by electrostatic contact potential. Two different materials in contact develop a contact potential caused by differences in the local energy states and electronic work functions. Electrons are transferred from one solid to the other until an equilibrium is reached and electron energy is the same across the interface between the two materials—the fermi level is constant across the interface. The resulting potential difference needed to establish a constant fermi level is called the contact potential difference, V, which generally ranges from 0 to about 0.5 V. This potential difference sets up a double-layer charge region at the interface. For semiconductors and insulators, these regions may extend into the bulk of the material, up to 1 μm or deeper; however, in metals, only the surface layer carries contact charges. For a spherical particle on a surface, this double-layer force can be calculated as (Bowling, 1985):

$$F_{el} = \frac{\pi \epsilon_o d_p V^2}{2 Z_o} . \tag{21-4}$$

For a maximum contact potential of 0.5 V and a separation distance Z_o = 4 A, the value of the force is about 1 mdyn for a 1-μm particle. This value may become comparable to the van der Waals force. This type of double-layer force is directly proportional to particle diameter, being greater for larger particles. The contact potential is a function of the materials involved and is sensitive to the presence of adsorbed impurities. A comparison of the magnitudes of the various forces will show that van der Waals forces predominate over electrostatic forces for very small particles (< 1 μm). Also, double-layer electrostatic forces are much greater than electrostatic image forces for small particles (Bowling, 1985).

Capillary Condensation

A particle/surface system that is exposed to humid conditions, or immersed and with-drawn from a liquid, can exhibit the presence of a liquid film around the region of contact (see Figure 21-4). The film, or meniscus, draws the bodies together due to surface tension and reduces the pressure of the liquid. Capillary condensation is an attractive force and is a function of the particle radius, the liquid surface tension, and the contact angles of the two materials. For wetting liquids, the contact angle approaches zero, and the capillary force is given by:

$$F_c = 2\pi d_p \gamma \quad . \tag{21-5}$$

where γ is the liquid surface tension.

For water, the surface tension is 72 dynes/cm, and for a 1-μm particle F_c is approximately $4.5\text{x}10^{-2}$ dyn, which is significant when compared with van der Waals forces.

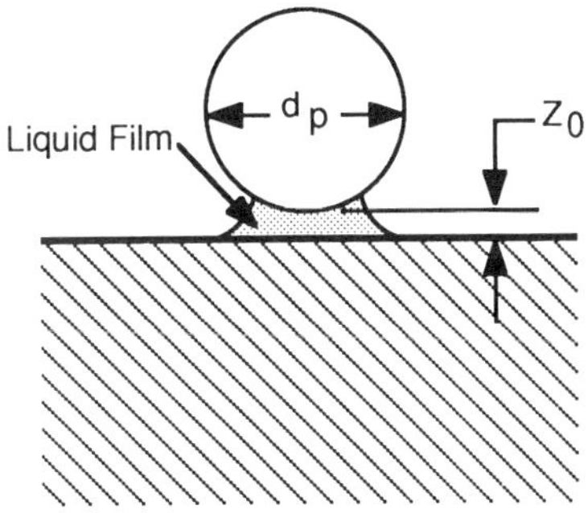

Source: Ranade, 1987

Figure 21-4. Capillary force.

The effects of capillary condensation may be prevalent even after baking the materials at temperatures above the boiling point of the liquid (Bhattacharya and Mittal, 1978). If the liquid contains species that can crystallize during evaporation, a solid bridge might form during drying and increase the force of adhesion. While immersion in a liquid generally lowers the force of adhesion (due to a decrease in Hamaker constant), it should be borne in mind that if particles are not removed from the substrate while it is immersed, the added capillary force upon removal from the liquid can increase the total adhesion force by an order of magnitude or more. Capillary condensation forces are generally not observed for surfaces that are hydrophobic.

Hydrophobic/Hydrophilic Interactions

An unusually strong attraction has been observed between nonpolar (hydrophobic) molecules and surfaces in water. This phenomenon is referred to as the hydrophobic interaction. Israelachvili (1985) attributes this type of interaction to an entropic phenomenon, which arises primarily from the structural rearrangement of water molecules in the overlapping solvation zones as two hydrophobic species come together. Such hydrophobic interactions are attractive because the approach of the two interacting surfaces close to each other will lead to the ejection of water molecules in the bulk, thus reducing the total free energy. Israelachvili and Pashley (1982) have experimentally measured these forces in the range of 0 to 10 nm between mica surfaces covered with hydrophobic monolayers. They found that hydrophobic interactions act over long distances (when compared to van der Waals interactions) and often result in the rapid coagulation of hydrophobic particles in water. In experiments conducted at RTI (Menon et al., 1988), hydrophobic polystyrene particles were found to agglomerate into clusters of 10 or more particles when a wafer with an initial deposit of single particles was immersed in water.

In contrast to the strong attraction exhibited by hydrophobic surfaces, certain molecules and groups tend to be hydrophilic and strongly repel each other in water. This is because hydrophilic (i.e., water-loving) groups prefer to be in contact with water rather than with each other. A silicon wafer with a layer of surface oxide is wetted by water and can exhibit repulsive hydrophilic interactions with other hydrophilic materials, including particles of SiO_2.

Hydrophobic and hydrophilic interactions are directly dependent on the wettability of the interacting bodies in water. Although contact angle measurements can help determine the existence of these forces, no theoretical models have as yet been developed to provide a quantitative understanding of these phenomena. Table 21-5 depicts the contact angles, measured through the aqueous phase, for materials that may be relevant to semiconductor processing. Silicon wafers with oxide layers generally exhibit lower adhesional forces in water and favor the detachment of hydrophilic and hydrophobic particles. Hydrophobic particles such as silicon dust or polystyrene prefer to agglomerate with each other rather than adhere on the hypdrophilic wafer surface. The agglomerates being larger in size are more easily removed during wafer cleaning. Hydrophilic particles such as SiO_2 prefer to be dispersed in the bulk aqueous phase rather than stay on the wafer surface because of hydrophilic repulsion. Thus, either hydrophobic or hydrophilic particles on a hydrophilic wafer surface will decrease the overall force of adhesion, while hydrophobic particles on a hydrophobic surface will increase the overall adhesional force.

Ionic Double-Layer Interactions in Liquids

In contrast to the contact potential double-layer interactions previously discussed, double-layer interactions in liquids owe their origin to the intrinsic charges that are on the contaminant particles and the wafer surface because of the presence of ions in the liquid phase. The double-layer interaction previously discussed originates with electronic redistribution; double-layer interactions in liquids arise because of ionic distributions. For a particle and substrate charge of the same sign, the ionic double-layer interaction is repulsive in

Table 21-5. Zeta Potential and Contact Angle of System Components in Water

Material	ξ (mV)	Contact angle
Bare wafer	-39.3	15° (with native oxide)
		87° (no native oxide)
Oxidized wafer	--	27°
Glass	-26.6	8°
Polystyrene latex	-22.2	91°

nature. Thc surface charge is often expressed in terms of the zeta potential, which is the electric potential measured at the shear boundary between (1) the particle or wafer surface including its ion cloud and (2) the liquid media through which it moves. Figure 21-5 is a schematic diagram of the electrostatic double-layer around a particle.

For interactions between a spherical particle and a planar surface, the force of repulsion is given by:

$$F_{ES} = \frac{2\pi\epsilon_o\epsilon_R d_p \varsigma_p \varsigma_s \kappa\, e^{-\kappa Zo}}{1 + e^{-\kappa Zo}} , \qquad (21\text{-}6)$$

where ϵ_R is the dielectric constant of the liquid, ϵ_o is the permittivity of free space, ς is the zeta potential of the particle, and κ is the reciprocal of the double-layer thickness (which can be estimated from the electrolyte concentration in solution and the ionic conductance). Table 21-5 gives typical values of zeta potential in DI-water of pH 7.0, for systems relevant to semiconductor processing. The zeta potentials are all of the same sign (negative) and have similar magnitudes, which implies the existence of repulsive forces of the same magnitude. Double-layer interactions in liquids can be quite large in comparison to van der Waals attraction, and they occur over greater distances of separation. For interactions between 1 μm glass or polystyrene particles and a bare silicon wafer, the force of repulsion can be of the order of 0.01 mdyn.

In addition to the interactions discussed above, specific systems may exhibit the presence of magnetic interactions, sintering, physical entrapment, and chemical bonding. However, for a dry system, the total forces of adhesion are a combination of van der Waals forces and electrostatic forces. For particles smaller than 50 μm, this total adhesional force greatly exceeds the force due to gravity. For a 1-μm silicon particle on a dry silicon surface, the total force of adhesion exceeds the gravitational force by a factor greater than 10^7 (Bowling, 1985).

CLEANING TECHNIQUES

Removal of particles from the wafer surface is accomplished by providing an external force that equals or exceeds the force of adhesion. External forces can be divided into three categories: vibratory or mechanical action, hydrodynamic or aerodynamic drag, and wet or dry chemical action. Wafer cleaning techniques include scrubbing, ultrasonics, megasonics, high-pressure air or liquid jets, ultraviolet (UV)/ozone cleans, and wet chemical cleans such as the RCA method. To select the right cleaning process, one must consider the following factors (Bhushan, 1979):

- Mechanical, chemical, and geometrical properties of the substrate
- Mechanical, chemical, and geometrical properties of the contaminants
- Required degree of cleanliness
- Effect of cleaning on the mechanical and chemical properties of the substrate

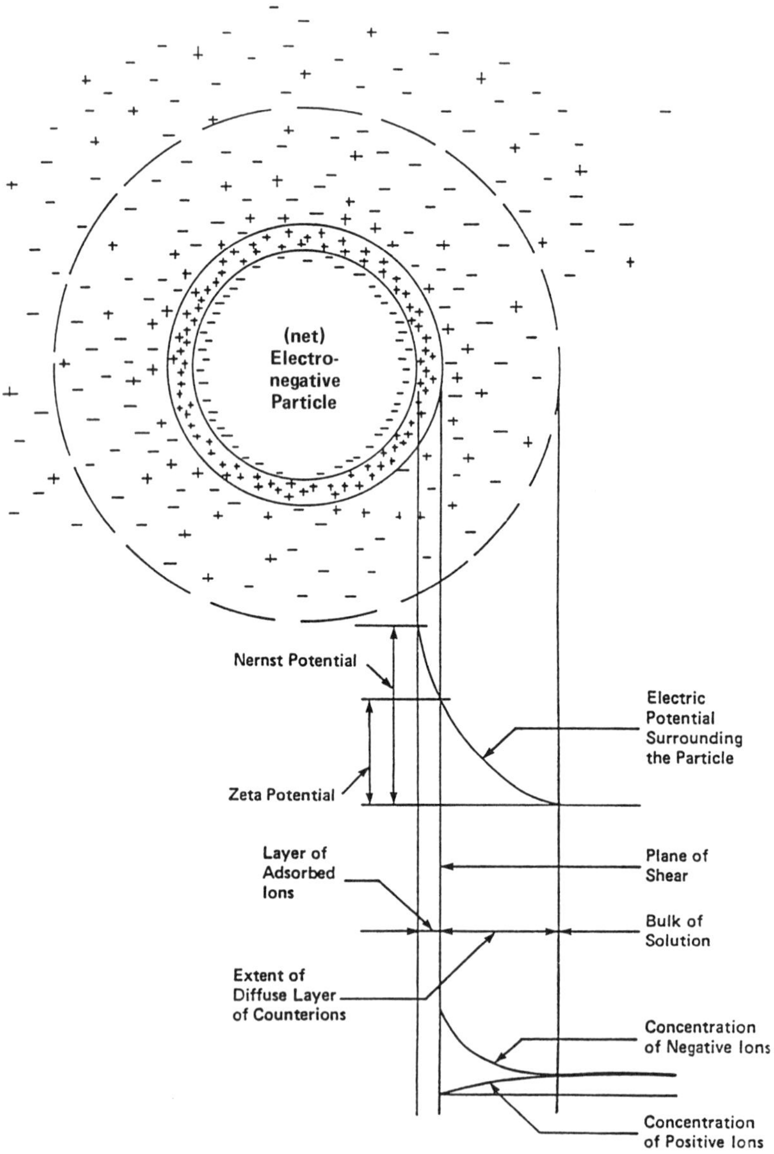

Source: Ranade, 1987

Figure 21-5. Electrostatic double-layer around a particle.

- Cost/advantage effectivness of the cleaning technique
- Quality control on the procedure.

Table 21-6 lists some of the common wafer cleaning techniques and the nature of the cleaning mechanism. In Table 21-7, the advantages and disadvantages of these cleaning techniques are briefly outlined. In the following paragraphs, some of these techniques are discussed in greater detail.

Wet Chemical Cleaning

Surface cleaning by immersing or spraying a wafer with liquid chemicals is one of the most common methods of decontamination. Such chemical cleans are good for removing soluble contaminants, but are very poor removers of insoluble particles. A very common prediffusion clean consists of the four-part RCA method. This method consists of the following operations:

- Removal of residual heavy organic contamination using a 4 H_2SO_4:1 H_2O_2 solution
- Stripping of the native oxide layer on silicon using 1 HF:100 H_2O, to remove mobile ions and metals trapped in the oxide
- Regrowth of native oxide on silicon surface with 1 NH_4OH:1 H_2O_2:5 H_2O (65 °C)
- Reduction of metal ions not trapped in the oxide with a solution of 1 HCl:1 H_2O_2:5 H_2O (65 °C).

Table 21-6. Mechanism of Action of Common Wafer Cleaning Techniques

Technique	Cleaning Action
RCA method and its variations	Chemical dissolution
Scrubbing	Mechanical force, often combined with hydrodynamic drag force
Pressurized fluid jets	Hydrodynamic and/or centrifugal drag force
Ultrasonics	Vibrational force
Megasonics	Vibrational force
UV/ozone	Chemical decomposition

Table 21-7. Advantages and Disadvantages of Common Wafer Cleaning Techniques

Cleaning method	Advantages	Disadvantages
Wet chemical cleaning	• Removes metal ions and soluble impurities	• Not good for particle removal • Requires wafer rinsing and drying • Known to add particles to wafer
Scrubbing	• Removes large particles (> 1 μm) effectively • Suitable for cleaning hydrophobic wafer surfaces	• Unsuitable for removing submicrometer particles from patterned wafers • Requires regular maintenance • Could damage wafers
Pressurized fluid jets	• Removes small particles from patterned wafers	• Possibility of wafer damage • Static buildup
Ultrasonics	• Removes small particles effectively	• Wafer damage possible • Control of cavitation intensity difficult • Unable to use reactive solutions
Megasonics	• Removes both soluble and insoluble contaminants • Removes small particles effectively • Can be used with chemical cleaning solutions • Less likelihood of wafer damage	• Contamination from chemical solutions possible • Wafer rinsing and drying required
Strippable polymer	• Capable of removing both soluble and insoluble impurities • Convenient for wafer storage and transport • Dry technique	• Possibility of depositing polymeric residue • Effectiveness not demonstrated adequately
UV/ozone	• Removes organic impurities • Dry technique	• Not proven for particle removal • Possibility of wafer damage

Some of the disadvantages of wet cleaning are given below (Ruzyllo, 1987).

- *Particles*: Wet cleaning methods, especially the chemical cleans, while being effective for removing soluble contaminants, are known to add particles to the wafer (especially the $HCl:H_2O_2$ step). This is mainly due to the inability of vendors to supply chemicals with zero particle counts and the chemical properties of the wafer surface subsequent to an HF etch. Even in the highest quality process chemicals, the count of particles 0.5 μm or larger is still in the range of tens to many hundreds per milliliter.

- *Wafer Drying*: Wet cleaning and deionized water rinse inherently must be followed by a wafer drying process. Particulate contamination during the drying operation is common and difficult to avoid.

- *Process Incompatibility*: More and more steps in the wafer processing sequence are conducted in the gas phase, and in the coming years, wafer cleaning may be the only wet process. Under such a scenario, there will be a strong impetus toward developing dry cleaning methods.

Wet chemical cleans would be attractive for detaching both soluble and insoluble impurities if they are combined with other cleaning operations designed specifically for particle removal.

Scrubbing

Scrubbing is a mechanical technique in which a brush rotates across the surface of a wafer, thereby removing inorganic particles and organic films. The brushes, usually made of nylon, polypropylene, camelhair, or mohair, never touch the wafer surface. Because the brushes are made of hydrophilic material, each bristle moves across the wafer like an automobile tire that hydroplanes on a wet street (Burggraaf, 1981). The hydrophilic brush material ensures that a cushioning film of scrubbing solution always remains between the two hard surfaces, i.e., the wafer and the brush bristles. The wafer does not always provide this film because it can be either hydrophilic or hydrophobic depending on its point in wafer processing. Also, with hydrophilic bristles, scrubbing works best when wafers are hydrophobic. As hydrophilic bristles move across a hydrophobic wafer, the film of solution surrounding each bristle is repelled from the wafer surface, thus sweeping suspended particles with it.

Most commercial scrubbers are sold either as components to automatic resist-coating systems or as stand-alone machines. Also, many scrubbers have the capability to clean both sides of a wafer simultaneously. A major disadvantage of the scrubber design is that it is unsuitable for removing submicrometer-size particles from patterned wafers. This is because the bristles that are several mils in diameter will ride over a several-micrometer-in-size feature on a wafer, missing debris lodged behind polysilicon or metal steps or in contact windows. An additional problem associated with scrubbing is the maintenance of scrubbers. The brushes load up with dirt particles, and wafers are periodically chipped or broken, showering debris over the brushes. If the brushes are not replaced frequently, they become a source of scratches and particle contamination (Shwartzman, Mayer, and Kern, 1985). Under such conditions, a high-pressure liquid jet, used alone or in conjunction with brushes, is preferred.

Brush scrubbing generally uses DI-water with some detergent to lower the surface tension of water. Typically, the scrubbing solution is a mixture of a low-ionic surfactant such as Triton-X, ammonium hydroxide, and water, and this mixture is dispensed from a pressurized canister. For example, a cleaning process may consist of a scrub with a low-ionic detergent, a continued scrub with DI-water, and then a final DI-water rinse before spinning dry. Other solvents such as isopropyl alcohol and methanol are also used to help remove organic films and speed up the drying process.

High-Pressure Fluid Jets

Air or nitrogen blowoff guns have been in use for many years for removing articles from surfaces. However, they are only suitable for removing large particles (> 10 μm) and have been known to cause static charge buildup. High-pressure liquid jets are capable of removing smaller particles. These jets are gaining popularity over brushes due to the inefficiency of the latter for submicrometer-size particles and due to the danger of wafer damage owing to the mechanical action of brushes.

Dislodging of particles from a surface by the mechanical action of a liquid jet is caused by the change in velocity of the particle relative to the surface. The force exerted on the particle is the momentum force and is proportional to the mass of the particle. When this force exceeds the adhesion force between the particle and the surface, the particle is dislodged.

The drag force, F, applied to a particle by a moving fluid is given by Stokes' law:

$$F = \frac{3 \pi \mu d_p V}{C_c}, \qquad (21\text{-}7)$$

where μ is the fluid viscosity, d_p the diameter of the particle, V the relative fluid-particle velocity, and C_c the slip correction. The force on the particle increases with fluid viscosity and velocity and with particle size. Thus, the drag force on particles in the path of a liquid jet is much higher than that for a gas jet because liquids have much higher viscosities than gases. Comparing air at one atmosphere pressure and liquid water at 20 °C, the drag force on a particle in water at comparable velocities is about 55 times that of the drag force in air (Grant et al., 1988).

However, there is a boundary layer of fluid on the surface of a wafer, so that particles on the surface are not subject to the full force of the fluid moving across the wafer surface. The boundary layer thickness is a function of the direction of fluid flow relative to the surface, but in general it is proportional to the square root of the kinematic viscosity, which is the fluid viscosity divided by the fluid density. Comparing air and water again, the square root of the kinematic viscosity of water is 0.25 times that of air, so that the boundary layer thickness of water is a factor of four less than that of air. Hence, the drag force experienced by a particle on a surface in a liquid is higher than that in air because the liquid boundary layer is thinner at comparable velocities.

The effect of the boundary layer is to shield particles from the full force of the fluid moving past the surface of the wafer. Smaller particles will be shielded more than larger particles because a larger fraction of their surface area will be inside the slower moving boundary layer (Grant et al., 1988). This implies that greater fluid velocities will have to be provided for detaching smaller particles from the surface. Figure 21-6 depicts the influence of the boundary layer on particle removal from a flat surface. Pressures as high as 3,000 psi may be required for removing submicrometer particles from a wafer. At such high pressures, the possibility of wafer damage is very real.

Several design features are important on high-pressure liquid jet cleaning systems. First, the nozzle should be located as close as possible to the wafer surface to ensure a minimum loss of fluid momentum. It is also important that the nozzle(s) be capable of sweeping across the entire surface of the wafer at various speeds and angles of incidence. A major problem with high-pressure scrubbing is static electricity, which is a universal problem with high-pressure jets and is a function of the pressure and resistivity of water. Most users of such systems add special additives to the water, such as CO_2, to reduce the water resistivity and static charge problem.

Ultrasonics

Ultrasonics uses sonic energy—20 kHz and up—to dislodge particles from the wafer surface. The mechanical disturbances that are caused by high-intensity sound waves produce pressure fluctuations above and below their ambient pressure. These pressure fluctuations result in the formation of cavitation bubbles. In the reduced-pressure cycle, bubbles tend to form and grow, whereas in the increased-pressure cycle, bubble growth stops and bubbles either oscillate or collapse. At the collapsing bubble interface, sufficient energy is released in a localized region to overcome the force of adhesion and to loosen particulate matter. To prevent the redeposition of particles from the liquid onto the wafer surface, the particles must be removed from the liquid through overflow or filtration after they are initially detached.

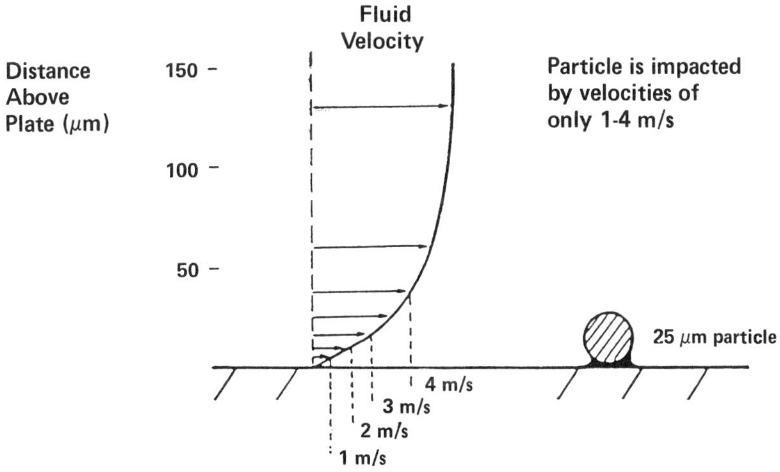

Source: Musselman and Yarbrough, 1987

Figure 21-6. Effect of boundary layer on flow profile.

The force required to dislodge a particle in an ultrasonic field, in the absence of any cavitation, is given by:

$$F = m \cdot a \quad , \tag{21-8}$$

where

$$a = 4\pi^2 f^2 A \quad , \tag{21-9}$$

and f is the frequency of vibration, A is the amplitude of vibration, and m is the mass of the particle. When cavitation is present, the above equation does not adequately represent the force imparted to the particle. There are no satisfactory theories to predict particle detachment by cavitation bubbles.

Wafer manufacturers use ultrasonics to remove large particles from wafers after sawing. It is also used after lapping and polishing to remove compounds used in these processes. Conventional ultrasonic units are not very popular in semiconductor fabrication because the frequency range tends to be near the natural frequency of some of the layers that are being cleaned, such as poly, nitride, and metallization layers. This increases the possibility of wafer damage. There are commercial ultrasonics cleaners specially designed for controlled wafer cleaning. One such unit is the acoustic cleaner manufactured by Estek of Rochester, New York, which is reportedly able to control the zone in which cavitation bubbles are created. This unit operates at a frequency of 20 kHz and consists of a tank shaped like a swimming pool. It has a deep end and a shallow end. Wafers are introduced one at a time, into the deep end, with the wafer face parallel to the surface of the cleaning tank. Cavitation bubbles are created by the transducer under which the wafers pass in a field of acoustical energy, and both sides of the wafer are cleaned uniformly (Skidmore, 1987).

Disadvantages of the ultrasonic or acoustic cleaning techniques include the possibility of damaging the wafer surface, the inability of processing large numbers of wafers simultaneously, and the difficulty of using reactive solutions such as those used during wet chemical cleaning.

Megasonics

Megasonics is similar to ultrasonics in that it employs sonic energy, but the operating frequency of 800 kHz is much higher than that of an ultrasonic unit. The megasonic cleaning system employs a noncontact apparatus for removing particles and films from both sides of a silicon wafer simultaneously. Sonic energy from the transducer array panels is directed parallel to the surface of the wafers which are immersed in the media liquid. The scrubbing action is provided by piezoelectric transducers that produce short sonic waves at a frequency of approximately 750 to 900 kHz. Particles are physically forced from the wafer surface by the very high-frequency sonic pressure waves generated in the bath liquid. Figure 21-7a is a photograph of a typical megasonic wafer cleaner. The chemical tank, with the transducer panel, is shown on the left, while the rinse tank is on the right. Figure 21-7b is a magnified view of the region around the transducer panel while it is vibrating.

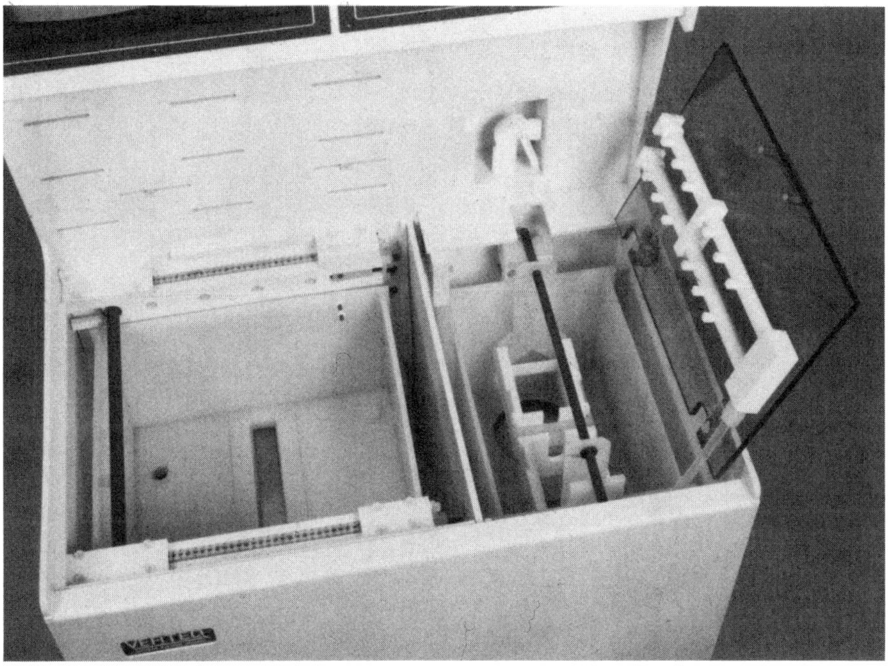

Source: Verteq Inc., Anaheim, CA.

Figure 21-7a. Photograph of a megasonic wafer cleaner.

For comparable cleaning efficiencies, the power intensities used in a megasonic unit are typically about 10 to 50 times lower than those used in ultrasonic cleaners. The reason for this increased efficiency lies in the mechanism of particle detachment. The higher frequency used in megasonics results in a high-pressure wave being set up in the solution rather than bubbles. The high-pressure waves push and tug at particles lodged on the wafer surface at a rate of approximately 800,000 times/s (Skidmore, 1987). Calculations based on a transfer of 7.5 W/cm^2 to the wafer surface suggest that the water molecules move only about 0.1 μm in each cycle, but that their acceleration is on the order of 10^5 g. The maximum instantaneous velocity is about 30 cm/s. It appears that the time between pulses, 1.25 μs, is too short for the formation of cavitation bubbles.

Shwartzman, Mayer, and Kern (1985) report that the best cleaning efficiencies were obtained when the liquid in the megasonic tank can wet the substrate and particle surfaces. They suggest the use of an ammoniacal hydrogen peroxide solution (one of the steps of the RCA method) as the medium. During the cleaning process, the particles appear to move sufficiently to permit the solvent to diffuse rapidly into the interface between the wetted substrate and the particle. The liquid film acts as a wedge and progressively reduces the contact area until the particle becomes dislodged, is free to move, and becomes suspended in the solution.

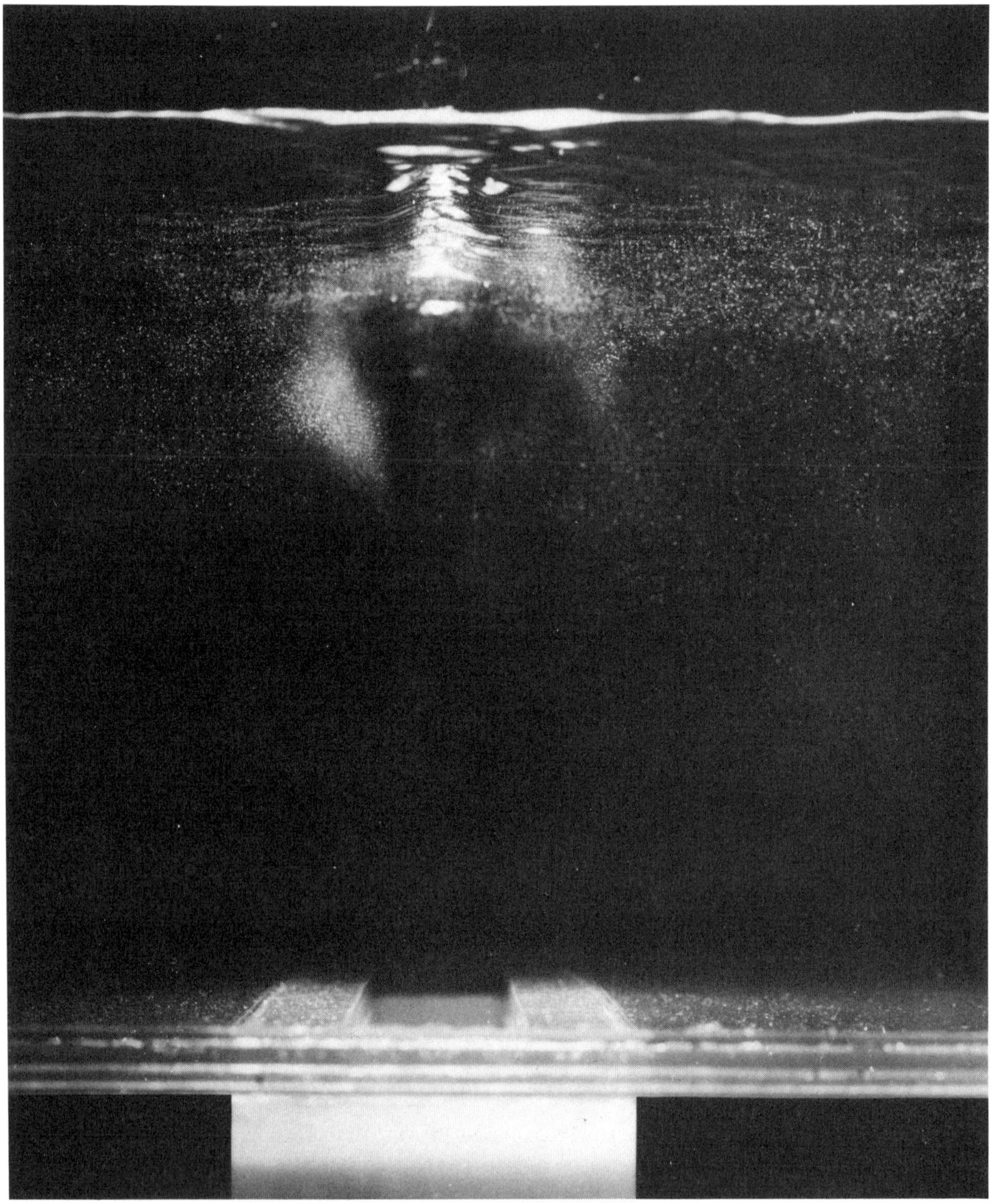

Source: Verteq Inc., Anaheim, CA.

Figure 21-7b. A magnified view of the megasonic transducer while it is vibrating in a chemical bath.

The megasonic cleaning technique possesses distinct advantages over the ultrasonic or acoustic cleaning systems. The absence of cavitation reduces the probability of wafer damage; the ability of using chemical solutions such as the ammoniacal hydrogen peroxide solution enables the simultaneous removal of particles and organic films; and the system design permits cleaning of multiple wafer carriers, thus resulting in high throughputs. Chapter 22 describes the application of megasonics to wafer cleaning in greater detail.

Other Cleaning Techniques

Besides the techniques previously discussed, other methods of wafer cleaning have been proposed. These include UV/ozone, plasma, and thermal cleaning. Each of these techniques is suitable for the removal of organic films and metal ions, but not particles. Because these are dry cleaning techniques, they do not possess some of the disadvantages of wet chemical cleaning; however, significant modifications will be required before they become widely applicable for particle removal.

An unconventional technique that has recently been suggested for wafer cleaning is particle removal using a strippable polymer (Bradford Laboratories, 1988). The procedure involves the spin coating of a contaminated wafer with a proprietary hydrophilic polymer, baking the polymer for a few minutes, and then peeling to clean the wafer. A hydration step is often suggested to help strip the polymer. In a limited number of cleaning experiments conducted at Research Triangle Institute (RTI), the polymer was observed to be quite effective in removing submicrometer particles from the wafer surface. Figure 21-8 depicts printouts from a wafer surface scanner before and after cleaning with the polymer. This cleaning technique, which could potentially be effective in cleaning patterned wafers, needs to be studied in greater detail.

SUMMARY

The ability of a cleaning technique to remove particles from surfaces is dependent on many factors, such as the composition of the substrate and particles, the wettability of both components, the physical properties of the cleaning medium, and the method of application of the force for particle removal.

The various interactions that contribute to the overall force of adhesion between particles and surfaces include van der Waals, electrostatic, capillary, and hydrophobic/hydrophilic interactions. The judicious selection of a cleaning technique requires an understanding of the relative magnitudes of these interactions for various systems. Also, the possibility of wafer damage, or subsequent contamination (e.g., particles have been known to deposit on wafers during the drying process after cleaning), needs to be gauged.

Of the many cleaning techniques available, scrubbing with water or surfactant solutions is effective during the sawing, lapping, and polishing steps, while megasonics appears to be the most suitable for cleaning a large number of wafers simultaneously after various steps of the device's manufacturing process.

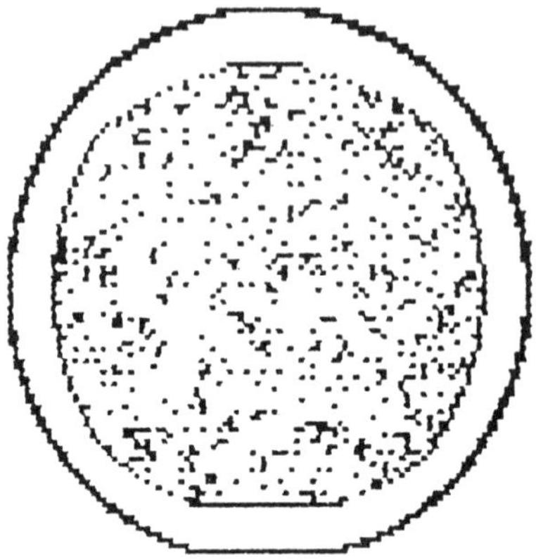

Size Range (μm)	Count
0.2 - 0.5	104
0.5 - 2.0	562
2.0 - 20	19

Before Cleaning

Wafer Size: 3.00 (75)

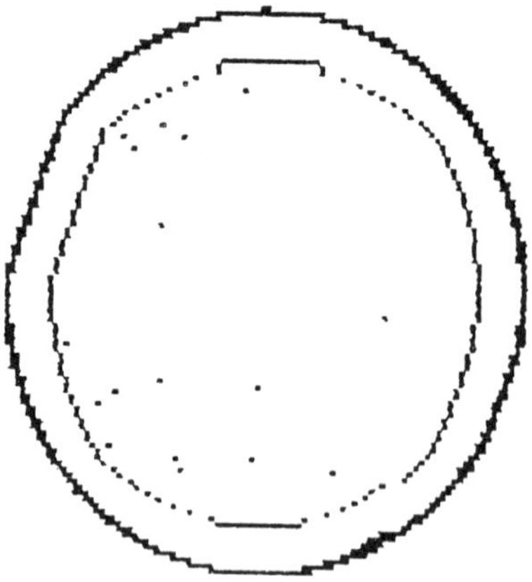

Size Range (μm)	Count
0.2 - 0.5	10
0.5 - 2.0	4
2.0 - 20	3

After Cleaning

Wafer Size: 3.00 (75)

Source: Menon and Clayton, 1988

Figure 21-8. Wafer scanner printouts before and after cleaning with polymer (no hydration before peeling).

REFERENCES

Bansal, I. 1984. "Particle Contamination During Chemical Cleaning and Photoresist Stripping of Silicon Wafers." *Microcontamination 2*:35.

Bhattacharya, S., and K.L. Mittal. 1978. *Surface Technol. 7*:413.

Bhushan, B. 1979. "Techniques for Removing Surface Contaminants in Thin Film Deposition." In: *Surface Contamination*, K.L. Mittal (ed.), Volume 1, pp. 393-406. Plenum Press, New York, NY.

Boehme, G., H. Krupp, and W. Schnabel. 1969. In: *Molecular Processes on Solid Surfaces*, E. Drauglis, R. Gretz, and R. Jaffee (eds.), McGraw Hill Book Publishing Co., New York, NY.

Bowling, R. A. 1985. "An Analysis of Particle Adhesion on Semiconductor Surfaces." *J. Electrochem. Soc: Solid-State Sci. Technol.* (Sept.):2208-2214.

Bradford Laboratories (Mountain View, CA). 1988. Personal Communication.

Brown, W. A., and T.I. Kamins. 1979. "An Analysis of LPCVD System Parameters for Polysilicon, Silicon Nitride and Silicon Dioxide Deposition." *Solid State Technol. 22*(7), 51.

Buettner, H., and E. Gerlach. 1970. *Chem. Phys. Lett. 5*:91.

Burggraaf, P.S. 1981. "Wafer Cleaning: Brush and High-Pressure Scrubbers." *Semiconductor International* (July):71-88.

Cooper, D.W. 1986. "Particulate Contamination and Microelectronics Manufacturing: An Introduction." *Aerosol Sci. Technol.* 5:287.

Fowkes, F.M. 1964. "Determination of Interfacial Tensions, Contact Angles and Dispersion Forces in Surfaces by Assuming Additivity of Intermolecular Interactions in Surfaces." *Ind. Eng. Chem. 56*:40.

Grant, D.C., B.Y.H. Liu, W.G. Fisher, and R. A. Bowling. 1988. "Particle Capture Mechanisms in Gases and Liquids: An Analysis of Operative Mechanisms." In: *Proceedings of the Institute of Environmental Sciences Annual Meeting*, King of Prussia, PA, May 2-6.

Gunawardena, S., R. Haven, U. Kaempf, M. Parikh, B. Tullis, and J. Vietor. 1984. "The Challenge to Control Contamination: A Novel Technique for the IC Process." *J. Environmental Sci.* (May):23-30.

Irwin, R.H. 1987. "The Needs and Requirements for Wafer Cleaning." In: *Proceedings of the Microcontamination Conference and Exposition, Santa Clara, October 27-30*, pp. 97-99.

Israelachvili, J.N. 1985. *Intermolecular and Surface Forces with Applications to Colloidal and Biological Systems*. Academic Press, New York, NY.

Israelachvili, J.N., and R.M. Pashley. 1982. *Nature 300*:341.

Kern, W., and D.A. Puotinen. 1970. "Cleaning Solutions Based on Hydrogen Peroxide for Use in Silicon Semiconductor Technology." *RCA Review 31*:187.

Kushner, L.M., and J.L. Hoffman. 1951. "Synthetic Detergents." *Scientific Am.* (Oct.):26.

Langbein, D. 1972. *Phys. Kondens. Materie, 15*, 61.

Menon, V.B., R.P. Donovan, L.D. Michaels, and M.B. Ranade. 1987. "Particle Removal from Semiconductor Wafers Using Cleaning Solvents." In: *Proceedings of the Symposium on Particles in Fluids: Detection, Characterization and Control*, Boston, MA, August 3-7.

Menon, V.B., and A.C. Clayton. 1988. "Evaluation of a Strippable Polymer for Silicon Wafer Cleaning." Unpublished report, Research Triangle Institute, Research Triangle Park, NC.

Menon, V.B., L.D. Michaels, R.P. Donovan, L. A. Hollar, and D.S. Ensor. 1988. "Removal Efficiencies of Organic and Inorganic Particles From Silicon Wafers," Proceedings of the Institute of Environmental Sciences, Annual Meeting, King of Prussia, PA, May 2-6.

Monkowski, J.R. 1987. "Particulate Surface Contamination and Device Failures." In: *Treatise on Clean Surface Technology*, K.L. Mittal (ed.), Volume 1, pp. 123-145. Plenum Press, New York, NY.

Mullins, M.E., B. Locke, L. D. Michaels, and V.B. Menon. 1988. "Effect of Particle Geometry on the Force of Adhesion on Surfaces." *Aerosol Sci. Technol.* (in press).

Musselman, R.P., and T.W. Yarbrough. 1987. "Shear Stress Cleaning for Surface Departiculation." *J. Environ. Sci.* (Jan./Feb.):51.

Ranade, M.B., V.B. Menon, M.E. Mullins, and V.L. Debler. 1986. "Adhesion and Removal of Particles: Effect of Medium." In: *Proceedings of the Symposium on Particles on Surfaces: Detection, Adhesion and Removal*, San Francisco, CA, July 28-August 2.

Ranade, M.B. 1987. "Adhesion and Removal of Fine Particles on Surfaces." *Aerosol Sci. Technol.* 7(2):161.

Ruzyllo, J. 1987. "Dry Cleaning of Silicon: Feasibility and Possible Solutions." In: *Proceedings of the Microcontamination Conference and Exposition, Santa Clara, CA, October 27-30*, pp. 61-68.

Shaw, D.J. 1980. *Introduction to Colloid and Surface Chemistry*. Butterworths, London.

Shwartzman, S., A. Mayer, and W. Kern. 1985. "Megasonic Particle Removal from Solid-State Wafers." *RCA Review 46*:81.

Skidmore, K. 1987. "Cleaning Techniques for Wafer Surfaces." *Semiconductor International* (Aug.):80-85.

Van den Tempel, M. 1972. "Interactions Between Condensed Bodies in Contact." *Advances in Colloid and Interface Science 3*:137.

Visser, J. 1976. "Adhesion of Colloidal Particles." *Surface and Colloid Science 8*:3.

Zimon, A.D. 1982. *Adhesion of Dust and Powder*. Plenum Press, New York, NY.

22. EXPERIMENTAL EVALUATION OF CLEANING TECHNIQUES

VENUGOPAL B. MENON Sematech, Austin, Texas

INTRODUCTION

In the previous chapter, the various factors that determine the adhesional force between particles and a surface were discussed. These factors include particle size and shape, particle and substrate composition, and the choice of cleaning medium. The force required to remove particles from a wafer is numerically equal and opposite to the force of adhesion. Therefore, if the magnitude of an externally applied force is measured at the instant of particle detachment from a surface, we may determine the force of adhesion. This chapter consists of a discussion of various experimental methods to measure the force of adhesion and several suggestions on conducting an evaluation of a wafer cleaning system.

It is very common to relate adhesional force using the ratio of the number of particles detached under the influence of a specific force to the initial number on a substrate. Because in most experimental investigations, it is not possible to use perfectly monosized, spherical particles, and substrates with no surface roughness, it is advantageous to estimate the force under which the majority of particles are removed. Kordecki and Orr (1960) and Zimon (1982) have shown that it is better to define the force of adhesion as being that value at which 50 percent of the particles are removed from a surface. This definition is now commonly used in experiments to estimate the force of adhesion.

Several methods have been used to measure the adhesional force of single particles on surfaces. These methods are based on the same external forces that are used in cleaning systems and include centrifugal, aerodynamic and hydrodynamic, vibrational, and microgravimetric methods. Most of these techniques have been used for measuring the adhesional force of particles with sizes larger than 1 μm, but they are capable of being used with submicrometer particles. Generally, these techniques will yield quantitative information on adhesion, provided the systems are sufficiently characterized and defined. It is particularly important to use particles with well-defined geometry and surfaces that are extremely smooth. A rigid control of experimental conditions is required, and often auxiliary measurements are necessary to interpret the results adequately (Visser, 1976). Table 22-1 gives a summary of the techniques used to measure the force of adhesion of micrometer-sized particles.

In most adhesion studies on individual particles, the number of particles on a well-defined surface area must be determined before and after a removal experiment. For silicon wafers, this can be done using a microscope or a wafer surface scanner. Both of these methods of particle counting have advantages and disadvantages. A wafer scanner such as the Estek 600 is very quick and accurate; however, it cannot distinguish between the particles that are being studied and other contaminants that deposit onto the wafer during the experiment. Also, the wafer scanner cannot differentiate between agglomerates and single particles. The microscopic technique (optical), while being cumbersome and limited to particles greater than 0.5 μm, is capable of separately identifying those particles that are being studied from other contaminants. Scanning electron microscopy (SEM) can be used with smaller particles;

Table 22-1. Summary of Techniques for Adhesion Measurement of Micrometer-Sized Particles ($F_{removed}$ = m•a)

Technique	Parameters	Maximum acceleration (g's)	F (1 μm, σ_g = 1) (dyn)
Centrifuge	75,000 rpm (commercial)	5 x 10^5	3 x 10^{-4}
	30,000 rpm (special)	10^6	6 x 10^{-4}
Vibration	kHz	10^5	6 x 10^{-5}
	MHz	10^6	6 x 10^{-4}
Hydrodynamic	1,000 rpm	--	9 x 10^{-6}(a)
	8,000 rpm (concentric cylinders)	--	2.26 x 10^{-4}(a)

[a]Visser (1970).
Source: Ranade, 1987.

however, it is very expensive to scan over a large surface area. The microscope enables one to conduct the experiment outside the clean room, if necessary. Also, the microscopic procedure enables the researcher to observe mechanistic phenomena, such as agglomeration, etc. Because the experimental procedure is statistical in nature, there must be a sufficiently large number of particles on the wafer to make the results meaningful. The wafer scanner provides reproducible counts with a lesser number of particles per square centimeter than a microscopic procedure. For investigations with patterned wafers, however, the now commercially available wafer surface scanners cannot detect particles smaller than 0.5 μm.

The deposition of single particles onto a clean wafer is an important step in the experimental procedure to measure the force of adhesion or cleaning efficiency. There are many ways of depositing particles on a wafer. Menon et al. (1987) have used aerosol deposition chambers, nebulizers, and vibrating membranes for particle deposition.

MEASUREMENT OF ADHESION

Particle adhesion in liquids is lower than in air by a factor of 2 to 4. This decrease in the force of adhesion is due to a multitude of factors, which include:

- Lowering of the Hamaker constant in liquids
- Presence of ionic, double-layer repulsive forces in liquids
- Occurrence of repulsive hydrophobic/hydrophilic interactions.

Chapter 21 provides a detailed discussion of these interactions. Although most of the experimental techniques for studying particle adhesion and cleaning are capable of being used with both air and liquid, care must be taken to understand the differences between the two systems.

The simplest procedure for investigating the efficacy of a commercial cleaning technique for silicon wafers is to measure cleaning efficiency on a variety of dirty wafers. This approach, although useful in the short-term, is empirical and does not provide answers to questions such as:

- Why do some wafer cleaners clean wafers from process A much better than those from process B?
- Why do wafers with thermally grown surface oxide layers clean better (or worse) than bare silicon wafers?
- Why are some wafer cleaners incapable of removing submicrometer particulate contaminants even at maximum cleaning intensity?
- Why do some wafer cleaning systems damage wafers more than others?
- Why do some contaminated wafers appear to become cleaner just by immersion in DI-water?

A systematic approach to answering some of these questions is to conduct experiments that elucidate the mechanisms and forces that contribute to particle adhesion and removal. The use of particles with known surface composition, size, and shape, and wafers with known orientation and surface properties, is important. The ability to estimate the energy imparted by the cleaning technique to the particle is also important, because it determines the capabilities and limitations of a particular cleaning system.

In the following paragraphs, experimental techniques are discussed for measuring the force of adhesion in different cleaning system configurations.

Centrifugal Method

As previously discussed, the methods used for studying adhesional behavior are the same as those used for wafer cleaning. Wafer rinsers/dryers, which are commonly used in the microelectronics industry, use a combination of hydrodynamic drag and centrifugal force to remove particles from wafers.

Hydrodynamic drag (pressurized water spray) and centrifugal force are used together in a rinser/dryer because centrifugal force alone would be inadequate for removing many types of particles.

The force of adhesion of particles to a solid surface can be determined by subjecting them to a centrifugal force. The force of adhesion, F_A, of such a particle is equal to the centrifugal force, F_C, when this force acts in the opposite direction to the adhesional force (Figure 22-1). For a spherical particle, the centrifugal force is given by:

$$F_C = \frac{4}{3} \pi R^3 (\rho_p - \rho_m) (\omega^2 x + g) \quad (22\text{-}1)$$

where R is the radius of the particle, ρ_p and ρ_m are the densities of the particle and the medium, respectively, ω is the rotational speed (in radians per second), and x is the distance between the center of gravity of the particle and the axis of rotation. For particles smaller than 100 μm, the adhesional force is greater than the particle weight, and the gravitational acceleration g can be neglected (Visser, 1976).

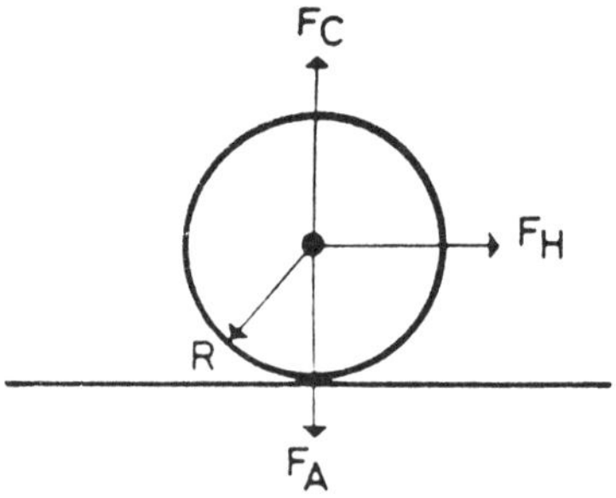

Figure 22-1. Direction of the forces acting on a sphere with radius R resting on a flat plane: F_A = force of adhesion, F_C = centrifugal force, and F_H = drag force.

This technique for determining the force of adhesion has been used by a great number of investigators, such as Bohme, Krupp, and Schnabel (1969), Rumpf (1972), Kordecki and Orr (1960), and Corn (1961). The maximum acceleration that can be attained is on the order of 10^5 to 10^6 g; hence, this method is reliable for large particles over 1 μm in diameter. Owing to the strong dependence on particle density, the centrifugal method is best suited to materials of high density (such as metals).

Van den Tempel (1972) used the centrifugal method to study the adhesion of gold and carbon particles on silicon wafers. In vacuum, the maximum force that can be applied to a 1.3-μm gold particle adhering to the outer rim of a spinning disk (ω = 10,000 s^{-1}) of 10 cm diameter is 0.09 dyn. Figure 22-2 depicts a typical result of adhesion experiments performed by van den Tempel (1972) with gold particles on silicon in ultrahigh vacuum. For gold particles with a radius of 1.3 μm, the experimentally measured force of adhesion, corresponding to 50 percent detachment, is about 2 x 10^{-5} dyn, which is well below the maximum force that can be imparted by the centrifuge. Hence, for this system, the centrifugal method appears to be adequate for particle removal.

The maximum force imparted by the centrifuge drops dramatically as the particle size and density decrease. For example, the maximum centrifugal force for carbon spheres (density = 2.1 g/cm^3 and 0.1 μm radius) on a silicon wafer (10 cm diameter) is about 3 x 10^{-6} dyn; whereas, the van der Waals force of adhesion (assuming a separation distance of 4 A) for 0.1 μm radius carbon spheres on silicon is 4.9 x 10^{-3} dyn. For this system, the centrifugal

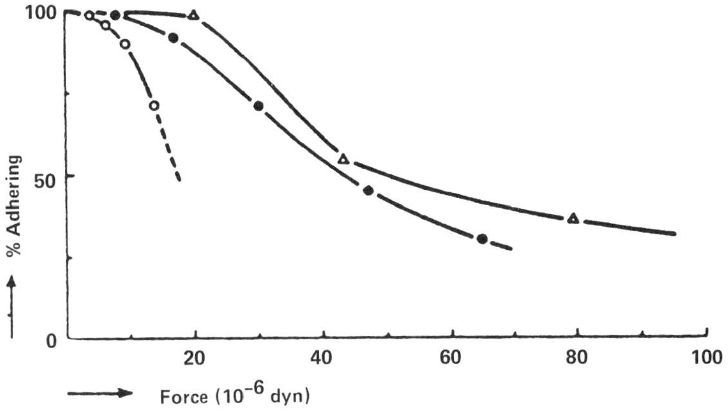

Source: Van den Tempel, 1972

Figure 22-2. Typical results of adhesion measurement for gold spheres on silicon in an ultrahigh vacuum. R = 1.3 μm (○); R = 2.1 μm (•); R = 2.9 μm (△).

method is completely inadequate for studying adhesion behavior. Although gold and carbon particles may generally not be encountered during wafer processing, they represent good model systems for investigating particle adhesion.

Aerodynamic and Hydrodynamic Methods

In aerodynamic and hydrodynamic methods, a stream of air or liquid is passed over particles adhering to a solid surface. When the drag and/or the lift force of the stream on the particles exceeds the force of adhesion, the particles will be dislodged. The hydrodynamic drag force, F_H, can be calculated from Stokes' Law (Chapter 2):

$$F_H = \frac{C_D \rho A_p V^2}{2} \quad , \tag{22-2}$$

where A_p is the projected area of the particle, C_D is the drag coefficient, ρ is fluid density, V is relative velocity between the fluid and the particle, and F_H is the hydrodynamic drag force. For a single sphere in an infinite sea of liquid, and a flow of sufficiently low Reynolds number, the force is given by the well-known Stokes' law (see Figure 22-3). The presence of a velocity gradient has no effect on the magnitude of the force if we take the liquid velocity in the streamline going through the center of the sphere (van den Tempel, 1972). The effect of a planar solid surface in contact with the sphere is to increase the drag force by a factor of 1.70 (Visser, 1976). The hydrodynamic drag (under laminar flow conditions) on an adhering particle is proportional to the square of the radius and does not depend directly on the fluid viscosity. The quantity that must be measured is the shear stress (τ) at the wall:

$$F_H = 32\ R^2\ \tau \quad . \tag{22-3}$$

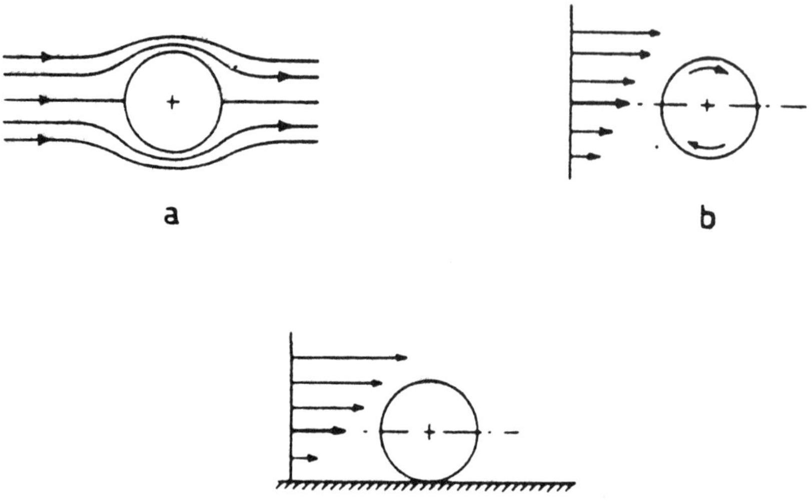

(a) $F = 6\pi\eta R\nu$; (b) $F = 6\pi\eta R\nu_c$; (c) $F = 1.7 \times 6\pi\eta R\nu_c$.

$$\nu_c = R\frac{d\nu}{dz} = \frac{\text{Wall shear stress } (\tau)}{\text{Viscosity } (\eta)}$$

$$\therefore F = 32\,R^2\,\tau.$$

Figure 22-3. Hydrodynamic force on a sphere adhering to a flat substrate.

Equations have been derived for the shear stress for various flow geometries and conditions (Visser, 1976; Wendt, 1933). The force of adhesion is generally related to the drag force by the following expression:

$$F_H = \mu F_A , \tag{22-4}$$

where μ is a proportionality constant. For submicrometer particles, the proportionality constant μ has been determined from removal experiments using a liquid flow between concentric cylinders (Zimon, 1982; Visser, 1976) to be close to unity. Hence,

$$F_A = 32\,R^2\,\tau . \tag{22-5}$$

The above relationship is valid only for small (less than 1 μm in diameter) particles. For larger particles, μ is on the order of 0.03 (Charles and Mason, 1960). In this case the thickness of the laminar boundary layer is comparable to particle size, so that the hydrodynamic theory is no longer applicable.

Within the category of aerodynamic and hydrodynamic methods, various flow configurations have been used for cleaning surfaces. Air or nitrogen blowoff guns that were common in the past are being replaced by liquid-based cleaning techniques because of the inadequacy of the former to remove small particles ($< 10\ \mu m$). The two most common methods are (1) liquid impingement from a nozzle onto the surface (also known as spray flow or shear stress departiculation), and (2) liquid flow between concentric cylinders.

Spray Flow: This type of fluid flow has been used for many years for cleaning surfaces (Stowers, 1978; McVey, Campuzano, and Fowler, 1981). Silicon wafer rinsers/dryers and most scrubbers use pressurized liquid sprays to remove contaminants from surfaces. When the spray impinges a surface, the fluid changes direction and flows parallel to the surface (see Figure 22-4). When a fluid passes over a surface, the velocity approaches zero immediately at the surface but increases to a maximum, known as the free stream velocity (U_s), at some distance above the surface. Fluid molecules at the solid surface are brought to rest, and those for a short distance above the surface are slowed because of viscous shear in the fluid. This region of retarded flow is called the "boundary layer" and for practical purposes extends to the point at which fluid velocity equals 99 percent of free stream velocity. For high-velocity spray applications, the flow within this boundary layer is usually turbulent over the entire surface. A thin laminar sublayer will exist between the solid surface and the turbulent portion of the boundary layer. Within this sublayer, velocities decrease rapidly to zero (Musselman and Yarbrough, 1987).

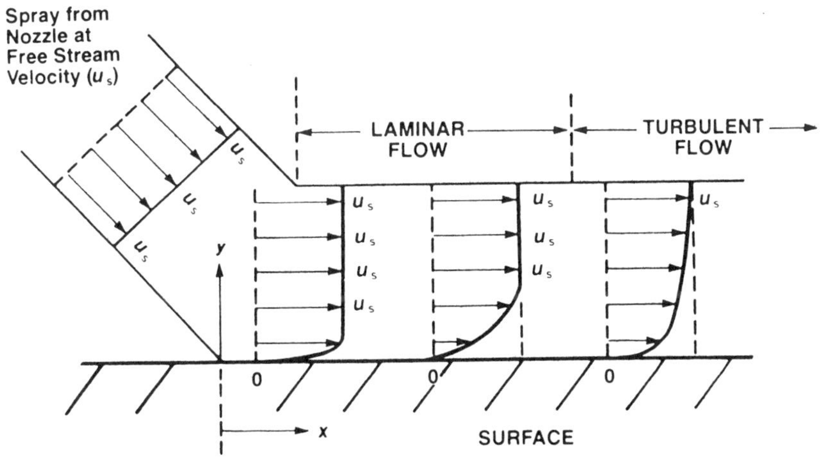

Source: Musselman and Yarbrough, 1987

Figure 22-4. Velocity profiles of spray flow.

To calculate the drag force acting on the particle, one needs to know the velocity at the center of the particle. From classical fluid mechanics, it is possible to determine the relative fluid velocities at any position above the surface of a smooth flat plate for a given distance downstream of the point of spray impact. Thus, knowing the nozzle or free stream velocity (which is usually provided by the nozzle vendor in terms of fluid pressure), the velocity at a particle's center can be ascertained, and the drag force can then be estimated. Table 22-2 gives the free stream velocity and boundary layer thickness at different pressures for freon spray at 70 °F (21 °C) (Musselman and Yarbrough, 1987). For particles smaller than 1 μm, the force of adhesion (based on van der Waals attractive potential) is typically so large that nozzle pressures in excess of 100 psig are required for particle removal. However, at such high pressures, the potential for wafer damage is great. In general, it is safe to conclude that surface cleaning at nozzle pressures in excess of a few hundred psig is not feasible for materials such as silicon wafers owing to the danger of wafer damage.

Figure 22-5 is a schematic diagram of the apparatus used at Research Triangle Institute (RTI) for studying wafer cleaning with a liquid spray. The apparatus consists of a stainless-steel chamber within which the wafer is placed. The roof of the chamber is mounted with two internal nozzles—one spraying perpendicular to the wafer surface, and the other at an angle of 45° to the wafer surface. The cleaning solvent enters the chamber through the nozzles (which can be operated one at a time or simultaneously), flows over the wafers, and collects in a solvent storage tank. The cleaning solvent is recirculated through the system by a high-pressure pump and a point-of-use filter.

Polystyrene latex (PSL) or glass microspheres (1 μm size) were deposited on clean wafers using a nebulizer. Each wafer was counted and placed in the cleaning chamber in a horizontal position. The wafers were cleaned one at a time, using either the top or side nozzle for 2 min. They were then dried within the chamber in a spray of filtered air. Particle counts of each wafer, taken at the conclusion of the experiment, allowed calculations of cleaning efficiency to be made. Figure 22-6 depicts the variation of drag force with nozzle pressure for water and Freon-TF, calculated using the data of Musselman and Yarbrough (1987). Also shown in Figure 22-6 are horizontal lines corresponding to the theoretical values based on

Table 22-2. Free Stream Velocity and Boundary Layer Thickness of Different Pressures

Nozzle Pressure (psig)	Nozzle Pressure (MPa)	Free Stream Velocity (m/s)	Turbulent Boundary Layer Thickness (Microns)	Laminar Sublayer Thickness (Microns)
1	6.9×10^{-3}	2.95	2000	16
10	6.9×10^{-2}	9.34	630	5.2
100	6.9×10^{-1}	29.5	200	1.6
1000	6.9×10^{0}	93.4	63	0.52
10000	6.9×10^{1}	295	20	0.16

Free Stream Velocity and the Reciprocal of Layer Thickness are Proportional to the Square Root of the Nozzle Pressure.

Source: Musselman and Yarbrough, 1987

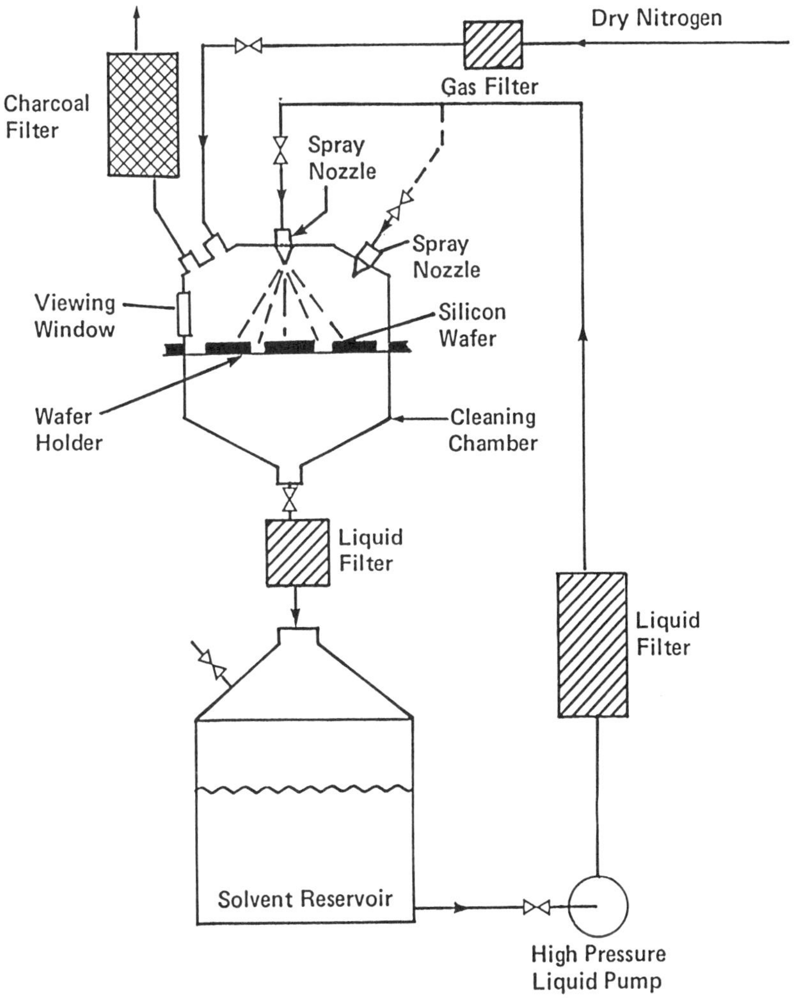

Figure 22-5. Schematic of experimental apparatus for shear stress cleaning.

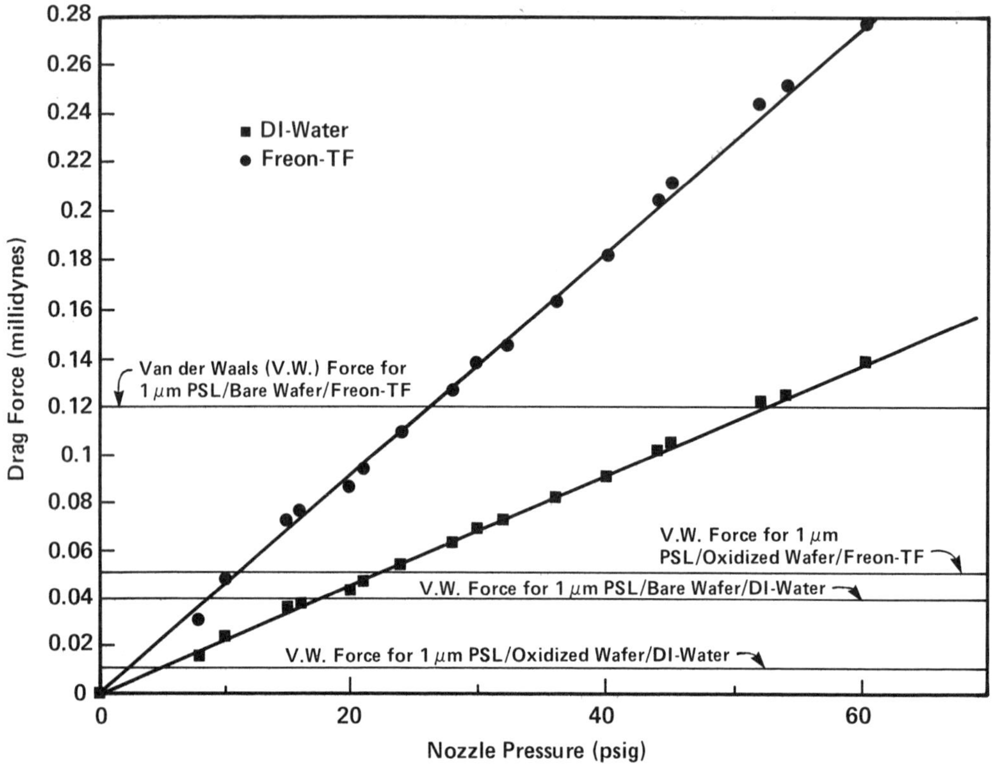

Figure 22-6. Comparison of drag forces for Freon-TF and DI-water as a function of nozzle pressure.

van der Waals interactions. Each point of intersection of the drag force-pressure curve with the horizontal line represents the minimum nozzle pressure predicted for particle removal. For 1 μm PSL and glass particles on bare or oxidized wafers, the liquid pressure required for cleaning is predicted to be less than 50 psig. Figure 22-7 shows the experimental measurements of cleaning efficiency with nozzle pressure for glass microspheres on a bare silicon wafer and a wafer with a 5,000-A surface oxide layer. Cleaning efficiencies approaching 99 percent were achieved at pressures below 40 psig. A similar plot was also obtained for PSL particles. The differences in adhesional force between bare and oxidized wafers predicted by van der Waals theory, and shown in Figure 22-6, are not observed in the experimental results (Figure 22-7). Also, cleaning efficiencies in excess of 50 percent were observed even at pressures below 10 psig. These differences between experimental results and van der Waals theory suggest the presence of other dominant particle-surface interactions that are repulsive in nature and contribute to a lowering of the overall force of adhesion. As will be seen later, electrostatic double-layer and hydrophobic/hydrophilic interactions play an important role in the adhesion process for such systems.

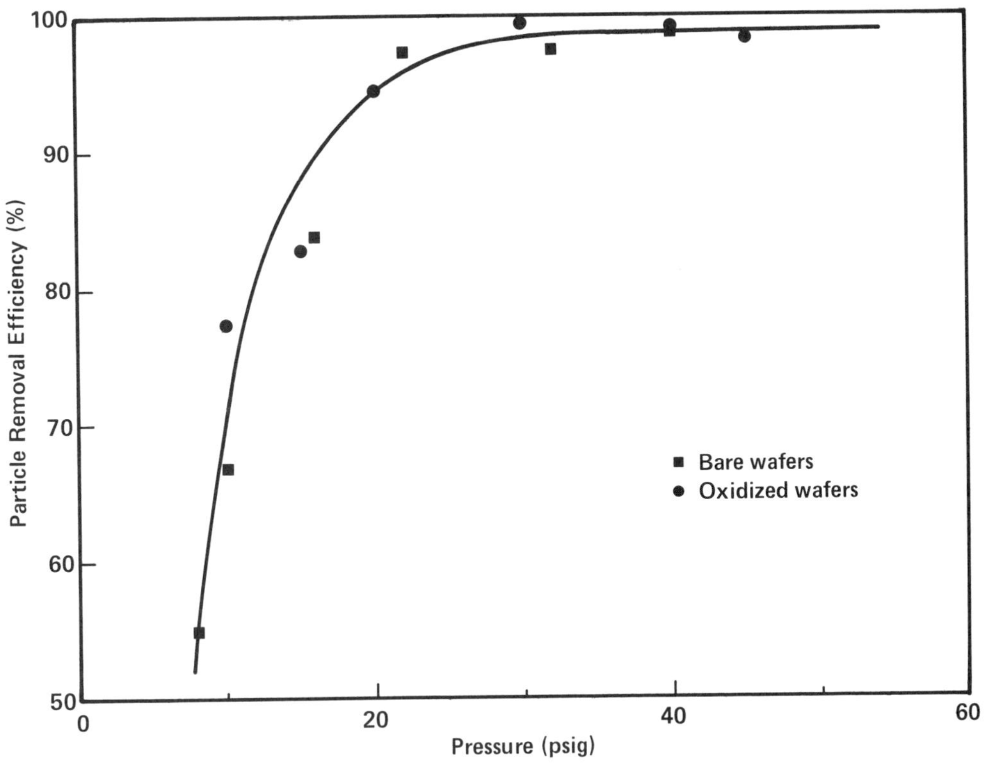

Figure 22-7. Effect of liquid pressure on cleaning efficiency for 1 μm glass microspheres on bare and oxidized silicon wafers.

Liquid Flow Between Concentric Cylinders: Many researchers have used this method to determine the force of adhesion (van den Tempel, 1972; Visser, 1976). In a typical experiment, particles are deposited on the outer wall of the inner cylinder, and the inner cylinder is rotated rapidly to produce hydrodynamic shear. The shear stress at the wall is obtained by measuring the torque on the outer cylinder. In a centrifuge experiment, the force of adhesion may be equated to the centrifugal force that is just sufficient to lift the particle from the substrate. This is not so for the viscous drag determined in a concentric cylinder apparatus, for two reasons. First, the particle is removed by liquid flow in a direction parallel to the surface, and it is subject to a torque. The force required to start the resulting rolling motion along the surface may be quite different from the adhesive force measured in a centrifuge. Second, the concept of a laminar boundary layer, although very successful in many applications, is in reality only a simplification of the actual situation in which some amount of turbulence may persist right up to the solid wall. This problem makes it necessary to calibrate the concentric cylinder using a different method, such as the centrifugal technique (van den Tempel, 1972).

Although the mechanism of particle detachment in the concentric cylinder apparatus is not clearly established, the force of adhesion will still be proportional to some function of the wall shear stress and particle diameter.

Vibration Method

The vibration method was first used for determining the adhesion of films (Zimon, 1982). The technique later became popular for studying particle adhesion to surfaces. To generate the detachment forces necessary for removing micrometer- and submicrometer-sized particles, the frequency of vibration must typically be in the ultrasonic region. In addition to ultrasonics, megasonics (with frequencies in the 850 to 900 kHz range) has also been applied for particle removal from surfaces.

Ultrasonics: For a vibrating surface, created by an ultrasonic transducer generating oscillations with frequencies around 10 to 25 kHz, the force of removal (F_V) is given by:

$$F_V = \frac{16\, \pi^3 f^2 A\, \rho\, R^3}{3} = m\,(a + g) \quad , \qquad (21\text{-}6)$$

where f is the frequency, A is the amplitude of the vibration, m is the particle mass, ρ is the particle density, R is the radius of the spherical particle, and a is the acceleration imparted by the vibrational motion.

To estimate the vibrational force imparted to a particle, the frequency and amplitude of oscillation should be known. The frequency is usually fixed by the generator, and the amplitude may be measured visually or with an oscillograph. Mullins et al. (1988) have used an ultrasonic vibrator to measure the force of adhesion in air of glass and carbon particles on glass substrates. Figure 22-8 is a schematic of their experimental apparatus. Spherical particles were deposited onto the substrate, which was then glued on the tip of the ultrasonic horn. A Climet 208 particle counter was placed below the vibrating horn to monitor the number of detached particles on a real-time basis. Ranade et al. (1986) extended this study to measure the force of adhesion for the same system in liquids. Figure 22-9 depicts the effect of medium on the particle removal efficiency. The acceleration (or force) required to remove 11.5 μm glass spheres from a glass substrate was significantly greater in air as compared to the liquid media. This trend may be attributed to a lowering of the Hamaker's constant in liquids as compared to air.

In comparison to ultrasonic cleaning in air, the mechanism of the action of ultrasound in liquids in not fully understood. This is mainly due to the difficulties associated with ultrasonic cavitation in liquids. The mechanical disturbances that are caused by high-intensity sound waves produce pressure fluctuations above and below their ambient pressure. In the reduced-pressure cycle, bubbles tend to form and grow; whereas, in the increased-pressure cycle, bubble growth stops, and bubbles either oscillate or collapse. At the collapsing bubble interface, high enough energy is released in a localized region to overcome the force of adhesion. Thus, the mechanism of ultrasonic cleaning involves the supply of (1) vibrational energy to the

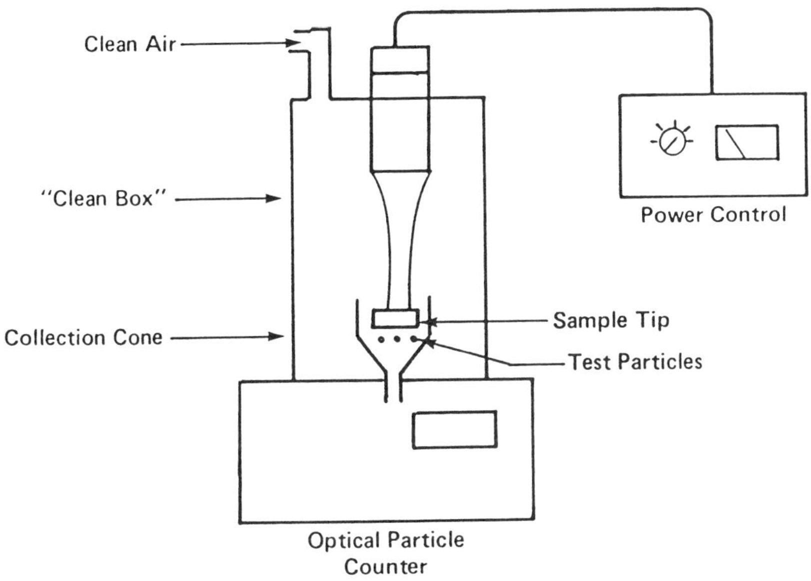

Figure 22-8. Schematic of ultrasonic adhesion test equipment.

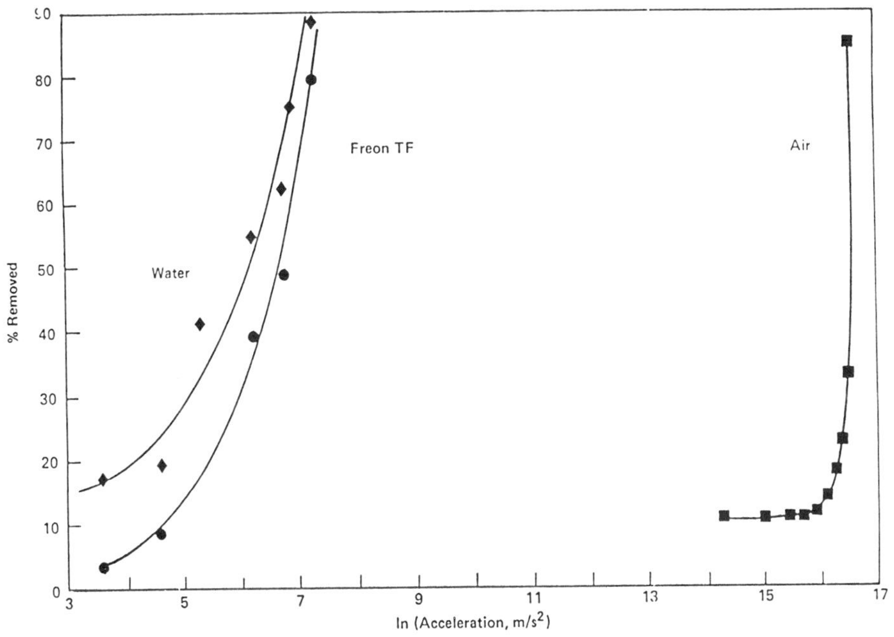

Figure 22-9. Removal of 11.5 μm glass particles from a flat glass surface by ultrasonic vibration.

surface being cleaned, and (2) cavitation energy to the particles. At low values of ultrasonic energy, the force applied to break the particle-surface bond is mostly supplied by the vibration of the surface; whereas, at high-energy inputs, both vibratory and cavitational forces are present. Cavitation intensities vary depending on the molecular nature of the liquid. Table 22-3 shows that the relative cavitational intensity for polar liquids such as water and alcohols is much higher than that for nonpolar liquids such as the freons (Niemczewski, 1980).

Table 22-3. Comparison of Cavitation Intensity in Various Liquids

Liquid	Maximum cavitation intensity (%)
Water	100
Toluene	71
Methanol	52
Ethanol-acetone	45
Freon-TMS	17.22
Freon-TF	15

Source: Niemczewski, 1980.

Figure 22-10 shows the cleaning efficiency using freon-TMS for 1 μm PSL microspheres on a bare (native oxide) silicon wafer. The ultrasonic force applied to the system is expressed in relative units as the log (1,000 x V_T), where V_T is the voltage corresponding to a specific energy input to the ultrasonic vibrator. Instead of the expected step function, where all the particles of a given size are removed at a certain value of the applied force, the plot shows a gradual increase in cleaning efficiency with an increase in applied force. This distribution of adhesive strength values is mainly due to surface roughness and variations in particle size.

Plots similar to Figure 22-10 were obtained by Menon et al. (1987) using water, ethanol-acetone (1:1 blend), freon-TF, and freon-TMS as the cleaning solvents. For all systems, the plots had similar profiles and varied only in the values of the force of adhesion (corresponding to 50 percent cleaning efficiency). Table 22-4 summarizes the experimentally measured values of the force of adhesion along with the values predicted from van der Waals adhesion theory. The force of adhesion obtained in voltage units was transformed to force units (dynes) using Equation 22-6. The amplitude of vibration as a function of voltage was provided by the manufacturer. To calculate the van der Waals force of attraction for a sphere-plate configuration, it is necessary to know the separation distance between the particles and the wafer. In studies of particle interactions with silicon substrates, van den Tempel (1972) found that values of Z_0 ranged from 5 Å to 25 Å. For the present calculations, a value of Z_0 = 15 Å was used.

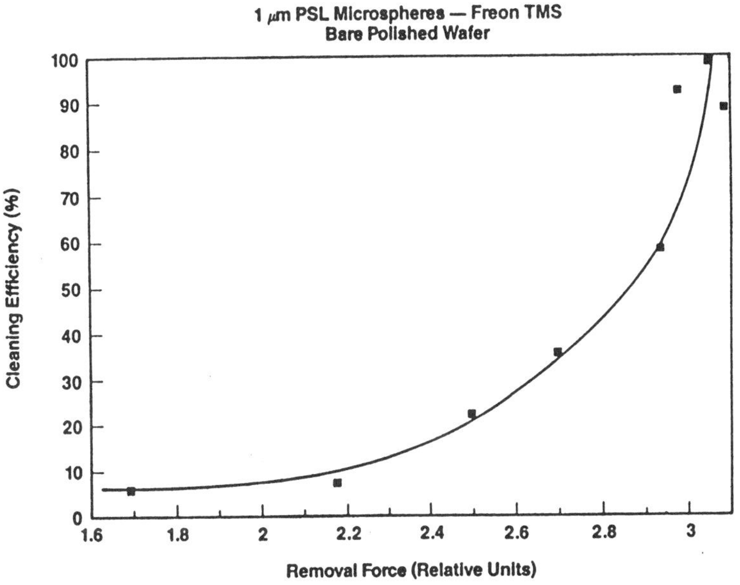

Figure 22-10. Variation of cleaning efficiency with ultrasonic force for 1 μm PSL spheres.

Table 22-4 reveals that the measured values of adhesional force for most systems were lower than those predicted theoretically. Also, the ease of particle removal from the wafer surface decreased as the cleaning solvent was varied from a polar liquid, such as water or ethanol-acetone, to the two freons. In some cases, cleaning efficiencies in excess of 50 percent were obtained even in the absence of any ultrasonic vibration. This was especially true for systems where water was the cleaning medium. In all cases, cleaning efficiencies approaching 99 percent were achieved. The discrepancy between experimental and theoretical results appears to occur due to the presence of repulsive forces between particles and substrate in the liquid. These repulsive forces may include contributions from ionic double-layer interactions, hydrophobic/hydrophilic interactions, and cavitation. For PSL particles on a wafer with a native surface oxide layer, significant agglomeration was observed after immersion of the wafer in water. This suggests the presence of strong hydrophobic interactions, which cause the hydrophobic PSL particles to slide along the wafer surface and agglomerate, thereby decreasing the overall free energy of the system. Among the four solvents (water, ethanol-acetone, freon-TF, and freon-TMS), freon-TF was the least effective cleaning solvent (see Figure 22-11). This may be attributed to the absence of hydrophobic/hydrophilic interactions and a low value of the cavitation intensity (Niemczewski, 1980).

As is evident from the results discussed above, ultrasonic wafer cleaning in liquids can achieve high degrees of cleanliness; however, the quantitative interpretation of experimental measurements of the force of adhesion is difficult due to the interplay of many attractive and

Table 22-4. Comparison of Theoretical (based on van der Waals) and Experimental Values of the Force of Adhesion for 1 μm Particles

Interacting Materials	Force of Adhesion (mdyn)									
	Intervening Medium									
	Air		Water		Ethanol-Acetone		Freon-TMS		Freon-TF	
	Experimental	Theoretical*	Experimental	Theoretical	Experimental	Theoretical	Experimental	Theoretical	Experimental	Theoretical
Glass-bare silicon	—	0.26[a]		−0.07		−0.05		0.009		0.015
	—	0.34[b]	3.5×10^{-4}	−0.03	7.7×10^{-5}	0.003	3.5×10^{-4}	0.067	0.8	0.07
	—	0.46[c]		0.04		0.08		0.16		0.16
Glass-silicon with 5,000 Å oxide layer	—	0.15[a]		−0.02		−0.016		0.004		0.007
	—	0.20[b]	$<10^{-5}$	−0.008	$<10^{-5}$	0.001	N/A[d]	0.03	N/A	0.03
	—	0.27[c]		0.012		0.03		0.068		0.071
Polystyrene-bare silicon	—	0.47	$<10^{-5}$	0.05	0.004	0.09	0.006	0.16	0.012	0.17
Polystyrene-silicon with 5,000 Å oxide layer	—	0.27	0.001	0.013	0.006	0.03	N/A	0.07	N/A	0.07

*Z_o = 15 Å.

[a]A_{11} for glass = 2.0×10^{-13} ergs.

[b]A_{11} for glass = 3.4×10^{-13} ergs.

[c]A_{11} for glass = 6.3×10^{-13} ergs.

[d]Data not available at present time.

Note: van der Waals force calculation based on A_{132}, the Hamaker constant of matrices 1 and 2 separated by medium 3.

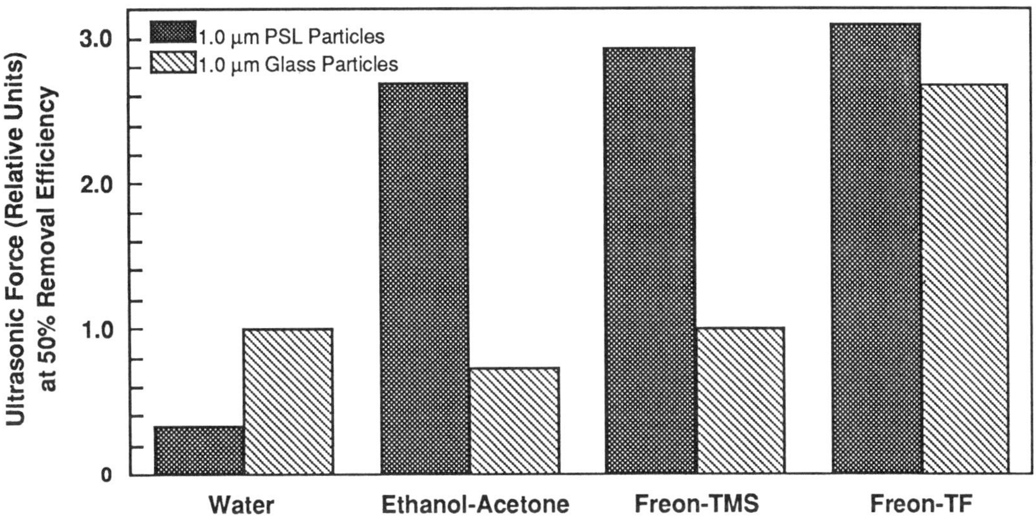

Figure 22-11. Comparison of adhesion force for glass and polystyrene latex particles using different cleaning solvents.

repulsive forces. When compared to hydrodynamic spray cleaning, ultrasonic cleaning appears to be more sensitive to the properties of the wafer surface.

Megasonics: Shwartzman, Mayer, and Kern (1985) described a megasonic cleaning system operating at frequencies around 750 to 900 kHz. At such high frequencies, the time between pulses is too short for the cavitation bubbles. The mechanism of cleaning with megasonics is attributed to the interaction between high-pressure waves in the liquid and the particles at the wafer surface. Shwartzman, Mayer, and Kern (1985) conducted a series of megasonic experiments with silicon wafers contaminated with particles to determine the efficiency of cleaning. The particles were deposited by dipping wafers into a slurry of 0.3 μm diameter aluminium oxide particles in water. The authors report that 4 min of megasonic action in an RCA cleaning solution was required to remove a majority of particles from the wafer surface.

A comparison of three cleaning media—DI-water, a solution of 1 part ammonium hydroxide/1 part hydrogen peroxide/5 parts water (often referred to as the SC-1 solution), and aqueous surfactant solution—showed that the SC-1 solution cleaned the best, followed by the surfactant solution, and water. A recent study (Menon et al., 1989a, 1989b) compares the effectiveness of various chemistries for megasonic cleaning of wafers with particles of different compositions. A commercially available megasonic wafer cleaner was used in these experiments. Figure 22-12 depicts the variation of particle removal efficiency with megasonic power for three cleaning media: DI-water, SC-1, and a choline-peroxide solution. The

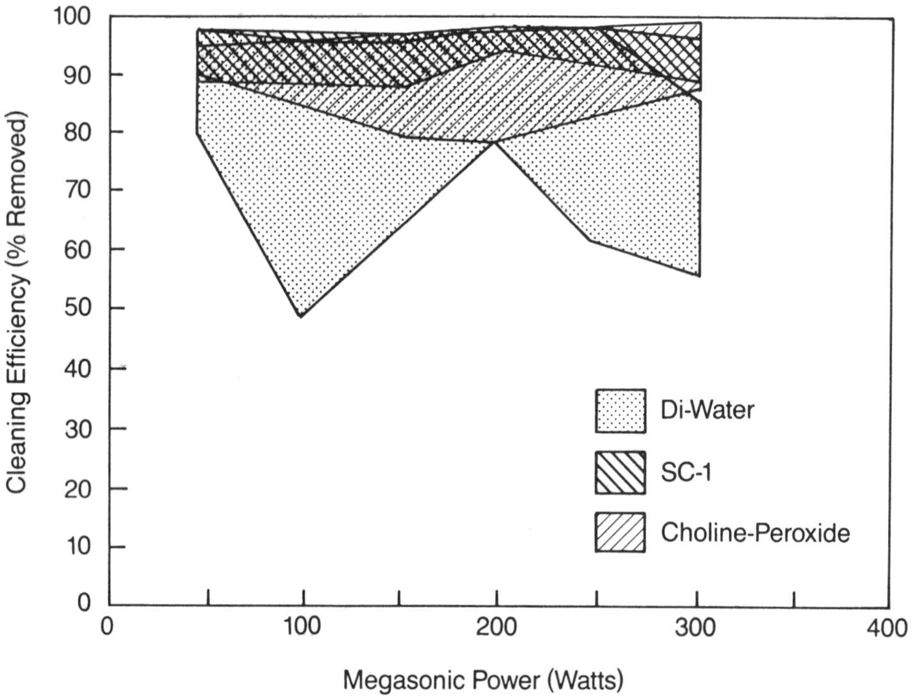

Figure 22-12. Variation of particle removal efficiency with megasonic power for different cleaning media.

choline-peroxide solution consisted of a blend of 7 parts DI-water, 1 part Summa-develop (a commercial brand name choline), and 1 part hydrogen peroxide. The data shown in Figure 22-12 represent the range over which cleaning efficiency varied when wafers with different types of particles, of size below 2 micrometers, were cleaned. The use of SC-1 consistently produced high cleaning efficiencies and relatively small variability for the various types of particles. The lowest cleaning efficiencies and largest variabilities were seen with DI-water. The results for choline-peroxide fell in between those for SC-1 and DI-water.

Figure 22-13 shows the number of particles remaining on a wafer after megasonic cleaning with SC-1. The wafers had approximately 5-8 particles per cm^2 in the specified size range, prior to cleaning. Under optimum conditions (corresponding to a megasonic power in the range 50-200 watts), the final particle count was less than 0.5 particles per cm^2. Figure 22-13 also shows that some removal of particles was observed even in the absence of megasonic energy, the particle counts being lower than 5-8 per cm^2. This phenomenon may be attributed to surface tension forces which come into play as the wafer passes through the air-liquid interface.

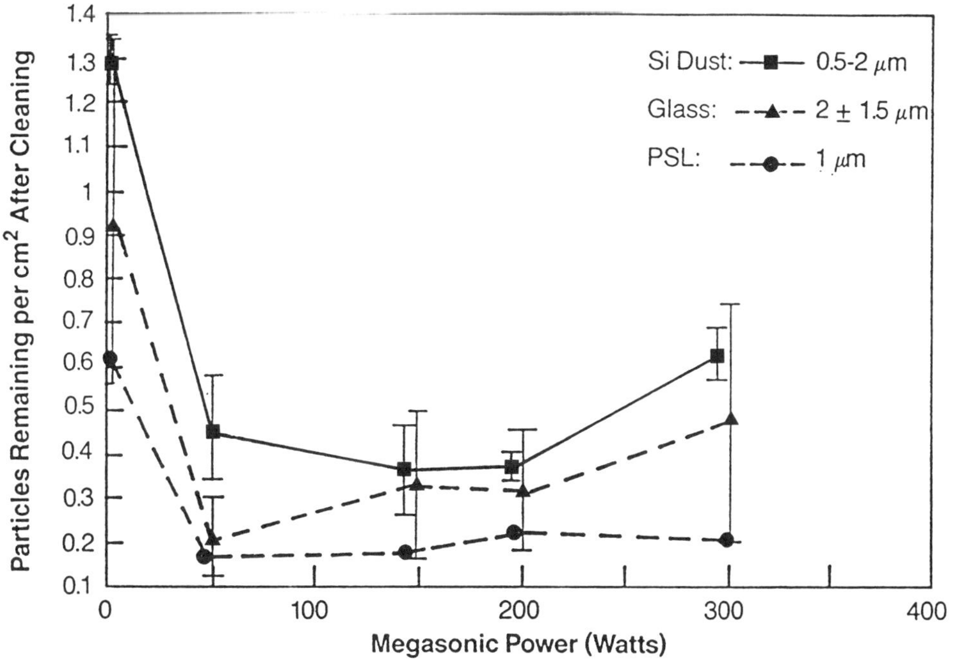

Figure 22-13. Particles left on bare silicon wafer after megasonic cleaning with SC-1 solution.

The precise mechanism of particle detachment from a wafer surface in an SC-1 solution is unclear. Shwartzman, Mayer, and Kern (1985) hypothesize that megasonic/SC-1 promotes wetting of the particles and the wafer, thus allowing the liquid to penetrate into the space between them. When this occurs, the van der Waals force of attraction between the two interacting materials is diminished, thus decreasing the force of adhesion. Indeed, the effect of SC-1 on the wettability of various particles and wafer surface may explain the observed differences in the cleaning profiles of PSL, glass particles, and silicon dust, shown in Figure 22-13. Further study is required to understand the mechanism of particle removal in various cleaning solutions.

Other Methods

Besides the aforementioned methods of measuring the particle adhesion and cleaning, there are other techniques based on the application of gravitational, electrical or mechanical forces to break the adhesive bond. These methods are, however, not suitable for studying micrometer- and submicrometer-sized particles, which represent the major contamination problem in the semiconductory industry; hence, they are not discussed here. For detailed discussions about this subject, the reader is referred to articles by Visser (1976) and Zimon (1982).

EVALUATION AND OPTIMIZATION OF A CLEANING PROCESS

The key components of a wet cleaning process are (1) the chemical tank, (2) the rinse tank, and (3) the spin dryer. A systematic evaluation of a cleaning process should include all three components. Figure 22-14 shows a diagram of a typical cleaning evaluation procedure. Particle test wafers are run through the entire clean, as well as through individual components. Liquid particle counting in the chemical tank and the rinse tank should be conducted in parallel with the wafer analysis. If possible, water analysis (TOC, colloidal silica, etc.) and surface analysis, such as scanning electron microscopy (SEM) and secondary ion mass spectroscopy (SIMS), should be included. Correlation of the results of such a comprehensive study as a function of the following variables can prove very useful in optimizing a cleaning process:

- wafer contamination source and level
- cleaning time and temperature
- cleaning energy
- rinse configuration (quick dump or overflow)
- rinse time
- dryer configuration (rinse and dry, or dry only)
- drying time.

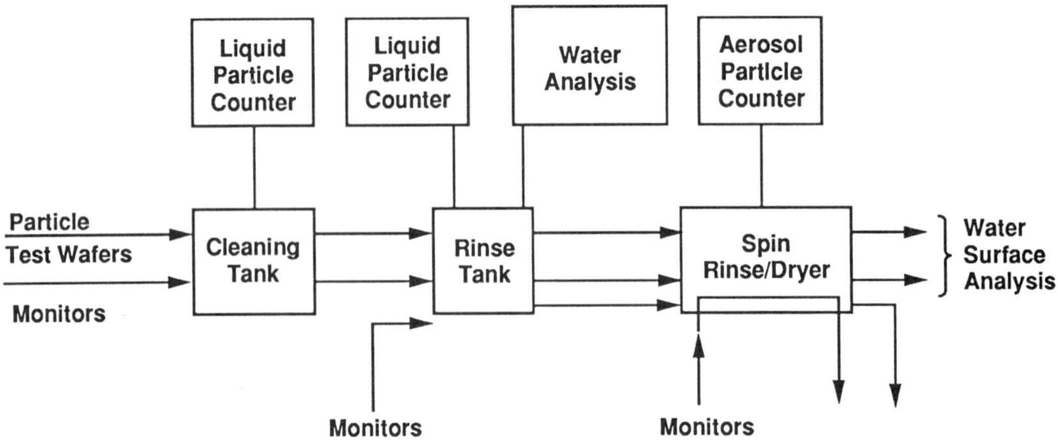

Figure 22-14. Cleaning evaluation procedure.

It is important to remember that each cleaning process is unique in its particle generation, deposition and removal behavior. Even units made by the same vendor and having the same model number, while possessing certain generic particle-related characteristics, can be very distinct. Hence, it is essential that each process, including each of its components, be evaluated and optimized. This may appear to be time consuming, but it will be beneficial in the long run.

SUMMARY

The selection of cleaning devices for removing particulate contaminants from wafer surfaces requires a fundamental understanding of the various forces that hold particles to surfaces. Although significant theoretical developments have been made in understanding particle adhesion, experimental evaluation of the force of adhesion and cleaning efficiency still represents the best approach. The techniques available for experimentally measuring the force of adhesion are the same as those used in wafer cleaning equipment; hence, it is feasible to conduct experimental studies in the actual cleaning devices or in a laboratory apparatus that is based on the same mechanism of cleaning.

The efficacy of a cleaning technique depends strongly on the chemical, physical, and morphological properties of the particles and the wafer. Experimental studies should therefore use well-characterized contaminants and wafers that span the entire range of situations encountered during actual processing.

Most commercial wafer cleaners are based on the application of centrifugal, hydrodynamic, or vibrational force to detach particles from the wafer surface. This chapter has discussed these forces and their applicability for various configurations of wafer, contaminant, and cleaning medium.

REFERENCES

Bohme, G., H. Krupp, and W. Schnabel. 1969. In: *Molecular Processes on Solid Surfaces*, p. 611. McGraw-Hill Book Publishing Co., New York, NY.

Charles, G.E., and S.G. Mason. 1960. *J. Colloid Sci.* 15:105.

Corn, M.J. 1961. *J. Air Pollution Control Assoc.* (JAPCA) *11*:253.

Kordecki, M.C., and C. Orr. 1960. *Archives Environ. Health 1*:13.

McVey, J.T., C. Campuzano, and D.E. Fowler. 1981. "Tools and Equipment: From Nuclear Waste to Reusable Items." *Nuclear and Chemical Waste Management 2*:197.

Menon, V.B., A.C. Clayton, and R.P. Donovan. 1989a. "Removing Particulate Contaminants from Silicon Wafers: A Critical Evaluation." *Microcontamination*, June, 31-34, 107-109.

Menon, V.B., R.P. Donovan, L.D. Michaels, and M.B. Ranade. 1987. "Particle Removal from Semiconductor Wafers Using Cleaning Solvents." In: *Proc. Symposium on Particles in Fluids: Detection, Characterization, and Control*, Boston, MA, August 3-7.

Menon, V.B., L.D. Michaels, A.C. Clayton, and R.P. Donovan. 1989b. "Megasonic Cleaning: A Comparative Evaluation of Cleaning Solutions." In: *Proc. Institute of Environ. Sci. Annual Meeting*, Anaheim, CA, May 4-9.

Menon, V.B., L.D. Michaels, R.P. Donovan, L.A. Hollar, and D.S. Ensor. 1988. "Removal Efficiencies of Organic and Inorganic Particles from Silicon Wafers." In: *Proc. Institute of Environ. Sci. Annual Meeting*, King of Prussia, PA, May 2-6.

Mullins, M.E., B. Locke, L.D. Michaels, and V.B. Menon. 1988. "Effect of Particle Geometry on the Force of Adhesion on Surfaces." *Aerosol Sci. Technol.* (in press).

Musselman, R.P., and T.W. Yarbrough. 1987. "Shear Stress Cleaning for Surface Departiculation." *J. Environ. Sci.* (Jan./Feb.):51.

Niemczewski, B. 1980. "A Comparison of Ultrasonic Cavitation Intensity in Liquids." *Ultrasonics* (May):107-110.

Ranade, M.B. 1987. "Adhesion and Removal of Fine Particles on Surfaces." *Aerosol Sci. Technol.* 7(2):161.

Ranade, M.B., V.B. Menon, M.E. Mullins, and V.L. Debler. 1986. "Adhesion and Removal of Particles: Effect of Medium." In: *Proc. Symposium on Particles on Surfaces: Detection, Adhesion and Removal*, San Fransisco, CA, July 28-August 2.

Rumpf, H. 1972. *Pharm. Ind. 34*:270.

Shwartzman, S., A. Mayer, and W. Kern. 1985. "Megasonic Particle Removal from Solid-State Wafers." *RCA Review 46*:81.

Stowers, I.F. 1978. *J. Vac. Sci. Technol. 15*:751.

Van den Tempel, M. 1972. "Interactions Between Condensed Bodies in Contact." *Advances in Colloid and Interface Sci. 3*:137.

Visser, J. 1970. *J. Colloid Interface Sci. 34*:26.

Visser, J. 1976. "Adhesion of Colloidal Particles." *Surface and Colloid Sci. 8*:3.

Wendt, F. 1933. *Ing. Arch. 4*:577.

Zimon, A.D. 1982. *Adhesion of Dust and Powder*. Plenum Press, New York, NY.

23. WAFER CLEANING: PRESENT AND FUTURE

KEITH DILLENBECK Ashland Chemical Company, Columbus, Ohio

INTRODUCTION

The purpose of a wafer-cleaning step is to remove surface particles as well as organic and inorganic contaminants, and to prepare a "clean" surface prior to specific process steps. To evaluate a wafer-cleaning technique, one must realize that a wafer-cleaning process is actually a chemical process designed to achieve a desired electrical result. Specific questions must be asked regarding the desired state of a wafer surface:

- What is the condition of the surface immediately prior to the cleaning treatment? Is there a native or chemical oxide layer that must be removed? What types of contaminants and particles appear on the surface? When do these contaminants appear? How do the types and quantities of surface impurities vary as a function of time?

- Is the wafer surface relatively flat, or are there many three-dimensional structures such as contact holes, vias, and trenches? What chemical treatments must be applied to the sides or bottom of these topographical structures?

- What is the required condition of the surface following the chemical treatment? Is a thin native or chemical oxide on the silicon surface desired? What level of particulate and molecular contaminants must be achieved? Is there a safe window of variability? To what degree must the surface, including the inner surfaces of three-dimensional structures, be dry?

SURFACE CHARACTERIZATION INSTRUMENTS

Many tools are available to the process technologist for assessing the condition of a surface or of a material interface that is affected by a wafer-cleaning treatment. Surface analysis techniques such as SIMS, AUGER, ESCA, TEM, and SEM/EDS (see Chapter 20) can provide useful information on the chemical state of a surface or an interface, such as the silicon/silicon dioxide interface within a metal oxide semiconductor (MOS) capacitor (Bakale, 1986; Phillips et al., 1983). Table 23-1 presents an overview of surface analysis tools that have been used in the past for evaluating the efficacy of a wafer cleaning treatment. It is not trivial to know which analysis tools are appropriate for a given situation, nor is it trivial to properly prepare the sample, conduct the analysis, or analyze the results. Particles on a surface can be detected by microscopic and light-scattering techniques. New technologies are available for detecting particles and certain process defects on and within topographical structures. To determine the elemental composition of a particle is possible with electron and ion beam analysis tools, but this can be very difficult in actual practice. All of these surface analysis techniques, when applied in a logical fashion by an experienced surface analyst, can enable the process technologist to design a chemical cleaning process that meets the electrical requirements of the devices being fabricated.

Table 23-1. Surface Analysis Tools for Evaluating a Wafer-Cleaning Process

Technique	Incident Signal	Annalyzed Signal	Reported Uses
SIMS	Ions	Ions	Metallic and ionic contaminant analysis
AUGER	Electrons	Auger electrons	Identification of small particles
ESCA	X-rays	Electrons	Measurement of contaminant layers
SEM/EDS	Electrons	Electrons/X-rays	Identification of gross contaminants
TEM	Electrons (backside)	Electrons/X-rays	Detecting contaminants at interfaces
Wafer particle counter	633 nm laser radiation	Scattered light	Counting particles on unpatterned surfaces
Patterned wafer particle counter	633 nm laser radiation	Scatter light (polarized)	Counting particles on patterns and thin films

RCA CLEANING TECHNOLOGY

The traditional method for cleaning silicon wafers prior to a thermal processing step involves a chemical process developed in the 1960s by RCA (Novak, 1987). The RCA cleaning process is based on oxidation and dissolution of residual organic and certain metallic impurities (Kern, 1984; Burkman, 1981). The steps of the RCA cleaning process and the function of each step are listed below:

- Sulfuric acid/hydrogen peroxide mixture—Removes any residual heavy organic contamination, such as photoresist.

- Dilute hydrofluoric acid (HF) (50:1 to 100:1)—Strips the native oxide, to expose the bare silicon surface. This is to remove any trapped impurities between the silicon and the thin oxide.

- Ammonium hydroxide/hydrogen peroxide/water mixture (1:1:5, SC1)—Regrows the native oxide and reduces the adhesion of particles onto the wafer surface.

- Hydrochloric acid/hydrogen peroxide/water mixture (1:1:5, SC2)—Dissolves metallic contaminants so they can be rinsed off the wafer surface. This chemistry also grows a native oxide, but not as thick as with the SC1 chemistry.

The exact volumetric ratios and temperatures of the chemicals involved vary depending on the specific applications. Additionally, some cleaning processes reverse the order of the SC1 and HF steps to maximize the removal of metallic impurities. In the RCA cleaning process, there is clearly a tradeoff between metallic removal rates and particle deposition rates. The sensitivity of the bare silicon surface to particles warrants the immediate regrowth of native oxide, but this can often trap metallic impurities at the silicon/oxide interface.

There are two commonly used ways in which to apply a wafer-cleaning process to silicon wafers (Kern and Puotinen, 1970). One involves immersing a cassette of wafers into a continually recirculating immersion tank containing the chemical solution. The wafers are then transferred to a rinse tank containing ultrapure water, as shown in Figure 23-1. Wafers are then transferred to the next chemical bath; this sequence is repeated until the final water rinse is complete. Next the wafers are transferred to some type of drying system. Centrifugal spin dryers are the most common way to dry wafers, but they can often be a serious source of particles or other process anomalies. New wafer drying techniques utilizing an isopropyl alcohol/water azeotrope are being used, with improved performance with regard to drying efficiency and particle deposition. An immersion cleaning system can be operated either manually or automatically, with the use of a reliable robotic mechanism.

The second technique for applying chemicals involves the use of spray technology in an automated spray cleaning system, as shown in Figure 23-2. In this type of system, filtered acids are pressure-fed into a mixing manifold and dispersed in a fine mist onto the wafer surface (Bansal, 1984). The cassettes of wafers rotate past the stationary acid spray column. Following the acid treatment, ultrapure water is sprayed to rinse the wafers. After the final rinse step, the wafers are dried at a relatively high spin speed while heated nitrogen flows into the chamber. This type of automated system can have advantages in terms of process consistency, as operator judgment does not come into play during the execution of the cleaning process. Another advantage is that chemicals are used only once, which can improve the purity of the cleaning process by avoiding particle and metallic contamination sources. One major disadvantage, however, is that centrifugal spray-cleaning systems are maintenance-intensive. Many wafer fabrication areas using spray-cleaning technology have encountered process problems in varying severity, either because of hardware malfunctioning or because of inadequate maintenance procedures.

METHODS FOR REMOVING PARTICLES

The RCA chemical cleaning process is not intended to remove particles from a wafer surface. Certain integrated circuit processes do require that particles be removed from a wafer surface. Two common methods are used: brush scrubbing and megasonic cleaning (Skidmore, 1987).

In a brush-scrubbing system, water or isopropyl alcohol is applied to the wafer surface while a brush moves over the surface (Burggraaf, 1981). The brush is made of a hydrophilic material (e.g., nylon or PVA foam), such that a thin layer of fluid resides dynamically between the wafer surface and the brush. The hydrodynamic force between the wafer and the brush is strong enough to remove deposited particles. Brush scrubbers are most effective for removing particles greater than 1 μm; their scrubbing efficiency is much greater on unpatterned wafers

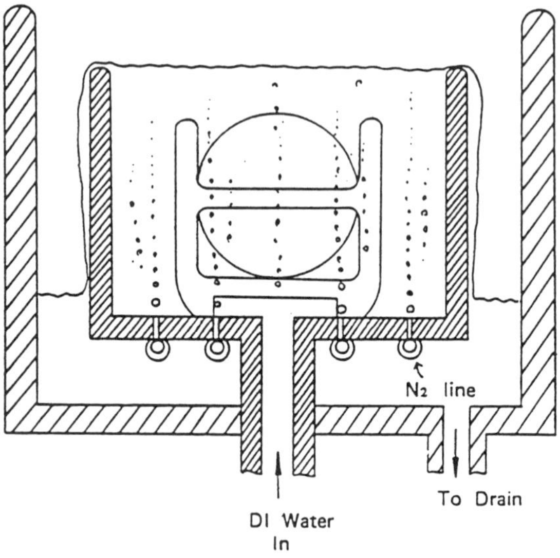

Figure 23-1. Schematic of overflow rinse tank.

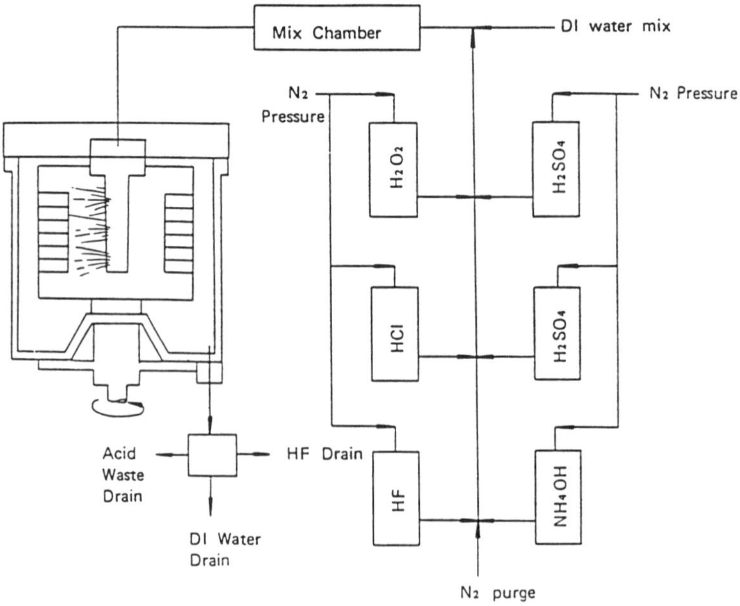

Figure 23-2. Spray acid processing system.

than on wafers with a topographical surface. Brush scrubbers are maintenance-intensive. One common problem associated with a brush-scrubbing unit is scratching of the wafer surface because of a lack of proper mechanical adjustment. Results can be disastrous if such a unit is left to operate without proper mechanical adjustments, brush replacement, etc.

Megasonic cleaning provides a 0.8 MHz sonic wave and 5 to 10 W/cm^3 of energy to the wafer surface to remove deposited particles (Shwartzman, Mayer, and Kern, 1985). Megasonic energy is provided by piezoelectric transducers, typically mounted at the bottom of a polypropylene or PVDF cleaning tank. A cassette of wafers is robotically moved past the megasonic wavefront several times during the cleaning cycle. The megasonic energy produces a high-pressure wave with a positive and negative vector force of frequency 850K to 900K times per second. Particle removal is aided by the use of an ammonium hydroxide:hydrogen peroxide:water solution (SC1) that has a surface wetting feature to help remove particles and prevent their redeposition onto the wafer surface. After the megasonic cleaning step, the wafers are moved to a rinse tank, then to some type of wafer-drying device.

The bar graph in Figure 23-3 shows typical results of a particle removal test with a megasonic cleaning system. Two particle sizes are shown: ≥0.3 μm and ≥1.5 μm. The wafers used for these tests were unpatterned silicon wafers with a 20 to 30 Å native oxide. The particles were a random mixture of inorganic and organic contaminants from an ion implant process operation. At particle sizes greater than 0.3 μm, the particle removal efficiency for the megasonic process averaged 75 percent for the specific type of contaminated wafers used in this test. The analogous average particle removal efficiency for a double-sided ultrapure water scrubber was 23 percent for particles greater than 0.3 μm. It must be noted that the wafer rinse and dry steps after a megasonic cleaning step can easily degrade the cleaning ability of the integrated megasonic process if sufficient care is not exercised (Dillenbeck, 1988).

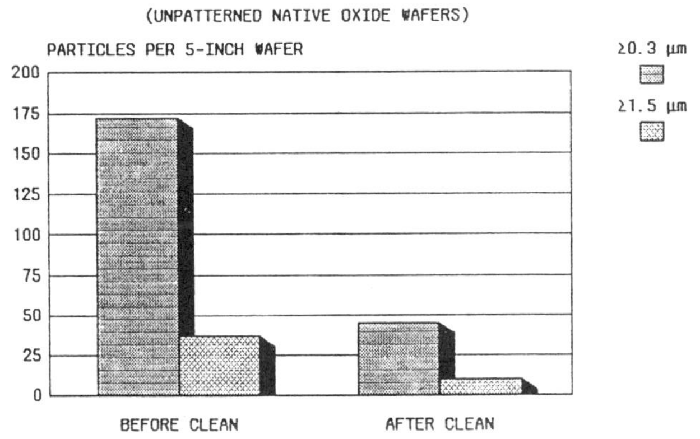

Figure 23-3. Typical megasonic process cleaning efficiency.

PROBLEMS ASSOCIATED WITH AQUEOUS CLEANING

There are several generic problems associated with wet wafer cleaning techniques. In an immersion bath system, the wafers themselves add ionic and organic impurities into the bath which redeposit on subsequently processed wafers (Jones et al., 1987). New technology involving continuous recirculation and distillation is proving to be effective for reducing ionic contamination onto wafers. This technology can be important for the specific case of hydrogen peroxide, as aluminum, tin, sodium, and other metallic impurities are typically present because of the stabilization compounds and the storage and transport vessels used.

In an immersion bath system, the particle concentration at the liquid/air interface can be very large due to buoyancy and surface tension effects at the surface. Wafers can be severely contaminated as they are pulled past this interface (Michaels et al., 1988). Properly designed recirculation systems can reduce this effect, but not eliminate it. Some chemicals are relatively difficult to filter for various reasons. The result is that these chemicals represent a significant particle source to the cleaning process. This is particularly true of the more viscous acids. Lastly, the wafer drying process can lead to several types of process anomalies that cause serious defect problems (Gill and Dillenbeck, 1988). These problems arise when water cannot be removed quickly, uniformly, and completely from the wafer surface, and are worsened as topographical complexity increases.

DEVELOPMENTS IN DRY WAFER CLEANING

Recent developments have been reported on the use of dry wafer-cleaning processes as a replacement for wet chemical cleaning processes (Ruzyllo, 1988). Dry cleaning processes can be of much higher purity than wet processes, in part because of the higher degree to which a gas can be purified and filtered (Ohmi and Kanno, 1987). There are several other potential advantages of wafer cleaning in the gas phase that are discussed in Ruzyllo (1988).

The ultimate requirement of an integrated dry cleaning process, which will almost certainly operate in a single-wafer mode, is to allow volatilization of all contaminants of concern, without adversely affecting other features of the circuit devices. A successful gas-phase cleaning process that addresses all contaminants of concern in sequence has yet to be developed; however, single dry processes that can remove specific impurities have been demonstrated.

To remove the native oxide from a silicon surface, a dry process has been developed using anhydrous HF gas in the presence of water vapor (Cleavelin and Duranko, 1987). This process has been reported to be able to uniformly remove silicon dioxide, yielding a partially fluorinated silicon surface that is ready for the subsequent process operation. The need to rinse or dry the wafers is thereby eliminated. The potential advantage of this treatment is that the deposition of particles on the bare (i.e., no native oxide) silicon surface is significantly reduced as compared to the aqueous case. It is well known that a silicon surface from which the native oxide has been removed is very chemically reactive, and hence sensitive to particle deposition during subsequent treatments. The mechanism of this phenomenon is not very well understood. While the usefulness of this approach is presently limited by the lack of an integrated dry cleaning process, it has been found to improve cleaning of contact holes prior to metallization (Cleavelin and Duranko, 1987).

A recognized method for removing organic contaminants from a surface is the use of ultraviolet (UV) radiation in the presence of oxygen (Vig, 1987). A 185 nm UV light source creates ozone in the presence of oxygen. The ozone/oxygen mixture reacts with 254 nm UV light to volatilize organic compounds on the wafer surface, releasing carbon dioxide. Additionally, the silicon surface is slightly oxidized as a result of the oxidizing environment. Successful attempts to use UV/ozone cleaning to remove residual organic impurities on silicon surfaces have been reported (Ruzyllo, Duranko, and Hoff, 1987; Norstrom et al., 1985; and Ruzyllo et al., 1987).

It has been demonstrated that metallic impurities can be removed through exposure of the surface to UV light in the presence of hydrogen chloride gas. This process was proven to reduce the deleterious electrical effects on silicon device performance of transition metal impurities. The effectiveness of thermally stimulating the UV/HCl environment by use of rapid thermal processing was demonstrated by Nulman (1987).

CONCLUSION

We are in an evolutionary period with regard to wafer cleaning process technology. New chemical treatments that significantly improve upon aqueous chemical cleaning processes are being developed and introduced commercially. The development of dry wafer cleaning techniques represents a new frontier in wafer surface preparation technology. The individual process elements to remove silicon dioxide and certain surface impurities in the gas phase exist and have been proven to be effective to some extent. Considerable process and hardware development activities are required to fully develop each component of an integrated dry wafer cleaning process. At least one significant gap remains: no one has found a suitable technique for removing small particles from a wafer surface in the gas phase. These developments in wafer-cleaning technology are certainly required if they are to keep pace with the increasingly complex integrated-circuit process technology.

REFERENCES

Bakale, D. 1986. Surface Analysis Techniques and their Applications. Microcontamination Conference Tutorial, November.

Bansal, I. K. 1984. Particle Contamination During Chemical Cleaning and Photoresist Stripping of Silicon Wafers. *Microcontamination* 2(8):35.

Burggraaf, P. S. 1981. Wafer Cleaning: Brush and High-Pressure Scrubbers. *Semiconductor International* (July):71-88.

Burkman. 1981. Optimizing the Cleaning Procedure for Silicon Wafers Prior to High-Temperature Operations. *Semiconductor International* (July).

Cleavelin and Duranko. 1987. Silicon Dioxide Removal in Anhydrous HF Gas. *Semiconductor International* 10(12):94-99.

Dillenbeck, Keith. 1988. Application of Contamination Engineering Techniques for Solving Particle Problems in Process Equipment. *Proceedings of the 34th Annual Technical Meeting of the Institute of Environmental Sciences*, May, pp. 532-539. IES, Mount Prospect, Illinois.

Gill, G. P., and K. D. Dillenbeck. 1988. The Use of Snake Patterns for Yield Enhancement. In: *Fifth International Symposium on Semiconductor Processing*, D. Gupta, ed. American Society for Testing and Materials, Philadelphia.

Jones, A. H., J. G. Hoffman, A. W. Jones, and W. Yuan. 1987. Improved Wafer Cleaning with Ultrapure Piranha. *Proceedings of the Microcontamination Conference*, October, pp. 69-80.

Kern, W. 1984. Purifying Si and SiO_2 Surfaces with Hydrogen Peroxide. *Semiconductor International* (April).

Kern, W., and D. A. Puotinen. 1970. Cleaning Solutions Based on Hydrogen Peroxide for Use in Silicon Semiconductor Technology. *RCA Review* 31(June):187.

Michaels, L. D., V. B. Menon, R. P. Donovan, and D. S. Ensor. 1988. Particle Deposition at the Solid-Liquid Interface. Presented at the Institute of Environmental Sciences Annual Meeting, May.

Norstrom, H., M. Ostling, R. Buchta, and C. S. Peterson. 1985. Dry Cleaning of Contact Holes Using Ultraviolet Light (UV) Generated Ozone. *Journal of the Electrochemical Society* 132:2285-2287.

Novak. 1987. Wafer Cleaning for VLSI/ULSI Processing. *Proceedings of the Microcontamination Conference*.

Nulman, J. 1987. In-Situ Rapid Thermal Cleaning and Growth of Thin Silicon Dioxide Films. Electronic Materials Conference, Abstract J-2, Santa Barbara, California.

Ohmi, T., and Y. Kanno. 1987. Ultra-Clean Gas Delivery for ULSI Fabrication and Its Evaluation. Presented at the ECS Symposium on ULSI Science and Technology, May.

Phillips, B. F., D. C. Burkman, W. R. Schmidt, and C. A. Peterson. 1983. The Impact of Surface Analysis Technology on the Development of Semiconductor Wafer Cleaning Processes. *J. Vac. Sci. Technol.* (April/June).

Ruzyllo, J. 1988. Evaluating the Feasibility of Dry Cleaning of Silicon Wafers. *Microcontamination* (March):39-43.

Ruzyllo, J., G. T. Duranko, and A. M. Hoff. 1987. Preoxidation UV Treatments of Silicon Wafers. *Journal of the Electrochemical Society* 134:2052-2055.

Ruzyllo, J., G. T. Duranko, J. T. Kennedy, and C. G. Pantano. 1987. Removal of Organic Contaminants from Silicon Surface after Reactive Ion Etching. Electrochemical Society, *Proceedings of the First International Symposium on ULSI Science and Technology*, Philadelphia, pp. 281-289.

Shwartzman, S., A. Mayer, and W. Kern. 1985. Megasonic Particle Removal from Solid-State Wafers. *RCA Review* 46(March):81.

Skidmore, K. 1987. Cleaning Techniques for Wafer Surfaces. *Semiconductor International* 10(9):80-85.

Vig, J. R. 1987. UV/Ozone Cleaning of Surfaces. In: *Treatise on Clean Surface Technology*, K. L. Mittal, ed. Vol. 1, pp. 1-26, Plenum Press, New York.

24. EQUIPMENT CLEANING TO MINIMIZE PARTICLE DEPOSITION

WAYNE G. FISHER Texas Instruments, Inc., Sherman, Texas

INTRODUCTION

This chapter is a discussion of practical ways to determine when processing equipment needs to be cleaned, what level of cleaning effort is needed, and how to accomplish the cleaning. Ways to determine the source of particle problems are discussed. Finally, a method for reducing the frequency of major equipment cleaning efforts by purging with inert gases is discussed.

DETERMINING THE NEED FOR EQUIPMENT CLEANING

Two general conditions determine when processing equipment should be cleaned. The first of these is when it is certain that a particle problem exists in the equipment. The problem may have occurred because of some catastrophe, such as a broken wafer or broken or worn part in the machine. Or, the particle problem may have been found by a particle measurement made according to a routine schedule, because a problem was suspected, or because a new machine was installed or an old one reconfigured for a new process.

The second condition that determines that equipment should be cleaned is an established schedule of periodic equipment maintenance. The time between such periodic cleanings may be as short as between each wafer in a single-wafer processing system, or as long as days or weeks when the only particle-generating mechanism is worn-out mechanical parts. Generally a great deal of experience with the equipment is required to determine the appropriate schedule for periodic cleaning. When the time between cleanings is too long, product is lost to particle contamination. When it is too short, equipment and personnel are lost to production while the machine is being unnecessarily serviced.

There are a variety of ways to measure the particle state of a piece of equipment, but the most common is to measure the number of particles added to a wafer that has passed through the equipment. Using unpatterned wafers, the number of "particles per wafer per pass" (PWP) may be measured quantitatively using a surface particle scanner. Equipment characterization using this method has been rigorously defined by Tullis (1985a, 1985b, 1986; see also Chapter 13). The minimum sizes of particles detected by surface particle scanners on bare polished silicon wafers are 0.1 to 0.3 μm.

A second, less rigorous way to number the particles on an unpatterned test wafer is to inspect its surface under an intense light in a black box. Microscope or slide projector lamps installed in the top of a plexiglas or metal box are suitable for these inspections. The sizes of particles that may be seen depend on the substrate, the type of particle, and the skill of the inspector, but particles with dimensions as small as 1 μm can typically be detected on bare silicon wafers. The disadvantages are that the particle size is undefined and the number of particles counted depends on the inspector. It is difficult to count the exact number of particles seen unless the total number of particles is very small. However, the method is very fast and the inspection equipment is inexpensive. Gross problems may be rapidly detected so that they can be corrected before much damage occurs.

Unpatterned wafers can be processed under conditions that closely approximate those used on product wafers, but often the conditions must be modified so that the resulting wafer surface is readable by a surface scanner. This is because surface particle scanners cannot generally accurately count particles on surfaces other than bare silicon. They also cannot be used on surfaces with a pattern. Real product wafers must be used to exactly duplicate the conditions experienced by product wafers during processing.

Patterned wafer inspection systems are now commercially available. Some instruments compare the pattern on the wafer to a data base; deviations of the pattern from the data base are interpreted as defects in the pattern. Other instruments detect the periodicity of the pattern either electronically or optically. In this case, deviations from the periodicity are interpreted as defects. The minimum size particle detected by this equipment is around 0.5 μm to 1 μm.

An advantage of the instrument inspection of patterned wafers is that the measurements can be made on product wafers that have undergone the full, unmodified process. A disadvantage is the high cost of the current generation of patterned wafer inspection systems. Also, the time required to inspect a patterned wafer is generally longer than the time to inspect an unpatterned wafer.

Patterned wafers may also be manually inspected with an optical microscope. Because the entire wafer cannot be inspected in this manner, it is best to define the specific parts of the wafer to be inspected. Preferably these parts should be whole integrated circuits so that the area inspected is unambiguous. The circuits should be located across the whole wafer, say, in the center and in each of the four quadrants, or along a diameter or two perpendicular diameters. The inspection is easiest if the pattern within the circuit is repetitive, as in a memory chip. Skilled inspectors can consistently locate particles only a few tenths of a micrometer across. The particles found are counted during the inspection.

An advantage of optical microscope inspection is that particle defects can be separated from defects of other types, although processing problems other than particle problems may often be solved at the same time that particle data are collected. Disadvantages are the possible inspector-to-inspector variability of the inspection. Also, patterns of particle deposition on the wafer might be missed because the whole wafer is not inspected. These patterns may be important clues to the origin of the particles. Finally, such inspections may be very tedious.

An inspection of unpatterned or patterned wafers will identify the number of particles added by the equipment or process to a wafer, a group of wafers, or to some fixed area on a wafer. To follow the performance of the equipment with time, inspectors may plot this number on a control chart (Western Electric Company, 1982; see Chapter 13). The appropriate control chart to use is a C-chart, a chart that uses the number of defects present in a sample of constant area. In the case at hand, the defects are particles on wafers, and the area of each sample is constant at one wafer, one group of wafers, or one group of circuits on a wafer.

A C-chart is constructed by collecting particle data from 10 or more samples, where each sample is a wafer, a group of wafers, or a group of circuits on a wafer. Each sample is plotted on a graph in the order the data was collected. The average of the number of particles found on each sample is calculated. This value becomes the centerline of the chart. Control limits are calculated and plotted on the graph on each side of the centerline. The values of the control limits are the average value, plus and minus three times the square root of the average value.

The value of control charts is that they allow one to rapidly determine whether the number of particles counted is changing outside of limits reasonably expected based on the history of the process. As each data point is added to the chart, simple rules are applied to make this determination. Detection of a change in the process can be used in two ways. First, if the number of particles added per unit area increases, the chart signals that there has been some change in the state of the equipment and that it is time to clean the machine. Second, the chart may be used to determine when changes to the equipment or process have reduced the number of particles added to wafers. When a change for the better can be made permanent, a new control chart is constructed and monitored to ensure that the improvement is sustained.

In practice, the number of particles added by each process or machine should be measured for each process in the whole manufacturing sequence. Efforts to reduce the number of particles added should be expended on those processes that add the most numbers of particles, or where it is known by experience that the particles are most likely to do damage. Processes that add fewer particles should be monitored to ensure that they do not change. When the cleanest processes add very few particles and are unchanging, fixed control limits should be set for their control charts.

The time between each particle measurement must be determined by experience. Some equipment may require a measurement each shift, each day, or before each lot is processed. For stable equipment and processes, or robust products, the interval between measurements may be extended. The idea is to catch changes in the particle state of the equipment before product is damaged without expending resources on unnecessary particle measurements.

At some point in the improvement of a process the number of particles found with each particle measurement will be so low that it is uncertain whether variability is due to the processing equipment or to the particle measurement apparatus (see Chapter 13). Generally, this number will be less than 10 particles per wafer. Then it is useful to magnify the number of particles seen. This may be done by looking at smaller particles, such as by resetting the threshold of detection of a surface scanner so that smaller particles are seen. One may also look at more wafers, or a larger number of circuits on a wafer, to increase the total surface area inspected. Also, wafers may be subjected to more than one pass through the equipment or to the process for a longer period of time so that they accumulate more total particles to detect. The total number of particles on the wafer surface will be the sum of the number of particles added with each pass.

Finally, it should be noted that control charts should be a means to control equipment rather than an end in themselves. As long as there is sound knowledge of the equipment and processes that the charts describe, they may be used in any manner that works. Control limits may be set according to good statistical principles or arbitrarily as the informed user sees fit.

When a method for monitoring a machine is established, and the normal particle performance of the machine has been determined, then the need for cleaning can be found by comparing the machine's present performance with its previous capability. Obviously, if a catastrophe like broken wafers or a broken machine part has occurred, some cleaning or repair is needed. Figure 24-1 illustrates the procedure to follow in all other cases. First, if the particle performance is satisfactory, no cleaning is required. If the number of particles added does not meet the specification or exceeds the control limit the first step is to verify the measurement. The verification ensures that a high particle count is not the result of an error in operation of the process or particle measurement machine or the result of careless wafer handling. If the particle count remains marginally above the specification, a group of clean test wafers should be cycled through the machine, then the particle count remeasured.

If particle levels are repeatedly found to be at very high levels above the machine's previous performance, some cleaning will be required. The level of cleaning effort must be determined by further tests.

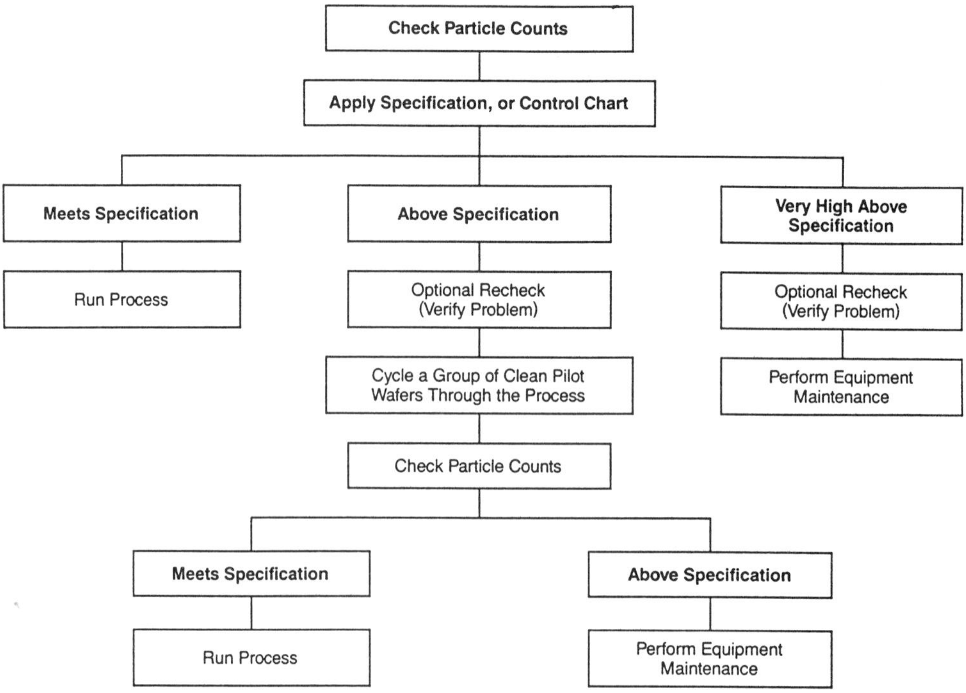

Figure 24-1. Flowchart for determining the need to perform equipment maintenance to reduce the number of particles added to wafers.

EQUIPMENT CLEANING METHODS

Two levels of cleaning effort may be applied to equipment. The first level requires no or only minor equipment disassembly and will be described first. The second level requires a major disassembly of the equipment. This level may include a detailed analysis of the contributions to the particle level by each component of the system. This procedure may be used not only to solve specific particle problems, but to improve the overall performance of equipment beyond its current capability.

The general philosophy of equipment cleaning should be that the minimum time and effort is spent to solve the particle problem. So, knowledge of the previous particle history of the equipment or process should always be used during attempts to locate and eliminate a particle problem. For example, if a plasma deposition reactor is known to accumulate deposits in a reaction chamber, and these deposits previously have been found to flake off the walls of the chamber, then the deposits should be suspected when the levels of particles added to wafers increase. Further, if during equipment inspection a failed part is found, say a broken or worn part or leak in a gas line, that part should be fixed before the investigation proceeds.

First, minor cleaning procedures that already have been established for the equipment should be performed. These procedures may include wiping down of interior and exterior surfaces; cleaning of o-ring grooves, gas ports, and other small openings that may accumulate particles; and reviewing plasma etching or ashing procedures in plasma equipment.

Chronic problem areas, such as hardware that has previously failed, should be inspected, and any problems found should be noted and repaired. These areas may include o-rings or other seals that wear out, flaking deposits on reactor walls, and cracks, grooves or seal areas that have been found to accumulate particles.

The equipment should be checked for broken wafers and wafer fragments that have escaped detection. If these are found, the affected area will probably have to be cleaned by vacuuming and wiping.

The equipment should be thoroughly checked for broken and worn parts, leaks, and any other malfunctions. Any indication that the equipment is not performing as it was previously should be investigated, even if wafers are processed correctly in all ways but particle addition.

The process gases should be cycled to full flow, one gas at a time, several times each, to flush particles from gas lines and from surfaces in the process chamber. The gases should be pumped from the process chamber at maximum pumping speed to remove entrained particles from the system.

The interior of the equipment can be cleaned with a vacuum cleaner. Special care should be taken to clean slots, grooves, corners, and cracks. Care should also be taken to use only clean tools. Disposable vacuum cleaner tool tips may be made from PVC or other plastic pipe or tubing.

The interior and exterior of the equipment, including all surfaces that are anywhere near wafers while they are in process, can be wiped down with lint-free cloths moistened with a solvent. The solvent may be water, isopropanol, chlorofluorocarbons, or mixtures. Foam swabs and razor blades may be used to clean grooves, slots, cracks, corners, and holes. Safety measures appropriate for the solvent should be rigorously followed. After wiping down surfaces it is generally advisable to purge the equipment with flowing gases, to perform a plasma etch or ash or other conditioning if possible, and to cycle a group of clean pilot wafers before rechecking particle levels.

At various points during the previous cleaning steps it is advisable to remeasure the levels of particles added to wafers to determine whether the particle problem has been fixed. Judgment must be applied to determine the likelihood that a particular procedure has solved the problem.

DETERMINING THE SOURCE OF PARTICLE PROBLEMS

Major-effort equipment cleaning methods generally involve disassembly of the equipment and inspection and cleaning of each of the individual parts. But before the disassembly it is best to troubleshoot the system to determine which parts of the hardware are causing the particle problem. This is done by analyzing the various parts of the process separately, and using particle detection methods to determine that part's contribution to the number of particles added by the process. The idea is to isolate each part of the process or each part of the hardware from the other parts and determine whether that part is contributing to the observed particle problem.

Wafer-handling hardware can be checked by cycling wafers into and out of a machine without flowing any of the processing gases, without pumping or venting reaction chambers or load locks, and without heating the wafers. The integrity of each of the processing gases may be checked by flowing one gas at a time. Pumping and venting hardware may be checked by performing only pump and vent cycles, while flowing only the minimum number of processing gases, or using inert gases. The temperature may be cycled without flowing process gases. In plasma-based processes the process may be run with and without the rf field.

The particle contribution of each part of the equipment can be measured in several ways. Unpatterned wafers may be cycled through the test, and the number of particles may be determined by before-and-after measurements with a surface scanner. Particles shed by mechanical equipment may be detected by collecting them on witness wafers placed near the mechanism while it is being cycled. It may be useful to repeat the abbreviated process a number of times to collect more particles. These particles may also be detected by an optical particle counter or condensation nucleus counter. A disadvantage of these methods is that the witness wafers or the particle counter inlet must either be placed in the area where it has been judged that particles are most likely to be found, or must be placed by trial and error. Also, these tests must be done in a very clean environment so that fallout on the wafers or particles counted by the particle counter from the air are an insignificant part of the total number of particles counted. An example of the use of both of these methods to test a wafer handling robot is detailed in the paper by Hardegen and Lane (1982). Chapter 14 also describes methods for measuring particulate emissions from equipment.

Another method of troubleshooting specific parts of equipment is to "vacuum clean" the suspected area, filtering the air and collecting the particles with a membrane filter. A convenient filter to use is the Nuclepore membrane filter, because these filters have a smooth, flat surface that is ideal for viewing collected particles, and they are available with pore sizes ranging from more than 10 μm to less than 0.1 μm. The filter is held in a stainless steel or plastic housing while air is drawn through the filter with house vacuum or a small vacuum pump. The airflow rate may be controlled by an orifice between the vacuum source and the filter (Hinds, 1982).

The membrane filter can be inspected for particles with an optical or scanning electron microscope (SEM). If particles are collected, they may be analyzed in the SEM by energy dispersive x-ray analysis, or by any of the other analysis techniques described in Chapter 20. The identity of the particles is an important clue to their origin. If aluminum particles are detected, for example, they can only come from aluminum parts of the system.

A final method of troubleshooting the various parts of a process is with an in situ particle detector. This method has an advantage over the previous ones because it measures in real time the number of particles emitted by the equipment. This capability is useful when it is difficult to separate the particle-generating parts of a process using other techniques. For example, to test a pumping and venting system, wafers may be placed in a chamber and the chamber pumped, then vented. If particles are detected on the wafers, it is not easy to tell whether they were deposited during the pumping portion or the venting portion of the cycle.

A commercially available in situ particle detector is the High-Yield Technology Particle Flux Detector. This instrument is similar to the particle-detection apparatus of a standard light scattering optical particle counter, but it has a novel design to increase the area that is sensitive to particles. A diode laser light beam is reflected back and forth between two mirrors to create a sheet of light. Particles that pass through the light beam scatter light out of the plane of the sheet. The scattered light is focused on detectors by mirrors. The amount of scattered light is proportional to the particle size. An electrical feedthrough provides power to the instrument when it is mounted inside equipment, and passes signals from the instrument to external signal processing hardware. The minimum particle size detected is about 0.5 μm, although the probability that particles of this size will be detected is quite low. Further details of the design, operation and performance of the High-Yield Technology monitor are discussed in Borden, Baron, and McGinley (1987a, 1987b); Borden and Knodle (1988); Caldow, Pui, and Liu (1987); and Bowling, Larrabee, and Fisher (1988).

Unlike a typical optical particle counter, no airflow is required to pass particles through the detection region of the High-Yield Technology detector. The instrument relies on the motion of particles through the ambient to move the particles through the light beam. This makes the instrument useful in reduced-pressure environments or in vacuum systems. The instrument provides a real-time measure of the number of particles that pass through the detector region so that particle events may be correlated with the activity in the process chamber. For example, the relative numbers of particles liberated into a vacuum system during pumping and venting cycles may be compared.

A similar system has been described by Bowling, Larrabee, and Fisher (1988), who modified a TSI laser particle counter for use as a particle detector in a vacuum system. The area between the counter's light source and detector was bored out to provide a 1-in. passage for particles. Although the particle detection region of this modified counter was smaller than that of the High-Yield Technology counter, useful data on the particle performance of vacuum systems were obtained. This instrument detects particles as small as 0.5 μm.

Another approach to in situ particle detection is the use of gas line particle monitors as detectors in pumping and venting lines. In contrast to the High-Yield Technology and modified TSI counters described above, these monitors use a light source outside the area of interest, the gas line, that looks into the line through a window. Gas line particle monitors may be installed in pumping lines to detect particles liberated in vacuum chambers during semiconductor processing. Commercially available systems made by PMS, Met One, and PPM detect particles as small as 0.1 μm.

MAJOR-EFFORT EQUIPMENT CLEANING METHODS

A major-effort equipment cleaning begins with disassembly of the equipment. In this type of cleaning, all parts should be inspected to detect failures, worn conditions, and particle or other residues. Particles and residues should be removed with an abrasive pad, swabs, razor blades, or other tools, then wiped clean with lint-free cloths and solvent. Small parts may be cleaned in an ultrasonic cleaner with a solvent such as isopropanol, then dried in an oven to remove solvent residue. A high-pressure solvent spray may also be used. After cleaning, it may be appropriate to passivate the surface of metal parts by anodization or other surface treatments. All the cleaned parts should again be carefully inspected, and worn and suspect parts should be replaced. The final cleaning steps should be done in a clean area so that the parts are not contaminated by the environment.

After the equipment is assembled, it should be purged with a high flow of inert gas to flush out residual particles. A series of high-flow-rate pump and vent cycles will help remove particles from surfaces inside the equipment. Then, etching, ashing, or other equipment conditioning steps may be appropriate. A group of clean wafers should be cycled through the process to test the final cleaning. When the equipment is judged to be clean, the process is ready to be tested for particles.

AVOIDING EQUIPMENT CLEANING

The frequency of major equipment cleaning may be reduced by in situ cleaning of the equipment by purging with inert gas. The object of the purging is to entrain particles into the inert gas from surfaces in the equipment and to pump them out of the system with the gas. If this procedure is done periodiccally, the system will remain clean longer and will be less likely to require major cleaning. The feasibility of this strategy has been demonstrated in experiments comparing the numbers of particles detected during slow and fast venting and pumping of a vacuum chamber (Bowling, Larrabee, and Fisher, 1988). During slow pumping and slow venting, few particles were detected by a particle monitor in the chamber. During fast pumping and fast venting, many particles were detected. The particles detected during the fast pumping and venting were dislodged from the walls of the chamber and the pumping and venting hardware by the rush of gas into and out of the chamber.

The probability that a particle on a surface will be entrained in a gas depends on the drag force on the particle, the boundary layer thickness of the gas over the surface, and whether the gas flow is laminar or turbulent. To maximize the entrainment of particles from a surface, the drag force should be large, the boundary layer should be thin, and the flow should be turbulent (see Chapters 21 and 22).

DRAG FORCE

For a free particle in a gas (and at Reynolds numbers less than 1000), the drag force is given by Stokes law,

$$F = \frac{3\pi u d_p v}{c} \tag{24-1}$$

where

- u = the gas viscosity,
- d_p = the particle diameter,
- v = the relative gas-particle velocity,
- c = a slip correction.

The drag force on a particle depends directly on the gas viscosity and velocity. To maximize the drag force on a particle, the viscosity and the velocity of gas moving past the surface should be maximized.

Because the drag force on a particle depends on the viscosity of the gas, the identity of the gases used for purging may be chosen to maximize the probability of entrainment. Helium, argon, and nitrogen are relatively inert and inexpensive and so are commonly used for purging. The viscosities of these three gases are listed in the table below.

Table 24-1. Viscosities of Three Inert Purge Gases

Gas	Viscosity (micropoise)
Helium	194
Argon	222
Nitrogen	175

Argon has the highest viscosity of the three gases, so it should produce the highest drag force on particles. The viscosity of helium is 13 percent lower than that of argon, and the viscosity of nitrogen is 21 percent lower than that of argon. Based on the drag force, argon is the best gas to use for cleaning by purging.

BOUNDARY-LAYER THICKNESS

The discussion of the drag force above assumed that the particle was free in the gas. For a particle on a surface, this is not the case, because the particle will be shielded from the full drag force by the boundary layer of the gas on the surface. The exact form of the equation describing the thickness of the boundary layer is a function of the orientation of the gas flow relative to the surface. Generally, though, the boundary-layer thickness is inversely proportional to the square root of the main stream velocity of the gas. It is directly proportional to the square root of the kinematic viscosity, which is the gas viscosity divided by the gas density. To minimize the boundary-layer thickness and increase entrainment of particles, the gas velocity must be high and the kinematic viscosity must be small.

The data below show that helium will have a boundary-layer thickness three times that of argon or nitrogen when the flow rates and system configurations are the same. This is because the square root of the kinematic viscosity of helium is about three times that of argon and nitrogen. To entrain particles during a purge cycle, argon is favored as it should have the thinnest boundary layer of the three gases. Nitrogen is a close second choice with a boundary-layer thickness only slightly greater than that of argon. Helium is a poor choice because it has a relatively thick boundary layer.

Table 24-2. Boundary-Layer Thicknesses of Three Inert Purge Gases

Gas	Relative Boundary-Layer Thickness
Helium	34.1
Argon	11.6
Nitrogen	12.3

TURBULENCE

A third factor to consider in the choice and handling of purge gases is whether the flow of the gas is laminar or turbulent. Turbulent flow will favor particle entrainment from surfaces because it reduces the boundary-layer thickness. Turbulence also creates ever-changing regions where the instantaneous fluid velocity is higher than the average velocity.

The Reynolds number of the flowing gas is a benchmark for the transition from laminar to turbulent flow. It is given by the formula

$$Re = \frac{vd\rho}{u} \qquad (24\text{-}2)$$

where

v = the gas velocity,
d = linear dimension characteristic of the system,
ρ = the gas density,
u = the gas viscosity.

For a given fixed fluid velocity and system configuration, the Reynolds number is determined by the ratio of the gas density to the gas viscosity. Note that this ratio is the inverse of the kinematic viscosity.

Flow is laminar if the Reynolds number is relatively low and turbulent if it is relatively high. The value of the Reynolds number where the transition from laminar to turbulent flow takes place depends on the configuration of the system. For example, flow through pipes is laminar for Reynolds numbers less than 2000, and turbulent for Reynolds numbers greater than 4000. For relative gas-particle motion, laminar flow occurs for Reynolds numbers less than 1, with increasing deviation from a smooth pattern of streamlines in the gas as the Reynolds number increases above 1 (see Chapter 2). In general, the transition point will be at lower Reynolds numbers when the gas must change direction sharply; i.e., the inertia of the gas is important. The transition point should be low for flow around particles suspended in the gas, or flow through orifices, around corners, or through tortuous passages in valves. The transition point will be at higher Reynolds numbers when the the gas must not change direction sharply, such as when the gas flows through a pipe or parallel to a surface.

In spite of the ambiguity in the value of the Reynolds number marking the transition between laminar and turbulent flow, the performance of various gases may be compared by examining their relative density/viscosity ratios.

Table 24-3. Density/Viscosity Ratios of Three Inert Purge Gases

Gas	Density/ Viscosity Ratio $\left(\frac{s}{cm^2}\right)$
Helium	0.86
Argon	7.49
Nitrogen	6.66

Helium has the lowest ratio, so the Reynolds number for a given flow rate and system configuration is lowest. Helium is more likely to be in the laminar flow regime than argon or nitrogen. The density/viscosity ratio for nitrogen is nearly seven times that of helium and that for argon is nearly eight times that of helium. The flow of nitrogen or argon is more likely to be turbulent than that of helium for a given gas flow rate and system configuration.

The Reynolds numbers of argon, nitrogen, and helium are shown in Figure 24-2 for flow velocities up to 5,000 cm^3/min in a 1/4-in. o.d., 0.035-in. wall thickness tube. The Reynolds number of helium is less than 200 for even the highest flow rate, so the flow is likely always laminar in such tubing. Argon and nitrogen have Reynolds numbers approaching 2000 for the highest velocities, so they are nearing the onset of turbulent flow. During the surges in flow when the gas stream is starting or stopping, the flow of these gases could easily be turbulent. It could also be turbulent if the gas must pass through bends, fittings, or other passages that change the direction of flow, or through valves or orifices that restrict the flow.

In summary, a purge cycle with an inert gas will remove particles from the interior surfaces of equipment by entraining the particles in the gas. The particles may then be pumped out of the system with the gas. The flow rate of the gas into the system, and the rate the gas is pumped out of the system, should both be as high as possible. Argon is the best gas to use for the purging because it will have both the highest drag force on particles and the thinnest boundary layer over surfaces, and is most likely to be in the turbulent flow regime. Nitrogen is a close second choice to argon, but helium is a poor choice for the purge gas.

The frequency required for such a purge cycle must be determined by experiment, comparing particle levels on test or product wafers as a function of the frequency of the purging. It is reasonable to perform a short purge cycle between every wafer in a single wafer processing system. The time required for the purge cycle will add to the overall process time of the wafer, but part of the time might otherwise be taken up solely in wafer handling.

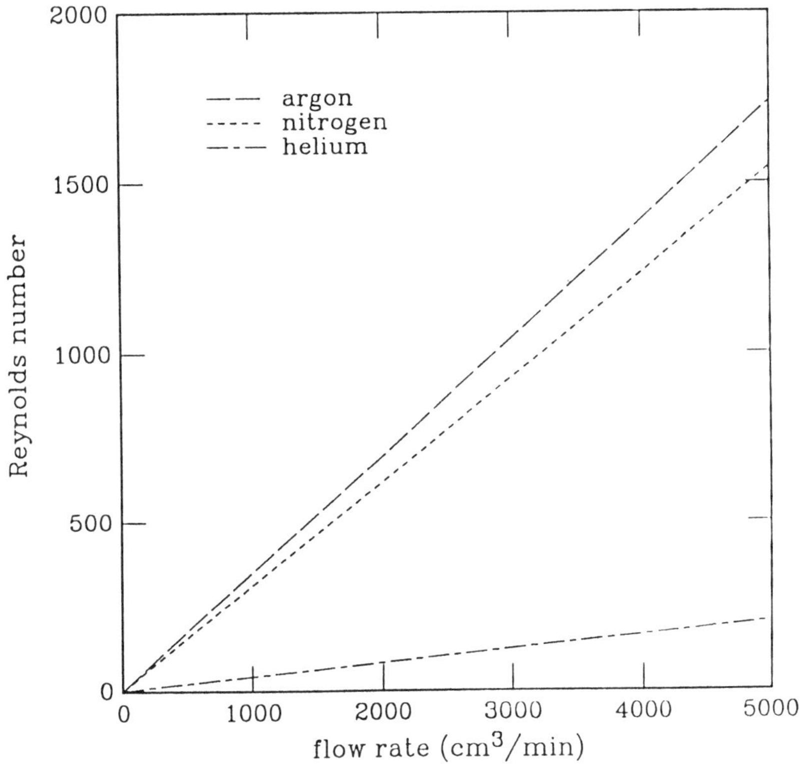

Figure 24-2. Reynolds number of argon, nitrogen, and helium as a function of flow rate in a ¼-in. od, 0.035-in. wall thickness tube.

SUMMARY

In this chapter, the cleaning of processing equipment has been discussed. Techniques to determine the need for cleaning have been described, including the use of control charts to track the particle condition of equipment. Procedures for carrying out minor and major equipment cleaning and for determining the sources of particles in equipment were detailed. Finally, a method was described for decreasing the frequency of equipment cleaning by way of periodic purging with an inert gas.

REFERENCES

Borden, P. G. In press. "An In-Situ Particle Flux Monitor for Semiconductor Processing Equipment Applications." *Proceedings of the 17th Annual Meeting of the Fine Particle Society, 1986.*

Borden, P. G. In press. "In-Situ Monitoring of Particulate in Semiconductor Processing Equipment." *Proceedings of the 18th Annual Meeting of the Fine Particle Society, 1987.*

Borden, P. G., Y. Baron, and B. McGinley. 1987a. "Real Time, In-Situ Studies of Particle Events in Vacuum Processing Equipment and Gas Streams." *Microcontamination* Conference Proceedings, pp. 25-34.

Borden, P. G., Y. Baron and B. McGinley. 1987b. "Monitoring Particles in Vacuum-Process Equipment." *Microcontamination* 5(10):30-34.

Borden, P., and W. Knodle. 1988. "Process Control Through the Measurement of Particle Flux to Wafer Surfaces in Vacuum Process Equipment." In: *Proceedings of the 34th Annual Technical Meeting of the Institute of Environmental Sciences, King of Prussia, PA, May 3-5*, pp. 429-432.

Bowling, R. A., G. B. Larrabee, and W. G. Fisher. 1988. "Status and Needs of In-Situ Real-Time Process Particle Detection." In: *Proceedings of the34th Annual Technical Meeting of the Institute of Environmental Sciences, King of Prussia, PA, May 2-6*, pp. 508-516.

Caldow, R., D. Y. H. Pui, and B. Y. H. Liu. 1987. "Performance of the High Yield Technology PM-100 In-Situ Particle Flux Monitor." Presented at Semicon/Southwest '87 Technical Symposium, Dallas, TX, October 13-15.

Hardegen, B., and A. P. Lane. 1982. "Testing Particle Generation by a Wafer Handling Robot." *Solid State Technology* 29(3):189-195.

Hinds, W. C. 1982. *Aerosol Technology*. Wiley, New York, NY.

Tullis, B. J. 1985a. "A Method of Measuring and Specifying Particle Contamination by Process Equipment." *Microcontamination* 3(11):66-73, 160-161.

Tullis, B. J. 1985b. "Measuring and Specifying Particle Contamination by Process Equipment: Part II, Statistical Considerations." *Microcontamination* 3(12):14-21.

Tullis, B. J. 1986. "Measuring and Specifying Particle Contamination by Process Equipment: Part III. Calibration." *Microcontamination* 4(1):50-55, 86.

Western Electric Company, Inc. 1982. *Statistical Quality Control Handbook*. Western Electric Company, New York, NY.

25. REVIEW OF FED-STD-209D

B. TULLIS Hewlett-Packard Company, Palo Alto, California

INTRODUCTION: HISTORY, SCOPE, AND APPLICATION

The original 209 standard was issued in December 1963, revised to 209A in August 1966, revised again to 209B in April 1973, and then 209B was amended in May 1976 to add a Class designation of 1000. The next revision was 209C dated October 1987 and finally, 209D, June 1988 which is simply a typographically corrected version of 209C. The purpose of this chapter is to review the content and interpretation of the new standard (209D) more completely than previous reviews have done (Cooper, 1986b, 1988; Peck, 1987; and Helander, 1987).

The main body of the text comprising the new standard is 13 pages long, up from 7 pages in 209B. It uses one figure and one table of class counts and a second table of minimum sampling volumes, all as a function of class limits and particle sizes. The 209B standard did not have this latter table of sampling volumes. The one 28-page appendix in 209B, which covered nonmandatory supplemental guidance information on the design and operation of clean rooms and clean work stations, has been replaced by five appendices numbering 27 pages and titled: A) Manual Counting and Sizing Methods [writer's note: for Classes 1000 and larger], B) Operation of Optical Particle Counters, C) Statistical Analysis, D) Sources of Supplemental Information, and E) Glossary.

Like its predecessors, 209D is organized in a typically bureaucratic fashion with a plethora of numbered sections and nested, similarly numbered, subsections. Several errors of a technical nature that this writer has found are listed in the section titled "Shortcomings," later in this chapter. The requirements and guidelines of 209D can be difficult to comprehend because they apply differently to different situations. This chapter may help readers to discriminate among these different applications.

This standard is still intended to establish classes of air cleanliness for airborne particulate levels in clean rooms and clean zones. (The new document uses the term "clean zones" as a more general term than the older term "clean stations.") Additionally, it is to prescribe verification and monitoring methods and class designating criteria.

Its limitations are that (1) the standard does not apply to equipment or supplies, only to clean rooms and clean zones; and (2) it does not characterize the physical, chemical, radiological, or viable nature of particles, other than by population and size.

The standard will certainly influence contractual agreements between government agencies purchasing clean rooms and their suppliers of clean rooms. It will also certainly influence regulations of the equipment and operating practices of the U.S. pharmaceutical industry, under the jurisdiction of the Food and Drug Administration (FDA). It will most likely influence contractual agreements between industrial purchasers of clean rooms and their consultants and suppliers. In all industries its effect will indirectly reach process equipment suppliers who supply equipment to operate in these clean rooms, if such equipment can have a detrimental effect on the maintenance of a clean room or clean zone to meet specified 209D requirements.

CHANGES RELATIVE TO FED-STD-209B

In addition to structural changes to the document organization and content as given above, basic changes to the standard include the following:

- The scope is broadened beyond clean rooms to include "clean zones," defined as those spaces to be controlled by specified limits. The older versions applied only to clean rooms and clean work stations where "work stations" were defined as enclosed areas having their own filtered air or gas supply.
- Differentiations are made among "as built," "at rest," and "operational" conditions or states of the clean room or clean zone under test. As specified by the application, the standard can be applied to any and all states. The earlier versions did not address these differences in application.
- Particle size is defined as the equivalent diameter of a standard reference spherical particle producing the same response in the sensing instrument as the particle being measured. (The index 1.6 presumes particles of latex that have a refractive index near 1.59.) The older standard 209B defined particle size only as the apparent maximum linear dimension of a particle, and the new standard still reserves this definition for particle sizes 5.0 μm and larger when they are measured optically with a microscope.
- The terms "laminar" and "turbulent" have been replaced with the terms "unidirectional" and "nonunidirectional." "Laminar and "turbulent" are scientific terms used in fluid dynamics and have precise definitions that do not apply accurately to clean rooms or clean zones. The newer terms are somewhat less restrictive and more accurate for this application.
- Classes "1" and "10" have been added along with count specifications at particle counts below the lower limit of 0.5 μm used in 209B. See the next section for more details.
- Provisions are included to allow other classes to be defined and concentrations to be measured at other particle sizes. More details and restrictions on defining classes and measuring concentrations are given in the next section. No such provisions were included in the earlier versions.
- Criteria for meeting class designations are carefully described, including calculations for statistical confidence (see the sections on "Class Definitions and Criteria" and "Statistical Analysis," below). The earlier versions did not address how count criteria were to be applied.
- Periodic calibration of particle counters is required, using reference spheres of index 1.6. Calibration for size detectability and sample volumes is required.

Calibration for count number is suggested only by way of the guidelines in Appendix B that refer to American Society for Testing and Materials (ASTM) standard practices. The earlier versions did not address calibration methods or requirements. See the section on calibration below for more details.

- Because the requirements for verification and for monitoring are applied differently, sorting them out can be a bit confusing. (See the next section for clarification.) Some requirements given for verification tests are not specified for monitoring purposes except by reference to a required user-specified "Monitoring Plan."

- Periodic verification and monitoring are both required with particle concentrations. For verifications, measurements must be taken initially and at periodic intervals or as specified. The earlier versions did not address requirements for frequency.

- Minimum sample volumes are set as a function of class and particle size, and are set for other classes and sizes that may be defined, within prescribed limits, by the user. This is done for verification tests only; such requirements for monitoring are left to the user of the standard to define in the required Monitoring Plan. These provisions are new. (See the next section for details.)

- Criteria for locations of sampling points, their number, and number of samples taken at each point are also new. This is the only requirement that discriminates between "unidirectional" and "nonunidirectional" airflow. There is one exception in Appendix A, which describes methodology and equipment for counting particles of size larger than 5.0 μm using a microscope. In that appendix, the requirements for probe orientation and flow rate are different for these two conditions. (See the next section for details.) According to Appendix B, requirements for probe orientation and flow conditions (e.g., isotropy) for particles smaller than 5.0 μm should be found in the main text, but in fact this is not the case. A reference to Appendix A, instead of to the main text, may have been intended.

- One section of the main text, on methods and equipment for both verification and monitoring, mentions that other particle-counting equipment may be used if it demonstrates accuracy and repeatability equal to or better than those for the methods listed. The methods listed are (1) that described in Appendix A, if the particles are greater than 5.0 μm and (2) that described in Appendix B, if the particles are greater than 0.1 μm. This mention of other particle counters refers to two footnotes. One footnote allows for use of condensation nuclei counters (CNCs) for counting particles 0.01 μm and larger. (Unfortunately for the application of CNCs, the standard specifies that particle sizes for defining class concentrations cannot be defined below 0.1 μm and that the concentration criterion must therefore be taken as that for the next larger size that is prescribed for the class in question, e.g., 0.1 μm for a Class 10). The other footnote allows for the use of sedimentation methods (with microscopic counting only), but only for monitoring, not for verification. From the wording here, a safe presumption may be that automatic particle counting may be allowed with the sedimentation technique if shown to have accuracy and repeatability equal to or better than those for microscopic methods. No mention is made of how to relate sedimentation counts to class counts or anything about sampling times and calculation procedures with sedimentation counts, but this is not inconsistent with the general requirements for monitoring, because under monitoring, the approach of the standard is to leave all of these detailed specifications to the user of the standard to define in the required Monitoring Plan.

- Both Appendix A and Appendix B add new subjects to the standard. Appendix A details manual (i.e., with microscope) counting and sizing methods for particle monitoring when the sizes to be counted are 5.0 μm and larger. Counting rules to meet statistical requirements are included. In Appendix A, two collection methods are described, both requiring a flow rate of 1 ft^3/min through a membrane filter, where one is used in what is called "the open filter holder method" and the other is used with an "aerosol adapter" and is called "the aerosol monitor method." Unfortunately the "aerosol adapter" and the "aerosol monitor" are pieces of hardware for which no description or reference is given. Appendix B discusses calibration and operation of optical particle counters and includes guidelines on permissible probe tube lengths, bends, and flow rates.

- Unlike its predecessors, 209D not only prescribes sampling procedures for verification such as frequency, sample volumes, sample location, number of locations, number of samples at each location, calibration, and equipment, but also the calculation procedure in step-by-step form to achieve some measure of statistical certainty in stating that the average sample concentration found meets a stated class count. Furthermore, Appendix C demonstrates the calculation steps with an example. Along with class definitions for the cleaner levels of Classes 1 and 10, this combination of improvements to the standard is a remarkable and much-needed step forward.

- A requirement is added to record environmental factors and use parameters—for example, air velocity, air volume change rate, room air pressure, make-up air volume, unidirectional airflow, temperature, humidity, vibration, equipment, and personnel activity.

- Appendix D is new, with seven pages of reference documents.

- Appendix E is a modest glossary of four terms.

CLASS DEFINITIONS AND CRITERIA

Figure 25-1 tells what classes are defined in 209D and how they differ from the earlier class definitions of 209B. The classes are depicted as straight line segments drawn on log-log graphs. Class designation is a function of particle concentration per cubic foot and of particle size. The count of particles per cubic foot is meant to indicate the count of all particle sizes in 1 ft^3 that are greater than the size indicated along the abscissas. The dots on the class lines mark the intersections of these class lines with particular particle sizes on the abscissa. The particle concentrations of the class lines at these particular sizes are used to compare measured values when testing the compliance of a clean room or clean zone. Figure 25-2 gives the same information (for the 209D definitions only) in tabular form. Note too, from Figure 25-1, that the numbers that give their values to the class designations are the intercepts of the class lines with the 0.5 μm particle size.

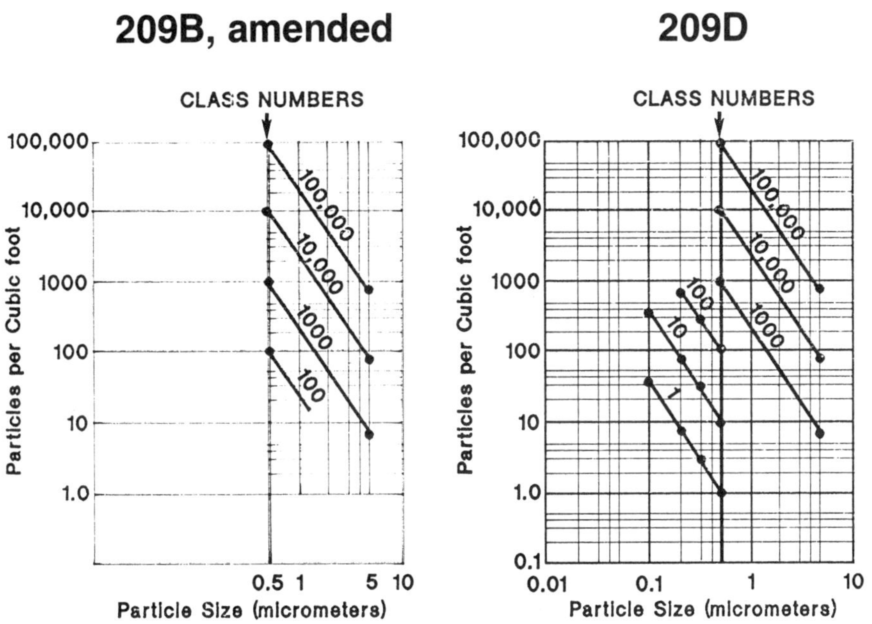

Figure 25-1. Graphs of class limit changes.

Particles/ft^3 with size $\geqslant$ size shown in micrometers

	Measured Particle Size (micrometers)				
Class	0.1	0.2	0.3	0.5	5.0
1	35	7.5	3	1	NA
10	350	75	30	10	NA
100	NA	750	300	100	NA
1000	NA	NA	NA	1,000	7
10,000	NA	NA	NA	10,000	70
100,000	NA	NA	NA	100,000	700

Figure 25-2. Tabular definition of class limits.

In Figure 25-1, note that in moving from the graph for 209B to the graph for 209D, three notable changes have occurred. The first is that in the graph for 209D, the new classes of 10 and 1 appear. Second is that the line defining Class 100 has flipped over into the region of particle sizes equal to or less than 0.5 μm. And third is that the dots on these three lines now reach to the smaller particle sizes of 0.1, 0.2, and 0.3 μm.

Provision is made in 209D for defining other classes and for making measurements at sizes other than those in Figure 25-2 (sizes corresponding to the dots on the class lines in Figure 25-1), but with the following limitations. Other classes can be defined by generating straight line segments that are parallel to those in Figure 25-1; however, these can only lie in a defined region of space described by Figure 25-3. This defined region is made up of three parts. The first part is that region swept out by sliding a line like the Class 1 line upward to coincide with the Class 10 line. The second part is that region swept out by sliding a line like the Class 100 line both downward to coincide with the right-hand-most part of the Class 10 line, or upward almost to where its right-hand end would meet the left-hand end of the Class 1000 line. The third part is that region swept out by sliding a line like the Class 1000 line upward to where it coincides with the Class 100,000 line. The class designation to be given any class is that concentration value where its line intersects the 0.5 μm particle size.

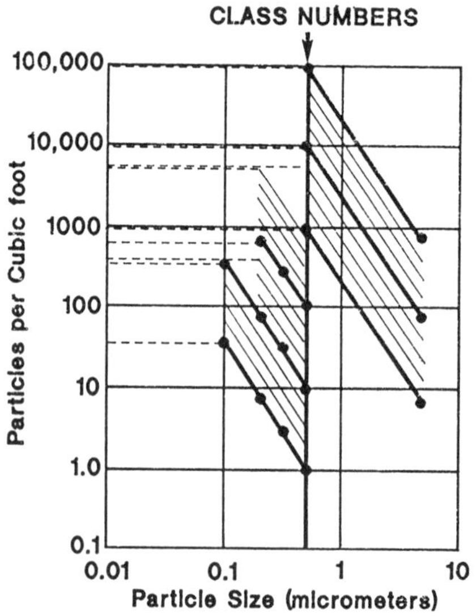

Dashed lines depict maximum concentrations for sizes to left of sloped lines. Shown only for some representative class lines.

Figure 25-3. Graphs of other allowable classes.

If concentrations are measured of particles above some size that is to the left of the left-hand end of a class line, then the criterion that must be met is that concentration value corresponding to the leftmost point on the allowed class line (see the dashed horizontal lines in Figure 25-3). If concentrations are measured at a size within the range of sizes spanned by the sloped line segment of a class, and this size is not one marked with a dot on Figure 25-1 (or found in Figure 25-2), then the concentration criterion to meet the class requirements is to be taken from that of the next larger size that is so marked (or found in Figure 25-2).

Specifications of how to determine whether a clean room or clean zone meets the class number criteria of a class are stated for verification, but for monitoring they are for the most part left for the user of the standard to spell out in a required Monitoring Plan. The specifications given for verification are for frequency of verification, minimum number of locations for measurements, spatial distribution of these locations, minimum number of samples to be taken at each location, volumes of samples, and counters that can be used.

Verifications should be done initially and at periodic intervals or as specified by the user of the standard. Measurements should be performed at a minimum of two locations and with a minimum total of five sample readings. In unidirectional flow, the minimum number of locations is the lesser of the cross-sectional area of the flow divided by 25 or divided by the square root of the class number, and where these measurements are to be uniformly spaced throughout the entrance plane. This entrance plane is defined as that plane perpendicular to the flow that is immediately upstream of the work activity. In nonunidirectional flow, the minimum number of locations is the floor area of the clean zone divided by the square root of the class number, and where these measurements are uniformly spaced horizontally throughout the clean zone (except as limited by equipment) and positioned vertically as specified by the user of the standard.

For monitoring, the standard specifies that measurements should be done during periods of operation and at locations throughout the clean zone, where critical, or where the larger counts were found during verification. Furthermore, it is stated that the air should be sampled "as it reaches the clean zone." "As it reaches the clean zone" is not further defined but probably means, for unidirectional flow, immediately upstream of the work activity. For nonunidirectional airflow, on the other hand, "as it reaches the clean zone" might be where the air enters the clean zone from the air-conditioning ducts, but this writer feels the specification for nonunidirectional flow should have read "adjacent to work activity zones."

Volumetric requirements for verification are specified by Figure 25-4. The specification attempts to assure that on average, samples will have counts of at least 20. At other classes or sizes, the minimum volume requirement is worded as that volume required at the next lower class or size. This writer believes this wording is incorrect and that 209D should have said, "at the next lower class or the next larger size." Only going to "the next larger size" would be consistent with attempts to assure a minimum count of 20 per sample on the average.

Equipment for both verification and monitoring must be in accordance with either (a) Appendix A for manually counting (with a microscope) particles that are equal to or larger than 5.0 μm, or (b) Appendix B for automatically counting (with optical particle counters) particles that are equal to or larger than 0.1 μm. Optical particle counters must count single particles and must be periodically calibrated. An infamous Footnote 1, however, allows users

Class	Measured Particle Size (micrometers)				
	0.1	0.2	0.3	0.5	0.6
1	0.6	3.0	7.0	20.0	NA
10	0.1	0.3	0.7	2.0	NA
100	NA	0.1	0.1	0.2	NA
1000	NA	NA	NA	0.1	3.0
10,000	NA	NA	NA	0.1	0.3
100,000	NA	NA	NA	0.1	0.3

- Use volume of next lower class or size* for other classes or sizes. [*209D error? Should be "larger" size.]
- Larger volumes improve precision but may make sample times impractical. Report volumes used.

Figure 25-4. Tabular specification of minimum sample volume (ft^3).

of the standard or their contracting agencies to specify their own methods, although it does not refer to equipment in the same manner. With reference to equipment, it does say that other methods and equipment may be used if demonstrated to have accuracy and repeatability equal to or better than those described in Appendixes A and B.

Another footnote allows for the use of CNCs to count particles of 0.01 μm or larger. But the criteria for a class, when measurements are used at sizes other than those listed in Figure 25-2, require that the concentration measured be numerically equal to or smaller than the concentration allowed at the next larger size of those given for the class in Figure 25-2. This next larger size for the case of a CNC counter would be the smallest size listed in Figure 25-3 for that class. If the class in question were one of those "other" classes depicted in Figure 25-3, then the concentration criterion to be met would be a concentration interpolated between the next higher and next lower classes that are found in Figure 25-2. Incidentally, this application of the rules for dealing with "measurements at other sizes" and "other classes" represents this writer's best interpretation of what is perhaps the most ambiguous wording found in 209D. The wording referred to is found in Section 4.4, titled "Classification by measurement at other particle sizes," and reads as follows: "A Classification of air cleanliness by measurement at particle sizes other than those specified herein may be performed with the following limitation: Particle counts may be interpolated between points but under no conditions may counts be extrapolated beyond the end points of Table I [same as Figure 25-2] or Figure 1 [same as Figure 25-1]. The particle count limit for the next larger particle size in Table I must not be exceeded."

Subtleties of optical counter performance are recognized to alert the user to use care in their use and data interpretation.

Equipment requirements for monitoring are to be specified in a Monitoring Plan. Sedimentation techniques can be used for monitoring, but only if the counting is done with a microscope (presumably, then, only where 5.0 μm counts are allowed).

VERIFICATION VERSUS MONITORING

Within the topic of "verification," unidirectional and nonunidirectional flow are specified by location and number of samples. Except for the aforementioned reference to Appendix A, in which probe orientation and isokinetic conditions are advised, this is the only place in which the two types of flows are differentiated in this way.

Monitoring requirements are identical in the sections of class, concentrations, sizes, other classes and sizes, and calibration, but are different in the sections of frequency, volumes, volumes at other sizes, location and number of samples, methods and equipment, and analysis. In monitoring, these are to be specified by the user in a Monitoring Plan. Two items unique to monitoring are mentioned explicitly. One, regarding locations and number of samples, is that measurements should be made throughout the clean zone or wherever critical or maximum counts were obtained during verification measurements. The other, regarding counter equipment and methods, is that (1) for those classes allowing counts at 5.0 μm, the microscope counting procedures of Appendix A are allowed; and (2) monitoring can be done using sedimentation techniques only if the counting is done with a microscope (presumably, then, only for classes in which 5.0-μm counts are allowed).

STATISTICAL ANALYSIS

Figure 25-5 presents the entire set of calculation steps prescribed by the new standard. The notable statistical technique on which the calculation is based is that of applying the "student" t-statistic in the calculation of an "upper confidence limit" (Gibra, 1973; Box, Hunter, and Hunter, 1978; Bzik, 1986a, b; Slooten, 1986; AT&T Technologies, 1956; Duncan, 1986). The procedure outlined in Figure 25-5 attempts to make the calculation as simple as possible and still apply statistics. This is a really notable contribution.

Unless samples are required to be taken randomly both in time and in location, however, the method prescribed will not represent a valid statistical estimate of the average concentration of particles in a clean zone. There are two other drawbacks to the method as prescribed. First, allowing only a single measurement at some of the measurement locations prevents "the standard deviation (SD) of means from the mean of averages (M)" from accurately representing the population of location means. In other words, unless the measurements at individual locations are averages of more than one measurement, the distribution of these averages will not be robust to the parent distribution of the measurements at any one location. Second, there is no explicit requirement that all the sampling volumes included in step 1 measurements of concentrations should be identical.

1. Average Particle Concentration at Each Location:

$$A = (C_1 + C_2 + \ldots + C_N) / N$$ (?presumes equal sample volumes.)

2. Mean of the Averages:

$$M = (A_1 + A_2 + \ldots + A_L) / L$$

3. Standard Deviation:

$$SD = \sqrt{\frac{(A1 - M)^2 + (A2 - M)^2 + \ldots + (AL - M)^2}{L-1}}$$

4. Standard Error of the Mean of the Averages (M):

$$SE = SD/\sqrt{L}$$

5. Upper Confidence Limit (UCL):

$$UCL = M + (SE) \times (UCL\ factor)$$

UCL factor for 95% upper control limit:

No. of locations (L)	2	3	4	5-6	7-9	10-16	17-29	29
95% UCL factor	6.3	2.9	2.4	2.1	1.9	1.8	1.75	1.65

Figure 25-5. Prescribed statistical analysis.

Without this requirement of identical sampling volumes, it is incorrect to average concentrations. If differing sample volumes are involved, then the average (A) to be calculated in step 1 should be obtained by dividing the sum of sample counts by the sum of sample volumes. Without this correction, sample volumes could conceivably be terminated any moment that the average concentration reads a low value, or a small sample with a rare high count could have disproportionate weight on the calculated average. In either case, a weakness in the requirement is that it could be interpreted to allow the user to skew the results away from being a truly representative average concentration of the clean zone. It should also have been stated that sample locations, volumes, and number of samples should be chosen so that they do not depend on where dirtier or cleaner areas are anticipated or are found to exist. Unfortunately, contamination problems within clean rooms and clean zones often are systematically located, so plans to randomly or even uniformly locate the points to sample offer little guarantee of catching the higher concentrations from these particle sources. A better technique is simply to sample here and there until one discovers the local zones of highest concentration, and then to evaluate the clean room or zone on the basis of these worst-case locations. Naturally, vendors of clean rooms would not appreciate this approach for finding the "hot spots."

The statistical analysis prescribed is intended to achieve a 95 percent chance that the class of the measured clean zone really is equal to or better than the calculated Upper Confidence Limit (UCL). Notwithstanding, the standard should caution that there is more than a 5 percent chance the class of the measured clean zone is really worse than this UCL. This increased probability of erroneous conclusions is because the assumption that the means taken at the sampling locations are Gaussian distributed is very likely optimistic due to the reproductive property of the Poisson distribution and to deviations from the reproductive property of the Poisson distribution and to deviations from necessary conditions on samples of randomness, independence, and uniformity. In addition, the 5 percent error probability may be misleading because the standard statistical method used can be argued to be wrong and that Bayesian analysis would provide a more objective analysis (Berger, and Berry, 1988).

CALIBRATION REQUIREMENTS

Recognizing that different optical particle counters can perform quite differently in counting particles, and that counters often significantly disagree in their counts (see Chapter 5), 209D requires that they be periodically calibrated against known reference standards. Calibrations are to include (but are not limited to) both sensitivity to particle size and airflow rate. The standard does not, however, properly recognize that counters of different designs may easily have different sensitivities to various size distributions (see Chapter 5). If it had done so, it would have required that these calibrations of counters all be performed at the same size(s)—i.e., those sizes at which measurements are to be made for use with this standard.

Although calibration is required, the methods to be used for calibration are only offered in Appendix B as guidelines:

- Sizes and sample volumes require primary calibration.
- Primary calibrations are to use spherical isotropic particles of index 1.6 (e.g., latex spheres).
- Secondary calibrations may be with atmospheric particles (i.e., natural dust).
- Counting and sizing stability and adjustability should be assured with built-in references or a secondary calibration system per ASTM's F-328 and F-649 (American Society for Testing and Materials, 1985a,c; see Figure 25-6), and IES-RP-CC-013 (Institute for Environmental Sciences, 1986). [Because F-328 references ASTM E-20, E-20 is referenced in this chapter as American Society for Testing and Materials (1985b). ASTM E-20 is a method of microscopically counting particles down to 0.2 μm in size, whereas Appendix A of 209D is only recommended for sizes as small as 5.0 μm.]
- Coincidence error should be less than 5 percent. Coincidence error is under-counting caused by two or more particles passing through the region of particle detection of the sensor, and spaced so closely in time that the counting circuitry cannot respond quickly enough to detect any more of them than the first one.
- Counter manufacturers should provide factory-available primary calibration facilities, should provide field primary capability and procedures, and should suggest recalibration intervals.
- A zero count check is required with a filter on the inlet and must register zero at 0.5 μm size. A tare count of 10 percent of class number is allowed for smaller sizes.

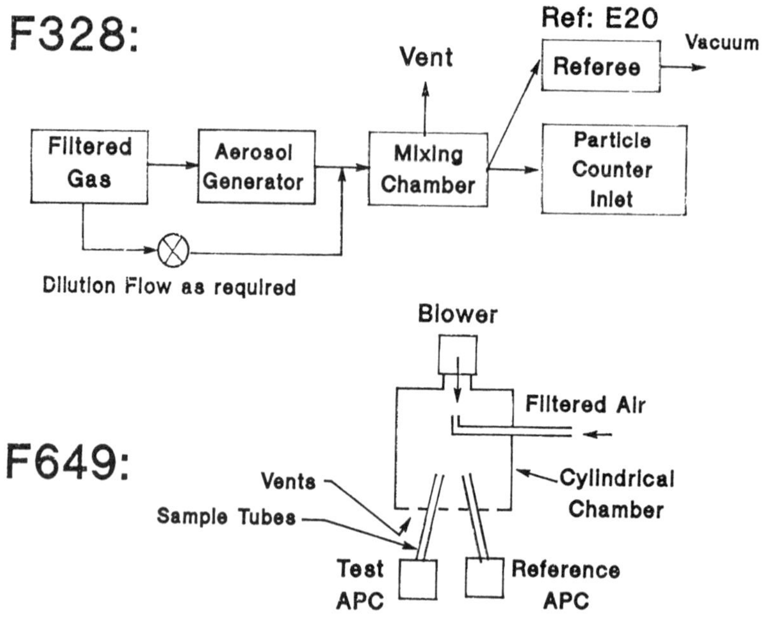

Figure 25-6. Instrumentation of referenced ASTM standards.

- Calibration data should be recorded, and should include sizes, volumes, counts, times, and locations.
- Tube lengths should be limited to 3 m if particles above 3 μm are to be counted; whereas it is presumed that within the particle range 0.1 to 1 μm, lengths up to 30 m may be acceptable. Generally, tubes should be configured to maintain flow Reynolds numbers in the range 5,000 to 10,000 and sample residence times below 5 s. No radius of curvature should be below 10 cm. For more on this subject, refer to Fissan and Schwientek (1987) and Cooper (1986a).

SHORTCOMINGS

The organization of 209D makes it hard for the reader to grasp its differentiations. Many other faults and shortcomings have been mentioned in the explanations above. These and a few others are collected below.

- In specifying sample volumes at sizes other than the main particle sizes, the standard says the minimum volumes specified at "either the lower class or size" should be met, when it should have said "lower class number or larger particle size." The committee was attempting to ensure that at least 20 counts would occur in a sample.

- The wording in the section on "Classification by measurement at other particle sizes" is ambiguous and confusing. One is left with the puzzle of why it is necessary to interpolate if the count at the next larger size is to be the limiting criterion anyway. It is not clear (1) whether class criteria counts or measured counts are being referred to as those that can be interpolated between points; (2) what "points" refers to; (3) whether this interpolation is to be made along the lines which define the classes, or between class lines as when defining an "other" class; or (4) whether interpolations should account for the log nature of the coordinates. An example of the allowed interpolation and the resulting count-limit criterion would have clarified this puzzle.

- Appendix B on "Operation and Use of Particle Counters" specifies that the orientation of the sample inlet tube in the airflow being sampled should be established in accordance with the requirements stated in the main body of the standard. Unfortunately, there is no mention of this subject in the main body. There is a requirement in Appendix A on "Particle Monitoring—Manual Counting Sizing Methods" that the sampler should be faced into the flow in the case of unidirectional flow, or upwards (or as specified otherwise) in the case of nonunidirectional flow. It also says in Appendix A that the airflow into the sampler should be adjusted to be isokinetic for unidirectional airflow. These are probably the requirements on orientation referred to in Appendix B, and the condition for isokinetic flow should have been specified for verification with optical counters as well.

- There is no requirement for calibration of the count performance of automatic particle counters, only for size detectivity and flow rate.

- The log-log descriptions for class-cumulative distributions offer little protection for the real distributions found in practice. The requirements in fact are that concentrations only need to meet the class criteria limits at one particle size point. Because many environments have distributions that fall off at smaller sizes, it would be easier to pass a test at a small size than at a large size. For this reason, it should be advised strongly that users of this standard first determine what sizes are critical to their products or occupants and require that the class measurements be met at those sizes specifically.

- Inadequate attention is given toward avoiding the pitfalls of using counters that are not sufficiently sensitive to those large particle sizes with which the user may be concerned.

- The number of sample locations and the number of samples at each location in verification testing is left too much to the user, with no protection against selections that may be made based on the data (Bayesian sampling) to artificially coerce the upper confidence limit to meet desired limits. For instance, if a low count is obtained on a first sample, the decision might be made not to take any more at that location in fear of the next reading being higher and raising the average. On the other hand, if the first reading is high, the sampler might decide to take more samples with the hope that the average of several will be lower. The sampling plan does not mention randomness, which is almost always required to ensure accuracy of statistical estimates such as the upper confidence limit.

- The specifications on sample locations for verification testing do not permit intentional checks around filter seams, which is where we know leaks of particles are apt to occur and where particles therefore tend to gather at relatively high concentrations. The requirement for uniformly distributed sample locations in verification measurements prevents users from selecting locations randomly and from specifically targeting critical locations to get a more functional measure of the cleanliness of a clean zone relative to the tasks which must be performed within it.

- Insufficient guidance is provided to help assure the validity and warranted interpretation of the 95-percent confidence figure of merit to the calculated results.

- No explicit requirement is made that all samples in verification testing should be of identical volume in order for the formula used in step 1 of the prescribed statistical calculations to be correct. And no alternative formula is prescribed when sample volumes differ from one another. Such a formula would be the following: $A = (c1 + c2 + ...)/(v1 + v2 + ...)$, where ci are counts of each sample, vi are the volumes of each sample, and the volume sum in the denominator would be required to be the same from location to location or become a weighting factor to be used in the subsequent steps of the computation.

- No specifications are included for viable particles, for surface contamination by particles, or for contamination by vapors that may leave films on products.

- The standard overuses its reference to Footnote 1 and thereby releases the user from just about every requirement except performing the prescribed statistical analysis against cleanliness acceptance criteria.

- English units (feet) and metric units (micrometers and meters) are mixed throughout.

- The words "count" and "concentration" are used synonymously; this could lead to confusion. In particular, it tends to overlook the involvement of volume sizes in the readings from counters that actually report counts and not concentrations and leave the user to divide counts by the product of flow rate and sample time.

REFERENCES

American Society for Testing and Materials. 1985a. Determining Counting and Sizing Accuracy of an Airborne Particle Counter Using Near-Monodisperse Spherical Particulate Materials (Standard Practice Designation 328-80). *Annual Book of ASTM Standards*, Vol. 15.03, pp. 676-680.

____. 1985b. Particle Size Analysis of Particulate Substances in the Range of 0.2 to 75 Micrometers by Optical Microscopy (Standard Practice Designation E-20-85). *Annual Book of ASTM Standards*. Vol. 14.02, pp. 18-28.

____. 1985c. Secondary Calibration of Airborne Particle Counter Using Comparison Procedures (Standard Practice Designation 649-80). *Annual Book of ASTM Standards*, Vol. 15.03, pp. 781-785.

AT&T Technologies. 1956. *Statistical Quality Control Handbook* (Western Electric Co.), Select Code 700-444 (1-800-432-6600).

Berger, James, and D. Berry. 1988. Statistical Analysis and the Illusion of Objectivity. *American Scientist*, 76(2):159-165.

Box, G., W. Hunter, and J. Hunter. 1978. *Statistics for Experimenters*. John Wiley & Sons, NY.

Bzik, T. 1986a. Statistical Management and Analysis of Particle Count Data in Ultraclean Environments: Part I. *Microcontamination*, (May):58-63, 98-99.

____. 1986b. Statistical Management and Analysis of Particle Count Data in Ultraclean Environments: Part II. *Microcontamination*, (June):34-41.

Cooper, D. 1986a. Minimizing Bias in Sampling Contaminant Particles. *Solid State Technology*, (January):73-79.

____. 1986b. Rationale for Proposed Revisions to Federal Standard 209B (Cleanrooms). *The Journal of Environmental Sciences*, (March/April):25-29.

____. 1988. Statistical Analysis Relating to Recent Federal Standard 209 (Cleanrooms) Revisions. *The Journal of Environmental Sciences*, (March/April):32-36.

Duncan, A. 1986. *Quality Control and Industrial Statistics*. Richard D. Irwin Press, Homewood, IL.

Fissan, H., and G. Schwientek. 1987. Sampling and Transport of Aerosols. *TSI Journal of Particle Instrumentation*, 2(2):3-10.

General Services Administration. 1973. Clean Room and Work Station Requirements, Controlled Environment. Federal Standard No. 209B, Washington, DC, April.

____. 1976. Clean Room and Work Station Requirements, Controlled Environment. Federal Standard No. 209B-Amended, Washington, DC, May.

____. 1987. Clean Room and Work Station Requirements, Controlled Environment. Federal Standard No. 209C, Washington, DC, October.

____. 1988. Clean Room and Work Station Requirements, Controlled Environment. Federal Standard No. 209D, Washington, DC, June.

Gibra, I. 1973. *Probability and Statistical Inference for Scientists and Engineers*. Prentice-Hall, NJ.

Helander, R. 1987. Certifying a Class 10 Cleanroom Using Federal Standard 209C. *Microcontamination*, (September):45-49, 76, 78.

Institute for Environmental Sciences. 1986. Recommended Practice for Equipment Calibration or Validation Procedures (IES-RP-CC-013).

Peck, R. 1987. FS 209 Requires More Work. *Microcontamination*, (August):8, 10-11.

Slooten, R. 1986. Statistical Treatment of Particle Counts in Clean Gases. *Microcontamination*, (February):32-38.

26. METHODS OF CIRCUIT DEFECTIVITY ANALYSIS

KEITH DILLENBECK Ashland Chemical Company, Columbus, Ohio

INTRODUCTION

A fundamental concern of contamination engineers is the effect that contamination has on the performance of electrical devices. Wafer particle counters and aerosol particle counters can be excellent tools for finding particle sources, but they do not indicate how the devices being fabricated will perform electrically. Of course, the final tests of the electrical functionality of the devices will occur after they have been stressed by temperature, voltage, humidity, and time. The contamination engineer, however, must have more first-order methods for assessing the effects of contamination, at any particular process step, on electrical function.

Short-flow electrical test structures are extremely useful for determining the effect of certain process conditions on process defectivity. To be useful, an electrical test structure should have relatively few process steps, and at the same time, assimilate actual device structures. A short process flow is important because it limits the number of variables that influence the end electrical effect. Thus, the test structure can be used to test the result of changes made to a process condition specifically to reduce the number of defects. Because they can be processed in a relatively short time, short-flow structures can also be used as real-time process-monitoring and process-control tools.

GATE OXIDE INTEGRITY

The gate oxide dielectric is the heart of a metal oxide semiconductor (MOS) or complementary MOS (CMOS) integrated circuit (Chou and Eldridge, 1979; Duffalo and Monkowski, 1984). A gate oxide can store electrons for prolonged periods, giving it the capability to store data, remember instructions, etc. This oxide must have superior dielectric properties so that electrons do not "leak" through any of the interfaces (Monkowski, 1984; Monkowski and Zahour, 1982). A simple MOS test capacitor is commonly used for short-flow tests and is shown schematically in Figure 26-1. The oxide and polysilicon thicknesses, as well as the poly doping level used for the test structure, should match those of the actual circuits being fabricated. The silicon dioxide and polysilicon thicknesses shown in Figure 26-1 are for example purposes only.

A common way to test this structure is to apply voltage to the doped polysilicon electrode and ramp that voltage until the oxide breaks down and begins to conduct electrons (Yamabe et al., 1983). The oxide breaks down when the current through it exceeds a predetermined value. A typical voltage ramp rate and current used to determine breakdown are 1 MV/cm-sec and 0.15 μA, respectively.

Gate oxide integrity tests can vary widely in terms of the specific capacitor structures and the criteria for evaluation (Smith, 1987). Examples of various types of capacitor structures are:

- Areas/perimeters
- Blocks or stripes (edge effects)

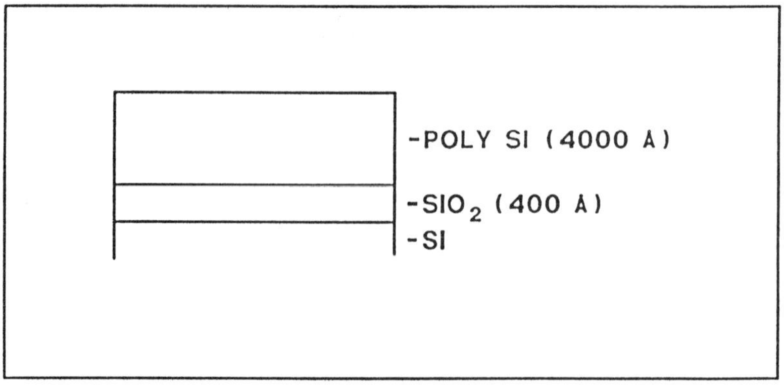

Figure 26-1. Simplified cross-section of an MOS capacitor.

- With or without field oxide
- With or without sidewall oxide.

The various methods for evaluating capacitor failure are:

- Histogram comparison of breakdown voltages
- Percent yield at a critical electric field (e.g., 5 MV/cm)
- Defect density, which is a function of capacitor yield and area
- Charge density to breakdown, which is the capacitor charge density per capacitor unit area.

The process engineer must carefully choose the capacitor features and evaluation criteria to be used for assessing certain process conditions. For example, oxide integrity tests are commonly used to evaluate the cleaning treatment of the silicon surface prior to gate oxidation. Conclusions to be drawn from wafer cleaning experiments can vary widely depending on whether time-zero or time-dependent evaluation criteria are used. The appropriate oxide integrity test can be very useful for assessing the effects of contamination forms on dielectric quality.

PATTERN INTEGRITY

It is important to determine the electrical integrity of the printed and etched patterns that make up an integrated circuit. Defects that lead to electrical opens and shorts are often caused by particles and other contaminants that adhere to the silicon substrate from sources throughout the fabrication process (Claudius, 1987; Wallis, 1985). A commonly used electrical monitor consists of a photolithographically defined conductive serpentine pattern on a dielectric film (Wallis, 1985). Such a structure is commonly referred to as a "snake." These patterns consist of two or three meandering conductive lines (Wallis, 1985; Gill and Dillenbeck, 1988). A meandering line (serpentine) is tested for continuity to account for open-related defects. An interdigitated comb structure can be used to test for electrical contact between two adjacent conductive lines; this accounts for short-related defects. A test array can either contain these two structures separately or combine them. Figure 26-2 shows a combined opens/shorts structure.

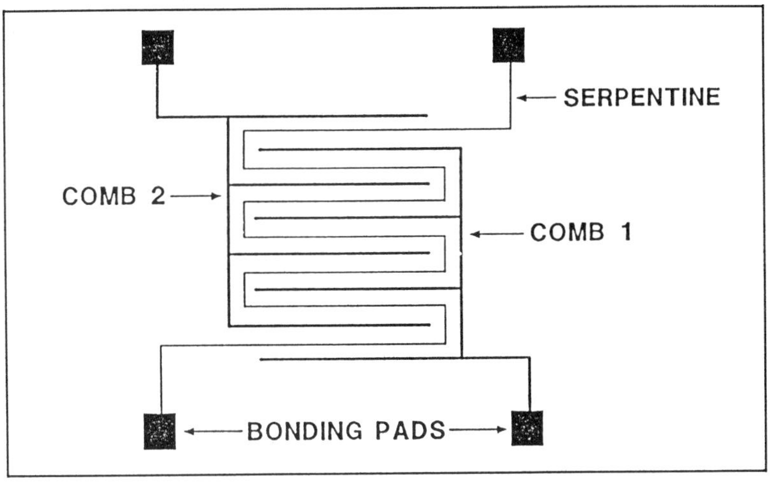

Figure 26-2. Simplified serpentine/comb defectivity array.

Several process variations can be used with the snake defectivity array. Such arrays have proven to be very flexible and useful for evaluating many conditions throughout the fabrication process (Gill and Dillenbeck, 1988). To maximize the usefulness of this defectivity-analysis tool, open- and short-related defects should be tested for and analyzed separately, as the defect mechanisms for opens and shorts can be quite different. A variety of conductive films--such as aluminum/silicon alloy and doped polysilicon--can be used to test different process modules. Additionally, a variety of underlying dielectric layers or sandwiches can be used to assess the cleanliness of the thermal oxidation or oxide deposition processes. Particles on or within the dielectric can extend through the conductive film to cause defects in the conductive pattern. Several pitch dimensions of the array can be used to account for various defect and particle sizes. The range of line and space dimensions will depend on the feature sizes of the devices being fabricated.

The following process modules can be effectively evaluated with this tool:

- Cleaning processes prior to the deposition of the conductive layer
- The photolithographic patterning process. This includes resist coat, expose, develop, and wafer transfer to the etch module (Durrant and Jenkins, 1985)
- The etch process itself. Particles before and during the etch can manifest themselves as defects of unetched metal or poly. Particles after the etch will cause short-related defects if they are conductive and have sufficient contact with the conductive lines.
- The deposition or sputtering process for the conductive film. For example, particles within the aluminum film will cause open and short-related defects during the photo and etch processes.

The snake defectivity array can be used both as a line monitor and as an engineering tool to evaluate specific process improvements. Careful experimental design can produce very powerful process experiments for determining the levels of defects of two or more process treatments.

FAILURE DEFECT ANALYSIS

Failure bit-map analysis is a technique that uses failed die information on memory devices as feedback to determine the cause of a particular die failure and at what step the defect occurred (Reyes, 1987). Bit mapping is an electrical testing technique that locates specific failed bit locations within the memory array of a memory circuit. These failed bit locations can then be scrutinized to better identify where in the process the defects originated. The major advantage of this technique is that only the most critical defects and particles, those that actually cause memory failure, are identified. The major disadvantage is that defects are only seen at the very end of the fabrication process, when it may be too late to correct the problem. While failure bit-map analysis should never be used as the sole yield-enhancement tool, it is one of the most effective techniques for identifying the "Red-X" source of defects. There are three components of this analysis technique:

- Bit-map electrical test
- Wafer deprocessing to sequentially strip off circuit layers
- Inspection of failed bit locations at each layer until the defect is found.

CRITERIA FOR RANKING OF CRITICAL PROCESS MODULES

There are many contamination and defect sources within an entire wafer fabrication process. For most semiconductor manufacturers, the number of particulate and dissolved contamination problems exceeds the manpower and monetary resources required to solve them. Within a wafer fabrication organization, it is important to identify the most critical defect sources from the standpoint of device yield and reliability. A pareto analysis can be effective for identifying the most critical defect sources. These defect sources are the ones that warrant the highest priority for expenditures of yield-enhancement resources. If yield-enhancement and contamination-engineering personnel are to be effective, identifying the most critical sources is the most critical step.

How do we identify "Red-X" defect sources? First, the tools of particle and contamination metrology and the methods of circuit defectivity analysis (both of which have been discussed) must be applied efficiently. The key is to develop and implement the proper yield-enhancement program, focusing on the problems that have proven to be critical, and not spending an inordinate amount of time on noncritical problems.

Figure 26-3 shows the types of killer defects found using the failure bit map analysis technique, as reported by Reyes (1987). The Red-X is clearly identified as "oxide particles." These large particles protrude through the deposited dielectric material beneath the aluminum interconnect layer, resulting in bridging lines of aluminum (Gill and Dillenbeck, 1988). These defects were not detected and identified until the wafers reached the very end of the process. Within certain steps throughout the fabrication process, very serious process problems could appear and disappear in the time required to process a wafer lot through the complete manufacturing process.

Figure 26-4 shows a typical distribution of the types of defects, using the snake/comb defectivity array. These data represent defects that were found at various points within the fabrication process, providing more timely process feedback.

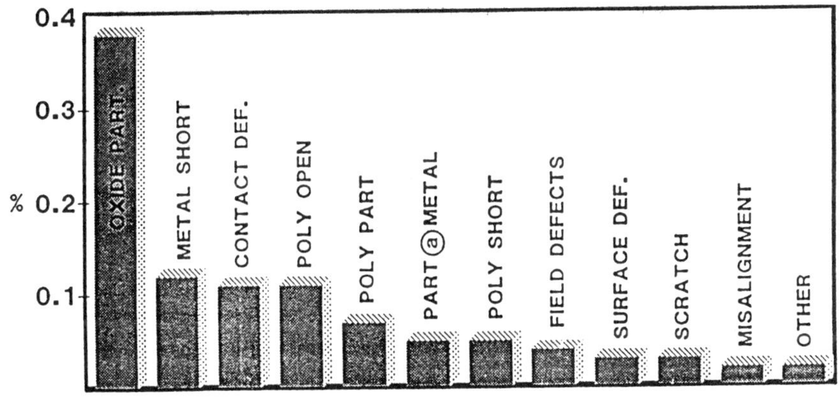

Figure 26-3. Types of defects identified using failure bit-map analysis technique.

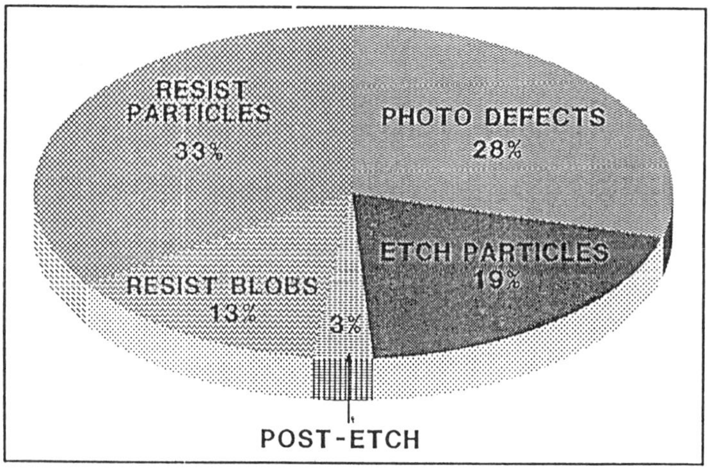

Source: Gill and Dillenbeck, 1988.

Figure 26-4. Identification of defect types: serpentine/comb defectivity array.

With both of these defect analysis techniques, emphasis must be placed on careful and consistent defect inspection and identification. A consistent set of criteria for defect classification must be established. The optical microscope can at times be misleading when process engineering personnel are trying to identify defects, because it represents the defect in only two dimensions; therefore the SEM should be used to identify defects having a questionable origin.

REFERENCES

Chou and Eldridge. 1979. Effects of Material and Processing Parameters on the Dielectric Strength of Thermally Grown SiO_2 Films. *Journal of the Electrochemical Society* (October).

Claudius, H. G. 1987. Practical Defect Reduction in an MOS I.C. Line. *Microcontamination* (April):47-52, 67.

Duffalo, J. M., and J. R. Monkowski. 1984. Particulate Contamination and Device Performance. *Solid State Technology* 27(3):109-114.

Durrant, N., and P. Jenkins. 1985. Defect Density Reduction Utilizing Wafer Edge Resist Removal. *Microcontamination* (April):45-51.

Gill, G. P., and K. D. Dillenbeck. 1988. The Use of Snake Patterns for Yield Enhancement. In: *Fifth International Symposium on Semiconductor Processing*, D. Gupta, ed. American Society for Testing and Materials, Philadelphia.

Monkowski, J. R. 1984. Significance of Submicron Particles to Device Performance. *Proceedings of the 30th Annual Technical Meeting of the Institute of Environmental Sciences*, May, pp. 81-84. IES, Mount Prospect, Illinois.

Monkowski, J. R., and R. T. Zahour. 1982. Failure Mechanism in MOS Gates Resulting from Particulate Contamination. J. T. Baker Chemicals for Electronics Series, Reprint 1; and in *Proceedings of the International Reliability Physics Symposium*, New York, Institute of Electrical and Electronics Engineers, pp. 224-248.

Reyes, J. D. 1987. Deprocessing Locates I.C. Failure Causes. *Semiconductor International* 10(6):108-110.

Smith, T. C. 1987. A User's Perspective on the Wafer Charging Problem. Motorola, Inc., internal communication, July.

Wallis, D. L. 1985. The Understanding and Uses of Electrical Monitors. *Proceedings of the Microcontamination Conference and exposition*, November, pp. 66-67. Canon Communications, Inc., Santa Monica, California.

Yamabe, Takanasha, and Mitsui. 1983. Thickness Dependence of Dielectric Breakdown Failure of Thermal SiO_2 Films. ECS Extended Abstracts, May.

INDEX